Oil Field Chemicals

Oil Field Chemicals

Johannes Karl Fink

Institut für Chemie der Kunststoffe
Montanuiversität Leoben

Gulf Professional Publishing
an imprint of Elsevier Science

Amsterdam Boston Heidelberg London New York Oxford Paris
San Diego San Francisco Singapore Sydney Tokyo

Gulf Professional Publishing is an imprint of Elsevier Science.

Copyright © 2003, Elsevier Science (USA). All rights reserved.

No part of this publication may be reproduced, stored in a retrieval system, or transmitted in any form or by any means, electronic, mechanical, photocopying, recording, or otherwise, without the prior written permission of the publisher.

∞ Recognizing the importance of preserving what has been written, Elsevier Science prints its books on acid-free paper whenever possible.

Library of Congress Cataloging-in-Publication Data

Fink, Johannes Karl
   Oil field chemicals / by Johannes Karl Fink.
     p. cm.
   Includes bibliographical references and index.
   ISBN 0-7506-7703-1
    1. Oil field chemicals.   I. Title.

TN871.F47 2003
622′.3382′028–dc21
                                                                                  2003044836

British Library Cataloguing-in-Publication Data
A catalogue record for this book is available from the British Library.

The publisher offers special discounts on bulk orders of this book.
For information, please contact:

Manager of Special Sales
Elsevier Science
200 Wheeler Road
Burlington, MA 01803
Tel: 781-313-4700
Fax: 781-313-4882

For information on all Gulf Professional Publishing publications available, contact our World Wide Web home page at: http://www.gulfpp.com

10 9 8 7 6 5 4 3 2 1

Printed in the United States of America

Contents

Preface, ix

CHAPTER 1

Drilling Muds..1
Classification of Muds, 1. Mud Compositions, 4. Additives, 11. Cuttings Removal by Sweep Materials, 30. Junk Removal, 30. Drilling Fluid Disposal, 31. Characterization of Drilling Muds, 31.

CHAPTER 2

Fluid Loss Additives..34
Mechanism of Action of Fluid Loss Agents, 34. Polysaccharides, 39. Synthetic Polymers, 44.

CHAPTER 3

Clay Stabilization..58
Properties of Clays, 58. Mechanisms Causing Instability, 61. Inhibitors of Swelling, 63. Chemicals in Detail, 64.

CHAPTER 4

Bit Lubricants..65
Refractory Metals, 65. Natural Compounds, 65.

CHAPTER 5

Bacteria Control..67
Mechanisms of Growth, 67. Treatments with Biocides, 69. Bactericides, 71. Various Biocides, 72. Bacterial Corrosion, 76. Assessment of Bacterial Corrosion, 79. Mechanisms of Microbial Corrosion, 80.

CHAPTER 6

Corrosion Inhibitors 82
History, 82. Classification of Corrosion Inhibitors, 82. Fields of Application, 82. Application Techniques, 85. Analytic Procedures, 85. Side Effects, 87. Amides and Imidazolines, 88. Nitrogenous Bases with Carboxylic Acids, 91. Nitrogen Quaternaries, 92. Polyoxylated Amines, Amides, and Imidazolines, 92. Nitrogen Heterocyclics, 98. Carbonyl Compounds, 99. Phosphate Esters, 100. Silicate-Based Inhibitors, 100. Miscellaneous Inhibitors, 100.

CHAPTER 7

Scale Inhibitors 103
Scale Inhibition, 103. Mathematical Models, 104. Chemicals in Detail, 104. Characterization, 106.

CHAPTER 8

Gelling Agents 108
Basic Mechanisms of Gelling Agents, 108.

CHAPTER 9

Filter-Cake Removal 120
Organic Acids, 120. Bridging Agents, 121. Enzymatic Breaker, 122. Peroxides, 123. Oligosaccharide, 124. Oscillatory Flow, 124.

CHAPTER 10

Cement Additives 125
Basic Composition of Portland Cement, 126. Special Cement Types, 130. Classification of Cement Additives, 135. Additives in Detail, 135.

CHAPTER 11

Transport 152
Pretreatment of the Products, 152. Corrosion Control, 156. Paraffin Inhibitors, 159. Pour Point Depressants, 159. Drag Reducers, 160. Hydrate Control, 162. Additives for Slurry Transport, 163. Additives for Odorization, 164. Cleaning, 164.

CHAPTER 12

Drag Reducers **166**
   Operating Costs, 166. Mechanism of Drag Reducers, 167. Drag
   Reducers in Detail, 171.

CHAPTER 13

Gas Hydrate Control **174**
   The Relevance of Gas Hydrates, 174. Inclusion Compounds,
   Clathrates, 174. Conditions for Formation, 177. Formation and
   Properties of Gas Hydrates, 178. Inhibition of Gas Hydrate
   Formation, 180. Hydrate Inhibitors for Drilling Fluids, 182.

CHAPTER 14

Antifreeze Agents **183**
   Theory of Action-Colligative Laws, 183. Overview of Antifreeze
   Chemicals, 184. Heat-Transfer Liquids, 185. Hydraulic Cement
   Additives, 191. Pipeline Transportation of Aqueous Emulsions of
   Oil, 191. Low-Temperature Drilling Fluids, 191.

CHAPTER 15

Odorization **192**
   Additives for Odorization, 192. Measurement and Odor
   Monitoring, 192. Uses and Properties, 194.

CHAPTER 16

Enhanced Oil Recovery **196**
   Waterflooding, 197. Caustic Waterflooding, 197. Acid
   Flooding, 199. Emulsion Flooding, 200. Chemical Injection, 203.
   Polymer Waterflooding, 205. Combination Flooding, 206. Foam
   Flooding, 208. Carbon Dioxide Flooding, 213. Steamflooding, 214.
   In situ Combustion, 215. Special Techniques, 215.
   Microbial-Enhanced Oil-Recovery Techniques, 217. Reservoir
   Properties, 228. Soil Remediation, 232.

CHAPTER 17

Hydraulic Fracturing Fluids **233**
   Stresses and Fractures, 233. Comparison of Stimulation
   Techniques, 234. Basic Constituents, 235. Types of Hydraulic

viii   Contents

    Fracturing Fluids, 236. Characterization of Fracturing Fluids, 238. Water-Based Systems, 240. Oil-Based Systems, 265. Foam-Based Fracturing Fluids, 267. Fracturing in Coal-Beds, 268. Propping Agents, 268. Acid Fracturing, 271. Special Problems, 272.

CHAPTER 18
Water Shutoff 276
Basic Principles, 276. Chemicals for Water Shutoff, 276.

CHAPTER 19
Oil Spill–Treating Agents 292
History, 292.

CHAPTER 20
Dispersants 309
Cement, 309. Aqueous Drilling Muds, 311. Miscellaneous, 315.

CHAPTER 21
Defoamers 316
Uses in Petroleum Technology, 316. Classification of Defoamers, 317. Theory of Defoaming, 319.

CHAPTER 22
Demulsifiers 325
Emulsions in Produced Crude Oil, 325. Waterflooding, 326. Oil Spill Treatment, 326. Desired Properties, 326. Mechanisms of Demulsification, 326. Performance Testing, 327. Classification of Demulsifiers, 328. Chemicals in Detail, 330.

References 345

Index 482

Preface

As crude oil resources decrease, the oil industry demands more sophisticated methods for the exploitation of natural resources. As a result, the use of oil field chemicals is becoming increasingly important.

When I started to research this topic seriously, I asked the experts for a monograph on oil field chemicals and they replied, "There is none." This book is the result of my efforts to create a definitive study on this important subject.

The material presented here is a compilation obtained by critically screening approximately 20,000 references from the literature (mainly from the Petroleum Abstracts Data Base and Patent Data Bases). Only materials that are accessible to the public have been included. The literature was screened from a chemist's point of view. Unfortunately several papers and patents did not disclose the chemical nature of the additives that are proposed for certain applications. In most cases it was not possible to learn the components by looking up cross-references and other sources. Papers of this kind are useless and have been omitted from the reference list. Research and procedures that are obviously not working, as well as wonder additives that are supposedly "good for everything you can imagine," have also been left out.

This book was originally intended to include a tutorial concerning the general chemistry, synthesis, and properties of oil field chemicals. In the course of the compilation I realized that the material is so extensive that the idea had to be abandoned. The material presented here is not complete. Only the recent developments of the last 10 years have been screened. Therefore additives that have been known and used as such for a long time (e.g., weighting materials) receive only brief mention. Attention is focused on the specific actions of oil field chemicals. Therefore patents dealing essentially with the same matter, but registered in multiple countries, are cited only once. The material presented here cannot replace a critical search in the patent literature. Longer quotations from original sources are marked as indented text. The text is ordered according to applications to parallel job processes. It starts with drilling, proceeds to productions, and ends (fortunately only occassionally in practice) with oil spills. Some of the chemicals are used in more than one main field.

x   Preface

For example, surfactants are used in nearly all of the applications. Separate chapters are devoted to these chemicals.

How to Use This Book

Utmost efforts have been undertaken to present reliable data. Because of the vast variety of material presented here, however, it cannot be complete in all relevant aspects, and it is recommended that the reader study the original literature for complete information. Therefore the author cannot assume responsibility for the completeness and validity of, nor for the consequences of the use of, the material presented here.

Index

There are two indices, an index of chemicals and a general index. In the chemical index, boldface page numbers refer to the sketches of structural formulas or to reaction equations.

Bibliography

The bibliography is sorted alphabetically by the first author. It is in this way an author index of the first author. Patents without authors are placed according to the patent numbers in the text.

Acknowledgments

I am indebted to our library, Dr. L. Jontes, Dr. J. Delanoy, and Mr. C. Slamenik for support in literature acquisition. Thanks are given to Professor I. Lakatos, University of Miskolc, who directed my interest to this topic, and to my wife Margit, who encouraged me to finalize the material when I felt exhausted.

Johannes Karl Fink

Chapter 1

Drilling Muds

Drilling fluids are mixtures of natural and synthetic chemical compounds used to cool and lubricate the drill bit, clean the hole bottom, carry cuttings to the surface, control formation pressures, and improve the function of the drill string and tools in the hole. They are divided into two general types: water-based fluids and oil-based fluids. The type of fluid base used depends on drilling and formation needs, as well as the requirements for disposition of the fluid after it is no longer needed. Drilling muds are a special class of drilling fluids used to drill most deep wells. *Mud* refers to the thick consistency of the formulation.

Following are the functions of a drilling mud:

1. To remove rock bit cuttings from the bottom of the hole and carry them to the surface
2. To overcome the fluid pressure of the formation
3. To avoid damage of the producing formation
4. To cool and lubricate the drill string and the bit
5. To prevent drill pipe corrosion fatigue
6. To allow the acquisition of information about the formation being drilled (e.g., electric logs, cutting analysis)

Classification of Muds

The classification of drilling muds is based on their fluid phase alkalinity, dispersion, and the type of chemicals used. We follow the classification of Lyons [1135]; see Table 1–1.

Dispersed Noninhibited Systems

Drilling fluids used in the upper hole sections are referred to as *dispersed noninhibited systems*. They are formulated from freshwater and may contain bentonite. The classification of bentonite-based muds is shown in Table 1–2.

Table 1–1
Classification of Drilling Muds

Class	Description
Freshwater muds—dispersed systems	pH from 7–9.5, include spud muds, bentonite-containing muds, phosphate-containing muds, organic thinned muds (red muds, lignite muds, lignosulfonate muds), organic colloid muds
Inhibited muds—dispersed systems	Water-based drilling muds that repress hydration of clays (lime muds, gypsum muds, seawater muds, saturated saltwater muds)
Low-solids muds—nondispersed systems	Contain less than 3%–6% v–v solids Most contain organic polymer
Emulsions	Oil in water and water in oil (reversed phase, with more than 5% water)
Oil-based muds	Contain less than 5% water; mixture of diesel fuel and asphalt

Table 1–2
Classification of Bentonite Fluid Systems

Solid–solid interactions	Inhibition level	Drilling fluid type
Dispersed	Noninhibited	Freshwater clay NaCl <1%, Ca^{2+} <120 ppm
Dispersed	Inhibited	Saline fluids Na^+, Ca^{2+} (seawater salt, saturated salt, gypsum, lime)
Nondispersed	Noninhibited	Freshwater low solids
Nondispersed	Inhibited	Salt and polymer fluids

The flow properties are controlled by a flocculant or thinner, and the fluid loss is controlled with bentonite and carboxymethylcellulose.

Phosphate-Treated Muds

Phosphates are effective only in small concentrations. The mud temperature must be less than 55° C. The salt contamination must be less than 500 ppm sodium chloride. The concentration of calcium ions should be kept as low as possible. The pH should be between 8 and 9.5. Some phosphates may decrease the pH, so adding more caustic soda is required.

Lignite Muds

Lignite muds are high-temperature resistant up to 230° C. Lignite can control viscosity, gel strength, and fluid loss. The total hardness must be lower than 20 ppm.

Quebracho Muds

Quebracho is a natural product extracted from the heartwood of the Schinopsis trees that grow in Argentina and Paraguay. Quebracho is a well characterized polyphenolic and is readily extracted from the wood by hot water. Quebracho is widely used as a tanning agent. It is also used as a mineral dressing, as a dispersant in drilling muds, and in wood glues. Quebracho is commercially available as a crude hot water extract, either in lump, ground, or spray-dried form, or as a bisulfite treated spray-dried product that is completely soluble in cold water. Quebracho is also available in a bleached form, which can be used in applications where the dark color of unbleached quebracho is undesirable [1622].

Quebracho-treated freshwater muds were used in shallow depths. It is also referred to as *red mud* because of the deep red color. Quebracho acts as a thinner. Polyphosphates are also added when quebracho is used. Quebracho is active at low concentrations and consists of tannates.

Lignosulfonate Muds

Lignosulfonate freshwater muds contain ferrochrome lignosulfonate for viscosity and gel-strength control. These muds are resistant to most types of drilling contamination because of the thinning efficiency of the lignosulfonate in the presence of large amounts of salt and extreme hardness.

Lime Muds

Lime muds contain caustic soda, an organic thinner, hydrated lime, and a colloid for filtrate loss. From this a pH of 11.8 can result, with 3 to 20 ppm calcium ions in the filtrate. Lime muds exhibit low viscosity, low gel strength, and good suspension of weighting agents. They can carry a larger concentration of clay solids at lower viscosities than other types of mud. At high temperatures, lime muds present a danger of gelation.

Table 1–3
Composition of Seawater

Component	Concentration (ppm)
Sodium	10,500
Potassium	400
Magnesium	1,300
Calcium	400
Chloride	19,000
Sulfate	3,000

Seawater Muds

The average composition of seawater is shown in Table 1–3. Seawater muds have sodium chloride concentrations above 10,000 ppm. Most of the hardness in seawater is caused by magnesium.

Seawater muds are composed of bentonite, thinner (lignosulfonate or lignosulfonate with lignite), and an organic filtration control agent.

Nondispersed Noninhibited Systems

In nondispersed systems no special agents are added to deflocculate the solids in the fluid. The main advantages of these systems are the higher viscosities and the higher yield point–to–plastics viscosity ratio. These altered flow properties provide a better cleaning of the bore hole, allow a lower annular circulating rate, and minimize wash out of the bore hole.

Low-Solids Freshwater Muds

Clear freshwater is the best drilling fluid in terms of penetration rate. Therefore it is desirable to achieve a maximal drilling rate using a minimal amount of solid additives. Originally low-solids mud formulations were used in hard formations, but they now also tend to find use in other formations. Several types of flocculants are used to promote the settling of drilled solids by flocculation.

Mud Compositions

Commercial products are listed in the literature. These include bactericides, corrosion inhibitors, defoamers, emulsifiers, fluid loss and viscosity control agents, and shale control additives [58–61, 65].

Table 1–4
Water-Based Drilling Muds

Compound	References
Glycol based	[1085]
Alkali silicates	[1276, 1777]
Acrylamide homopolymer, carboxymethylcellulose	[999]
Carboxymethylcellulose zinc oxide	[661]
Acrylamide copolymer, polypropylene glycol (water-based mud)	[1409]

Water-Based Muds

Components for water-based muds are shown in Table 1–4. Various modification methods for lignosulfonates have been described, for example, by condensation with formaldehyde [1170] or modification with iron salts [885]. It has been found that chromium-modified lignosulfonates, as well as mixed metal lignosulfonates of chromium and iron, are highly effective as dispersants and therefore are useful in controlling the viscosity of drilling fluids and in reducing the yield point and gel strength of the drilling fluids. Because chromium is potentially toxic, its release into the natural environment and the use thereof is continuously being reviewed by various government agencies around the world. Therefore less toxic substitutes are desirable. The lignosulfonates are prepared by combining tin or cerium sulfate and an aqueous solution of calcium lignosulfonate, thereby producing a solution of tin or cerium sulfonate and a calcium sulfate precipitate [1406].

Oil-Based Muds

Oil-based muds are being replaced now by synthetic muds. Diesel oil is harmful to the environment, particularly the marine environment in offshore applications. The use of palm oil derivative could be considered as an alternative oil-based fluid that is harmless to the environment [1869]. Hydrated castor oil can be used as a viscosity promoter instead of organophilic quaternized clays [1285].

An oil-based drilling mud can be viscosified with maleated ethylene-propylene elastomers [919]. The elastomers are ethylene-propylene copolymers or ethylene-propylene-diene terpolymers. The maleated elastomers are far more effective oil mud viscosifiers than the organophilic clays used. On the other hand, specific organophilic clays can provide a drilling fluid composition less sensitive to high temperatures [491].

Poly-α-olefins (PAOs) are biodegradable and nontoxic to marine organisms; they also meet viscosity and pour point specifications for formulation into oil-based muds [78].

The hydrogenated dimer of 1-decene [1208] can be used instead of conventional organic-based fluids, as can n-1-octene [1105].

Polyethercyclicpolyols

Polyethercyclicpolyols possess enhanced molecular properties and characteristics and permit the preparation of enhanced drilling fluids that inhibit the formation of gas hydrates; prevent shale dispersion; and reduce the swelling of the formation to enhance wellbore stability, reduce fluid loss, and reduce filter-cake thickness. Drilling muds incorporating the polyethercyclicpolyols are substitutes for oil-based muds in many applications [195–197, 1906, 1907]. Polyethercyclicpolyols are prepared by thermally condensing a polyol, for example, glycerol to oligomers and cyclic ethers.

Electric Conductive Nonaqueous Mud

A wellbore fluid has been developed that has a nonaqueous continuous liquid phase that exhibits an electrical conductivity increased by a factor of 10^4 to 10^7 compared with conventional invert emulsion. 0.2% to 10% by volume of carbon black particles and emulsifying surfactants are used as additives. Information from electrical logging tools, including measurement while drilling and logging while drilling, can be obtained [1563].

Water Removal

Water can be removed from oil-based muds by the action of magnesium sulfate [1652].

Synthetic Muds

Synthetic-based muds are mineral oil muds in which the oil phase has been replaced with a synthetic fluid, such as ether, ester, PAO, or linear alkylbenzene, and are available from major mud companies. The mud selection process is based on the mud's technical performance, environmental impact, and financial impact. Synthetic muds are expensive. Two factors influence the direct cost: unit or per-barrel cost and mud losses. Synthetic muds are the technical equivalent of oil-based muds when drilling intermediate hole sections. They are technically superior to all water-based systems when drilling reactive shales in directional wells. However, with efficient solids-control equipment, optimized drilling, and good housekeeping practices, the cost of the synthetic mud can be brought to a level comparable with oil-based mud [1308].

By the end of 2000, rig operators in the U.K. Continental Shelf had to reduce to zero the discharge of their synthetic-based drilling fluid-contaminated cuttings. To comply with government requirements, operators had to reduce discharges by at least 20% each year until the 2001 deadline. Most of the rest of Europe had already adopted a zero or limited discharge approach. In the United States, the Environmental Protection Agency (EPA) has taken a more pragmatic approach by including the potential polluting effects of not only the discharge of cuttings into the sea but also any fuel use or air emissions from barging, reinjection, or cuttings disposal on land [1666].

Skeletally isomerized linear olefins exhibited a better high-temperature stability in comparison with a drilling fluid prepared from a conventional PAO. The fluid loss properties are good, even in the absence of a fluid loss additive [684–686, 1852]. Although normal α-olefins (NAOs) are not generally useful in synthetic hydrocarbon-based drilling fluids, mixtures of mostly linear olefins are minimally toxic and highly effective as the continuous phase of drilling fluids [683, 685]. Acetals as mineral oil substitutes exhibit good biodegradability and are less toxic than mineral oils [827, 828]. Acrylic acid salts are formed by the neutralization reaction of acrylic acid in aqueous solution [1613].

Alginates are hydrocolloids, which are extracted from brown marine microalgae. Water-soluble alginates are prepared as highly concentrated, pumpable suspensions in mixtures of propylene glycol and water by using hydroxypropylated guar gum in combination with carboxymethylated cellulose, which is used as a suspending agent [945].

Inverted Emulsion Drilling Muds

Inverted emulsion muds are used in 10% to 20% of all drilling jobs. Historically, first crude oils, then diesel oils and mineral oils, have been used in formulating invert drilling fluids. Considerable environmentally damaging effects may occur when the mud gets into the sea. Drilling sludge and the heavy mud sink to the seabed and partly flow with the tides and sea currents to the coasts. All of these hydrocarbons contain no oxygen and are not readily degraded [826].

Because of problems of toxicity and persistence, which are associated with these oils, in particular for offshore use, alternative drilling oils have been developed. Examples of such oils are fatty acid esters and branched chain synthetic hydrocarbons such as PAOs. Fatty acid ester–based oils have excellent environmental properties, but drilling fluids made with these esters tend to have lower densities and are prone to hydrolytic instability. PAO-based drilling fluids can be formulated to high densities and have good hydrolytic stability and low toxicity. They are, however, somewhat less biodegradable than esters; they are expensive; and the fully weighted, high-density fluids tend to be overly viscous [1105].

8   Oil Field Chemicals

Esters

Esters of C_6 to C_{11} monocarboxylic acids [1288–1292], acid-methyl esters [1282], and polycarboxylic acid esters [1287], as well as oleophilic monomeric and oligomeric diesters [1293], have been proposed as basic materials for inverted emulsion muds. Natural oils are triglyceride ester oils [1844] and are similar to synthetic esters. Diesters also have been proposed [1293–1297].

Acetals

Acetals and oleophilic alcohols or oleophilic esters are suitable for the preparation of inverted emulsion drilling muds and emulsion drilling muds. They may replace the base oils, diesel oil, purified diesel oil, white oil, olefins, and alkylbenzenes [825, 826]. Examples are isobutyraldehyde; di-2-ethylhexyl acetal; dihexyl formal; and mixtures with coconut alcohol, soya oil, and α-methyldecanol.

Inverted emulsion muds are more advantageous in stable, water-sensitive formations and in inclined boreholes. They are stable up to very high temperatures and provide excellent corrosion protection. Disadvantages are the higher price, the greater risk if gas reservoirs are bored through, the more difficult handling for the team at the tower, and the greater environmental problems. The high setting point of linear alcohols and the poor biologic degradability of branched alcohols limit their use as an environment-friendly mineral oil substitute. Higher alcohols, which are still just somewhat water-soluble, are eliminated for use in offshore muds because of their high toxicity to fish. Esters and acetals can be degraded anaerobically on the seabed. This possibility minimizes the environmentally damaging effect on the seabed. When such products are used, rapid recovery of the ecology of the seabed takes place after the end of drilling. Acetals, which have a relatively low viscosity and in

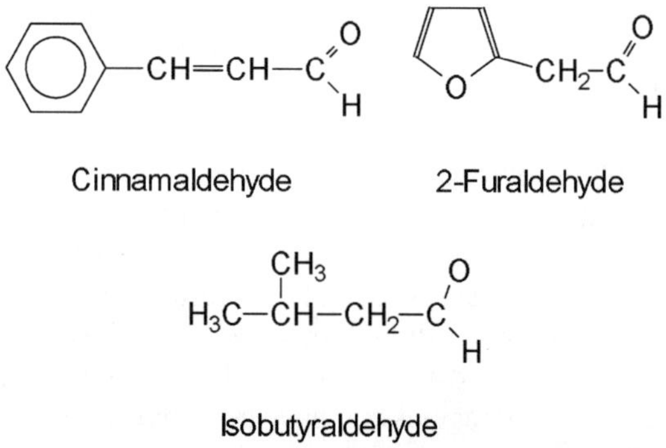

Figure 1–1. Cinnamaldehyde, 2-furaldehyde, isobutyraldehyde.

particular a relatively low setting point, can be prepared by combining various aldehydes and alcohols [826, 1877].

Anti-Settling Properties

Ethylene-acrylic acid copolymer neutralized with amines such as triethanol amine or N-methyl diethanol amine enhances anti-settling properties [1198, 1554].

Glycosides

The advantage of using glycosides in the internal phase is that much of the concern for the ionic character of the internal phase is no longer required. If water is limited in the system, the hydration of the shales is greatly reduced. The reduced water activity of the internal phase of the mud and the improved efficiency of the shale is an osmotic barrier if the glycoside interacts directly with the shale. This helps lower the water content of the shale, thus increasing rock strength, lowering effective mean stress, and stabilizing the wellbore [760].

Methylglucosides also could find applications in water-based drilling fluids and have the potential to replace oil-based drilling fluids [801]. The use of such a drilling fluid could reduce the disposal of oil-contaminated drilling cuttings, minimize health and safety concerns, and minimize environmental effects.

Others

Other base materials proposed are listed in Table 1–5. Quaternary oleophilic esters of alkylolamines and carboxylic acids improve the wettability of clay

Table 1–5
Other Materials for Inverted Emulsion Drilling Fluids

Base material	References
Ethers of monofunctional alcohols	[1299]
Branched di-decyl ethers	[710, 711]
α-Sulfofatty acids	[1301]
Oleophilic alcohols	[1281, 1283, 1300]
Oleophilic amides	[1284]
Hydrophobic side chain polyamides from N,N-didodecylamine and sodium polyacrylate or polyacrylic acid	[1240]
Polyether polyamine	[1820]
Phosphate ester of a hydroxy polymer	[237]

[1443, 1444]. Nitrates and nitrites can replace calcium chloride in inverted emulsion drilling muds [612].

Biodegradable Formulations

Biodegradable drilling fluid formulations have been suggested. These are formulations of a polysaccharide in a concentration insufficient to permit a contaminating bacterial proliferation, namely a high-viscosity carboxymethylcellulose sensitive to bacterial enzymes produced by the degradation of the polysaccharide [1419].

On the other hand, the biodegradability of mud additives is a problem [752]. The biodegradability of seven kinds of mud additives was studied by determining the content of dissolved oxygen in water, a simple biochemical oxygen-demand testing method. The biodegradability is high for starch but is lower for polymers of allyl monomers and additives containing an aromatic group.

Foam Drilling

While drilling low-pressure reservoirs with nonconventional methods, it is common to use low-density dispersed systems, such as foam, to achieve underbalanced conditions. To choose an adequate foam formulation, not only the reservoir characteristics but also the foam properties need to be taken into account. Parameters such as stability of foam and interactions between rock–fluid and drilling fluid–formation fluid are among the properties to evaluate while designing the drilling fluid [13].

A foaming composition having a specific pH and containing an ionic surfactant and a polyampholytic polymer whose charge depends on the pH is circulated in a well. By varying the pH, it is possible to destabilize the foam in such a way as to more easily break the foam back at the surface and possibly to recycle the foaming solution [76].

Chemically Enhanced Drilling

Chemically enhanced drilling offers substantial advantages over conventional methods in carbonate reservoirs. Coiled tubing provides the perfect conduit for chemical fluids that can accelerate the drilling process and provide stimulation while drilling [1471]. The nature of the chemical fluids is mainly acid that dissolves or disintegrates the carbonate rock.

Supercritical Carbon Dioxide Drilling

The efficiency of drilling operations can be increased using a drilling fluid material that exists as supercritical fluid or a dense gas at temperature and pressure conditions occurring in the drill site, such as carbon dioxide.

A supercritical fluid exhibits physical–chemical properties intermediate between those of liquids and gases. Mass transfer is rapid with supercritical fluids. Their dynamic viscosities are nearer to those in normal gaseous states. In the vicinity of the critical point the diffusion coefficient is more than 10 times that of a liquid. Carbon dioxide can be compressed readily to form a liquid. Under typical borehole conditions, carbon dioxide is a supercritical fluid.

The viscosity of carbon dioxide at the critical point is only 0.02 cP, increasing with pressure to about 0.1 cP at 70 MPa (about 10,000 psi). Because the diffusivity of carbon dioxide is so high, and the rock associated with petroleum-containing formations is generally porous, the carbon dioxide is quite effective in penetrating the formation. This penetration is beneficial. Carbon dioxide is commonly used to stimulate the production of oil wells, because it tends to dissolve in the oil, reducing the oil viscosity while providing a pressure gradient that drives the oil from the formation.

Carbon dioxide can be used to reduce mechanical drilling forces, to remove cuttings, or to jet erode a substrate. Supercritical carbon dioxide is preferably used with coiled-tube drilling equipment. The very low viscosity of supercritical carbon dioxide provides efficient cooling of the drill head and efficient cuttings removal. Furthermore, the diffusivity of supercritical carbon dioxide within the pores of petroleum formations is significantly higher than that of water, making jet erosion using supercritical carbon dioxide much more effective than jet erosion using water. Supercritical carbon dioxide jets can be used to assist mechanical drilling, for erosion drilling, or for scale removal. Spent carbon dioxide can be vented to the atmosphere, collected for reuse, or directed into the formation to aid in the recovery of petroleum [986].

Additives

Thickeners

A variety of compounds useful as thickeners is shown in Table 1–6.

Polymers

Thickener polymers include polyurethanes, polyesters, polyacrylamides, natural polymers, and modified natural polymers [510].

pH Responsive Thickeners. The viscosity of ionic polymers is dependent on the pH. In particular, pH responsive thickeners can be prepared by copolymerization of acrylic or methacrylic acid ethyl acrylate or other vinyl monomers and tristyrylpoly(ethyleneoxy)$_x$ methyl acrylate. Such a copolymer provides a stable aqueous colloidal dispersion at an acid pH lower than 5.0 but becomes an effective thickener for aqueous systems upon adjustment to a pH of 5.5 to 10.5 or higher [1512, 1513].

Table 1–6
Thickeners

Compound	References
A water-soluble copolymer of hydrophilic and hydrophobic monomers, acrylamide–acrylate of silane or siloxane	[1219]
Carboxymethylcellulose, polyethylene glycol	[1130, 1131]
Combination of a cellulose ether with clay	[1481]
Amide-modified carboxyl-containing polysaccharide	[152]
Sodium aluminate and magnesium oxide	[1405]
Thermally stable hydroxyethylcellulose 30% ammonium or sodium thiosulfate and 20% hydroxyethylcellulose (HEC)	[1129]
Acrylic acid copolymer and oxyalkylene with hydrophobic group	[541]
Copolymers acrylamide–acrylate and vinyl sulfonate–vinylamide	[1809]
Cationic polygalactomannans and anionic xanthan gum	[1873]
Copolymer from vinyl urethanes and acrylic acid or alkyl acrylates	[1843]
2-Nitroalkyl ether–modified starch	[724]
Polymer of glucuronic acid	[404]
Ferrochrome lignosulfonate and carboxymethylcellulose	[1001]
Cellulose nanofibrils[a]	[1063, 1065]
Quaternary alkyl amido ammonium salts	[1689]
Chitosan[b]	[853]

a) Stable up to temperatures of about 180° C.
b) Solubilized in acidic solution.

Mixed Metal Hydroxides

By addition of mixed metal hydroxides, typical bentonite muds are transformed to an extremely shear thinning fluid [1060]. At rest these fluids exhibit a very high viscosity but are thinned to an almost waterlike consistency when shear stress is applied. In theory, the shear thinning rheology of mixed metal hydroxides and bentonite fluids is explained by the formation of a three-dimensional, fragile network of mixed metal hydroxides and bentonite. The positively charged mixed metal hydroxide particles attach themselves to the surface of negatively charged bentonite platelets. Typically, magnesium aluminum hydroxide salts are used as mixed metal hydroxides.

Mixed metal hydroxides demonstrate the following advantages in drilling:

- High cuttings removal
- Suspension of solids during shutdown
- Lower pump resistance
- Stabilization of the borehole
- High drilling rates
- Protection of the producing formation [579]

Mixed metal hydroxide drilling muds have been successfully used in horizontal wells; in tunneling under rivers, roads, and bays; for drilling in fluids; for drilling large-diameter holes; with coiled tubing; and to ream out cemented pipe.

Mixed metal hydroxides can be prepared from the corresponding chlorides treated with ammonium [276]. Experiments done with various drilling fluids showed that the mixed metal hydroxides system, coupled with propylene glycol [469], caused the least skin damage of the drilling fluids tested.

Thermally activated mixed metal hydroxides, made from naturally occurring minerals, especially hydrotalcites, may contain small or trace amounts of metal impurities besides the magnesium and aluminum components, which are particularly useful for activation [946]. Mixed hydroxides of bivalent and trivalent metals with a three-dimensional spaced-lattice structure of the garnet type ($Ca_3Al_2[OH]_{12}$) have been described [275, 1279].

Lubricants

During the drilling of oil and gas wells, a drill bit at the end of a rotating drill string, or at the end of a drill motor, is used to penetrate through geologic formations. During this operation, drilling mud is circulated through the drill string, out of the bit, and back to the surface via the annular space between the drill pipe and the formation. The drilling mud serves numerous functions including cooling and lubricating the drill string and drill bit, counterbalancing the pressures encountered in the formation using hydrostatic pressure, providing a washing action to remove the formation cuttings from the wellbore, and forming a friction-reducing wall cake between the drill string and the wellbore. During drilling, the drill string may develop an unacceptable rotational torque or, in the worst case, become stuck. When this happens, the drill string cannot be raised, lowered, or rotated. Common factors leading to this situation include:

- Cuttings or slough buildup in the borehole
- An undergauge borehole
- Irregular borehole development embedding a section of the drill pipe into the drilling mud wall cake
- Unexpected differential formation pressure

Differential pressure sticking occurs when the drill pipe becomes imbedded in the mud wall cake opposite a permeable zone. The difference between the hydrostatic pressure in the drill pipe and the formation pressure holds the pipe in place, resulting in a sticking pipe. Differential sticking may be prevented, and a stuck drill bit may be freed, using an oil–mud or an oil- or water-based surfactant composition.

Such a composition reduces friction, permeates drilling mud wall cake, destroys binding wall cake, and reduces differential pressure. Unfortunately, many of such compositions are toxic to marine life. Synthetic PAOs are nontoxic and effective in marine environments when used as lubricants, return-of-permeability enhancers, or spotting fluid additives for water-based drilling muds. A continuing need exists for other nontoxic additives for water-based drilling muds, which serve as lubricants, return-of-permeability enhancers, and spotting fluids.

Olefin isomers containing between approximately 8 and 30 carbon atoms are suitable. However, isomers having fewer than 14 carbon atoms are more toxic, and isomers having more than 18 carbon atoms are more viscous. Therefore olefin isomers having between 14 and 18 carbon atoms are preferred [767].

Alcohols

Both polyalkylene glycols [45] and side chain polymeric alcohols such as polyvinyl alcohol have been suggested. These substances are comparatively environmentally safe [1420, 1553].

Polyvinyl alcohols may be applied as such or in crosslinked form [90]. Crosslinkers can be aldehydes (e.g., formaldehyde, acetaldehyde, glyoxal, glutaraldehyde), to form acetals, maleic acid or oxalic acid to form crosslinked ester bridges, or others (e.g., dimethylurea, polyacrolein, diisocyanate, divinyl sulfonate) [89, 91].

An amine-terminated polyoxyalkylene having an average molecular weight from about 600 to about 10,000 can be acylated with a succinic acylating agent (e.g., hexadecenyl succinic anhydride or a Diels-Alder diacid) obtained from an unsaturated fatty acid [628, 629]; similarly, alkyl–aryl sulfonate salts [1319] can be used in lubrication.

The pendant hydroxy groups of ethylene oxide–propylene oxide copolymers of dihydroxy and trihydroxy alcohols may be sulfurized to obtain a sulfurized alcohol additive. This is effective as a lubricant in combination with oils and fats [387, 533]. The sulfurized alcohols may be obtained by the reaction of sulfur with an unsaturated alcohol. Furthermore, fatty alcohols and their mixtures with carboxylic acid esters as lubricant components [1286] have been proposed.

Ether Lubricants

2-Ethylhexanol can be epoxidized with 1-hexadecene epoxide. This additive also helps reduce or prevent foaming. By eliminating the need for traditional oil-based components, the composition is nontoxic to marine life, biodegradable, environmentally acceptable, and capable of being disposed of at the drill site without costly disposal procedures [44].

Ester-Based Oils

Several ester-based oils are suitable as lubricants [532,690], as are branched chain carboxylic esters [1588]. Tall oils can be transesterified with glycols [1536] or condensed with monoethanolamine [51].

The ester class also comprises natural oils, such as vegetable oil [75]; spent sunflower oil [940, 941, 992, 993]; and natural fats, for example, sulfonated fish fat [161]. In water-based mud systems no harmful foams are formed from partially hydrolyzed glycerides of predominantly unsaturated C_{16} to C_{24} fatty acids. The partial glycerides can be used at low temperatures and are biodegradable and nontoxic [1280]. A composition for high-temperature applications is available [1818]. It is a mixture of long chain polyesters and polyamides.

In the case of esters from, for example, neopentylglycol, pentaerythrite, and trimethylolpropane with fatty acids, tertiary amines, such as triethanol amine, together with a mixture of fatty acids, improve the efficiency [74].

Nitrogen-Containing Additives

The group of nitrogen-containing additives comprises phenolic mannich bases, phosphoric acid [1776], and oxalkylated alkylphenols with nitrogen-containing additives [996].

Polymers

Synthetic polymers and natural polymers suitable for drilling muds are listed in Tables 1–7 and 1–8, respectively. Polyacrylamides are eventually hydrolyzed in the course of time and temperature. This leads to a lack of tolerance toward electrolyte contamination and to a rapid degradation inducing a loss of their properties. Modifications of polyacrylamide structures have been proposed to postpone their thermal stability to higher temperatures. Monomers such as AMPS or sulfonated styrene/maleic anhydride can be used to prevent acrylamide comonomer from hydrolysis [92].

Friction Loss by Ellipsoidal Glass Granules

The use of ellipsoidal glass granules instead of spherical glass beads is substantiated by the effort to increase the contact surface of antifrictional particles, reduce their ability to penetrate deeply into the mud cake, and increase their breaking strength [1034, 1035, 1037, 1038].

Graphite. Graphite is a classical lubricant [1879].

Paraffins. Purified paraffins are nontoxic and biodegradable [765].

Table 1–7
Synthetic Copolymers

Polymer	References
2-Acrylamido-2-methylpropane sulfonic acid (AMPS), diallyldimethylammonium chloride (DADMAC), N-vinyl-N-methylacetamide (VIMA), acrylamides and acrylates[a]	[824, 1101, 1376]
AMPS/AM (acrylamide)/VAC (vinylacetate) copolymer	[1177, 1825]
Acrylamide styrene sulfonate copolymer[b]	[1399]
Acrylamide; vinylpyrrolidone N-vinyl lactam	[1408]
Copolymer from acrylamide and 2-acrylamide-2-methyl propane sulfonic acid, with methylene-bis-acrylamide as the crosslinker	[1398]
Vinylamide morpholine derivative acrylic acid copolymer	[1773]
Styrene-butadiene copolymer latex and styrene-acrylate-methacrylate terpolymer latex	[125, 126]
Polymers of amido-sulfonic acid	[1660]
Acrylic polymer	[1587]
N-Vinyl-2-pyrrolidone, acrylamidopropane sulfonic acid, acrylamide, and acrylic acid copolymer	[1679]
Acrylamidopropenylmethylenesulfonic acid and N-vinyl amides of acrylics and methacrylics or N-vinylcaprolactam	[808]
Sulfonated chromium humate	[1716]
Sulfonated phenolic resin and hydrolytic ammonium polyacrylate	[1716]
Polyamide and polyimide	[1819]
Hydrolyzed polyacrylonitrile and cyan-ethylate carboxymethylcellulose[c]	[1117]
Poly-α-olefins (PAOs)[d]	[1207]
Polymers of hydroxy carboxylic acids as a rheologic additive	[1298]
Dimethyl silicone fluids	[1397]

a) Deep-drilling additives.
b) The polymer additive is characterized by increased viscosity at low shear rates and enhanced fluid loss control.
c) Salt tolerance (above 10%).
d) The additive also reduces drill string drag.

Differential Sticking Reducer

Various additives have been proposed to assist in freeing a stuck drill pipe, the most common of which is diesel oil that is added directly to the drilling mud as a spotting fluid. However, this is not always successful. An additive comprising an oil-in-water microemulsion has been proposed. Sodium dodecyl

Table 1–8
Natural and Natural Modified Polymers

Polymer	References
Amylopectin[a]	[983, 984]
Polyanionic cellulose sulfonate–containing polymer[b]	[812]
Hydroxyethyl and hydroxypropyl cellulose	[1437]
Hydroxyethylcellulose, hydrophobically modified	[95, 97]
Carboxymethylcellulose	[1542]
Gellan	[515]
Diutan	[1326, 1328]
Cornstarch, carboxylated methyl, crosslinked hydroxypropyl cornstarch	[50, 173]
Graft copolymer of starch, acrylamide, and polyvinyl alcohol	[671]
Waxy maize starch, epichlorohydrin crosslinked	[561]
Crosslinked starches	[394, 1626]
Amine-derivatized potato starch	[50]
Sulfonated chromium humate, sulfonated phenolic resin, and hydrolytic ammonium polyacrylate	[1716]
Gellan scleroglucan xanthan gum	[516]
Hydrophobically modified guars	[94]
Hydroxypropylguar gum, hydrophobically modified	[93]
Deacetylated xanthan gum	[1064]
Vinyl grafted lignite[c]	[873, 874]

a) The amylopectin starch may be crosslinked with epichlorohydrin to stabilize the starch molecule. The molecule may also be stabilized by hydroxypropylation, carboxymethylation, or both.
b) Combination for high temperature/high pressure.
c) For example, with dimethylaminopropyl methacrylamide, methacrylamido propyltrimethyl ammonium chloride, N-vinylformamide, N-vinylacetamide, diallyl dimethyl ammonium chloride, and diallylamine.

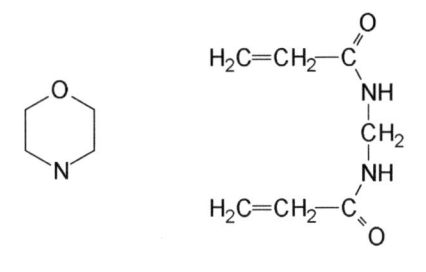

Morpholine   Methylene-bis-acrylamide

Figure 1–2. Morpholine, methylene-bis-acrylamide.

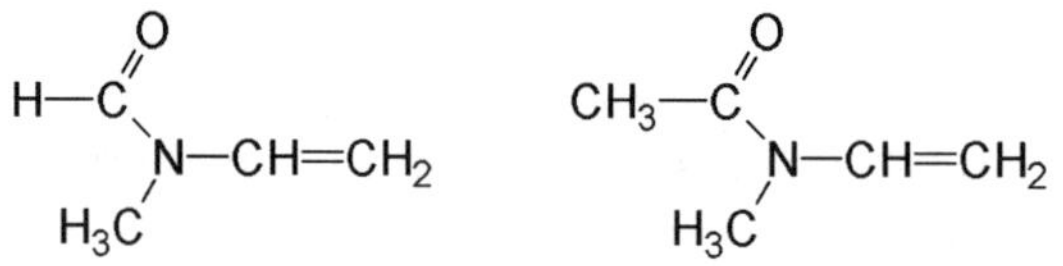

18   Oil Field Chemicals

N-Vinylformamide   N-Vinyl-N-methylacetamide

Figure 1–3. N-Vinylformamide, N-vinyl-N-methylacetamide.

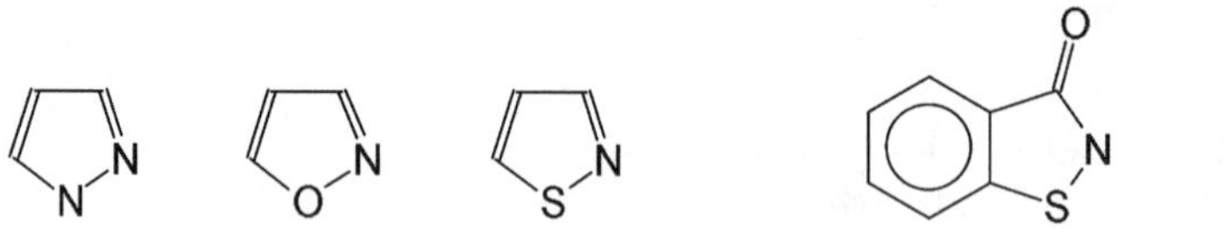

Pyrazol   Isooxazol   Isothiazol   1,2-Benzoisothiazolin-3-one

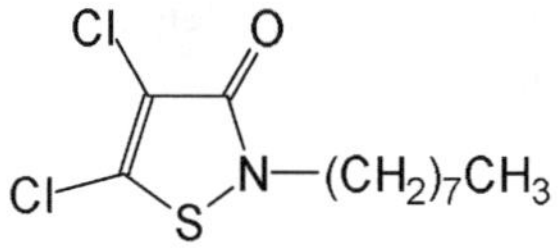

4,5-Dichloro-2-N-octyl-isothiazolin-3-one

Figure 1–4. Pyrazol, isooxazol, isothiazol, 1,2-benzoisothiazolin-3-one, 4,5-dichloro-2-N-octyl-isothiazolin-3-one.

benzene sulfonate may be used as a surfactant. Ethylene glycol or diethylene glycol act as cosurfactants [453].

Studies on Pipe Sticking. A study of the effect of various additives on pipe sticking is available [1391]. The effect of various available oil field additives to reduce down-hole friction and their optimal concentration has been studied. As previously mentioned, the frictional forces present at the string–borehole interface are of prime importance. The friction at the string–borehole interface can be reduced through various chemicals incorporated in the drilling fluid system. To obtain a mud cake in which sensitivity of various chemicals could be studied, a highly sticky cake was prepared from mud containing gypsum, kaolinite, sand, and shale powder. The sensitivity of various mud additives in minimizing the friction at the string–borehole interface, and thereby reducing the sticking coefficient tendency, was evaluated systematically with time. The study was extended to several mud systems.

Bacteria

Bacterial contamination of drilling fluids contributes to a number of problems. Many of the muds contain sugar-based polymers in their formulation that provide an effective food source to bacterial populations. This can lead to direct degradation of the mud. In addition, the bacterial metabolism can generate deleterious products. Most notable among these is hydrogen sulfide, which can lead to decomposition of mud polymers, formation of problematic solids such as iron sulfide, and corrosive action on drilling tubes and drilling hardware [550]. Moreover, hydrogen sulfide is a toxic gas.

Many polymers are used in drilling fluids as fluid loss control agents or viscosifiers. Because of the degradation of the polymers by bacteria in drilling fluids, an increase in fluid loss can occur. All naturally occurring polymers are capable of being degraded by bacterial action. However, some polymers are more susceptible to bacterial degradation than others. One solution, besides using bactericides, is replacing the starch with low-viscosity polyanionic cellulosic polymer, polyanionic lignin, or other enzyme-resistant polymers [838].

Certain additives are protected from biodegradation while drilling deep wells by quaternary ammonium salts [1482]. This results in a considerably reduced consumption of the additives needed.

Bacteria control is important not only in drilling fluids, but also for other oil and gas operations. The topic is treated more extensively in Chapter 5. Some bactericides especially recommended for drilling fluids are summarized in Table 1–9.

Table 1–9
Bactericides for Drilling Fluids

Bactericide	References
Tetrakis-hydroxymethyl-phosphonium sulfate[a]	[550]
Dimethyl-tetrahydro-thiadiazine-thione	[937]
2-Bromo-4-hydroxyacetophenone[b]	[1369]
Thiocyanomethylthio-benzothiazole[c]	[1367]
Dithiocarbamic acid; hydroxamic acid[d]	[117]
1,2-Benzoisothiazolin-3-one, 3-(3,4-dichlorophenyl)-1,1-dimethylurea, and di-iodomethyl-4-methylphenyl sulfone[e]	[1258]
Isothiazolinones	[513, 860, 863, 1257]

a) Absorbed on solid.
b) Synergistically effective with organic acids.
c) Synergistically effective with organic acids.
d) Fungicide.
e) Algicide.

Corrosion Inhibitors

Corrosion inhibitors are the subject of several topics in petroleum industries, such as transport and completion. They are detailed in Chapter 6.

Bentonite

Bentonites are highly colloidal and swell in water to form thixotropic gels. This property results from their micaceous sheet structure. Because of these viscosity-building characteristics, bentonites find major use as viscosity enhancers or builders in such areas as drilling muds and fluids, concrete and mortar additives, foundry and molding sands, and compacting agents for gravel and sand, as well as cosmetics. Most bentonites that are found in nature are in sodium or calcium form. The performance of a calcium bentonite as a viscosity builder often can be enhanced by its conversion to the sodium form. Crude bentonite can be upgraded to prepare a variety of solutions that have unusually high aqueous viscosities [153]. The crude material is sheared and dried. Sodium carbonate is then dry-blended with the material and pulverized. These bentonite clays are self-suspending, self-swelling, and self-gelatinizing when mixed with water.

The modification of bentonite with alkylsilanes improves the dispersing properties [991]. Incorporation of phosphonate-type compounds in bentonites for drilling mud permits the blockage of free calcium ions in the form of soluble and stable complexes and the preservation or restoration of the initial fluidity of the mud [1222]. The phosphonates also have dispersing and fluidizing effects on the mud.

Clay Stabilization

Selected clay stabilizers are shown in Table 1–10. Thermal-treated carbohydrates are suitable as shale stabilizers [1609–1611]. They may be formed by heating an alkaline solution of the carbohydrate, and the browning reaction product may be reacted with a cationic base. The inversion of nonreducing sugars may be first effected on selected carbohydrates, with the inversion catalyzing the browning reaction.

Formation Damage

Polyacrylates are often added to drilling fluids to increase viscosity and limit formation damage. The filter-cake is critical in preventing reservoir invasion by mud filtrate. Polymer invasion of the reservoir has been shown to have a great impact on permeability reduction [98]. The invasion of filtrate and solids in drilling in fluid can cause serious reservoir damage.

Table 1–10
Clay Stabilizers for Drilling Fluids

Additive	References
Modified poly-amino acid[a]	[265]
Acrylamide homopolymer	[134]
Amphoteric acetates and glycinates	[901]
Caproyloamphoglycinate, cocoamphodiacetate, disodium cocoamphodiacetate, lauroamphoacetate, sodium capryloamphohydroxypropyl sulfonate, sodium mixed C_8 amphocarboxylate, and alkylamphohydroxypropyl sulfonate	[43]
Polyvinylpyrrolidone, polyvinyl alcohol, starches, cellulosic material, or partially hydrolyzed polyacrylamide, and polypropylene glycol, or a betaine	
Quaternized trihydroxy alkyl amine	[1402]
Polyfunctional polyamine	[1192]

a) Water-sensitive smectite or illite shale formations.

Shale Stabilizer

Swelling due to shale hydration is one of the most important causes for borehole instability. Three processes contributing to shale instability are considered [127]:

1. Movement of fluid between the wellbore and shale, which is limited to flow from the wellbore into the shale
2. Changes in stress and strain, which occur during the interaction of shale and filtrate
3. Softening and erosion, caused by invasion of mud filtrate and consequent chemical changes in the shale

Adding a shale stabilizer to drilling fluids is an effective way to control clay swelling [653]. A copolymer of acrylamide and acrylonitrile has been found to be effective as a shale hydration swelling retarder. Experimental results showed that the inhibitors developed have good properties to inhibit shale hydration swelling, especially their quaternized product. 2-Hydroxybutyl ether and polyalkyl-ether–modified polygalactomannans are described as useful shale inhibitors [492]. A copolymer of styrene and maleic anhydride with alkylene oxide–based side chains is effective as a shale stabilizer [1643]. A variety of polyoxyalkyleneamines may serve as shale inhibition agents. It was found that polyoxypropylenediamine $H_2N-CH(CH_3)CH_2[-OCH_2CH(CH_3)]_x-NH_2$ [1403] is the best, with $x < 15$.

Table 1–11
Surface-Active Agents for Drilling Muds

Compound	References
Alkylpolyglycosides	[1082]
Amphoteric surfactants	[438]
Acetal or ketal adduct hydroxy polyoxyalkylene ether[a]	[578]
Amphoteric anion ethoxy and propoxy units	[797]
Alkanolamine	[796]

a) Controlling foam formation, drilling muds.

Table 1–12
Lost Circulation Additives

Material	References
Encapsulated lime	[1811]
Encapsulated oil-absorbent polymers	[470, 1812, 1813]
Hydrolyzed polyacrylonitrile Divinylsulfone, crosslinked	[1864]
Polygalactomannan gum	[982]
Polyurethane foam	[709]
Partially hydrolyzed polyacrylamide 30% hydrolyzed, crosslinked with Cr^{3+}	[1704]

Surfactants are used to change the interfacial properties. Suitable surfactants are given in Table 1–11.

Fluid Loss Additives

Fluid loss additives are detailed in Chapter 2.

Lost Circulation Additives

There are a number of methods that have been proposed to help prevent the loss of circulation fluid [1214]. Some of these methods use fibrous, flaky, or granular material to plug the pores as the particulate material settles out of the slurry. Examples are oat hulls [855], rice products [294, 295], waste olive pulp [519], nut cork [654, 1532], pulp residue waste [750], petroleum coke [1840], and shredded cellophane [296].

Other methods propose to use materials that interact in the fissures of the formation to form a plug of increased strength. Lost circulation additives are summarized in Table 1–12.

Water-Swellable Polymers. Certain organic polymers absorb comparatively large quantities of water, for example, alkali metal polyacrylate or crosslinked polyacrylates [734]. Such water-absorbent polymers, insoluble in water and in hydrocarbons, can be injected into the well with the objective of encountering naturally occurring or added water at the entrance to and within an opening in the formation. The resulting swelling of the polymer forms a barrier to the continued passage of circulation fluid through that opening into the formation. The hydrocarbon carrier fluid initially prevents water from contacting the water-absorbent polymer until such water contact is desired. Once the hydrocarbon slug containing the polymer is properly placed at the lost circulation zone, water is mixed with the hydrocarbon slug so that the polymer will expand with the absorbed water and substantially increase in size to close off the lost circulation zone [194, 471, 1812, 1814]. The situation is similar to an oil-based cement.

The opposite mechanism is used by a hydrocarbon-swellable elastomer [1857].

Anionic Association Polymer. Another type of lost circulation agent is a combination of an organic phosphate ester and an aluminum compound, for example, aluminum isopropoxide. The action of this system as a fluid loss agent seems to be that the alkyl phosphate ester becomes crosslinked by the aluminum compound to form an anionic association polymer, which serves as the gelling agent [1488].

Aphrons. Other lost circulation additives can be encapsulated. The encapsulation is dissolved and the material swells to close fissures. Microbubbles in a drilling fluid can be generated by certain surfactants, and polymers known as *aphrons* are a different approach to reduce the fluid loss [891]. An aphron drilling fluid is similar to a conventional drilling fluid, but the drilling fluid system is converted to an energized air-bubble mud system before drilling [967].

Permanent Grouting. Lost circulation also can be suppressed by grouting permanently, either with cement [37, 409] or with organic polymers that cure in situ.

Scavengers

Oxygen Scavenger

Oxygen corrosion is often underestimated. Studies have shown that the corrosion can be limited when proper oxygen scavengers are used. Hydrazine leads the group of chemicals that are available for oxygen removal. Because of its special properties, it is used for corrosion control in heating systems and in drilling operations, well workover, and cementing [1628].

Hydrogen Sulfide Removal

It is sometimes necessary to remove hydrogen sulfide from a drilling mud. Techniques using iron compounds that form sparingly soluble sulfides have been developed, for example, with iron (II) oxalate [1695] and iron sulfate [1456]. The sulfur is precipitated out as FeS. Ferrous gluconate is an organic iron-chelating agent, stable at pH levels as high as 11.5 [452].

Zinc compounds have a high reactivity with regard to H_2S and therefore are suitable for the quantitative removal of even small amounts of hydrogen sulfide [1831]. However, at high temperatures they may negatively affect the rheology of drilling fluids.

Surfactants

Surfactant in Hydrocarbon Solvent

Methyl-diethyl-alkoxymethyl ammonium methyl sulfate has high foam extinguishing properties [563].

Biodegradable Surfactants

Alkylpolyglucosides (APGs) are highly biodegradable surfactants [1344]. The addition of APGs, even at very low concentrations, to a polymer mud can drastically reduce the fluid loss even at high temperatures. Moreover, both fluid rheology and temperature resistance are improved.

Deflocculants and Dispersants

Deflocculants have a relatively low molecular weight. Polymers composed of sodium styrene sulfonate, maleic anhydride, and a zwitterionic functionalized maleic anhydride [738, 1411, 1412, 1415] are suitable.

Complexes of tetravalent zirconium with organic acids, such as citric, tartaric, malic, and lactic acids, and a complex of aluminum and citric acid have been claimed to be active as dispersants. The dispersant is especially useful in dispersing bentonite suspensions [288]. Polymers with amine sulfide terminal moieties are synthesized by using aminethiols as chain transfer agents in aqueous addition polymerizations. The polymers are useful as mineral dispersants [1182].

Defoamers

Defoamers are covered in Chapter 21.

Shale Stabilizing Surfactants

There are special shale stabilizing surfactants consisting of nonionic alkanolamides [900, 902], for example, acetamide monoethanolamines and diethanolamines.

Toxicity

Alkylphenol ethoxylates (APEOs) are a class of surfactants that have been used widely in the drilling fluid industry. The popularity of these surfactants is based on their cost-effectiveness, availability, and range of obtainable hydrophilic–lipophilic balance values [693]. Studies have shown that APEOs exhibit oestrogenic effects and can cause sterility in some male aquatic species. This may have subsequent human consequences, and such problems have led to a banning of their use in some countries and agreements to phase out their use. Alternatives to products containing APEOs are available, and in some cases they show an even better technical performance.

Hydrate Inhibitors

Hydrate inhibitors for drilling fluids are summarized in Chapter 13.

Weighting Materials

There are many weighting materials, including barite and iron oxides, to increase the specific weight of a slurry. Conversely, the specific weight can be reduced by foaming or hollow glass particles.

Barite

Barite has been used as a weighting agent in drilling fluids since the 1920s. It is preferred over other materials because of its high density, low production costs, low abrasiveness, and ease of handling. Other weighting materials have been used, but they are problematic or costly. Finished barite producers sometimes blend ores from different sources to obtain the desired average density to meet API specifications. Some barite ores contain alkaline-soluble carbonate minerals that can be detrimental to a drilling fluid, such as iron carbonate (siderite), lead carbonate (cerussite), and zinc carbonate (smithsonite) [1020].

Details of how to characterize barite have been worked out [64]. Barite can be modified to become oleophilic [1607, 1608].

To recover barite from drilling muds, a direct flotation without prior dewatering and washing of the drilling muds has been described [809]. An alkylphosphate is used as a collecting and frothing reagent.

Ilmenite

Environmental aspects suggest replacing barite with ilmenite. However, the use of ilmenite as weighting material can cause severe erosion problems. Using ilmenite with a narrow particle size distribution around $10\,\mu$ can reduce the erosion to a level experienced with barite [1543].

Carbonate

It is reasonable to replace barite and iron-based weighting material with carbonate if a high degree of weighting the muds is not required by the drilling conditions. Besides being cheaper than barite, a carbonate weighting material is less abrasive, which is especially important when drilling is performed in producing formations, and is readily soluble in hydrochloric acid. The main shortcomings of carbonate powders are due to the presence of a coarsely divided fraction and noncarbonate impurities [1112].

Zinc Oxide, Zirconium Oxide, and Manganese Tetroxide

Zinc oxide [ZnO] is a particularly suitable material for weighting because, it has a high density (5.6 g/ml versus 4.5 for barite); it is soluble in acids (e.g., HCl); and its particle size can be designed so that it does not invade the formation. Acid solubility is particularly desirable because dissolved ZnO can be produced through a production screen without plugging it. A high density means less weighting material is needed per unit mud volume to achieve a desired density. The particle size (around $10\,\mu$) is such that the ZnO particles do not invade the formation core with the filtrate. On the other hand, the particle size it is not large enough to settle out of suspension. Zirconium oxide possesses similar properties as ZnO. It has a density of 5.7 and is soluble in nitric acid and hot concentrated hydrochloric, hydrofluoric and sulfuric acids. Therefore, a filter-cake formed from zinc oxide or zirconium oxide then can be dissolved. The high acid solubility of ZnO makes it particularly suitable as weighting material [1071].

On the other hand, manganese tetroxide is so fine that it invades the formation with the filtrate.

Hollow Glass Microspheres

Initially glass microspheres were used in the 1970s to overcome severe lost circulation problems in the Ural Mountains. The technology has been used in other sites [1189]. Hollow glass beads reduce the density of a drilling fluid and can be used for underbalanced drilling [1199–1201]. Field applications have been reported [73].

Organoclay Compositions

It has long been known that organophilic clays can be used to thicken a variety of organic compositions. Such organophilic clays are prepared by the reaction of an organic cation with a clay in various methods known in the art. If the organic cation contains at least one alkyl group containing at least 8 to 10 carbon atoms, then such organoclays have the property of increasing viscosity in organic liquids and thus providing rheologic properties to a wide variety of such liquids including paints, coatings, adhesives, and similar products. It is also well known that such organoclays may function to thicken polar or nonpolar solvents, depending on the substituents on the organic salt. The efficiency of organophilic clays in nonaqueous systems can be further improved by adding a low–molecular-weight polar organic material to the composition. Such polar organic materials have been called *dispersants, dispersion aids*, and *solvating agents*. The most efficient polar materials have been found to be low–molecular-weight alcohols and ketones, particularly methanol and acetone.

There is a synergistic action of two or more types of organic salts in the presence of an organic anion. The combination of hydrophobic and hydrophilic organic salts and an organic anion provides an organophilic clay gellant, which exhibits improved gelling properties in nonaqueous systems [1312]. Examples are dimethyl dihydrogenated tallow quaternary ammonium chloride and methyl bis-polyoxyethylene (15 units) cocoalkyl quaternary ammonium chloride and the salts from stearic acid, succinic acid, and tartaric acid [1166, 1310, 1311, 1313].

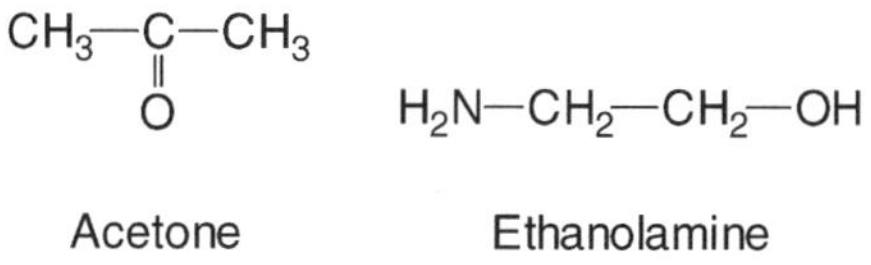

Acetone Ethanolamine

Figure 1–5. Acetone, ethanolamine.

Reticulated Bacterial Cellulose

Reticulated bacterial cellulose may be used in place of a conventional gellant or in combination with conventional gellants to provide enhanced drilling muds [1837]. The addition of relatively small quantities of reticulated bacterial cellulose to wellbore drilling muds enhances their rheologic properties.

Scleroglucan

Scleroglucan is a polysaccharide secreted by the mycelia of certain microorganisms. It is produced by aerobic fermentation of D-glucose by such microorganisms. Scleroglucan has been proposed as a better alternative to xanthan gum for drilling fluid compositions [666].

For drilling fluid applications, scleroglucan can be used unrefined. It is an effective thickener for water [1795] and enhances the lubricating and cleaning power of water-based muds [506, 507]. In drilling of deviated wells, scleroglucan permits better cleaning of the well [508, 509, 1793, 1794]. A further application is in drilling jobs concerning large-diameter wells [1045–1047].

Uintaite

Uintaite is a naturally occurring hydrocarbon mineral classified as an asphaltite. It is a natural product whose chemical and physical properties vary and depend strongly on the uintaite source. Uintaite has also been called "Gilsonite" although this usage is incorrect; Gilsonite is a registered trademark of American Gilsonite Co., Salt Lake City, Utah.

General purpose Gilsonite brand resin has a softening point of about 175° C, and Gilsonite HM has a softening point of about 190° C, and Gilsonite Select 300 and Select 325, which have softening points of about 150° C and 160° C, respectively. The softening points of these natural uintaites depend primarily on the source vein that is mined when the mineral is produced.

Uintaite is described in Kirk–Othmer [1329]. Typical uintaite used in drilling fluids is mined from an area around Bonanza, Utah and has a specific gravity of 1.05 with a softening point ranging from 190° C to 205° C, although a lower softening point (165° C) material is sometimes used. It has a low acid value, a zero iodine number, and is soluble or partially soluble in aromatic and aliphatic hydrocarbons, respectively.

For many years uintaite and other asphaltic-type products have been used in water-based drilling fluids as additives assisting in borehole stabilization. These additives can minimize hole collapse in formations containing water-sensitive, sloughing shales.

Uintaite and asphalt–type materials have been used for many years to stabilize sloughing shales and to reduce borehole erosion. Other benefits derived from these products include borehole lubrication and reduction in filtration.

Uintaite is not easily water wet with most surfactants. Thus, stable dispersions of uintaite are often difficult to achieve, particularly in the presence of salts, calcium, solids and other drilling fluid contaminants and/or in the presence of diesel oil. The uintaite must be readily dispersible and must remain water wet; otherwise it will coalesce and be separated from the drilling fluid, along with cuttings at the shale shaker or in the circulating pits. Surfactants and emulsifiers are often used with uintaite drilling mud additives.

Uintaite is not water-wettable. Loose or poor bonding of the surfactant to the uintaite will lead to its washing off during use, possible agglomeration, and the removal of uintaite from the mud system with the drilling wastes. Thus, the importance of the wettability, rewettability, and storage stability criteria is evident [382].

An especially preferred product comprises about 2 parts Gilsonite HM, about 1 part Gilsonite Select, about 1 part causticized lignite and about 0.1 to 0.15 part of a nonionic surfactant [378–382].

Sodium Asphalt Sulfonate

Neutralized sulfonated asphalt (i.e., salts of sulfonated asphalt and their blends with materials such as Gilsonite, blown asphalt, lignite, and mixtures of the latter compounds) are commonly used as additives in drilling fluids. These additives, however, cause some foaming in water or water-based fluids. Furthermore, these additives are only partially soluble in the fluids. Therefore, liquid additives have been developed to overcome some of the problems associated with the use of dry additives. On the other hand, with liquid compositions containing polyglycols, stability problems can arise. Stable compositions can be obtained by special methods of preparation [1407]. In particular first the viscosifier is mixed with water, then the polyglycol, and finally the sulfonated asphalt is added.

Formation Damage by Gilsonite and Sulfonated Asphalt

Laboratory experiments have been conducted with a chromium lignite–chromium lignosulfonate mud system both without and with solid lubricants. These studies include filtration loss, cake quality, and their impact on the formation. A comparative evaluation has led to the conclusion that Gilsonite is a better additive compared with sulfonated asphalt as it results in less filtration loss and compact cake formation, thereby reducing formation damage. The flow studies have indicated that the addition of these solid lubricants can be used in drilling fluids without adverse impact on the producing zones [672].

Cuttings Removal by Sweep Materials

In drilling of deviated and horizontal wells, gravity causes deposits of drill cuttings and especially fines, or smaller sized cuttings, to build up along the lower or bottom side of the wellbore. Such deposits are commonly called *cuttings beds*. Buildup of cuttings beds can lead to undesirable friction and possibly to sticking of the drill string.

Removing the drill cuttings from a deviated well, in particular when drilled at a high angle, can be difficult. Limited pump rate, eccentricity of the drill pipe, sharp build rates, high bottom hole temperatures, and oval-shaped wellbores can contribute to inadequate hole cleaning. Well treatments or circulation of fluids specially formulated to remove such cuttings beds are periodically necessary to prevent buildup to the degree that the cuttings or fines interfere with the drilling apparatus or otherwise with the drilling operation.

Commonly, the drilling operation must be stopped while such treatment fluids are swept through the wellbore to remove the fines. Alternatively, or additionally, special viscosifier drilling fluid additives have been proposed to enhance the ability of the drilling fluid to transport cuttings, but such additives at best merely delay the buildup of cuttings beds and can be problematic if they change the density of the drilling fluid. A mechanical operation for removing cutting beds has also been used wherein the drill string is pulled back along the well, pulling the bit through the horizontal or deviated section of the well.

Barium sulfate can be used as a sweep material. The material should be ground and sieved to a size range sufficiently small to be suspendable in the drilling fluid. After adding the composition to the drilling fluid, the fluid is circulated in the wellbore and allowed to remove the small cuttings or cuttings beds from the borehole and deliver them to the well surface. The composition and the cuttings are then removed from the drilling fluid in a manner that prevents a significant change of the density of the drilling fluid. The composition is removed from the drilling fluid by sieving or screening the drilling fluid, which may preferably be accomplished by the principal shale shaker in the drilling operation [1833].

Junk Removal

Drilling equipment that is broken or stuck in the hole can be dissolved by means of a mixture of nitric and hydrochloric acids in the proportion of 1:3. To accelerate the dissolving of the metal, a mixture containing 1.1 parts by weight of sodium nitrate and 1.0 part of monoethanolamine is added initially to the acids in the amount of 0.05 to 13.0 parts by weight per 100 parts of acid mixture. The acidic residue in the hole is neutralized by addition of alkali and converted into drilling fluid by addition of polymer solution [504].

Drilling Fluid Disposal

Conversion into Cements

Water-based drilling fluids may be converted into cements using a hydraulic blast furnace slag [163, 410, 411, 413, 1897]. Hydraulic blast furnace slag is a unique material that has low impact on rheologic and fluid loss properties of drilling fluids. It can be activated to set in drilling fluids that are difficult to convert to cements with other solidification technologies. Hydraulic blast furnace slag is a more uniform and consistent quality product than are Portland well cements, and it is available in large quantities from multiple sources. Fluid and hardened solid properties of blast furnace slag and drilling fluids mixtures used for cementing operations are comparable with properties of conventional Portland cement compositions.

Environmental Regulations

In response to effluent limitation guidelines promulgated by the EPA for discharge of drilling wastes offshore, alternatives to water– and oil–based muds have been developed. Thus synthetic–based muds are more efficient than water-based muds for drilling difficult and complex formation intervals and have lower toxicity and smaller environmental impacts than diesel or conventional mineral oil–based muds. Synthetic drilling fluids may present a significant pollution prevention opportunity because the fluids are recycled, and smaller volumes of metals are discharged with the cuttings than for water–based muds. A framework for a comparative risk assessment for the discharge of synthetic drilling fluids has been developed. The framework will help identify potential impacts and benefits associated with the use of drilling muds [1205].

Characterization of Drilling Muds

Important parameters to characterize the properties of a drilling mud are specific weight, viscosity, gel strength, and filtration performance.

Viscosity

Viscosity is measured by means of a Marsh funnel. The funnel is dimensioned so that the outflow time of 1 qt (926 ml) freshwater at 21° C (70° F) is 26 s.

Viscosity also is measured with a rotational viscometer. The mud is placed between two concentric cylinders. One cylinder rotates with constant velocity. The other cylinder is connected with a spring. The torque on this cylinder results in a deviation of the position from rest, which may serve as a measure of viscosity.

Gel strength is obtained with the rotational viscometer when the maximal deflection of the pointer is monitored when the motor is turned on with low speed, the liquid being at rest for a prolonged time before, for example, for 10 minutes. This maximal deflection is referred to as a 10-minute gel.

API Filtration

A filter press is used to determine the wall-building characteristics of a mud. This press consists of a cylindrical chamber, which may withstand alkaline milieu. A filter paper is placed on the bottom of the chamber. The mud is placed into the chamber and a pressure of 0.7 MPa is applied. After 30 minutes the volume of filtrate is reported. The filter-cake is inspected visually and the consistency is noted as hard, soft, tough, rubbery, or firm.

There is another procedure suitable for oil–based muds under high-pressure and high-temperature conditions. Filtration is performed at 100 psi (7 MPa) and temperatures of 93° C (200° F).

It should be noted that research has shown that there may be significant differences between static and dynamic filtering.

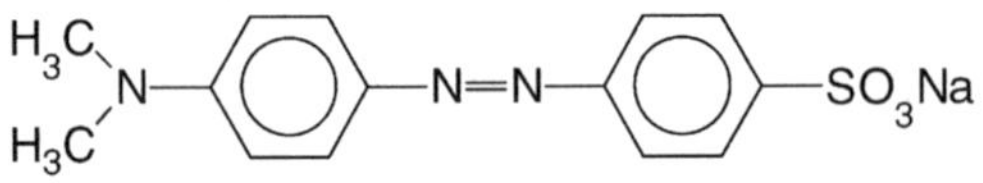

Methylorange

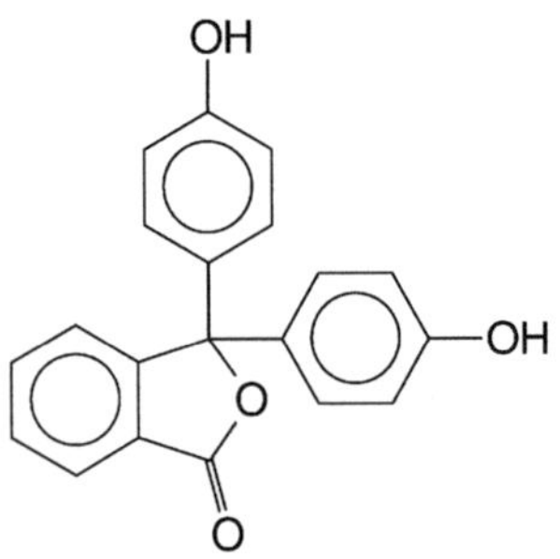

Phenolphthalein

Figure 1–6. Methylorange, phenolphthalein.

Alkalinity and pH

Alkalinity is measured by acid–base titration with methylorange or phenolphthalein as indicator. Phenolphthalein changes color at pH 8.3, whereas methylorange changes color at pH 4.3. At pH 8 the neutralization of the strong alkali ingredients like NaOH is essentially complete. Further reduction of the pH to 4 will also measure carbonates and bicarbonates. Colorimetric tests and glass electrode systems are used to determine pH.

Total Hardness

The sum of calcium and magnesium ions in the mud determines the total hardness. These ions are analyzed with complexometric titrations using EDTA (versenate).

Chapter 2

Fluid Loss Additives

Fluid loss additives are also called *filtrate-reducing agents*. Fluid losses may occur when the fluid comes in contact with a porous formation. This is relevant for drilling and completion fluids, fracturing fluids, and cement slurries.

The extent of fluid loss is dependent on the porosity and thus the permeability of the formation and may reach approximately 10 t/hr. Because the fluids used in petroleum technology are in some cases quite expensive, an extensive fluid loss may not be tolerable. Of course there are also environmental reasons to prevent fluid loss.

Mechanism of Action of Fluid Loss Agents

Reduced fluid loss is achieved by plugging a porous rock in some way. The basic mechanisms are shown in Table 2–1.

Action of Macroscopic Particles

A monograph concerning the mechanism of invasion of particles into the formation is given by Chin [375].

One of the basic mechanisms in fluid loss prevention is shown in Figure 2–1. The fluid contains suspended particles. These particles move with the lateral flow out of the drill hole into the porous formation. The porous formation acts like a sieve for the suspended particles. The particles therefore will be captured near the surface and accumulated as a filter-cake.

The hydrodynamic forces acting on the suspended colloids determine the rate of cake buildup and therefore the fluid loss rate. A simple model has been proposed in literature [907] that predicts a power law relationship between the filtration rate and the shear stress at the cake surface. The model shows that the cake formed will be inhomogeneous with smaller and smaller particles being deposited as the filtration proceeds. An equilibrium cake thickness is achieved when no particles small enough to be deposited are available in the suspension. The cake thickness as a function of time can be computed from the model.

Table 2–1
Mechanisms of Fluid Loss Prevention

Macroscopic particles	Suspended particles may clog the pores, forming a filter-cake with reduced permeability.
Microscopic particles	Macromolecules form a gel in the boundary layer of a porous formation.
Chemical grouting	A resin is injected in the formation, which cures irreversibly; suitable for bigger caverns.

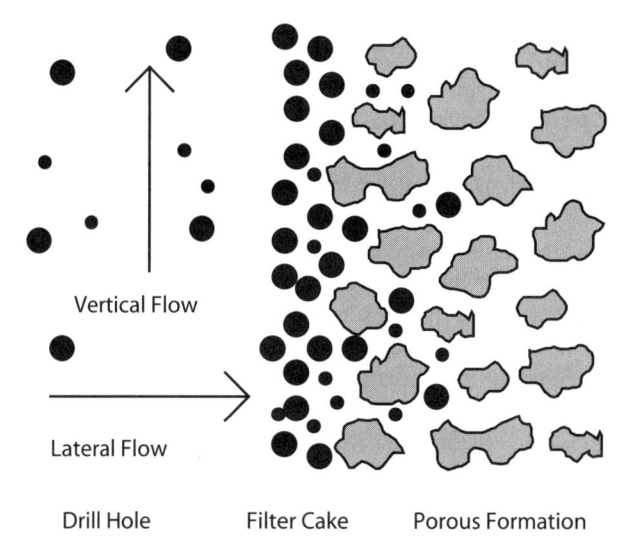

Figure 2–1. Formation of a filter-cake in a porous formation from suspension (•) in a drilling fluid. The suspended particles can invade the formation to some extent.

For a given suspension rheology and flow rate there is a critical permeability of the filter, below which no cake will be formed. The model also suggests that the equilibrium cake thickness can be precisely controlled by an appropriate choice of suspension flow rate and filter permeability.

Action of Cement Fluid Loss Additives

Two stages are considered with respect to the fluid loss behavior of a cement slurry [140]:

1. A dynamic stage corresponding to placement
2. A static stage, awaiting the setting of the cement

During the first period, the slurry flow is eroding the filter-cake as it is growing; thus a steady state, in which the filtration occurs through a cake of constant thickness, is rapidly reached. At the same time, because the slurry is losing water but no solid particles, its density is increasing in line with the fluid loss rate.

During the second period, the cake grows because of the absence of flow. It may grow to a point at which it locally but completely fills the annulus: Bridging takes place and the hydrostatic pressure is no longer transmitted to the deeper zones. From the typical mudcake resistance it can be estimated that under both dynamic and static conditions, the fluid loss could require reduction to an American Petroleum Institutue (API) value lower than what is generally considered a fair control of fluid loss.

Testing of Fluid Loss Additives

Predictions on the effectiveness of a fluid loss additive formulation can be made on a laboratory scale by characterizing the properties of the filter-cake formed by appropriate experiments. Most of the fluids containing fluid loss additives are thixotropic. Therefore the apparent viscosity will change when a shear stress in a vertical direction is applied, as is very normal in a circulating drilling fluid. For this reason, the results from static filtering experiments are expected to be different in comparison with dynamic experiments.

Static fluid loss measurements, which are the present standardized testing method, provide inadequate results for comparing fracturing fluid materials or for understanding the complex mechanisms of viscous fluid invasion, filter-cake formation, and filter-cake erosion [1806]. On the other hand, dynamic fluid loss studies have inadequately addressed the development of proper laboratory methods, which has led to erroneous and conflicting results.

Results from a large-scale, high-temperature, high-pressure simulator were compared with laboratory data, and significant differences in spurt loss values were found [1125].

Static experiments with pistonlike filtering can be reliable, however, to obtain information on the fluid loss behavior in certain stages of a cementation process, in particular when the slurry is at rest.

Organic Additives

The properties of the filter-cake formed by macroscopic particles can be significantly influenced by certain organic additives. The overall mechanism of water-soluble fluid loss additives has been studied by determining the electrophoretic mobility of filter-cake fines. Water-soluble fluid loss additives are divided into four types according to their different effects on the negative electrical charge density of filter-cake fines [1891]:

1. Electrical charge density is reduced, such as polyethylene
2. Glycol and pregelatinized starch
3. Electrical charge density is not changed, such as carboxymethyl starch
4. Electrical charge density is increased, such as sulfonated phenolic resin, carboxymethylcellulose, and hydrolyzed polyacrylonitrile

The properties of filtrate reducers contribute to their different molecular structures. Nonionic filtrate reducers work by completely blocking the filter-cake pore, and anionic ones work by increasing the negative-charge density of filter-cakes and decreasing pore size. Anionic species cause further clay dispersion, but nonionic species do not, and both of them are beneficial to colloid stability [1890].

The change of properties of the filter-cake due to salinity and polymeric additives has been studied by scanning electron microscope (SEM) photography [1438]. Freshwater muds with and without polymers such as starch, polyanionic cellulose (PAC), and a synthetic high-temperature–stable polymer, were prepared, contaminated with electrolytes (NaCl, $CaCl_2$, $MgCl_2$), and aged at 200° to 350° F. Static API filtrates before and after contamination and aging were measured. The freeze-dried API filter-cakes were used for SEM studies. The filter-cake structure was influenced by electrolytes, temperature, and polymers. In an unaged, uncontaminated mud, bentonite forms a cardhouse structure with low porosity. Electrolyte addition increases the average filter-cake pore size. Temperature causes coagulation and dehydration of clay platelets. Polymers protect bentonite from such negative effects.

Formation Damage

The damage of the formation resulting from the use of a filtration loss agent can be a serious problem for certain fields of application. Providing effective fluid loss control without damaging formation permeability in completion operations has been a prime requirement for an ideal fluid loss control pill.

Filter-cakes are hard to remove and thus can cause considerable formation damage. Cakes with very low permeability can be broken up by reverse flow. No high-pressure spike occurs during the removal of the filter-cake. Typically a high-pressure spike indicates damage to the formation and wellbore surface because damage typically reduces the overall permeability of the formation. Often formation damage results from the incomplete back-production of viscous, fluid loss control pills, but there may be other reasons.

Reversible Gels

Another mechanism for fluid loss prevention is caused by other additives, which are able to form gels on a molecular mechanism.

Ultrafine Filtrate–Reducing Agents

Methods are available for reducing the fluid loss and reducing the concentration of polymer required to provide a desired degree of fluid loss control to a drilling fluid and to a well servicing fluid, respectively [500]. The fluids contain, as usual polymeric viscosifiers, a polymeric fluid loss additive and a water-soluble bridging agent suspended in a liquid in which the bridging agent is not soluble. It is important to add to the fluids a particulate, water-soluble, ultrafine filtrate–reducing agent. The particle size distribution should be such that approximately 90% of the particles are less than 10 μ, the average particle size being between 3 and 5 μ and the ultrafine filtrate–reducing agent being insoluble in the liquid.

Bentonite

Bentonite is an impure clay that is formed by weathering of volcanic tuffs. It contains a high content of montmorillonite. Bentonites exhibit properties such as ability to swell, ion exchange, and thixotropy. Properties can be modified by ion exchange, for example, exchange of earth alkali metals to alkali metals. The specific surface can be modified with acid treatment. Organophilic properties can be increased by treatment with quaternary ammonia compounds.

Polydrill

Polydrill is a sulfonated polymer for filtration control in water-based drilling fluids [1775]. Tests demonstrated the product's thermal stability up to 200° C and its outstanding electrolyte tolerance. Polydrill can be used in NaCl-saturated drilling fluids as well as in muds containing 75,000 ppm of calcium or 100,000 ppm of magnesium. A combination of starch with Polydrill was used successfully in drilling several wells. The deepest hole was drilled with 11 to 22 kg/m^3 of pregelatinized starch and 2.5 to 5.5 kg/m^3 of Polydrill to a depth of 4800 m. Field experience with the calcium-tolerant starch/Polydrill system useful up to 145° C has been discussed in detail [1774].

In dispersed muds (e.g., lignite or lignosulfonate), minor Polydrill additions result in a significantly improved high-temperature, high-pressure filtrate. Major benefits come from a synergism of polymer with starch and polysaccharides. The polymer exerts a thermal stabilizing effect on those polymers. In conventional or clay-free drilling and completion fluids, Polydrill can be used by itself or in combination with other filtrate reducers for various purposes [1436]. Handling and discharge of the product as well as the waste mud created no problem in the field.

Bacteria

Instead of using polymers, the addition of bacteria cultures, which may form natural polymers and could then prevent fluid loss, has been suggested. In one study, a bacterial culture selected for its abundant exopolymer production was added to drilling mud to determine whether the PAC component could be replaced without sacrificing viscosity or fluid retention [50]. Drilling mud performance was tested using a standard API test series. The bacterial inoculum was not as effective in maintaining viscosity or preventing fluid loss as was the PAC. However, the inoculum was capable of reducing the amount of PAC required in the drilling mud.

The combination of the bacterial inoculum with less expensive start sources, for example, carboxylated methyl cornstarch, crosslinked hydroxypropyl cornstarch, and amine-derivatized potato starch, gave viscosity and fluid loss control as good as or better than PAC alone. The bacterial strain tested was effective over a wide range of drilling mud conditions with growth at varying pH (3 to 11), varying salinities (0% to 15%), and a wide range of temperatures.

Polysaccharides

Cellulose-Based Fluid Loss Additives

Polyanionic Cellulose

A composition containing polyanionic cellulose and a synthetic polymer of sulfonate has been tested for reducing the fluid loss and for the thermal stabilization of a water-based drilling fluid for extended periods at deep well drilling temperatures [812].

Sulfonate. When a sulfonate-containing polymer is added to a drilling fluid containing PAC, the combination reduces fluid loss. Improved fluid loss is obtained when PAC and the sulfonate-containing polymer, which has a molecular weight of 300,000 to 10 million Dalton, are combined in a water-based drilling mud after prolonged aging at 300° F.

Carboxymethylcellulose

Certain admixtures of carboxymethylhydroxyethylcellulose or copolymers and copolymer salts of N,N-dimethylacrylamide and 2-acrylamido-2-methylpropane sulfonic acid (AMPS), together with a copolymer of acrylic acid, may provide fluid loss control to cement compositions under elevated temperature conditions [252].

Hydroxethylcellulose

Hydroxyethylcellulose with a degree of substitution of 1.1 to 1.6 has been described for fluid loss control in water-based drilling fluids [1473]. An apparent viscosity in water of at least 15 cP should be adjusted to achieve an API fluid loss of less than 50 ml/30 min. Crosslinked hydroxyethylcellulose is suitable for high-permeability formations [344, 346].

A derivatized hydroxyethylcellulose polymer gel exhibited excellent fluid-loss control over a wide range of conditions in most common completion fluids. This particular grated gel was compatible with the formation material and caused little or no damage to original permeability [1341]. Detailed measurements of fluid loss, injection, and regained permeability were taken to determine the polymer particulate's effectiveness in controlling fluid loss and to assess its ease of removal. Hydroxyethylcellulose can be etherified or esterified with long chain alcohols or esters. An ether bond is more stable in aqueous solution than is an ester bond [96].

Starch

Crosslinked Starch

A crosslinked starch was described as a fluid loss additive for drilling fluids [632, 1626]. The additive resists degradation and functions satisfactorily after exposure to temperatures of 250° F for periods of up to 32 hours. To obtain crosslinked starch, a crosslinking agent is reacted with granular starch in an aqueous slurry. The crosslinking reaction is controlled by a Brabender

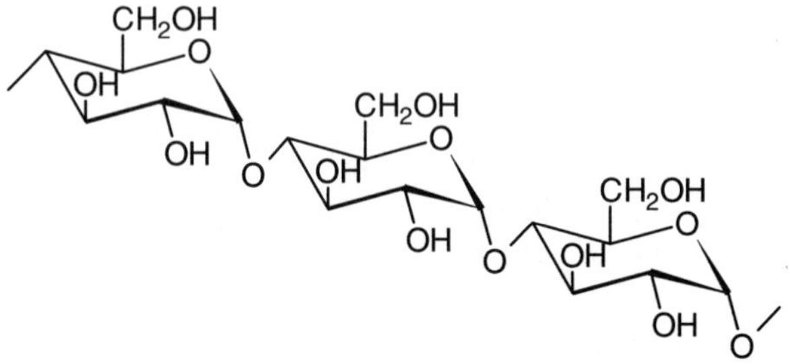

Starch (Amylose)

Figure 2–2. Starch.

Viscometer test. Typical crosslinked starches are obtained when the initial rise of the viscosity of the product is between 104° and 144° C, and the viscosity of the product does not rise above 200 Brabender units at temperatures less than 130° C. The crosslinked starch slurry is then drum-dried and milled to obtain a dry product. The effectiveness of the product is checked by the API Fluid Loss Test after static aging of sample drilling fluids containing the starch at elevated temperatures. The milled dry product can then be incorporated into the oil well drilling fluid of the drill site.

Granular Starch and Mica

A fluid loss additive is described that consists of granular starch composition and fine particulate mica [337]. An application comprises a fracturing fluid containing this additive. A method of fracturing a subterranean formation penetrated by a borehole comprises injecting into the borehole and into contact with the formation, at a rate and pressure sufficient to fracture the formation, a fracturing fluid containing the additive in an amount sufficient to provide fluid loss control.

Depolymerized Starch

Partially depolymerized starch provides decreased fluid losses at much lower viscosities than the corresponding starch derivatives that have not been partially depolymerized [498].

Controlled Degradable Fluid Loss Additives

A fluid loss additive for a fracturing fluid comprises a mixture of natural and modified starches plus an enzyme [1850]. The enzyme degrades the α-linkage of starch but does not degrade the β-linkage of guar and modified guar gums when used as a thickener. Natural or modified starches are utilized in a preferred ratio of 3:7 to 7:3, with optimum at 1:1, and the mix is used in the dry form for application from the surface down the well. The preferred modified starches are the carboxymethyl and hydroxypropyl derivatives. Natural starches may be those of corn, potatoes, wheat, and soy, and the most preferred is corn starch. Blends include two or more modified starches, as well as blends of natural and modified starches. Optionally, the starches are coated with a surfactant, such as sorbitan monooleate, ethoxylated butanol, or ethoxylated nonylphenol, to aid dispersion into the fracturing fluid.

A fluid loss additive is described [1849] that helps achieve a desired fracture geometry by lowering the spurt loss and leak-off rate of the fracturing fluid into the surrounding formation by rapidly forming a filter-cake with low permeability. The fluid loss additive is readily degraded after the completion of the fracturing process. The additive has a broad particulate size distribution that is ideal for use in effectively treating a wide range of formation porosities and is easily dispersed in the fracturing fluid. The fluid loss additive comprises a blend of modified starches or blends of one or more modified starches and one or more natural starches. These blends have been found to maintain injected fluid within the created fracture more effectively than natural starches. The additive is subject to controlled degradation to soluble products by a naturally proceeding oxidation reaction or by bacterial attack by bacteria naturally present in the formation. The oxidation may be accelerated by adding oxidizing agents such as persulfates and peroxides.

Guar

A hydrophobically modified guar gum can be used as an additive for drilling, completion, or servicing fluids [93,94]. The modified gum is used together with polymers or reactive clay.

Hydroxypropylguar Gum

Hydroxypropylguar gum gel can be crosslinked with borates [1227], titanates, or zirconates. Borate-crosslinked fluids and linear hydroxyethylcellulose gels are the most commonly used fluids for high-permeability fracture treatments. This is for use for hydraulic fracturing fluid under high-temperature and high-shear stress.

Succinoglycan

Succinoglycan is a biopolymer. It has been shown to possess a combination of desirable properties for fluid loss control [1069,1070]. These include ease of mixing, cleanliness, shear-thinning rheology, temperature-insensitive viscosity below its transition temperature (T_m), and an adjustable transition temperature (T_m) over a wide range of temperatures. Being a viscous fluid, succinoglycan relies solely on viscosity to reduce fluid loss. It does not form a hard-to-remove filter-cake, which can cause considerable formation damage. Based on these findings, succinoglycan has been used successfully as a fluid loss pill before and after gravel packing in more than 100 offshore wells. Calculations based on laboratory-measured rheology and field experience have shown succinoglycan to be effective in situations in which hydroxyethylcellulose is not. Fluid loss, even over 40 barrels/hr, was reduced to several barrels per hour after application of a properly designed succinoglycan pill. Most wells experienced no problem in production after completion.

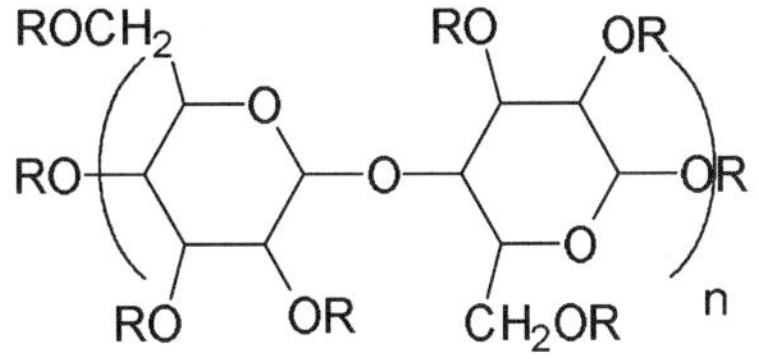

Hydroxyethylcellulose R=(CH$_2$CH$_2$O)$_m$OH

Carboxymethylcellulose R=(CH$_2$CO)OH

Figure 2–3. Hydroxyethylcellulose, carboxymethylcellulose.

Succinoglycan can be degraded with an internal acid breaker [222]. The formation damage that results from the incomplete back-production of viscous fluid loss control pills can be minimized if a slow-acting internal breaker is employed. In particular, core-flow tests have indicated that combining a succinoglycan-based pill with a hydrochloric acid internal breaker enables a fluid loss system with sustained control followed by delayed breakback and creates only low levels of impairment. To describe the delayed breaking of the succinoglycan/hydrochloric acid system, a model, based on bond-breaking rate, has been used. With this model, it is possible to predict the change of the rheologic properties of the polymer as a function of time for various formation temperatures, transition temperatures of the succinoglycan, and acid concentrations. The model can be used to identify optimal formulations of succinoglycan and acid breaker on the basis of field requirements, such as the time interval over which fluid loss control is needed, the overbalance pressure a pill should be able to withstand, and the brine density required.

Polyether-Modified Polysaccharides

Compositions containing mixtures of metal hydroxides a polysaccharide, partially etherified with hydroxyethyl and hydroxypropyl groups, are used as fluid loss additives for aqueous, clay-mineral–based drilling muds [1437].

Scleroglucan

A combination of graded calcium carbonate particle sizes, a nonionic polysaccharide of the scleroglucan type, and a modified starch, has been claimed for use as a fluid loss formulations [915]. It is important that the calcium carbonate particles are distributed across a wide size range to prevent filtration or fluid loss into the formation. Because the filter-cake particles do not invade the wellbore due to the action of the biopolymer and starch, no high-pressure spike occurs during the removal of the filter-cake.

The rheologic properties of the fluid allow it to be used in a number of applications in which protection of the original permeable formation is desirable. The applications include drilling, fracturing, and controlling fluid losses during completion operations, such as gravel packing or well workovers.

Gellan (Xanthan, Scleroglucan, or Wellan)

It has been found that gellan has good characteristics as a filtrate reducer in water-based drilling fluids [515, 516]. Preferential use is made of native gellan, which has a considerable gelling capacity and good solubility. It should be noted that native gellan contains cellular debris or other insoluble residue. Xanthan gum has been used extensively in the oil industry as a viscosifier for various applications [1327]. Deacetylated xanthan gum is used in guar-free compositions instead of guar [1064].

Synthetic Polymers

Polyhydroxyacetic Acid

A low–molecular-weight condensation product of hydroxyacetic acid with itself or compounds containing other hydroxy acid, carboxylic acid, or hydroxycarboxylic acid moieties has been suggested as a fluid loss additive [164]. Production methods of the polymer have been described. The reaction products are ground to 0.1 to 1500 µ particle size. The condensation product can be used as a fluid loss material in a hydraulic fracturing process in which the fracturing fluid comprises a hydrolyzable, aqueous gel. The hydroxyacetic acid condensation product hydrolyzes at formation conditions to provide hydroxyacetic acid, which breaks the aqueous gel autocatalytically and eventually provides the restored formation permeability without the need for the separate addition of a gel breaker [315–317, 329].

Lignosulfonates

Polymer of Monoallylamine

A water-soluble polymer of monoallylamine can be used in conjunction with a sulfonated polymer, such as a water-soluble lignosulfonate, condensed naphthalene sulfonate, or sulfonated vinyl aromatic polymer, to minimize fluid loss from the slurry during well cementing operations [1510, 1511]. The polymer

$$H_2C=CH-CH_2-NH_2$$

Allylamine

Figure 2–4. Monoallylamine.

of monoallylamine may be a homopolymer or a copolymer and may be crosslinked or uncrosslinked. These components react with each other in the presence of water to produce a gelatinous material that tends to plug porous zones and to minimize premature water loss from the cement slurry to the formation.

Polyphenolic Materials for Oil-Based Drilling Fluids

Organophilic polyphenolic materials for oil-based drilling fluids have been described [407]. The additives are prepared from a polyphenolic material and one or more phosphatides. The phosphatides are phosphoglycerides obtained from vegetable oils, preferably commercial lecithin. Humic acids, lignosulfonic acid, lignins, phenolic condensates, tannins; the oxidized, sulfonated, or sulfomethylated derivatives of these polyphenolic materials may serve as polyphenolic materials.

A fluid loss additive is described that uses graded calcium carbonate particle sizes and a modified lignosulfonate [917]. Optionally, a thixotropic polymer, such as a wellan or xanthan gum polymer, is used to keep the $CaCO_3$ and lignosulfonate in suspension. It is important that the calcium carbonate particles are distributed across a wide size range to prevent filtration or fluid loss into the formation. Furthermore, the lignosulfonate must be polymerized to an extent effective to reduce its water solubility. The modified lignosulfonate (lignin sulfonate) is necessary for the formation of a filter-cake essentially on the surface of the wellbore. Because the filter-cake particles do not invade the wellbore due to the action of the modified lignosulfonate, no high-pressure spike occurs during the removal of the filter-cake, which would indicate damage of the formation and wellbore surface. The additive is useful in fracturing fluids, completion fluids, and workover fluids.

Tests showed that a fluid loss additive on a base of a sulfonated tannicphenolic resin is effective for fluid loss control at high temperature and pressure, and it exhibits good resistance to salt and acid [868].

Grafting to Lignin and Lignite

In hydraulic cement slurries, fluid loss additives based on sulfonated or sulfomethylated lignins have been described.

Sulfonated or sulfomethylated lignins are reacted with phenol-blocking reagents, such as ethylene oxide, propylene oxide, or butylene oxide [1571].

The fluid loss and thickening time characteristics of the cement slurry is altered, either by increasing the molecular weight of the lignin by cross-linking with formaldehyde or epichlorohydrin or by adding agents such as sodium sulfite, sodium metasilicate, sodium phosphate, and sodium naphthalene sulfonate.

Another method is amination with a polyamine and an aldehyde [1567]. The formulation also contains sodium carbonate, sodium phosphate, sodium sulfite, sodium metasilicate, or naphthalene sulfonate. The sulfonated or sulfomethylated aminated lignin shows less retardation (shorter thickening time) than a sulfonated or sulfomethylated lignin without the attached amine.

Lignite can be grafted with synthetic comonomers to obtain lignite fluid loss additives [873]. Comonomers can be AMPS, N,N-dimethylacrylamide, acrylamide, vinylpyrrolidone, vinylacetate, acrylonitrile, dimethylaminoethyl methacrylate, styrene sulfonate, vinyl sulfonate, dimethylaminoethyl methacrylate methyl chloride quaternary, and acrylic acid and its salts.

Various polymers, for example, lignin, lignite, derivatized cellulose polyvinyl alcohol, polyethylene oxide, polypropylene oxide, and polyethyleneimine, can be used as a backbone polymer onto which some other groups can be grafted [650]. The grafted pendant groups can be AMPS, acrylonitrile, N,N-dimethylacrylamide, acrylic acid, N,N-dialkylaminoethylmethacrylate, and their salts. The backbone polymer makes 5% to 95% by weight of the graft polymer, and consequently the pendant groups are in the range of 5% to 95% by weight of the graft polymer.

A polymeric composition for reducing fluid loss in drilling muds and well cement compositions is obtained by the free radical–initiated polymerization of a water-soluble vinyl monomer in an aqueous suspension of lignin, modified lignins, lignite, brown coal, and modified brown coal [705, 1847]. The vinyl monomers can be methacrylic acid, methacrylamide, hydroxyethyl acrylate, hydroxypropyl acrylate, vinylacetate, methyl vinyl ether, ethyl vinyl ether, N-methylmethacrylamide, N,N-dimethylmethacrylamide, vinyl sulfonate, and additional AMPS. In this process a grafting process to the coals by chain transfer may occur.

Graft copolymers and other polymers are prepared in a way that is common in polymerization techniques [1894]. For example, they are made by providing a foamed, aqueous solution of water-soluble monomeric material, initiating polymerization by adding an initiator, exothermically polymerizing

$H_2C{=}CH{-}O{-}CH_3$ $H_2C{=}CH{-}O{-}CH_2{-}CH_3$

Methylvinylether   Ethylvinylether

Figure 2–5. Methyl vinyl ether, ethyl vinyl ether.

the monomeric material to form a foamed gel, and comminuting the gel. Preferably, the polymerization temperature is held below 60° C for at least the first 10 minutes of the polymerization and then rises exothermically to a higher temperature. Graft copolymers that can be made by this technique and that are of particular value as fluid loss additives are formed from a polyhydroxy polymer, a sulfonate monomer, acrylamide, and acrylic acid.

The vinyl grafted wattle tannin [872] comprises a wattle tannin grafted with AMPS and small amounts of acrylamide. The wattle tannin is present in an amount between 2% and 14% by weight. The AMPS is present in an amount between 98% and 84% accordingly.

Latex

A thermally stable drilling fluid system includes an additive that comprises styrene-butadiene copolymers having an average molecular weight greater than approximately 500,000 Dalton, wherein the drilling fluid system exhibits fluid loss control at high-temperature (350° F) and high-pressure conditions. The drilling fluid may be either all-oil– or water-in-oil–based. In each case, gas oil is preferably used, although mineral oils with low aromatic content, synthetic oils, or vegetable oils are suitable alternatives. The all-oil fluid further comprises an organophilic clay (hectorite, bentonite, and mixtures thereof).

Optionally, a surfactant that behaves as an emulsifier and a wetting agent and a weighting agent (calcium carbonate, barite, hematite, and mixtures thereof) are included. Suitable emulsifiers and wetting agents include surfactants; ionic surfactants such as fatty acids, amines, amides, and organic sulfonates; and mixtures of any of these with nonionic surfactants such as ethoxylated surfactants. The water-in-oil emulsion may consist of an oil phase, a water phase (salt or fresh), a surfactant, a weighting agent, and salts or electrolytes [125, 819, 820].

Polyvinyl Alcohol

Partially hydrolyzed polyvinylacetate polymer (PVA), a crosslinker for the polymer, and other additives such as calcium sulfate can be used in cementing casing strings [1253]. PVA is not totally water-soluble below 50° C, but is, instead, water swellable. It is believed that the individual PVA particles swell and soften to form small gel-balls in the slurry. These gel-balls deform by flattening and become a part of the filter-cake, greatly reducing the filter-cake permeability, thus giving good fluid loss control. Because PVA is not totally water soluble, it does not significantly increase the slurry viscosity. Furthermore, PVA does not delay the setting of the cement, and it has high-temperature properties that are relatively insensitive to external conditions.

PVA can be crosslinked with a crosslinker present in a molar concentration, relative to monomer residues, of 0.01% to 1.0%. The crosslinker may be formaldehyde, acetaldehyde, glyoxal, glutaraldehyde, maleic acid, oxalic acid, dimethylurea, polyacrolein, diisocyanate, divinyl sulfonate, or a chloride of a diacid [89–91].

Humic Acid Derivatives

Polysulfonated humic acid resin is a drilling fluid filtrate loss additive composed of three mud additives: sulfonated chromium humate, sulfonated phenolic resin, and hydrolytic ammonium polyacrylate [1716]. Field application and effectiveness of polysulfonated humic acid resin, especially in extra-deep wells and in sylvite and undersaturated salt muds, have been described. Polysulfonated humic acid resin resists high temperature, salt concentration, and calcium contamination, as well. A polysulfonated humic acid resin–treated drilling fluid has stable properties and good rheologic characteristics and can improve cementing quality.

Oil-Based Well Working Fluids

Adducts of aminoethylethanolamine and polyethylenepolyamines with humic acid–containing materials and fatty acids [1400] are useful as fluid loss additives in oil-based drilling fluids [854].

In addition, a fluid loss additive for oil-based drilling fluids, which consists of fatty acid compounds and lignite or humic acid, an oil-soluble or oil-dispersible amine or amine salt with phosphoric acid, or an aliphatic amide or hydroxyamide [392], has been described.

Polyethyleneimine

A liquid fluid loss–reducing additive for well cementing compositions consists of water, polyethyleneimine, an alkali metal salt of alkylbenzene sulfonic acid, and an alkali metal salt of naphthalene sulfonic acid, condensed with formaldehyde [231]. The polyethyleneimine has a molecular weight of 40,000 to 60,000 Dalton and is present in an amount of 50% to 55% by weight of the additive. The alkali metal salt of the alkylbenzene sulfonic acid is sodium dodecylbenzene sulfonate and is present in an amount of 3% to 4% by weight of the additive. The alkali metal salt of the naphthalene sulfonic acid that is condensed with formaldehyde is sodium naphthalene sulfonate. The sodium naphthalene sulfonate–formaldehyde condensation product has a molecular weight of 1400 to 2400 Dalton, and the condensation product is present in an amount of 3% to 4% by weight of the additive. The alkyl group of the alkylbenzene sulfonic acid salt contains from 8 to 16 carbon atoms.

Acrylics

Copolymers of N-Vinyl-2-pyrrolidone and Sodium-2-acrylamido-2-methylpropane Sulfonate

Homopolymers and copolymers from amido-sulfonic acid or salt containing monomers can be prepared by reactive extrusion, preferably in a twin screw extruder [1660]. The process produces a solid polymer. Copolymers of acrylamide, N-vinyl-2-pyrrolidone, and sodium-2-acrylamido-2-methylpropane sulfonate have been proposed to be active as fluid loss agents. Another component of the formulations is the sodium salt of naphthalene formaldehyde sulfonate [207]. The fluid loss additive is mixed with hydraulic cements in suitable amounts.

A fluid loss additive for hard brine environments has been developed [1685], which consists of hydrocarbon, an anionic surfactant, an alcohol, a sulfonated asphalt, a biopolymer, and optionally an organophilic clay, a copolymer of N-vinyl-2-pyrrolidone and sodium-2-acrylamido-2-methylpropane sulfonate. Methylene-bis-acrylamide can be used as a crosslinker [1398]. Crosslinking imparts thermal stability and resistance to alkaline hydrolysis.

Terpolymers and Tetrapolymers

Terpolymers and tetrapolymers have been proposed as fluid loss additives for drilling fluids [1676, 1679]. The constituent monomers are a combination of nonionic monomers and ionic monomers.

The nonionic monomer can be acrylamide, N,N-dimethylacrylamide, N-vinyl-2-pyrrolidone, N-vinyl acetamide, or dimethylamino ethyl methacrylate. Ionic monomers are AMPS, sodium vinyl sulfonate, and vinylbenzene sulfonate. The terpolymer should have a molecular weight between 200,000 to 1,000,000 Dalton.

A formulation consisting of 2-acrylamido-2-methylpropane sulfonic acid, acrylamide, and itaconic acid has been proposed [676]. Such polymers are used as fluid loss control additives for aqueous drilling fluids and are advantageous when used with lime- or gypsum-based drilling muds containing soluble calcium ions.

For seawater muds, another example [139] is a copolymer of 10% by weight AMPS and 90% by weight acrylic acid in its sodium salt form. The polymers have an average molecular weight between 50,000 and 1,000,000 Dalton.

A terpolymer from a family of intramolecular polymeric complexes (i.e., polyampholytes), which are terpolymers of acrylamide–methyl styrene sulfonate–methacrylamido propyltrimethylammonium chloride [106, 1418], has been reported.

A terpolymer formed from ionic monomers AMPS, sodium vinyl sulfonate or vinylbenzene sulfonate itaconic acid, and a nonionic monomer, for example, acrylamide, N,N-dimethylacrylamide, N-vinylpyrrolidone, N-vinyl acetamide, and dimethylaminoethyl methacrylate, is used as a fluid loss agent in oil well cements [1562]. The terpolymer should have a molecular weight between 200,000 and 1,000,000 Daltons. The terpolymer comprises AMPS, acrylamide, and itaconic acid. Such copolymers also serve in drilling fluids [1892].

A tetrapolymer consisting of 40 to 80 mole-percent of AMPS, 10 to 30 mole-percent of vinylpyrrolidone, 0 to 30 mole-percent of acrylamide, and 0 to 15 mole-percent of acrylonitrile was also a suggested as a fluid loss additive [1061]. Even at high salt concentrations, these polymers yield high-temperature–stable protective colloids that provide for minimal fluid loss under pressure.

Similar copolymers with N-vinyl-N-methylacetamide as a comonomer have been proposed for hydraulic cement compositions [669]. The polymers consist of AMPS in an amount of 5% to 95%, vinylacrylamide in an amount of 5% to 95%, and acrylamide in an amount of 0% to 80%, all by weight. The polymers are effective at well bottom-hole temperatures ranging from 200° to 500° F and are not adversely affected by brine. Terpolymers of 30 to 90 mole-percent AMPS, 5 to 60 mole-percent of styrene, and residual acrylic acid are also suitable for well cementing operations [253].

Copolymer of Acrylamide/Vinyl Imidazole

A fluid loss additive useful in cementing oil and gas wells is a blend [423, 424, 1015] of a copolymer of acrylamide/vinyl imidazole. The second component in the blend is a copolymer of vinylpyrrolidone and the sodium salt of vinyl sulfonate. Details are given in Table 2–2. The copolymers are mixed together in the range of 20:80 to 80:20. Sodium or potassium salts or a sulfonated naphthalene formaldehyde condensate can be used as a dispersant.

Table 2–2
Copolymer Blends for Fluid Loss [423, 424, 1015]

Copolymer	Composition	Molecular weight (Dalton)
Acrylamide/vinyl imidazole	95:5 to 5:95	100,000 to 3,000,000
Vinylpyrrolidone and the sodium salt of vinyl sulfonate	80:20 to 20:80	100,000 to 3,000,000

N-Vinylpyrrolidone/Acrylamide Random Copolymer

An N-vinylpyrrolidone/acrylamide random copolymer (0.05% to 5.0% by weight) is used for cementing compositions [371, 1076]. Furthermore, a sulfonate-containing cement dispersant is necessary. The additive can be used in wells with a bottom-hole temperature of 80° to 300° F. The fluid loss additive mixture is especially effective at low temperatures, for example, below 100° F and in sodium silicate–extended slurries.

Copolymer of N-Vinylpyrrolidone and a Salt of Styrenesulfonic Acid

For aqueous cement slurries a copolymer of N-vinylpyrrolidone and a salt of styrenesulfonic acid has been proposed [1585]. A naphthalene sulfonic acid salt condensed with formaldehyde serves as a dispersant.

Copolymer of Acrylamide and N-Vinylamide

The fluid loss control of aqueous, clay-based drilling mud compositions is enhanced by the addition of a hydrolyzed copolymer of acrylamide and an N-vinylamide [402]. The copolymer, which is effective over a broad range of molecular weights, contains at least 5 mole-percent of the N-vinylamide units, which are hydrolyzed to N-vinylamine units. The copolymers can be made from various ratios of N-vinylamide and acrylamide by using common radical-initiated chain growth polymerization techniques.

N-vinylamide can be polymerized by the inverse emulsion polymerization technique [1050]. The polymers of that monomer are used in cementing compositions for oil and gas wells. The method for preparing the inverse, or water-in-oil, emulsion involves colloidally dispersing an aqueous solution containing 10% to 90% by weight water-soluble N-vinylamide in a hydrocarbon liquid, using a surfactant with a hydrophilic-lipophilic balance value between 4 and 9. The weight ratio of a monomer-containing aqueous solution to hydrocarbon liquid is preferably from 1:2 to 2:1. To initiate the polymerization, an azo-type free radical initiator is used. The resultant high–molecular-weight polymer emulsion has a low viscosity ranging from 2 to less than 10 cP at 15% solids (60 rpm Brookfield and 20° C), thus eliminating problems of solution viscosity that arise when the polymer is prepared by a solution polymerization process.

Copolymer of Styrene with 2-Acrylamido-2-methylpropane Sulfonic Acid

Copolymers of styrene with AMPS having fluid loss capabilities for use in well cementing operations have been described [254]. The styrene is present

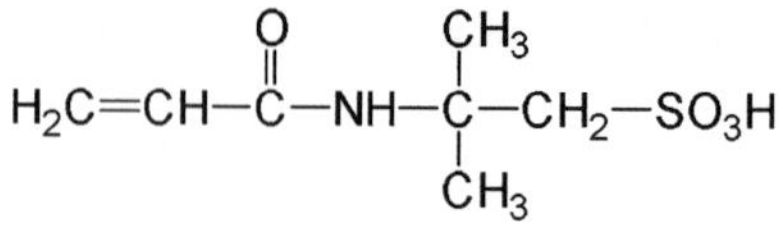

$$H_2C{=}CH-\overset{\overset{O}{\|}}{C}-NH-\underset{\underset{CH_3}{|}}{\overset{\overset{CH_3}{|}}{C}}-CH_2-SO_3H$$

Figure 2–6. 2-Acrylamido-2-methylpropane sulfonic acid.

in an amount of 15 to 60 mole-percent, and the AMPS is present in an amount of 40 to 85 mole-percent. The polystyrene units are not hydrophilic, so AMPS will affect solubility in water.

Copolymers of Acrylic Acid and Itaconic Acid

Copolymers of mainly acrylic acid and 2% to 20% by weight of itaconic acid are described as fluid loss additives for aqueous drilling fluids [138]. The polymers have an average molecular weight between 100,000 and 500,000 Dalton and are water dispersible. The polymers are advantageous when used with muds containing soluble calcium and muds containing chloride ions, such as seawater muds.

Copolymers of 2-Acrylamido-2-methylpropane Sulfonic Acid

Copolymers from the monomers AMPS, diallyldimethylammonium chloride (DADMAC), N-vinyl-N-methylacetamide (VIMA), acrylamides, and acrylates are particularly useful for fluid loss additives [824]. The molecular weights of the copolymers range from 200,000 to 1,000,000 Dalton. The copolymers are used in suspensions of solids in aqueous systems, including saline, as water binders. In these systems, the water release to a formation is substantially reduced by the addition of one or more of these copolymers.

A suitable formulation for filtration reducers, good temperature resistance (over 200° C), and good tolerance to salts and calcium compounds is a copolymer of AMPS and other vinyl monomers [1101].

Poly-N-vinyl Lactam

An additive described as reducing the water loss and enhancing other properties of well-treating fluids in high-temperature subterranean environments consists of polymers or copolymers from N-vinyl lactam monomers or vinyl-containing sulfonate monomers. Organic compounds like lignites, tannins, and asphaltic materials are added as dispersants [175].

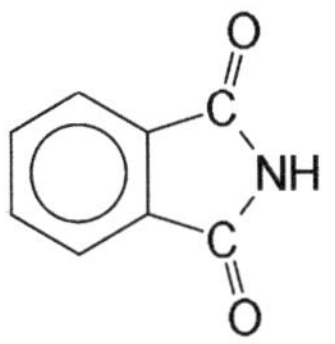

Phthalimide

Figure 2-7. Phthalimide.

Phthalimide as a Diverting Material

Phthalimide has been described as a diverting material, or fluid loss additive, for diverting aqueous treating fluids, including acids, into progressively less permeable portions of a subterranean formation [485]. The additive also reduces the fluid loss to the formation of an aqueous or hydrocarbon treating fluid utilized, for example, in fracturing treatments. The use of the material depends on the particle size of the material that is used. Phthalimide will withstand high formation temperatures and can be readily removed from the formation by dissolution in the produced fluids or by sublimation at elevated temperatures. The material is compatible either with other formation permeability–reducing materials or formation permeability–increasing materials. The phthalimide particles act by sealing off portions of a subterranean formation by blocking off the fissures, pores, channels, and vugs that grant access to the formation from the wellbore penetrating the formation.

Tall Oil Pitch

A fluid loss additive for well drilling fluids consists of air-blown tall oil pitch, which has a softening point (ring and ball) of 100° to 165° C. Tall oil pitch is available as the residue from the distillation of tall oil. Its solubility is low in fatty acids and high in fatty esters, higher alcohols, and sterols. Blowing air through tall oil pitch at an elevated temperature partially oxidizes and polymerizes the material and drives off volatiles. Blowing reduces the volume of the pitch by 30% and increases the viscosity and the softening point. The softening point of the resultant blown pitch is therefore a measure of the degree of oxidation-polymerization that has occurred. It has been found that optimal properties as a fluid loss additive are given by blown tall oil pitches that have a softening point between 125° and 130° C [1851].

54   Oil Field Chemicals

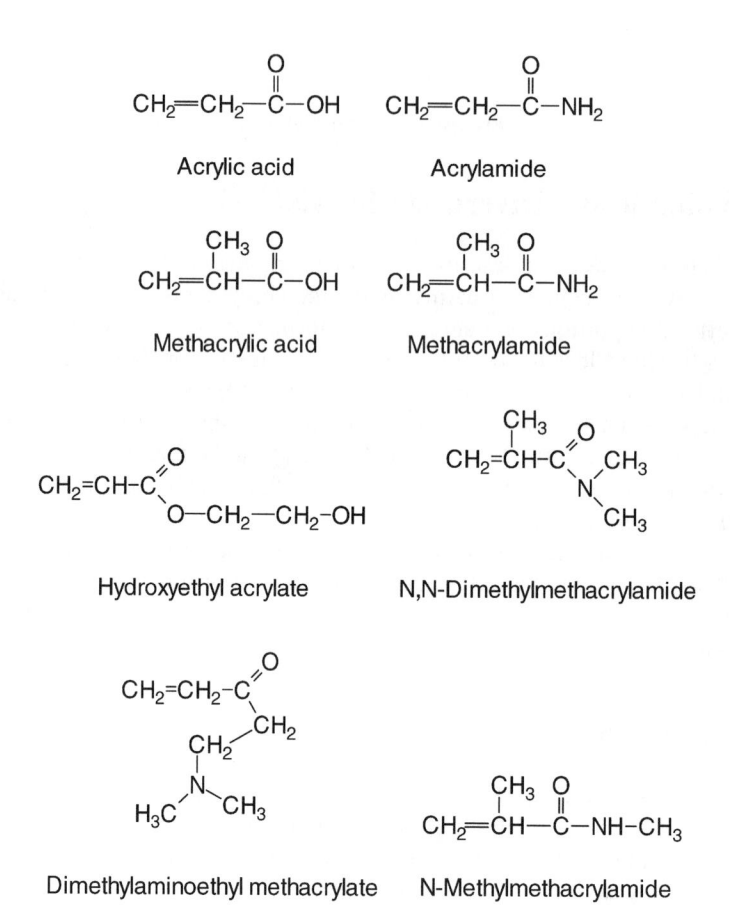

Figure 2–8. Acrylic acid, acrylamide, methacrylic acid, methacrylamide, hydroxyethyl acrylate, N,N-dimethylmethacrylamide, dimethylaminoethyl methacrylate, N-methylmethacrylamide.

Table 2–3
Summary of Formulations of Fluid Loss Additives

Composition	References
Lignite or humic acid oil-soluble or oil-dispersible amine or amine salt with phosphoric acid, or an aliphatic amide or hydroxyamide, and fatty acid compounds	[392]
Adducts of aminoethylethanolamine and polyethylenepolyamines with humic acid–containing materials and fatty acids[Ob,D]	[854]
Polysulfonated humic acid resin–sulfonated chromium humate, sulfonated phenolic resin, and hydrolytic ammonium polyacrylate[C,D,HT]	[1716]
Polyvinylacetate partially hydrolyzed crosslinker: formaldehyde, acetaldehyde, glyoxal, glutaraldehyde, maleic acid, oxalic acid, dimethylurea, polyacrolein, diisocyanate, divinyl sulfonate, calcium sulfate[C,HT]	[89, 90, 1253]
Sodium dodecylbenzene sulfonate and naphthalene sulfonic acid salt condensed with formaldehyde	[231]
Hydrocarbon, an anionic surfactant, an alcohol, a sulfonated asphalt, biopolymer, organophilic clay, copolymer of N-vinyl-2-pyrrolidone, and sodium-2-acrylamido-2-methylpropane sulfonate[SB]	[1685]
Copolymers of acrylamide, N-vinyl-2-pyrrolidone, and sodium-2-acrylamido-2-methylpropane sulfonate; dispersant is sodium salt of naphthalene formaldehyde sulfonate[C]	[207]
Combination of nonionic monomers and ionic monomers acrylamide, N,N-dimethylacrylamide, N-vinylpyrrolidone, N-vinyl acetamide, or dimethylamino ethyl methacrylate, 2-acrylamido-2-methylpropane sulfonic acid, sodium vinyl sulfonate, and vinylbenzene sulfonate	[1676, 1679]
Copolymer of 2-acrylamido-2-methylpropane sulfonic acid, acrylamide, and itaconic acid[Aq,D]	[676]
Copolymer of 2-acrylamido-2-methylpropane sulfonic acid, vinylphosphonic acid, diallyldimethylammonium chloride[S]	[1376]
Copolymer 2-acrylamido-2-methylpropane sulfonic acid and acrylic acid sodium salt	[139]
Copolymers of acrylamide-methyl styrene sulfonate-methacrylamido propyltrimethylammonium chloride	[1418]
Copolymer from 2-acrylamido-2-methylpropane sulfonic acid, itaconic acid, and acrylamide[C]	[1562]
Copolymer of 2-acrylamido-2-methylpropane sulfonic acid, vinylpyrrolidone, acrylamide, and acrylonitrile[HT,SB]	[1061]

Continued.

Table 2–3 (continued)

Composition	References
Copolymers with N-vinyl-N-methylacetamide, acrylamide, and 2-acrylamido-2-methylpropane sulfonic acid[C,HT,SB]	[669]
Terpolymers of 2-acrylamido-2-methylpropane sulfonic acid, styrene, and acrylic acid[C]	[253]
Copolymer of acrylamide and vinyl imidazole, copolymer of N-vinylpyrrolidone, and the sodium salt of vinyl sulfonate; dispersants are sodium or potassium salts or a sulfonated naphthalene formaldehyde condensate[C]	[423, 424, 1015]
Copolymer of N-vinylpyrrolidone-acrylamide[C,LT]	[371]
Copolymer of N-vinylpyrrolidone and a salt of styrene-sulfonic acid; dispersant is naphthalene sulfonic acid salt[C]	[1585]
Hydrolyzed copolymer of acrylamide and an N-vinylamide[Aq,Cb,D]	[402]
Copolymers of styrene with 2-acrylamido-2-methylpropane sulfonic acid[C]	[254]
Styrene sulfonate polymers[HT]	[105]
Copolymers of acrylic acid and itaconic acid[Aq,D,S]	[138]
Copolymers 2-acrylamido-2-methylpropane sulfonic acid, diallyldimethylammonium chloride, N-vinyl-N-methyl-acetamide (VIMA), acrylamides, and acrylates	[824]
Polyanionic cellulose and sulfonate-containing polymer	[812]
Copolymer of 2-acrylamide-2-methyl propane sulfonic acid and other vinyl monomers[HT,SB]	[1101]
Polymers or copolymers from N-vinyl lactam monomers or vinyl-containing sulfonate monomers; lignites, tannins, and asphaltic materials are added as dispersants[HT]	[175]
Tall oil pitch	[1851]
Admixtures of carboxymethylhydroxyethylcellulose or copolymers and copolymer salts of N,N-dimethylacryl-amide, AMPS, together with a copolymer of acrylic acid/AMPS[C,HT]	[252]
Hydroxyethylcellulose, degree of substitution of 1.1 to 1.6[Aq,D]	[1473]
Starch, crosslinked[Aq,D,HT]	[632]
Starch, granular mica, fine particulate starches, mixture of natural and modified plus an enzyme modified: carboxymethyl and hydroxypropyl derivatives[FF]	[337]
Natural corn, potatoes, wheat, and soy coating with a surfactant sorbitan monooleate, ethoxylated butanol, or ethoxylated nonylphenol, to aid dispersion[FF]	[1850]
Guar gum, hydrophobically modified; used together with polymers or reactive clay[Aq,D]	[93]
Succinoglycan	[1069, 1070]

Table 2–3 (continued)

Composition	References
Polyether-modified polysaccharides; mixtures of metal hydroxides polysaccharide partially etherified with hydroxyethyl and/or hydroxypropyl groups[Aq,Cb,D]	[1437]
Scleroglucan, starch, modified calcium carbonate particles	[915]
Gellan[Aq,D]	[515]
Hydroxyacetic acid, low–molecular-weight condensation product[FF]	[164]
Polymonoallylamine, crosslinked or uncrosslinked lignosulfonate, condensed naphthalene sulfonate, or sulfonated vinyl aromatic polymer; components react with each other in the presence of water to produce a gelatinous material[C]	[1510, 1511]
Polyphenolic materials prepared from a polyphenolic material (humic acids, lignosulfonic acid, lignins, phenolic condensates, tannins) and phosphatides (lecithin)[Ob,D]	[407]
Tannic-phenolic resin, sulfonated[HT,HP,SB]	[868]
Lignins, sulfonated or sulfomethylated; reaction products with ethylene oxide, propylene oxide, butylene oxide[C]	[1571]
Lignite grafted with synthetic comonomers; comonomers can be AMPS, dimethylacrylamide, acrylamide, vinylpyrrolidone, vinylacetate, acrylonitrile, dimethylaminoethyl methacrylate, styrene sulfonate, vinyl sulfonate, dimethylaminoethyl methacrylate methyl chloride quaternary, and acrylic acid and its salts	[873]
Lignin, brown coal; polymer of methacrylic acid, methacrylamide, hydroxyethyl acrylate, hydroxypropyl acrylate, vinyl acetate, methyl vinyl ether, ethyl vinyl ether, N-methylmethacrylamide, N,N-dimethylmethacrylamide, vinyl sulfonate, or 2-acrylamido-N-methylpropane sulfonic acid; free radical polymerization of a water-soluble vinyl monomer in an aqueous suspension of coals[D,C]	[705, 1847]
Wattle tannin, vinyl grafted AMPS and acrylamide	[872]
1S-endo-borneol, camphor, iodine, β-carotene, lycopene, cholesterol, lanosterol, and agnosterol	[1048, 1049]
Copolymer with vinylamide morpholine[Aq,D]	[1773]
Lignite and Gilsonite[HT,HP]	[419]
Peanut hulls	[625–627]

Aq, Aqueous; C, cementing; Cb, clay based; D, drilling fluids; FF, fracturing fluids; HP, high pressure application; HT, high-temperature application; LT, low temperature; Ob, oil based; S, seawater mud; SB, salt and brine tolerant.

Chapter 3

Clay Stabilization

The problems caused by shales in petroleum activities are not new. At the beginning of the 1950s, many soil mechanics experts were interested in the swelling of clays. It is important to maintain wellbore stability during drilling, especially in water-sensitive shale and clay formations. The rocks within these types of formations absorb the fluid used in drilling; this absorption causes the rock to swell and may lead to a wellbore collapse. The swelling of clays and the problems that may arise from these phenomena are reviewed in the literature [528, 529, 1788, 1900]. Various additives for clay stabilization are shown in Table 3–1.

Properties of Clays

Clay minerals are generally crystalline in nature. The structure of the clay crystals determines its properties. Typically, clays have a flaky, mica-type structure. Clay flakes are made up of a number of crystal platelets stacked face-to-face. Each platelet is called a unit layer, and the surfaces of the unit layer are called basal surfaces. A unit layer is composed of multiple sheets. One sheet is called the octahedral sheet; it is composed of either aluminum or magnesium atoms octahedrally coordinated with the oxygen atoms of hydroxyl groups. Another sheet is called the tetrahedral sheet. The tetrahedral sheet consists of silicon atoms tetrahedrally coordinated with oxygen atoms. Sheets within a unit layer link together by sharing oxygen atoms. When this linking occurs between one octahedral and one tetrahedral sheet, one basal surface consists of exposed oxygen atoms while the other basal surface has exposed hydroxyl groups. It is also quite common for two tetrahedral sheets to bond with one octahedral sheet by sharing oxygen atoms. The resulting structure, known as the Hoffman structure, has an octahedral sheet that is sandwiched between the two tetrahedral sheets. As a result, both basal surfaces in a Hoffman structure are composed of exposed oxygen atoms.

Table 3–1
Clay Stabilizers

Additive	References
Polymer latices	[1686]
Partially hydrolyzed poly-(vinylacetate)[a]	[1014]
Polyacrylamide[b]	[1878, 1880]
Copolymer of anionic and cationic monomers	
Acrylic acid, methacrylic acid, 2-acrylamido-2-methyl-propane sulfonic acid	
Dimethyl diallyl ammonium chloride	[119, 1649–1651]
Nitrogen[c]	[1640, 1641]
Partially hydrolyzed acrylamide-acrylate copolymer, potassium chloride, and polyanionic cellulose[MA]	[100, 768]
Aluminum/guanidine complexes with cationic starches and polyalkylene glycols[MA]	[232]
Hydroxyaldehydes or hydroxyketones	[1834]
Polyols and alkaline salt	[762]
Tetramethylammonium chloride and methyl chloride quaternary salt of polyethyleneimine[SF]	[8–10]
Pyruvic aldehyde and a triamine	[420]
Quaternary ammonium compounds	
In situ crosslinking of epoxide resins	[405, 406]
Oligomer (methyl quaternary amine containing 3 to 6 moles of epihalohydrin)	[832]
Quaternary ammonium carboxylates[BD,LT]	[829]
Quaternized trihydroxyalkyl amine[LT]	[1401]
Polyvinyl alcohol, potassium silicate, and potassium carbonate	[32]
Copolymer of styrene and substituted maleic anhydride	[1643]
Potassium salt of carboxymethylcellulose	[1390]
Water-soluble polymers with sulfosuccinate derivative–based surfactants, zwitterionic surfactants[BD,LT]	[42, 46]

BD, biodegradable; *LT*, low toxicity; *SF*, well stimulation fluids.
a) 75% hydrolyzed; 50,000 Dalton.
b) Shear-degraded; for montmorillonite clay dispersed in sandpacks.
c) Injection of unreactive gas.

The unit layers stack together face-to-face and are held in place by weak attractive forces. The distance between corresponding planes in adjacent unit layers is called the c-spacing. A clay crystal structure with a unit layer consisting of three sheets typically has a c-spacing of about 9.5×10^{-7} mm.

In clay mineral crystals, atoms having different valences commonly will be positioned within the sheets of the structure to create a negative potential at the crystal surface. In that case, a cation is adsorbed on the surface. These adsorbed cations are called exchangeable cations because they may chemically trade places with other cations when the clay crystal is suspended in water. In addition, ions may also be adsorbed on the clay crystal edges and exchange with other ions in the water.

The type of substitutions occurring within the clay crystal structure and the exchangeable cations adsorbed on the crystal surface greatly affect clay swelling, a property of primary importance in the drilling fluid industry. Clay swelling is a phenomenon in which water molecules surround a clay crystal structure and position themselves to increase the structure's c-spacing thus resulting in an increase in volume.

Swelling of Clays

Two types of swelling may occur. Surface hydration is one type of swelling in which water molecules are adsorbed on crystal surfaces. Hydrogen bonding holds a layer of water molecules to the oxygen atoms exposed on the crystal surfaces. Subsequent layers of water molecules align to form a quasi-crystalline structure between unit layers which results in an increased c-spacing. All types of clays swell in this manner.

Osmotic swelling is a second type of swelling. Where the concentration of cations between unit layers in a clay mineral is higher than the cation concentration in the surrounding water, water is osmotically drawn between the unit layers and the c-spacing is increased. Osmotic swelling results in larger overall volume increases than surface hydration. However, only certain clays, like sodium montmorillonite, swell in this manner.

Exchangeable cations found in clay minerals are reported to have a significant impact on the amount of swelling that takes place. The exchangeable cations compete with water molecules for the available reactive sites in the clay structure. Generally cations with high valences are more strongly adsorbed than ones with low valences. Thus, clays with low valence exchangeable cations will swell more than clays whose exchangeable cations have high valences.

Clay swelling during the drilling of a subterranean well can have a tremendous adverse impact on drilling operations. The overall increase in bulk volume accompanying clay swelling impedes removal of cuttings from beneath the drill bit, increases friction between the drill string and the sides of the borehole, and inhibits formation of the thin filter cake that seals formations. Clay swelling can also create other drilling problems such as loss of circulation or stuck pipe that slow drilling and increase drilling costs [1403].

Guidelines

The literature offers several papers that may serve as guidelines for issues such as selecting a proper clay stabilizing system or completing wellbore stability analyses of practical well designs [367, 427–429, 562, 1565].

Mechanisms Causing Instability

Shale stability is an important problem faced during drilling. Stability problems are attributed most often to the swelling of shales. It has been shown that several mechanisms can be involved [680, 681]. These can be pore pressure diffusion, plasticity, anisotropy, capillary effects, osmosis, and physicochemical alterations. Three processes contributing to the instability of shales have to be considered [127]:

1. Movement of fluid between the wellbore and shale (limited to flow from the wellbore into the shale)
2. Changes in stress (and strain) that occur during shale–filtrate interaction
3. Softening and erosion caused by invasion of mud filtrate and consequent chemical changes in the shale

The major reason for these effects is of a chemical nature, namely the hydration of clays. Borehole instabilities were observed even with the most inhibitive fluids, that is oil-based mud. This demonstrates that the mechanical aspect is also important. In fact, the coupling of both chemical and mechanical mechanisms has to be considered. For this reason, it is still difficult to predict the behavior of rock at medium-to-great depth under certain loading conditions.

The stability of shales is governed by a complicated relationship between transport processes in shales (e.g., hydraulic flow, osmosis, diffusion of ions, pressure) and chemical changes (e.g., ion exchange, alteration of water content, swelling pressure).

Clays or shales have the ability to absorb water, thus causing the instability of wells either because of the swelling of some mineral species or because the supporting pressure is suppressed by modification of the pore pressure. The response of a shale to a water-based fluid depends on its initial water activity and on the composition of the fluid. The behavior of shales can be classified into either deformation mechanisms or transport mechanisms [1765]. Optimization of mud salinity, density, and filter-cake properties is important in achieving optimal shale stability and drilling efficiency with water-based mud.

Kinetics of Swelling of Clays

Basic studies on the kinetics of swelling have been performed [1699]. Pure clays (montmorillonite, illite, and kaolinite) with polymeric inhibitors were investigated, and phenomenologic kinetic laws were established.

Hydrational Stress

Stresses caused by chemical forces, such as hydration stress, can have a considerable influence on the stability of a wellbore [364]. When the total pressure and the chemical potential of water increase, water is absorbed into the clay platelets, which results either in the platelets moving farther apart (swelling) if they are free to move or in generation of hydrational stress if swelling is constrained [1715]. Hydrational stress results in an increase in pore pressure and a subsequent reduction in effective mud support, which leads to a less stable wellbore condition.

Borehole Stability Model

A borehole stability model has been developed that takes into account both the mechanical and the chemical aspects of interactions between drilling fluid and shale [1231]. Chemically induced stress alteration based on the thermodynamics of differences in water molar free energies of the drilling fluid and shale is combined with mechanically induced stress. Based on this model, it should be possible to obtain the optimal mud weight and salt concentration for drilling fluids.

Further stability models based on surface area, equilibrium water-content–pressure relationships, and electric double-layer theory can successfully characterize borehole stability problems [1842]. The application of surface area, swelling pressure, and water requirements of solids can be integrated into swelling models and mud process control approaches to improve the design of water-based mud in active or older shales.

Shale Inhibition with Water-Based Muds

One potential mechanism by which polymers may stabilize shales is by reducing the rate of water invasion into the shale. The control of water invasion is not the only mechanism involved in shale stabilization [133]; there is also an effect of the polymer additive. Osmotic phenomena are responsible for water transport rates through shales.

Inhibiting Reactive Argillaceous Formations

Argillaceous formations are very reactive in the presence of water. Such formations can be stabilized by bringing them in contact with a polymer solution with hydrophilic and hydrophobic links [101–104]. The hydrophilic portion consists of polyoxyethylene, with hydrophobic end groups based on isocyanates. The polymer is capable of inhibiting the swelling or dispersion of the argillaceous rock resulting from its adsorptive and hydrophobic capacities.

Thermal Treatment to Increase the Permeability

To increase the permeability of a certain region of the reservoir, the liquid-absorbed water is evaporated by heating the portion to a temperature above the boiling point of water, taking into account the ambient pressure [897, 1487]. The liquid water is evaporated by injecting a water-undersaturated gas, such as heated nitrogen, into the reservoir.

Formation Damage in Gas Production Shut-In

Sometimes it may become necessary to shut-in a gas well when the demand for gas is low. In such instances, the well is shut-in for an indefinite period, after which it is reopened and production is resumed. It often has been found that the production rate of gas from the reopened well is substantially less than it was before the well was shut-in. During production, the inner wall of the production tubing will be coated with a film of condensed freshwater because of the geothermal gradient. This water flows down when production is interrupted and can cause formation damage. This may occur because clays are normally saturated with brine water and not with freshwater. This swelling can be prevented with the injection of some additive, for example, sodium chloride, potassium chloride, calcium chloride, or an alcohol or a similar organic material [1853].

Inhibitors of Swelling

Inhibitors of swelling act in a chemical manner rather than in a mechanical manner. They change the ionic strength and the transport behavior of the fluids into the clays. Both the cations and the anions are important for the efficiency of the inhibition of swelling of clays [503].

Chemicals in Detail

Saccharide Derivatives

A drilling fluid additive, which acts as a clay stabilizer, is the reaction product of methylglucoside and alkylene oxides such as ethylene oxide, propylene oxide, or butylene oxide. Such an additive is soluble in water at ambient conditions, but becomes insoluble at elevated down-hole temperatures [386]. Because of their insolubility at elevated temperatures, these compounds concentrate at important surfaces such as the drill bit cutting surface, the borehole surface, and the surfaces of the drilled cuttings.

Anionic Polymers

Anionic polymers may be active as the long chain with negative ions attaches to the positive sites on the clay particles or to the hydrated clay surface through hydrogen bonding [768]. Surface hydration is reduced as the polymer coats the surface of the clay. The protective coating also seals, or restricts, the surface fractures or pores, thereby reducing or preventing the capillary movement of filtrate into the shale. This stabilizing process is supplemented by polyanionic cellulose. Potassium chloride enhances the rate of polymer absorption onto the clay.

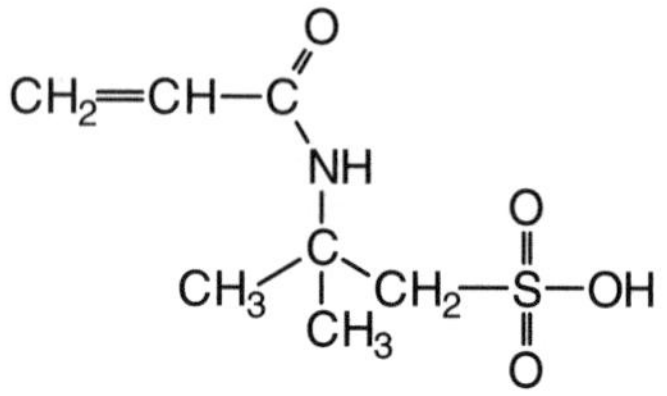

2-Acrylamido-2-methylpropane sulfonic acid

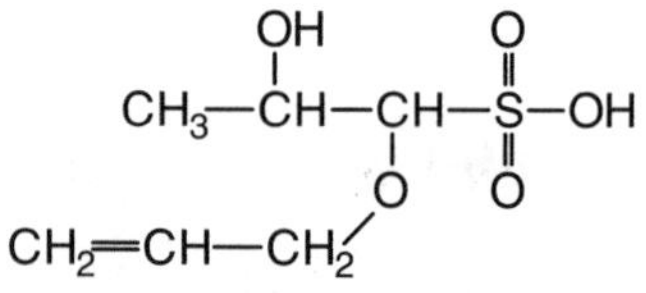

1-Allyloxy-2-hydroxypropyl sulfonic acid

Figure 3–1. Monomers for clay stabilizers: 2-acrylamido-2-methylpropane sulfonic acid, 1-allyloxy-2-hydroxypropyl sulfonic acid.

Chapter 4

Bit Lubricants

Refractory Metals

Molybdenum disulfide is used traditionally in greases for bit lubrication. In addition, polymers of 2-methylpropene (i.e., isobutene) and metal soaps are used to formulate synthetic greases [475]. A viscosity of 600 to 750 cP at 120° C is desirable. However, in the severe environment of a rock bit bearing, the viscosity of the composition should be at least 200 cP at 100° C [472]. Other heavy-duty greases based on molybdenum sulfide also contain calcium fluoride [1058, 1059] and metal soaps as thickeners. Specialized lubricating greases were developed for the bearing assemblies of roller bits. The greases are prepared from petroleum oils that are thickened with alkali and alkaline-earth metal soaps [1136]. The greases contain additives and fillers, such as synthetic dichalcogenides of refractory metals, which exhibit the necessary service characteristics. Tests have shown that such greases exceed the initial grease by 7 to 12 times with respect to performance time.

Natural Compounds

Phosphatides or phospholipids are environmentally safe lubricating additives [678].

Table 4–1
Compounds Suitable as Lubricants

Compound	References
Carbon black glycol esters of synthetic 21–25C fatty acids	[1537]
Sodium ethyl siliconate, diethylene glycol, graphite powder, and fatty acids[a]	[712]
Polyacrylamide, carboxymethylcellulose, gypsum[b]	[928]
Olefins[c]	[767, 987, 988]
2,4,8,10-tetra-oxaspiro-5, 5-undecane[d]	[648]
Phosphatides or phospholipids[e]	[678]
Polypropylene glycol[f]	[555, 573, 1553]
Fluoropolymers and zinc dioctyl-phenyl-dithiophosphate[g]	[349]

a) Sealing lubricant down to −40° C.
b) Sealing lubricant.
c) With metal soaps and aerosil as thickener.
d) Reaction product from pentaerythrite and paraformalehyde.
e) Environmentally safe lubricating additives.
f) As mud additive for lubrication.
g) Antioxidizing and anticorrosion additive to lubricating oils.

Chapter 5

Bacteria Control

Major problems in oil and gas operations result from the biogenic formation of hydrogen sulfide (H_2S) in the reservoir. The presence of H_2S results in increased corrosion, iron sulfide formation, higher operating costs, and reduced revenue and constitutes a serious environmental and health hazard.

In secondary oil recovery, which involves waterflooding of the oil-containing formation, biofilms can plug the oil-bearing formation. Severe corrosion also can result from the production of acids associated with the growth of certain bacterial biofilms. These bacterial biofilms are often composed of sulfate-reducing bacteria, which grow anaerobically in water, often in the presence of oil and natural gases. Once biofilms are established, it is extremely difficult to regain biologic control of the system.

When biofilms are formed on metallic surfaces, they can seriously corrode performance oil production facilities, chemical processing plants, paper mills, ships, and water distribution networks. Microbiologically influenced corrosion (MIC) represents the most serious form of that degradation.

It is estimated that MIC may be responsible for 15% to 30% of failures caused by corrosion in all industries.

Thus an effective control of bacteria responsible for these undesired effects is mandatory. Several biocides and nonbiocidal techniques to control bacterial corrosion are available, and procedures and techniques to detect bacteria have been developed.

Mechanisms of Growth

Growth of Bacteria Supported by Oil Field Chemicals

Growth experiments were conducted using bacteria from oil installations with several chemicals normally used in injection water treatment. The studies revealed that some chemicals could be utilized as nitrogen sources, as phosphorus sources, and as carbon sources for the bacteria [1696]. Therefore it is concluded that the growth potential of water treatment additives may be substantial and should be investigated before selection of the respective chemicals.

In other experiments it was established that the cultures of sulfate-reducing bacteria isolated from waters of several oil fields have a greater capacity to form H_2S than the collection culture of sulfate-reducing bacteria used as the standard in the investigations. The stimulating effect of the chemical products on individual elective cultures of sulfate-reducing bacteria can vary considerably depending on the species, activity, and adaptation of bacteria to the chemical products. Elective cultures of sulfate-reducing bacteria as a result of adaptation acquire relative resistance to toxic compounds. Thus when bactericides are used for complete suppression of the vital activity of sulfate-reducing bacteria in the bottom-hole zone and reservoir, higher doses of the bactericide than those calculated for laboratory collection cultures are necessary [1011].

It has been shown that sulfidogenic bacteria injected into a reservoir with floodwater may survive higher temperatures in the formation and can be recovered from producing well fluids [1546]. These organisms may colonize cooler zones and sustain growth by degrading fatty acids in formation waters.

Mathematical Models

A mathematical model for reservoir souring caused by the growth of sulfate-reducing bacteria is available. The model is a one-dimensional numerical transport model based on conservation equations and includes bacterial growth rates and the effect of nutrients, water mixing, transport, and adsorption of H_2S in the reservoir formation. The adsorption of H_2S by the rock was considered.

Two basic concepts for microbial H_2S production were tested with field data:

- H_2S production in the mixing zone between formation water and injection water (mixing zone model)
- H_2S production caused by the growth of sulfate-reducing bacteria in a biofilm in the reservoir rock close to the injection well (biofilm model)

Field data obtained from three oil producing wells on the Gullfaks field correlated with H_2S production profiles obtained using the biofilm model but could not be explained by the mixing zone model.

Detection of Bacteria

Microbiologically influenced souring (MIS) is the production of H_2S through the metabolic activities of microorganisms. A better chance for mitigating MIS in some down-hole environments using biocides may be possible if the problem is detected early in the souring process [1259]. However, if the

H$_2$S-producing creatures are allowed to spread into subsurface regions that are less accessible to biocides (profuse-stage MIS), the problem becomes less mitigable by conventional means.

API Serial Dilution Method. The API serial dilution method is the most widely used method for the detection of microorganisms. Field test methods for estimating bacterial populations have been standardized. A standard method dealing with the dose-response (time-kill) testing for evaluating biocides has been established. Sampling methods are of special importance because effective sampling is essential to any successful analysis.

Enzymatic Assay. The enzymatic (luciferase) assay for adenosine triphosphate (ATP) is one of the methods applied to areas of biocidal control in oil production operation [1454]. A reliable method for the determination of ATP is the measurement of bioluminescence produced by the luciferin luciferase system.

Electrochemical Determination. An electrochemical method has been developed to allow on-line monitoring of biofilm activity in aqueous environments.

Colorimetry. Laboratory data on the persistence of biocides formulated in glutaraldehyde and acrolein are available [1260]. A colorimetric general aldehyde detection method based on m-phenylenediamine was used. Such studies follow the demand for better understanding of ecologic systems in the aspect of environmental protection.

In another study, a mathematical model was constructed that incorporated experimentally determined glutaraldehyde persistence, rates of water production, and other factors. The model was used to calculate the levels of glutaraldehyde in a specified environment (lagoons) as a function of time, based on the amount of glutaraldehyde applied down hole [476].

Treatments with Biocides

Previously Fractured Formations

A special problem is the refracturing of a previously fractured formation that is contaminated with bacteria. In such a case the fracturing fluid must be mixed with an amount of biocide sufficient to reach and to kill the bacteria contained in the formation. The refracturing of the formation causes the bactericide to be distributed throughout the formation and to contact and kill bacteria contained therein [1181].

Intermittent Addition of Biocide

The intermittent addition technique [805–807] consists of

- The addition of a slug dose of a biologically effective amount of a quick-kill biocide.
- Intermittent addition of biologically effective amounts of a control biocide. This means that the control biocide is dosed for a certain period of time, followed by a period of much lower or zero dosing. This cycle is repeated throughout the treatment.

This process reduces the amount of control biocide employed in the control of contamination of oil production system waters by sessile bacteria. The biocide may be applied at intervals of 2 to 15 days. The duration of biocide application is preferably from 4 to 8 hours [1244].

Nonbiocidal Control

Chemical treatments for bacteria control represent significant cost and environmental liability. Because the regulatory pressure on the use of toxic biocides is increasing, more environmentally acceptable control measures are being developed.

Biocompetitive Exclusion Technology

Besides adding biocides to wells, another approach seems to be promising—modifying the reservoir ecology. The production of sulfide can be decreased, and its concentration is reduced by the establishment and growth of an indigenous microbial population that replaces the population of sulfate-reducing bacteria.

The technology is based on the addition of low concentrations of a water-soluble nutrient solution that selectively stimulates the growth of an indigenous microbial population, thereby inhibiting the detrimental sulfate-reducing bacteria population that causes the generation of H_2S. This deliberate and controlled modification of the microflora and reservoir ecology has been termed *biocompetitive exclusion* [835, 1548].

Inhibitors for Bacterial Films

Laboratory tests with quaternary amine additives showed a very low surface colonization and lower corrosion rates [556]. On the other hand, the biocidal effect of quaternary amines in the test fluids appeared to be minimal. These results suggest that quaternary amines may prevent MIC by mechanisms other than killing bacteria and that treatments preventing colonization on the surface may persist longer than most biocides.

Periodic Change in Ionic Strengths

For effective control of microorganisms, it is necessary to take into account the mechanism of formation of bacteria and the ecologic factors affecting it. The process of vital activity of bacteria begins with their adsorption on the enclosing rocks and adaptation to the new habitat conditions. Pure cultures of sulfate-reducing bacteria are not active in crude oil. Their development in oil reservoirs depends entirely on hydrocarbon-oxidizing bacteria, which are the primary cause of breakdown of oil. If the ecologic conditions in the reservoir are changed during formation of the microorganisms, the established food chains are disrupted and the active development of microflora ceases. It was experimentally established that the periodic injection of waters markedly differing in mineralization, taking into account the ecologic characteristics of the formation of the microorganisms, makes it possible to control biogenic processes in an oil reservoir without disturbing the sanitary state of the environment [187].

Bactericides

Biocides are often misapplied in the petroleum industry. Many of the misapplications occur because the characteristics of the biocides are not considered before use. Some guidelines of biocide selection are outlined in a review in the literature [201]. Early detection of microbiologic problems is imperative, and reparative actions must be taken as soon as possible. These measures should include changes in operating methods to prevent degradation of the operating environment. This might include the rejection of untreated waters for cleaning deposits in vessels and lines. In general, biocides are needed to control the activity of the bacteria in a system. However, biocides alone usually will not solve a microbiologic problem. Five requirements for the bactericide selection are emphasized [1898]:

- Wide bacteria-killing ability and range
- Noncorrosive property, good inhibiting ability, and convenience for transportation and application
- Nontoxic or low-toxicity property that causes no damage to human beings and is within environmental control regulations
- Good miscibility, with no damage or interference to drilling fluid or its chemical agents
- Bacteria killing effect that is not affected by environmental adaptation of the bacteria

Various Biocides

Formaldehyde

Tables 5–1 to 5–3 summarize some biocides proposed for bacteria control. Core flood experiments were used to evaluate the efficacy of periodic formaldehyde injection for the control of in situ biogenic reservoir souring. Formaldehyde treatments were demonstrated to control souring in both

Table 5–1
Biocides Proposed for Bacteria Control

Biocide	References
Thiocyanomethylthio-benzothiazole[a]	[1368]
1-(2-Hydroxyethyl)-2-methyl-5-nitroimidazole =(metronidazole)	[1113]
2-Bromo-4-hydroxyacetophenone	[1370]
Di-(tri-N-butyl)-(1,4-benzodioxan-6,7-dimethyl) diammonium dichloride	[1272]
Dimethyl-tetrahydro-thiadiazine-thione	[937]
Formaldehyde	[1011]
Glutaraldehyde	
Nitrate[b]	[1584]
Diammonium salts of tetrahydrophthalic acid or methyl-tetrahydrophthalic acid[c]	[955]
Tetrakis-hydroxymethyl-phosphonium sulfate	[1147]
Zinc slurry[d]	[1764]
o-Phthalaldehyde	[1730]
Monochloroamine	[203]

a) Drilling lubricant.
b) 5 to 50 ppm.
c) 25 to 75 ppm.
d) Waste from the production of 1-naphthol-3,6-disulphonic acid.

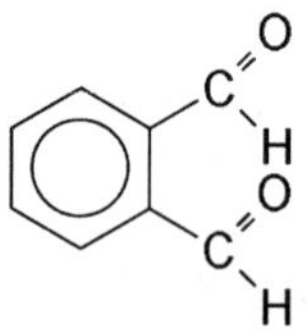

o-Phthalaldehyde

Figure 5–1. o-Phthalaldehyde.

environments; if the formaldehyde can be transported through the reservoir, in situ biogenic souring should be mitigated.

Glutaraldehyde

Glutaraldehyde is a useful antimicrobial agent, but it is dangerous and unpleasant to handle and is thermally unstable. Despite these disadvantages, glutaraldehyde is specified for use against bacteria in cooling towers of air-conditioning systems in buildings and to control anaerobic sulfate-reducing bacteria in oil wells.

Bisulfite Adduct. A bisulfite addition complex of an aldehyde or dialdehyde has been proposed for use as an antimicrobial agent [1858, 1859]. The complex is less toxic than free glutaraldehyde. In oil wells, its digestion by the sulfate-reducing bacteria releases the free dialdehyde that controls the bacteria. In these ways, a more economic and environmentally safer use of antimicrobial additives is likely.

Combined Chlorine-Aldehyde Treatment. A combined chlorine-aldehyde treatment that has two stages, that is, chlorination and subsequent biocide application, has been suggested. Short-residence–time shock doses of glutaraldehyde have been applied after chlorination [1180]. It has been established that a primary chlorination in overall bacterial control is useful.

Tetrakis-hydroxymethyl Phosphonium Salts

Offshore oil production is one area in which environmental pollution has been highlighted as an issue, particularly in the use of biocides in production waters. Tetrakis-hydroxymethyl phosphonium salts exhibit acceptable environmental profiles [1120], and they are regarded as the preferred products for bacterial control within the oil production industry.

Chlorine Dioxide

Chlorine dioxide has been evaluated as a replacement for chlorine [1630]. Gaseous chlorine as a biocide for industrial applications is declining because of safety and environmental and community impact considerations. Various alternatives have been explored, for example, bromo-chorodimethyl hydantoin (BCDMH), nonoxidizing biocides, ozone, and chlorine dioxide. Chlorine dioxide offers some unique advantages because of its selectivity, effectiveness over a wide pH range, and speed of kill. Safety and cost considerations have restricted its use as a viable replacement.

Table 5-2
Biocides Proposed for Bacteria Control

Biocide	References
Bromo-2-nitropropane/1,3-diol (Bronopol)	[1195]
1-N-Hexadecyl-1,2,4-triazole bromide, 1-N-octadecyl-1,2,4-triazole bromide	[473]
3-Hydroxy-4-methylthiazol-2(3H)-thione, 3-hydroxy-4-phenylthiazol-2(3H)-thione, 3-acetoxy-4-methylthiazol-2(3H)-thione, 1-hydroxy-4-imino-3-phenyl-2-thiono-1, 3-diazaspiro(4,5)decane, 1-hydroxy-5-methyl-4-phenylimidazoline-2-thione	[115]
Iodoacetone	[1483]
Glutaraldehyde	[330, 537]
2,3-Dibromo-1-chloro-4-thiocyanato-2-butene	[116]
Isothiazolin-3-ones[a]	[1090, 1175]
Tetrakis-(hydroxymethyl)-phosphonium salts[b]	[266, 1796]
1,2-Benzoisothiazolin-3-one and 3-(3,4-dichlorophenyl)-1,1-dimethylurea	[1258]
Tributyltetradecyl phosphonium chloride and pentanedial	[1057]
1,2-Dimethyl-5-nitro-1H-imidazole and glutaraldehyde or formaldehyde	[849]
Ethylene glycol monoacetate	[959]
5-Chloro-2-methyl-4-isothiazolin-3-one or 2-methyl-4-isothiazolin-3-one	[860–862]
2-N-Octyl-4-isothiazolin-3-one with 1,2-dibromo-2,4-dicyanobutane	[863, 864]
2,2-Dibromo-2-nitroethanol	[1083, 1084]
Chlorine dioxide	[388]
Lactic acid and sodium chlorite	[1171–1173]
Polyoxymethylene[c]	[580]
Isothiazolin-3-one[d]	[708]
4,5-Dichloro-2-N-octyl-isothiazolin-3-one and 2-methylthio-4-tert-butylamino-6-cyclopropylamino-S-triazine or N,N-dimethyl-N'-phenyl-N'-fluorodichloromethylthio)sulfamide[e]	[513]
Anthraquinones	[280, 1832]
Free halogen sources with 5,5-dimethylhydantoin, 5-ethyl-5-methylhydantoin	

Table 5–2 (continued)

Biocide	References
Cyanuric acid, succinimide, urea, 4,4-dimethyl-2-oxazolidinone, glycouril[f]	[1702]
2-Bromo-2-bromomethylglutaronitrile with 2,2-dibromo-3-nitrilopropionamide	[896]
Dimethylaminopropyl methacrylamide or dimethylaminopropyl acrylamide	[1784]
Iodine	[477]

a) Stabilized with orthoesters.
b) Quaternary phosphonium.
c) In situ biocide, formaldehyde release.
d) pH 4 to 5 with bromate buffer.
e) Broad control of microorganisms.
f) Sodium hypochlorite and chlorine are more efficient with N-hydrogen compounds.

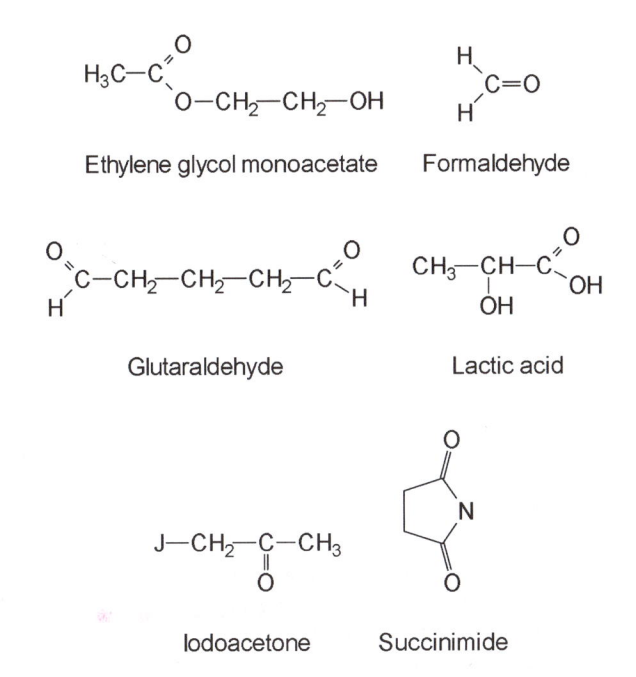

Figure 5–2. Ethylene glycol monoacetate, formaldehyde, glutaraldehyde, lactic acid, iodoacetone, succinimide.

Table 5–3
Biocides Proposed for Bacteria Control

Biocide	References
Chlorine dioxide	[977]
Alkyl amine carboxylic acid and metronidizole	[483]
Sodium hypochlorite[a]	[633, 751]
Toxic alloys[b]	[818]

a) Continuous or intermittent chlorination; high concentrations increase corrosion.
b) Passive layers enriched with toxic elements.

Bacterial Corrosion

Bacterial corrosion is often referred to as *microbiologically influenced corrosion*. MIC involves the initiation or acceleration of corrosion by microorganisms. The metabolic products of microorganisms appear to affect most engineering materials, but the more commonly used corrosion-resistant alloys, such as stainless steels, seem to be particularly susceptible.

The importance of MIC has been underestimated, because most MIC occurs as a localized, pitting-type attack. In general this corrosion type results in relatively low rates of weight loss, changes in electrical resistance, and changes in total area affected. This makes MIC difficult to detect and to quantify using traditional methods of corrosion monitoring [1447].

To adequately address MIC problems, interdisciplinary cooperation of specialists in microbiology, metallurgy, corrosion, and water chemistry is required. Because the complexities of MIC are so great, one single technique cannot provide all the answers in terms of corrosion mechanisms.

The problem of and importance of MIC was not fully realized until recently. Even in the mid-1980s the statement was made that "The major problem encountered by the petroleum microbiologist working in the North Sea oil fields is that of convincing the oil field engineer that bacterial corrosion is a subject worthy of serious attention."

Reviews on Bacterial Corrosion

A reference guide on recognizing, evaluating, and alleviating corrosion problems caused by microorganisms has been compiled [57]. This manual provides a guide, training manual, and reference source for field and engineering personnel for dealing with corrosion problems caused by microorganisms. The trends in the 1990s of MIC have been reviewed in the literature [575]. The basic goal of a practicing corrosion engineer should be not to identify, count, or even kill the microorganisms, but to effectively control corrosion in an oil field.

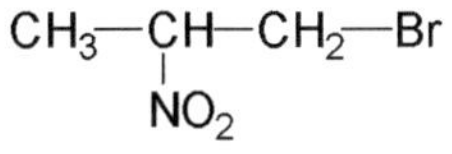

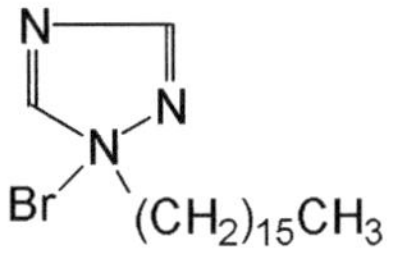

Bromo-2-nitropropane 1-N-hexadecyl-1,2,4-triazole bromide

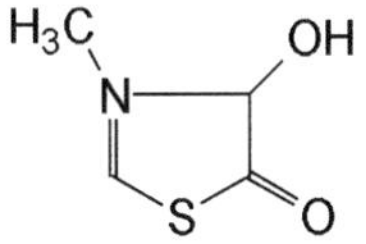

3-Hydroxy-4-methylthiazol-2(3H)-thione

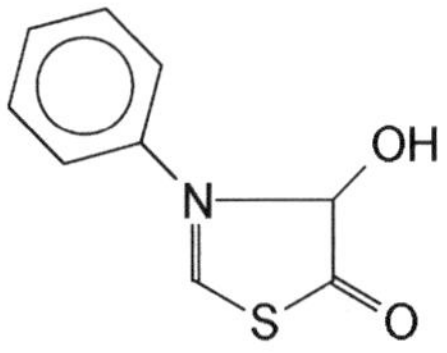

3-Hydroxy-4-phenylthiazol-2(3H)-thione

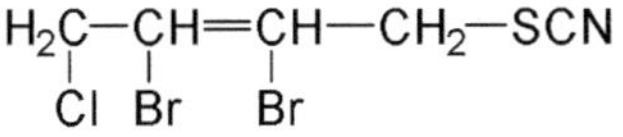

2,3-Dibromo-1-chloro-4-thiocyanato-2-butene

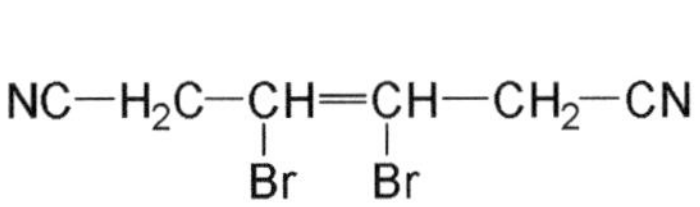

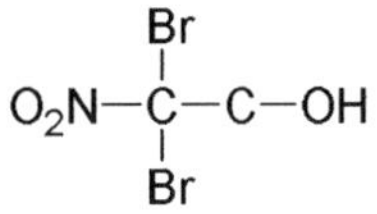

1,2-Dibromo-2,4-dicyanobutane 2,2-Dibromo-2-nitroethanol

Figure 5–3. Biocides: bromo-2-nitropropane, 1-N-hexadecyl-1,2,4-triazole bromide, 3-hydroxy-4-methylthiazol-2(3H)-thione, 3-hydroxy-4-phenylthiazol-2(3H)-thione, 2,3-dibromo-1-chloro-4-thiocyanato-2-butene, 1,2-dibromo-2,4-dicyanobutane, 2,2-dibromo-2-nitroethanol.

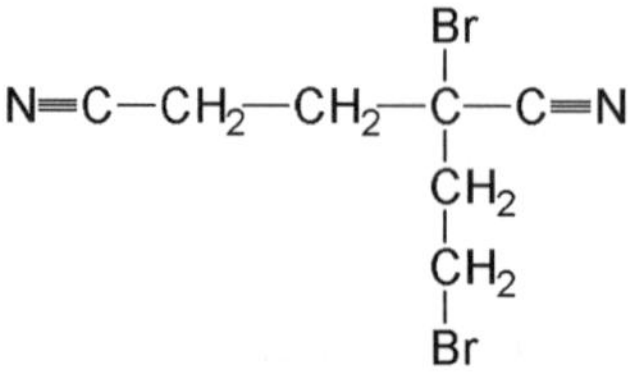

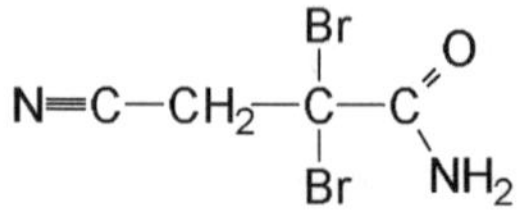

2-Bromo-2-bromomethylglutaronitrile 2,2-Dibromo-3-nitrilopropionamide

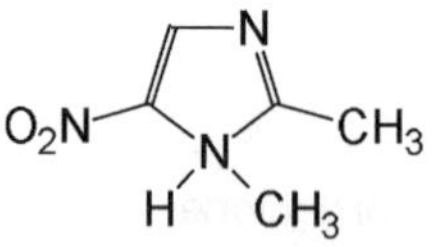

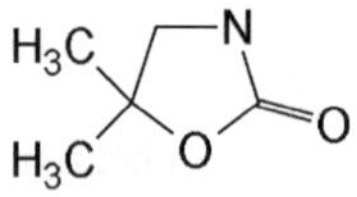

1,2-Dimethyl-5-nitro-1H-imidazole 4,4-Dimethyl-2-oxazolidinone

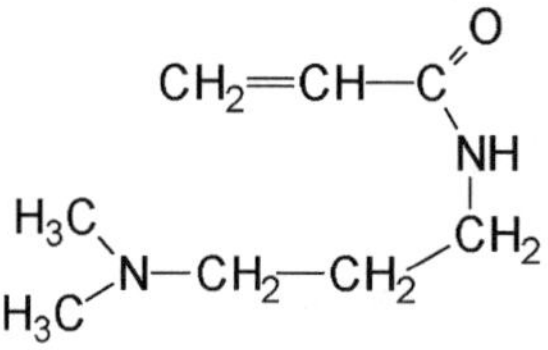

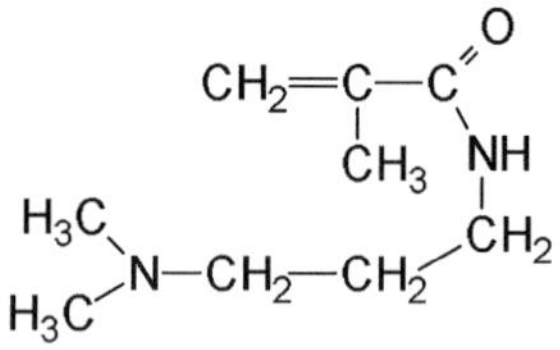

Dimethylaminopropylacrylamide Dimethylaminopropylmethacrylamide

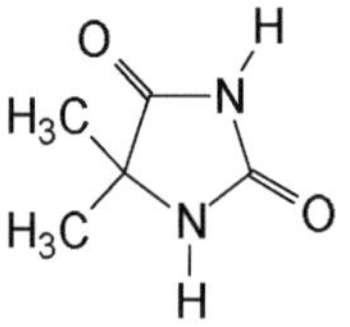

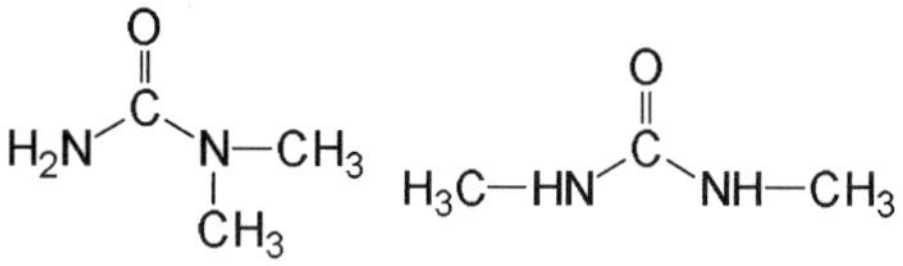

5,5-Dimethylhydantoin Dimethylurea (asymmetric and symmetric)

Figure 5–4. 2-Bromo-2-bromomethylglutaronitrile, 2,2-dibromo-3-nitrilopropionamide, 1,2-dimethyl-5-nitro-1H-imidazole, 4,4-dimethyl-2-oxazolidinone, dimethylaminopropyl acrylamide, dimethylaminopropyl methacrylamide, or dimethylaminopropyl acrylamide, 5,5-dimethylhydantoin, dimethylurea.

Assessment of Bacterial Corrosion

A critical review of the technical literature concerned with monitoring techniques for the study of MIC has been presented in the literature [212]. The monitoring techniques in this review include measurements of electrochemical properties, measurements of physical metal loss, and enumeration of sessile organisms. The procedures for the study of MIC, as well as the advantages and the disadvantages of each technique, are discussed.

MIC can be misdiagnosed as attack caused by conventional chloride crevice or as pitting corrosion unless specialized techniques are used during the failure analysis [213]. These techniques include in situ bacteria sampling of residual water; bacterial analysis of corrosion products using analytical chemistry, culture growth, and scanning electron microscopy; as well as nondestructive examination using ultrasonic and radiographic techniques. Metallographic examination can reveal MIC characteristics such as dendritic corrosion attack in weld metal.

Bacterial Hydrogenase

Theoretical and experimental studies have shown that the removal of molecular hydrogen from cathodic surfaces is a primary driving force in MIC. A rapid (1 to 4 hours) test has been developed for the presence of bacterial hydrogenase [202] that detects the presence of a wide range of corrosion-causing bacteria in water, sludge, and adherent bacterial biofilms. This test can be used to monitor oil and gas systems for the development of potentially corrosive bacterial populations and to assess the efficacy of control measures, including biocide treatment, because the hydrogenase test yields negative results when this pivotal, corrosion-causing enzyme has been denatured.

Lipid Biomarkers

Microbes of differing physiologic types, acting in consortia, appear to be more destructive than monocultures. Methods for examining consortia are based on the detection of lipid biomarkers that are characteristic for different classes of microbes. These can be analyzed by gas chromatography coupled with mass spectrometry [512].

Electron Microscopy

Side stream sampling devices can be used to collect biofilm and corrosion samples. The biofilm, inorganic passive layers, and metal attacked samples can be characterized with scanning electron microscopy and energy dispersive x-ray analysis. Results of such a study showed a correlation between biofouling and corrosion attack of carbon steel samples [1800].

Electrochemical Impedance Spectroscopy

Electrochemical impedance, weight loss, and potentiodyne techniques can be used to determine the corrosion rates of carbon steel and the activities of both sulfate-reducing bacteria and acid-producing bacteria in a water injection field test. A study revealed that the corrosion rates determined by the potentiodyne technique did not correlate with the bacterial activity, but those obtained by electrochemical impedance spectroscopy (EIS) were comparable with the rates obtained by weight loss measurements [545].

Other electrochemical techniques covered include measurements of the corrosion potential, the redox potential, the polarization resistance, the electrochemical impedance, electrochemical noise, and polarization curves, including pitting scans. A critical review of the literature concerned with the application of electrochemical techniques in the study of MIC is available [1164].

Mechanisms of Microbial Corrosion

The role of microorganisms can be direct in microbially induced corrosion if an electrochemical cell is created. Alternatively, the role can be indirect, in that it maintains a preexisting electrochemical cell by stimulating either the cathodic or the anodic reaction [771]. Various microorganisms and mechanisms are thought to be involved. Most commonly, a differential aeration cell is created with a low concentration of oxygen, shielded beneath slime or colony growth, as compared with the high concentration externally in the bulk environment. Under these conditions, the surface in the low-concentration area becomes anodic with the dissolution of metal, while the electrons react at the cathodic region with the high concentration of oxygen, giving rise to hydroxide. Ultimately, metal oxides and hydroxides are characteristic for aerobic corrosion.

Mechanisms whereby microbes influence the corrosion rate are described as follows [1446]:

1. Cathodic depolarization
2. Formation of occluded area on metal surface
3. Fixing the anodic sites
4. Underdeposit acid attack

Simultaneous Mechanisms of Corrosion

MIC almost always acts in concert with other corrosion mechanisms and may, at times, appear to be crevice corrosion, underdeposit acid attack, oxygen-concentration cell corrosion, ion-concentration cell corrosion, and CO_2

corrosion [1445]. In most cases in which MIC is found on external surfaces, it is associated with disbonded coatings or other areas that are shielded from the potentially protective action of cathodic protection. Furthermore, pipelines often are in contact with wet clays, which have little scaling potential.

pH Regulation

Weak acids are products of bacterial metabolisms. Sulfate-reducing bacteria regulate the pH of their environment at levels that depend on the potential secondary reactions:

- Precipitation of iron sulfide
- Oxidation of sulfide ions to thiosulfate by traces of oxygen
- Metabolism of this thiosulfate or of other sulfur compounds

In this way, it is possible to explain the initiation and growth of bacterial corrosion pits.

Chapter 6

Corrosion Inhibitors

History

The history of corrosion inhibitors and neutralizers and their invention, development, and application in the petroleum industry is documented by a review of Fisher [605]. Early corrosion inhibitor applications in each of the various segments of the industry, including oil wells, natural gas plants, refineries, and product pipelines, are reviewed.

Corrosion and scale deposition are the two most costly problems in oil industries. Corrodible surfaces are found throughout production, transport, and refining equipment. The *Corrosion and Scale Handbook* gives an overview of corrosion problems and methods of corrosion prevention [159].

Classification of Corrosion Inhibitors

Corrosion inhibitors, which are used for the protection of oil pipelines, are often complex mixtures. The majority used in oil production systems are nitrogenous and have been classified into the following broad groupings:

- Amides and imidazolines
- Salts of nitrogenous molecules with carboxylic acids (fatty acids, naphthenic acids)
- Nitrogen quaternaries
- Polyoxylated amines, amides, and imidazolines
- Nitrogen heterocyclics

Fields of Application

Corrosion problems may occur in numerous systems within the petroleum industry. These include:

- Acid stimulation jobs
- Cooling systems

- Drilling muds
- Oil production units
- Oil storage tanks
- Oil well
- Protection of pipelines
- Refinery units
- Scale removal treatments using acids
- Steam generators
- Technologic vessel

Many compositions involve environmentally dangerous products, such as chromates, fatty amines of high molecular weights, imidazolines, etc. The use of some of the alternatives, for instance, polyphosphate or polyphosphonate, is limited because they precipitate in the presence of the salts of alkaline earth metals or because of their high costs.

Acidization

Acidization is an oil reservoir stimulation technique for increasing well productivity. Stainless steels are being used successfully to combat hydrogen sulfide (H_2S) and carbon dioxide (CO_2) corrosion; however, they are proving susceptible to hydrochloric acid (HCl). HCl is used in oil and gas production to stimulate the formation. The down-hole temperature may be in excess of 200° C in deep wells. The acid treatment occurs through steel tubes. This process requires a high degree of corrosion inhibition. Electrochemical measurements are nonpredictive in inhibited concentrated HCl at high temperatures [798].

Oil Storage Tanks

Storage tank bottoms have historically been protected from corrosion through the use of cathodic protection. In general, this method is successful. However, problems arise when there is not complete contact with the soil. This occurs when the bottom buckles slightly, leaving air spaces with filling or emptying of the tank. Over time, a portion of the base may erode away. In either case, the electrical continuity is lost. Other methods of protection, such as protective coatings, are not suitable.

When the bottom plates are welded together, the coating is partially destroyed. Research and field work showed that protection can be achieved using volatile corrosion inhibitors under the tank [688]. This works alone or in combination with cathodic protection. Double tank bottoms for leakage monitoring are often specified for new tanks. However, the same problem of coating destruction occurs. Volatile corrosion inhibitors are an excellent solution from both a technical and an economic standpoint. This type of corrosion inhibitor has a long history of corrosion protection under these conditions—wet, corrosive environments in void spaces.

Pipelines

The normal industrial practice for controlling the internal corrosion of petroleum pipelines is to use coatings, nonmetallic pipeline materials, or corrosion inhibitors. Corrosion inhibitors, which are used for the protection of oil pipelines, are often complex mixtures.

The consequences of pipeline failure caused by corrosion can include inventory loss, production shutdown, environmental damage, safety risks, and excessive repair and replacement costs [950]. Chemical treatment can delay or inhibit internal corrosion of a pipeline so that the line can fulfill its operating requirements over its design life. Using pigs for corrosion inhibitor application is particularly useful in gas and gas-condensate transmission pipelines, especially in multiphase flow service. Selecting a pig for inhibitor batching is based on the pig's ability to create a good seal between the pig cups and the pipe wall. The thickness of the film deposited during inhibition must be known to correctly size the slug inhibitor.

Coatings

Epoxide resins with aromatic amines are used as coatings for pipelines [305–307].

Production Wells

Unalloyed or low-alloyed steels of various strength are generally used in the production of oil and gas. Inhibitors must be injected into the borehole to increase the life of well casing, flow lines, and equipment of unalloyed and low-alloyed steels in corrosive media. If the inhibitor is improperly chosen, considerable corrosion damage may result, such as:

- Damage without hydrogen influence
- Hydrogen-induced damage in the presence of H_2S

Agitator autoclave tests can be used as screening tests despite the more intensive localized corrosion attack and the generally greater erosion rates. This test method elucidates the influences of certain test parameters including temperature, H_2S/CO_2 ratio, and flow [564].

Scale Removal Treatments Using Acids

Acids injected down hole for scale removal treatments are extremely corrosive to the production tubing and casing liners. Inhibitors are added to the stimulation fluids to minimize this corrosion. The effectiveness of inhibitors can be estimated with laboratory screening methods [279].

Application Techniques

The application techniques include batch application and continuous application.

Batch Application Versus Continuous Application

Batch treatment of pipelines with liquid or gel slugs of inhibitor, with continuous injection as a backup (or vice versa), are accepted methods of corrosion prevention [951]. Batching liquid or gel inhibitors using pigs is more likely to attain complete coverage of the internal surface of the pipe wall than is continuous injection. The film laid down is quite resilient and of long duration. Important factors to optimize the application include determining film thickness and selecting an appropriate pigging system and program. Cleaning of the pipeline before inhibitor pigging is recommended.

Emulsions

Corrosion inhibitors are often emulsions that are able to form an organic film on the parts to be protected.

Application in Solid Form

The preparation of a corrosion inhibitor in the solid form allows the development of a new technique of continuous intensive anticorrosive protection for gas and oil pipelines, as well as for acidizing operations of oil wells [746]. The controlled dissolution of the solid inhibitor creates a thin protective layer on the metallic surface that prevents or minimizes the undesirable corrosion reactions.

Analytic Procedures

The common method of treating rod-pumped wells is to periodically batch inhibitor into them. The treatment period for a given well is selected using empirical rules based on well production volumes. A successful and economic corrosion inhibition program must carefully control the inhibitor concentration in the well fluids. Environmental aspects and efficacious inhibitor usage necessitate the measurement of very low corrosion inhibitor concentrations. Inhibitor concentrations as low as one part per million are significant, thus requiring an analytic technique with a detection limit of a fraction of a part per million.

Accurate monitoring of inhibitor residual concentrations is most important in systems in which the volume of water is unknown or is highly variable. Frequent monitoring of the inhibitor concentration in the water exiting the pipeline is the simplest method, and sometimes the only method that can be used, to ensure that the line in fact is being protected.

Dye Transfer Method

The classic method for the determination of corrosion inhibitors in oil field brines is the dye transfer method. This method is basically sensitive to amines. Within this method, there are many variations that the analyst may use to determine the amount of corrosion inhibitor in either water or crude oil. Unfortunately these methods detect all amines present as corrosion inhibitors [1174].

Pressure Liquid Chromatography

Improved high-pressure liquid chromatography (HPLC) methods have been developed for the analysis of quaternary salt type corrosion inhibitors in brine waters [400]. However, these methods are not suitable for imidazolines and amido-amines. A method based on fluorescence detection has been described for the quantitative analysis of the imidazoline– and amido-amine–type corrosion inhibitors in both oil field water and crude oil samples by HPLC [1174].

Another analytic procedure based on HPLC has been developed for the quantitative determination of nitrogen-containing corrosion inhibitors [1194]. The method was primarily developed for the analysis of certain oil pipeline condensate samples.

A fully automated instrumental procedure has been developed for analyzing residual corrosion inhibitors in production waters in the field. The method uses ultraviolet (UV) and fluorescence spectrophotometric techniques to characterize different types of corrosion inhibitors. Laboratory evaluations showed that fluorescence is more suitable for field application because errors from high salinity, contamination, and matrix effect are minimized in fluorescence analysis. Comparison of the automated fluorescence technique with the classic extraction–dye transfer technique showed definite advantages of the former with respect to ease, speed, accuracy, and precision [1658].

Thin Layer Chromatography

With thin layer chromatography, attempts have been made to analyze amounts of residual inhibitors down to below the parts-per-million region [272].

Ultraviolet Spectroscopy

UV spectroscopy can be used to detect low levels of organic corrosion inhibitors in produced water. An analytic method has been developed using a diode array UV spectrophotometer [630].

Side Effects

Stabilizer for Emulsions

Some corrosion inhibitors have a side effect of stabilizing emulsions. This is sometimes undesirable.

Antisynergism with Alcohols

In stimulation fluids containing concentrated HCl, the partial substitution of water by alcohols such as methanol, ethanol, and glycerol increases the corrosivity of the acid fluids and reduces the efficiency of the corrosion inhibitors [1148]. This effect is especially important for fatty amine–based inhibitors. For products containing acetylenic-type inhibitors the detrimental effect is less important and a weight loss may be maintained within acceptable limits using slightly higher, but still reasonable, levels of inhibitor.

Synergism with Surfactants

Surfactants greatly improve the performance of trans-cinnamaldehyde as a corrosion inhibitor for steel in HCl [741, 1590, 1591]. They act by enhancing the adsorption at the surface. Increased solubility or dispersibility of the inhibitor is an incidental effect. N-dodecylpyridinium bromide is effective in this aspect far below its critical micelle concentration, probably as a result of electrostatic adsorption of the monomeric form of N-dodecylpyridinium bromide. This leads to the formation of a hydrophobic monolayer, which attracts the inhibitor. On the other hand, an ethoxylated nonylphenol, a nonionic surfactant, acts by incorporating the inhibitor into micelles, which themselves adsorb on the steel surface and facilitate the adsorption of trans-cinnamaldehyde.

Effect of Flow on Inhibitor Film Life

Experiments with low- and high-velocity conditions were performed in standard laboratory tests [539]. It was found that corrosion is governed by the flow of the reactants and the products to and from the corroding surface. Corrosion in oxygenated fluids is increased with the velocity of the fluid because a greater amount of oxygen is made available to the surface. Corrosion of steel in fluids containing CO_2 produces a protective iron carbonate film that initially results in decreased corrosion. However, at high velocities the protective layers are broken off, thus exposing the bare metal to the aggressive medium and increasing the corrosion rate. Inhibitor films are protective, because they reduce transfer rate of the corrosants. Inhibitor films can become ineffective because of aging, removal, and dilution. In all of the previous examples the velocity is an important variable, governing the ability of the inhibitor to control the corrosion rate.

Amides and Imidazolines

Amides

An amide-type corrosion inhibitor is prepared as follows: Methylmethacrylate is converted with tallow triamine or tallow tetramine at 80° to 90° C into the corresponding amides. After completion, the temperature is raised to initiate the polymerization reaction [1350]. The polymerization reaction is performed at temperatures up to 200° C. The polymer controls the corrosion of metal surfaces in contact with a corrosive hydrocarbon-containing medium.

Ammonium salts of alkenyl succinic half-amides have been described for use as corrosion inhibitors in oil and gas production technology to combat corrosion by media containing CO_2, H_2S, and elemental sulfur [1366]. The inhibitor composition may contain a dispersing agent, such as a low molecular weight or polymeric anionic surfactant like an alkylsulfonic acid or an alkylaryl sulfonic acid.

Ethoxylated and propoxylated alkylphenol amines, converted into the amides with a fatty acid or similar long chain diacids, are effective in controlling sour and sweet corrosion [1782, 1783, 1785–1787].

Table 6–1
Fatty Acids

Name	Formula	mp[a] (°C)
Palmitic acid	$H_3C-(CH_2)_{14}-COOH$	63
Stearic acid	$H_3C-(CH_2)_{16}-COOH$	69–71
Oleic acid	$C_{18}H_{34}O_2$	16
Lauric acid	$H_3C-(CH_2)_{10}-COOH$	44
Myristic acid	$H_3C-(CH_2)_{12}-COOH$	59
Linoleic acid	Z,Z–9,12–Octadecadienoic acid	−5
Linolenic acid	Z,Z,Z–9,12,15–Octadecatrienoic acid	−11

a) Melting point.

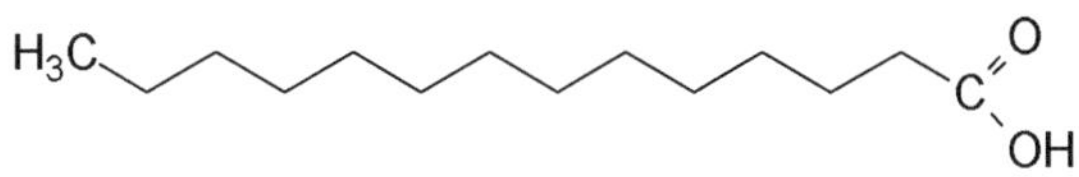

Myristic acid

Figure 6–1. Myristic acid.

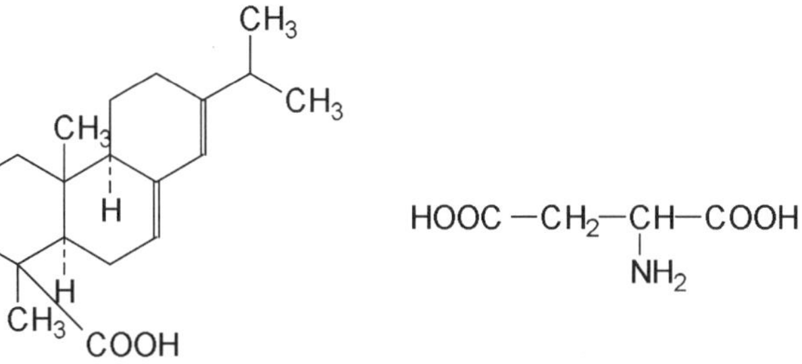

abietinic acid aspartic acid

Figure 6–2. Abietinic acid, aspartic acid.

Tall oil fatty acids consist of resin acids (25% to 30%) and of a mixture of linolic acid, conjugated C_{18} fatty acids (45% to 65%), oleic acid (25% to 45%), 5,9,12-octadecatrienic acid (5% to 12%), and saturated fatty acids (1% to 3%). Resin acids are abietinic acid, dehydroabietic acid, and others. Properties of fatty acids are shown in Table 6–1.

Polyimidoamines

Corrosion inhibiting compositions for metals subjected to highly acidic environments may be produced by reacting in a condensation reaction a styrene/maleic anhydride copolymer with a polyamine to produce a polyimidoamine inhibitor [1568]. These inhibitors exhibit film-forming and film-persistency characteristics. Some relevant polyamines are listed in Table 6–2.

Polypeptides

Polypeptides have been under consideration as corrosion inhibitors because of environmental concerns [1358]. Polyaspartate is the most efficient corrosion inhibitor known among the polypeptides [1196]. The molecular weight (1000 to 22,000 Dalton) does not affect the efficiency, but both high calcium ion and

Table 6-2
Polyamines

Name	Formula	mp[a] (° C)
Tetraethylene-pentamine	$H_2N-(CH_2-CH_2-NH)_4H$	−30
Ethylenediamine	$H_2N-CH_2-CH_2-NH_2$	9
1,2-Propylenediamine		—
Trimethylenediamine		−12
1,4-Butanediamine	$H_2N-(CH_2)_4-NH_2$	27

a) Melting point.

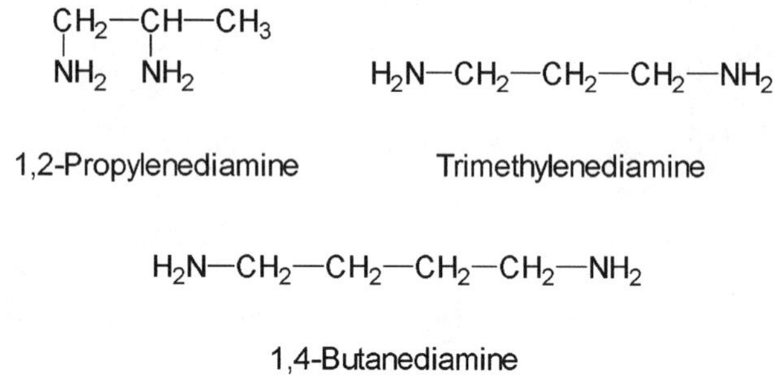

$$CH_2-CH-CH_3$$
$$\;|\quad\;|$$
$$NH_2\;\;NH_2$$

1,2-Propylenediamine

$$H_2N-CH_2-CH_2-CH_2-NH_2$$

Trimethylenediamine

$$H_2N-CH_2-CH_2-CH_2-CH_2-NH_2$$

1,4-Butanediamine

Figure 6-3. 1,2-Propylenediamine, trimethylenediamine, 1,4-butanediamine.

high pH enhance the effectiveness. The performance is particularly good in batch treatment tests.

In another study, polyaspartic acid was examined as a corrosion inhibitor for steel as a function of pH and temperature [1629]. At low to neutral pH values, polyaspartic acid increases the corrosion rate of steel. At pH values above 10, polyaspartic acid is a reasonably robust corrosion inhibitor.

Ampholytes

Corrosion inhibitors used in offshore oil production are highly cationic. However, the use of such cationic-based corrosion inhibitors for offshore oil platforms is becoming less acceptable for environmental reasons. Cationic inhibitors are attracted to metal surfaces, thereby controlling the acid-type corrosion. When these cationic corrosion inhibitors enter seawater, they are attracted to a particular type of algae, diatomes. These algae are part of a food-chain for mussels. The cationic inhibitors inhibit the growth of these algae. Betaines and ampholytes [1067] can be used instead of cationic inhibitors or

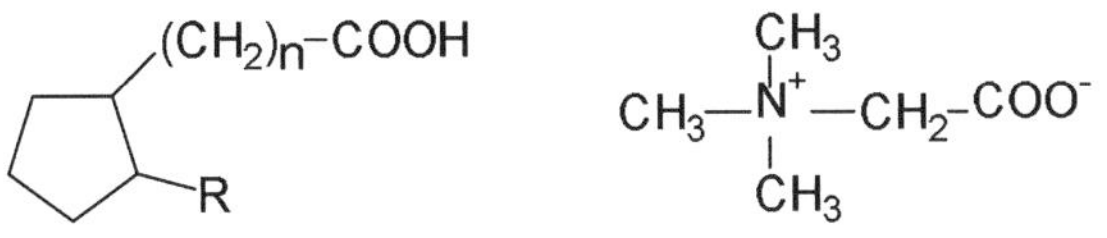

Naphthenic acids

Betaine

Figure 6–4. Naphthenic acids, betaine.

can be neutralized with acids such as acetic acid, adipic acid, sebacic acid, naphthenic acids, paraffinic acids, tall oil acids, and free sulfur dioxide. They are claimed to prevent CO_2 corrosion.

Slow-Release Formulation

An amido-amine (e.g., from the reaction of tetraethylenepentamine with stearic acid) is modified with propylene oxide [792]. The product is dispersed in a polymer matrix such as an acrylic or methacrylic polymer. The inhibitor is slowly released into the surrounding environment, such as in an oil or gas well, to prevent corrosion of metal equipment in the well.

Nitrogenous Bases with Carboxylic Acids

Diels–Alder Adducts

A corrosion inhibitor with excellent film-forming and film-persistency characteristics is produced by first reacting C_{18} unsaturated fatty acids with maleic anhydride or fumaric acid to produce the fatty acid Diels–Alder adduct or the fatty acid-ene reaction product [31]. This reaction product is further reacted in a condensation or hydrolyzation reaction with a polyalcohol to form an acid-anhydride ester corrosion inhibitor. The ester may be reacted with amines, metal hydroxides, metal oxides, ammonia, and combinations thereof to neutralize the ester. Surfactants may be added to tailor the inhibitor formulation to meet the specific needs of the user, that is, the corrosion inhibitor may be formulated to produce an oil-soluble, highly water-dispersible corrosion inhibitor or an oil-dispersible, water-soluble corrosion inhibitor. Suitable carrier solvents may be used as needed to disperse the corrosion inhibitor formulation.

Similarly, a salt of an ethoxylated amine and a reaction product of an alcohol and a fatty acid maleic anhydride adduct produced by a reaction between maleic anhydride and an unsaturated fatty acid has been described [511].

Nitrogen Quaternaries

Quaternary ammonium iodides were tested alone and in combination with propargyl alcohol with several steels in 15% HCl. The quaternary ammonium iodides showed superior inhibitor performance to that of propargyl alcohol (propargyl := $-CH_2-C\equiv CH$) at identical dosage levels. Mixtures of propargyl alcohol and quaternary ammonium iodide showed a synergistic effect [1330], as did formic acid [246] and thiols [1808].

It has been shown that the corrosion rates of various steels can be reduced to less than 1 mg/cm^2/hr using ternary inhibitor mixtures containing quaternary ammonium salts, trans-cinnamaldehyde, and potassium iodide in amounts of 0.2% of each component [1758].

Thio-Substituted Quaternary Ammonium Salts

A thio-substituted, quaternary ammonium salt can be synthesized by the Michael addition of an alkyl thiol to acrylamide in the presence of benzyl trimethyl ammonium hydroxide as a catalyst [793–795]. The reaction leads to the crystallization of the adducts in essentially quantitative yield. Reduction of the amides by lithium aluminum hydride in tetrahydrofuran solution produces the desired amines, which are converted to desired halide by reaction of the methyl iodide with the amines. The inhibitor is useful in controlling corrosion such as that caused by CO_2 and H_2S.

Synergism of Thiosulfate with Quaternary Ammonium Salts

Laboratory observations have shown that a combination of thiosulfate with cationic nitrogenous inhibitors has a significant effect on improving their performance [1432].

Polyoxylated Amines, Amides, and Imidazolines

Polyoxyalkylene Amines

A mixture of C-alkyl-ethylene diamine and di-(C-alkyl)-diethylenetriamine, with an alkyl side chain of 8 to 26 carbon atoms is suitable [1876] as a corrosion inhibitor. This product can be further reacted with an alkylating agent or an alkylene oxide [836, 837].

The inorganic nitrite used as a corrosion inhibitor in aqueous alkylene glycol or polyoxyalkylene glycol solutions can be replaced with polyoxyalkylene amines [1263,1264]. Such polyoxyalkylene amines impart corrosion inhibition to the liquid in contact with the metal and the metal in contact with the vapors of the aqueous composition. Aqueous compositions containing the glycol and the polyoxyalkylene amine also exhibit a low foaming tendency.

Alkylene Amine Compounds with Mercaptan

In highly acidic environments a reaction product of an isobutyraldehyde and an alkylene amine compound with an alkylsulfopropionic amide group is recommended [1888, 1889]. The alkylene amine compound can be the product of a reaction of equimolar amounts of N-dodecylmercaptan, methyl methacrylate, and diethylenetriamine.

Polyamine Derivatives

Fatty Amine Adducts

Dimerized fatty acid thioesters (with a dithiol) in combination with fatty amines are sulfur-containing corrosion inhibitors [888]. The corrosion inhibitor solvent is preferably a hydrocarbon.

Adducts of a fatty amine adduct to unsaturated acid in which the product contains only secondary or tertiary amine groups that have a lower toxicity to the environment [391].

Adducts to Polymers

Polymeric polyolefins, such as polybutadiene, secondary amines, and synthesis gas, are reacted in the presence of a catalyst system comprising a ruthenium-containing compound, a rhodium-containing compound, a sterically hindered phosphine, and a solvent [1191]. Preferred polybutadiene feedstocks are those with a predominance of main chain, rather than pendant olefin groups and in particular, those polymers containing both the 1,2-polybutadiene and 1,4-polybutadiene units. These polymers of high amine content are useful as down-hole corrosion inhibitors.

A low–molecular-weight, polyfunctional polymer can be formed by polymerizing a vinyl monomer in the presence of a mercaptan chain transfer agent [1861]. The vinyl monomer may be an unsaturated acid, acrylonitrile,

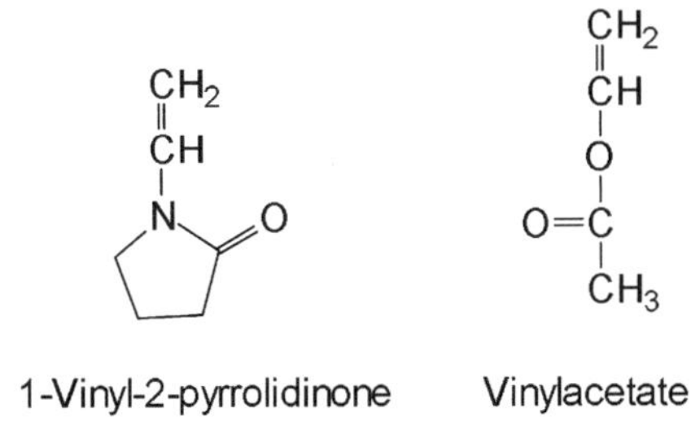

1-Vinyl-2-pyrrolidinone   Vinylacetate

Figure 6–5. 1-Vinyl-2-pyrrolidinone, vinylacetate.

vinylester, a variety of acrylamides, or N-vinyl-2-pyrrolidone. The molar ratio of the vinyl monomer to the mercaptan is preferably in the range of 2 to 40 moles of the vinyl monomer to 1 mole of the mercaptan. The composition and methods are useful for inhibiting corrosion of down-hole metal surfaces present in oil and gas wells.

Formaldehyde Condensates with Amines

Corrosion inhibitor compositions useful for oil and gas well applications are prepared by reacting, for example, 2,5-dimethylpyridine or 2,4,6-collidine with formaldehyde or acetone and an amine such as 1-dodecanamine [1761, 1762]. A hydrocarbon-soluble corrosion inhibitor is obtained by the acid-catalyzed oligomerization of an alkylaniline and formaldehyde [123]. These oligomers exhibit good initial inhibition of metal corrosion in aqueous environments, and this effect is more persistent than that observed for the corresponding monoamine starting material. Moreover, in an acidic environment, the products show superior persistence in inhibiting corrosion when compared with known monoamine corrosion inhibitors, such as tallow amine. The oligomers can be formulated to be both hydrocarbon soluble and water dispersible. The water dispersibility can be controlled by varying the type and amount of the additional aromatic compound, such as ethoxylated alkylphenol, included in the oligomerization reaction mixture [122].

A corrosion inhibitor that is the adduct of a carbonyl compound, an amine, and a thiocyanate has been described [1431]. The product provides protection against ferrous corrosion in severe environments. 500 ppm by weight is sufficient. The inhibitor is employed in wells producing both oil and water and in high-temperature environments around 120° C.

Lignin Amines

Lignin amines with high nitrogen content are water soluble at both alkaline and acidic pH values. The lignin amines have various useful properties. For example, they are active as flocculants, filtration aids, scale inhibitors, fluid loss additives, oil well cement additives, and corrosion inhibitors among other potential uses. The nitrogen is introduced into the lignins with the Mannich reaction [1570].

Fatty Acid Amides with Propargylalcohol

Propargylalcohol has been found to be active in corrosion control, and a variety of formulations with propargylalcohol have been proposed. A condensate of a polyamine, such as diethylenetriamine, triethylenetetramine, or

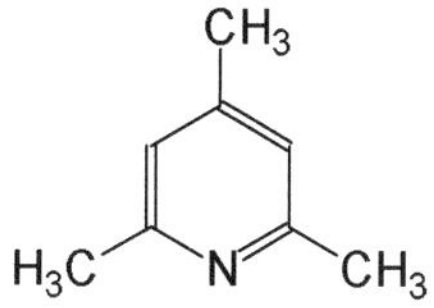

$H-C\equiv C-CH_2-OH$

Collidine   Propargyl alcohol

Figure 6–6. Collidine, propargyl alcohol.

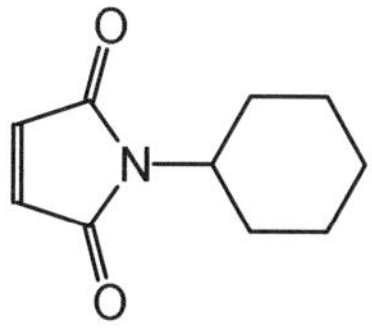

N-Cyclohexyl maleimide

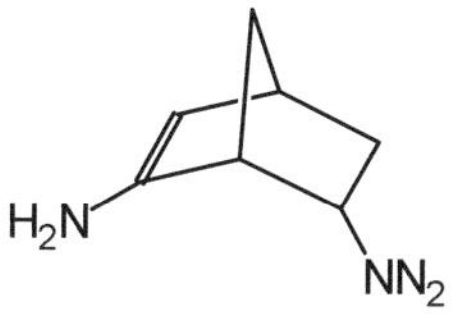

2,5-Diaminonorbornylene

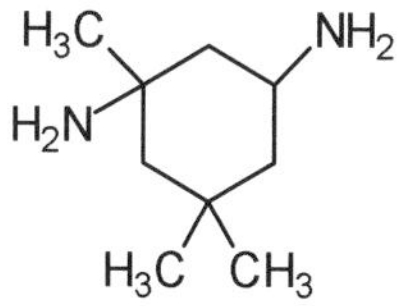

Isophorone diamine

Figure 6–7. N-Cyclohexyl maleimide, 2,5-diaminonorbornylene, isophorone diamine.

aminoethylethanol amine, with C_{21} or C_{22} carbon fatty acids or tall oil fatty acids can be used as corrosion inhibitor base [1569]. Propargyl alcohol has been found to enhance the anticorrosive effects of the composition.

Most simply, a mixture of mainly (80% to 90% by weight) propargyl alcohol and cellosolve, with minor amounts of polyglycol, amine derivatives, a phenol-formaldehyde resin, and tar bases, has been described [248, 249].

Instead of propargyl alcohol, propargyl ether has been proposed to be used as a corrosion inhibitor. Propargyl alcohol is added to olefins to form the corresponding ether [936].

Fatty acid amides of isophorone diamine, 2,5-diaminonorbornylene, and 2,2,4-trimethyl-1,6-diaminohexane are particularly suitable for high-temperature and high-pressure applications [971].

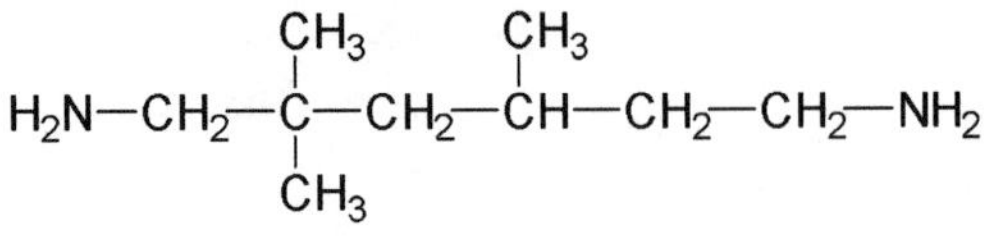

$$H_2N-CH_2-\underset{\underset{CH_3}{|}}{\overset{\overset{CH_3}{|}}{C}}-CH_2-\underset{}{\overset{\overset{CH_3}{|}}{CH}}-CH_2-CH_2-NH_2$$

2,2,4-Trimethyl-1,6-diaminohexane

Figure 6–8. 2,2,4-Trimethyl-1,6-diaminohexane.

Unsaturated Alcohols with Hexamethylenetramine

Acetylenic alcohols such as propargyl alcohol, 1-hexyn-3-ol and 5-decyne-4,7-diol have been used as corrosion inhibitors in HCl for the protection of ferrous metals. However, acetylenic alcohols are expensive and their use at high temperatures, that is, temperatures in the range of from approximately 180° to 350° F has been limited by the high concentrations of acetylenic alcohols needed to achieve the desired corrosion protection [656]. The presence of a small amount of hexamethylenetetramine with acetylenic alcohols dramatically improves the performance of the acetylenic alcohols in reducing corrosion and enables the use of these alcohols at lower concentrations or higher temperatures than when used alone. Hexamethylenetetramine also acts as a sulfide scavenger, whereby the formation of free sulfur or the formation of ferrous sulfide precipitate is prevented. In addition to acetylenic alcohols and hexamethylenetetramine, the metal corrosion inhibiting compositions of this invention can also include solvents. The formation of ferric hydroxide precipitate, free sulfur, and ferrous sulfide precipitate; the precipitation of sludge from a sludging oil; and the corrosion of metal surfaces can be prevented.

Imidazolines

Quaternized imidazolines with an amido moiety are suitable formulations for general oil and gas field applications. The synthesis of such compounds is detailed in the literature [1218]. For aqueous systems that contain sulfide compounds, a mixture has been described [262] that consists of an aqueous solution of an alcohol such as diethylene glycol monobutyl ether, butyl cellosolve, additional orthophosphoric acid, a fatty acid (from tall oil), substituted imidazoline, an ethoxylated fatty diamine (polyamines such as ethylenediamine, diethylenetriamine, etc.), and a molybdate compound.

A modification of the previous formulation uses amine products containing preferably only tertiary amino groups [1845]. These amines have favorable ecotoxicity levels in marine or freshwater environments. The ecotoxicity decreases with increasing substitutions on the N atoms present; that is, it appears that tertiary groups are less toxic than secondary groups, which are less toxic than primary groups. Combinations of imidazoles with wetting agents also have been described [227].

Water-soluble corrosion inhibitors are necessary to prevent corrosion of the pipe walls, joints, pumps, and collection stations. An ampholytic, substituted imidazoline has been described for inhibiting corrosion in such systems [297]. This type of corrosion inhibitor is intended for continuous treatment.

Pourable emulsions comprising up to 50% of a kerosene-containing corrosion inhibiting compound, such as an imidazoline, have been claimed to allow longer treatment intervals [637, 638]. A formulation that is resistant to sludge formation and does not tend to stabilize oil/water emulsions when added to oil and water systems has been described in the literature. An imidazoline derivate, prepared from a long-chain fatty acid and a polyamine, is dissolved in an aromatic solvent and dispersed with glycolic acid and hexylene glycol [1188].

In water systems, sulfate-reducing bacteria and sulfides are present. To prevent their growth, chlorine dioxide (ClO_2) is added. However, ClO_2 is highly corrosive to the metallic components used in oil field equipment. Chromates are successful ClO_2 corrosion inhibitors. However, chromates are also undesirable because of their high toxicity. As an alternative for chromates, a mixture of an alcohol, an acid, a fatty imidazoline, an ethoxylated fatty diamine, and water can be used [1360]. Such a composition has proven more effective than chromates at inhibiting the corrosion caused by ClO_2, without the serious toxicologic effects caused by the use of chromates.

It has been reported that the corrosion rate of mild steel in the presence of H_2S is greatly decreased by the addition of an imidazoline derivative or a quaternary ammonium salt into the drilling fluids [908]. When the concentration of inhibitor is 2 g/liter, an inhibition efficiency of 70% to 90% is achieved. If it is used together with the H_2S scavenger–alkaline zinc carbonate, the inhibition becomes even more effective. A synergistic effect is found between imidazoline inhibitors and calcium oxide. They also inhibit dioxide corrosion to some extent in H_2S-free drilling muds. Moreover, imidazoline can improve the rheologic properties of drilling mud.

Polyesters may be used [27–30, 223] instead of a fatty acid modifier for imidazoline. Thus a corrosion inhibitor with film-forming and film-persistency characteristics can be produced by first reacting, in a condensation reaction, a polybasic acid with a polyalcohol to form a partial ester. The partial ester is reacted with imidazoline or fatty diamines to result in a salt of the ester. Oil-soluble, highly water-dispersible corrosion inhibitor or oil-dispersible, water-soluble corrosion inhibitor modification can be achieved by the addition of suitable surfactants.

Nitrogen Heterocyclics

Pyridinium Compounds

Aliphatic pyridinium salts or aliphatic quinolinium salts in the presence of a sulfur-containing compound have been claimed to be active as corrosion inhibitors [952].

N-(p-dodecylphenyl)-2,4,6-trimethylpyridinium sulfoacetate is suitable as an inhibitor in aqueous media [606]. Such pyridinium compounds exhibit greater thermal stability than N-aralkyl pyridinium compounds or N-alkyl pyridinium compounds. Therefore the desired properties are retained both during and after exposure to elevated temperatures.

Furthermore, an α,β–ethylenically unsaturated aldehyde together with organic amines will form intermediate products, which are further reacted with a carboxylic acid, an organic halide, or an epoxide-containing compound [1760]. The final products are suitable corrosion inhibitors for preventing corrosion of steel in contact with corrosive brine and oil and gas well fluids.

Still bottom residues produced in the distillation of quinoline from coal tar can be oxidized and are suitable as a metal corrosion inhibitor for use in aqueous acid solutions [244].

Azoles

The effectiveness of various chemicals such as 1H-benzotriazole, 2-methylbenzotriazole, and 2-phenylbenzimidazole as a corrosion inhibitor for mild steel in 15% HCl was investigated by weight loss and electrochemical techniques [1547]. Among different azoles, 2-phenylbenzimidazole has shown the best performance. A synergism of iodide and 2-phenylbenzimidazole was observed.

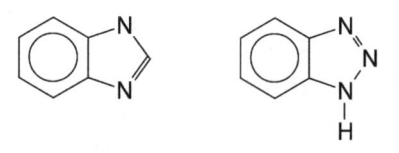

Benzimidazole   1H-Benzotriazole

Figure 6–9. Benzimidazole, 1H–benzotriazole.

Aminopyrazine with Epoxide Compound

An inhibitor during drilling and servicing of oil and gas wells is a condensate of aminopyrazine and an epoxide compound, such as the glycidyl ether of a mixture of C_{12} to C_{14} alkanols [604].

Carbonyl Compounds

Aldehydes with Surfactants

Mixtures of aldehydes with surfactants are active in preventing corrosion, in particular in the presence of mineral or organic acids [646]. The aldehyde may be trans-cinnamaldehyde. The surfactant may be N-dodecylpyridinium bromide or the reaction product of trimethyl-1-heptanol with ethylene oxide [645]. Such aldehyde and surfactant mixtures provide greater and more reliable corrosion inhibition than the respective compositions containing aldehydes alone.

Aldose Group Antioxidants

Corrosion inhibitors have been described that can be used with calcium-free drilling, completion, and workover fluids in carbonate- or sulfate-containing wells [435–437]. These inhibitors are sodium, ammonium, and/or calcium thiocyanate alone, or in combination with specific aldose group antioxidants. Aldose group antioxidants include arabinose, ascorbic acid, isoascorbic acid, gluconic acid, and their corresponding salts. In addition, ammonium thioglycolate may be incorporated as another corrosion inhibitor. Thio groups and aldose group antioxidants exhibit synergistic properties [1614].

Similarly, a high-density brine, useful as a drilling fluid for deep wells, is made corrosion resistant by adding an aliphatic or aromatic aldehyde and thiocyanates [817]. The aldehyde can be reacted with a primary amine before use.

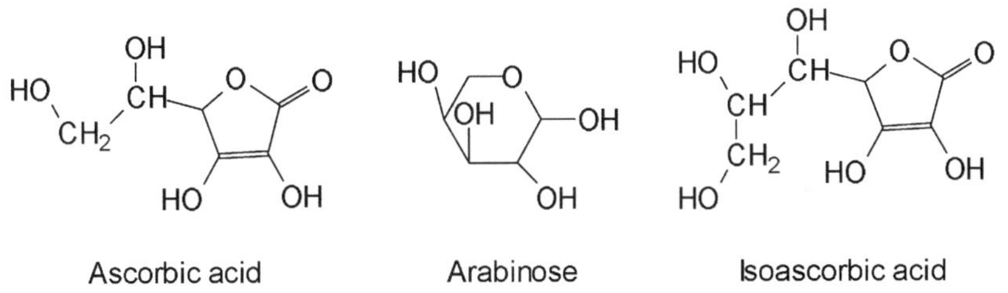

Figure 6–10. Ascorbic acid, arabinose, isoascorbic acid.

Phosphate Esters

Phosphate ester–type inhibitors are the reaction products of ethoxylated, propoxylated, or butoxylated alcohols or phenols with phosphating agents [1317, 1318, 1816]. Inhibitors for both general corrosion and cracking-type corrosion are obtained by the reaction of a nitrogen base and a phosphate ester [1167, 1168]. Although the nitrogen bases and phosphate esters have good general corrosion inhibition properties, neither the nitrogen base nor the phosphate ester provide suitable inhibition for cracking-type corrosion. However, the neutralization product of the nitrogen base and phosphate ester provides inhibition of both general and cracking-type corrosion. This inhibitor type is safe for aquatic organisms and is biodegradable.

Silicate-Based Inhibitors

Silicates [1149] offer advantages with respect to low costs, low toxicity, and low environmental impact.

Miscellaneous Inhibitors

Group VA Halides

Antimony tribromide and other group VA halides minimize the corrosion rate effectively [1799]. Unfortunately, antimony tribromide is toxic.

Aldol-amine Adduct

The corrosion of metal surfaces and the precipitation of a metal sulfide by an aqueous acid solution can be prevented by an aldol-amine adduct. Aldol (from acetaldehyde) $CH_3CH(OH)CH_2CHO$ has been utilized as a H_2S scavenger that prevents the precipitation of metal sulfides from aqueous acid solutions. However, when the aldol or an aqueous solution of the aldol is stored, the solution separates quickly into two layers, with all of the aldol concentrated in the bottom layer. The bottom layer is not redispersible in the top layer or in water or acid. In addition, the aldol in the bottom layer has very little activity as a sulfide scavenger. Thus the use of aldol as a H_2S scavenger in aqueous acid solutions can result in unsatisfactory results [245, 247]. However, the aldol can be reacted with an amine, such as monoethanoleamine (=aminoethanol), to form an aldol-amine adduct to overcome these difficulties. The amine utilized to prepare the aldol-amine adduct must be a primary amine. The aldol-amine adduct preferentially reacts with sulfide ions when they are dissolved in the

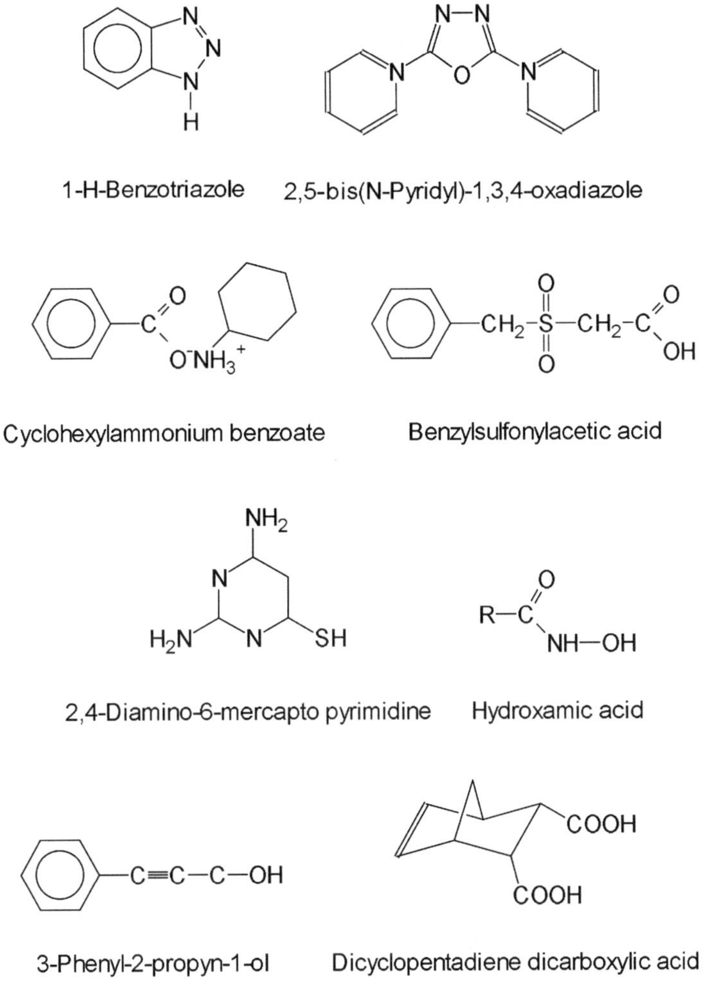

Corrosion Inhibitors 101

1-H-Benzotriazole

2,5-bis(N-Pyridyl)-1,3,4-oxadiazole

Cyclohexylammonium benzoate

Benzylsulfonylacetic acid

2,4-Diamino-6-mercapto pyrimidine

Hydroxamic acid

3-Phenyl-2-propyn-1-ol

Dicyclopentadiene dicarboxylic acid

Figure 6–11. 1-H-Benzotriazole, 2,5-bis(N-pyridyl)-1,3,4-oxadiazole, cyclohexylammonium benzoate, benzylsulfonylacetic acid, 2,4-diamino-6-mercapto pyrimidine, hydroxamic acid, 3-phenyl-2-propyn-1-ol, dicyclopentadiene dicarboxylic acid.

acid compositions, thereby preventing the dissolved sulfide ions from reacting with dissolved metal ions and precipitating.

Some formulations that cannot be readily classified into any of the previous sections are summarized in Table 6–3.

Table 6–3
Corrosion Inhibitors

Corrosion inhibitor	References
Acetylinic alcohol[a]	[1728]
Tall oil fatty acid anhydrides	[602, 603]
3–Phenyl-2-propyn-1-ol (PPO)[b]	[742]
Dicyclopentadiene dicarboxylic acid salts[c]	[444, 445]
Hydroxamic acid	[618]
Cyclohexylammonium benzoate	[912, 913]
Acyl derivatives of tris-hydroxy-ethyl-perhydro-1,3,5-triazine	[86, 87]
2,4-Diamino-6-mercapto pyrimidine sulfate (DAMPS) combined with oxysalts of vanadium, niobium, tantalum or titanium, zirconium, hafnium	[1476]
Aqueous alkanol amine solution[d]	[1581, 1797]
Quaternized fatty esters of alkoxylated alkyl-alkylene diamines	[1854]
Mercaptoalcohols	[15]
Polysulfide[e]	[679]
Polyphosphonohydroxybenzene sulfonic acid compounds[f]	[1009]
1-Hydroxyethylidene-1,1-diphosphonic acid[g]	[1586]
2-Hydroxyphosphono-acetic acid[h]	[1881, 1882]
Water-soluble 1,2-dithiol-3-thiones[i]	[1377, 1378]
Sulfonated alkylphenol[j]	[121]
Polythioether	[887]
Thiazolidines	[35]
Substituted thiacrown ethers pendent on vinyl polymers	[1228]
Benzylsulfinylacetic acid or benzylsulfonylacetic acid	[1110]
Halohydroxyalkylthio-substituted and dihydroxyalkylthio-substituted polycarboxylic acids[k]	[1109]
Alkyl-substituted thiourea	[1719]
2,5-bis(N-Pyridyl)-1,3,4-oxadiazoles	[169]

a) In combination with ClO_2 treatment for bacteria control.
b) Aqueous HCl.
c) 0.1 to 6% with antifreezers such as glycols.
d) Gas stream containing H_2S or CO_2.
e) Forms a film of iron disulfide.
f) Relatively nontoxic, substitution of chromate-based corrosion inhibitors, conventional phosphate, and organophosphonate inhibitors and the zinc-based inhibitors.
g) CO_2 environment.
h) Calcium chloride brine.
i) 10 to 500 ppm.
j) 5 to 200 ppm to inhibit naphthenic acid corrosion.
k) In drilling equipment.

Chapter 7

Scale Inhibitors

Some risk of scale deposition occurs in many operations in the petroleum industry. Scale deposition happens particularly in production, stimulation, and transport. Scaling can occur when a solution becomes supersaturated, which occurs mostly if the temperature changes in the course of injection operations. Also, if two chemicals that will form a precipitate are brought together, a scale is formed (e.g., if a hydrogen fluoride solution meets calcium ions). From a thermodynamic perspective, there is a stable region, a metastable region, and an unstable region, separated by the binodale curve and the spinodale curve, respectively.

The scale may consist of calcium carbonate, barium sulfate, gypsum, strontium sulfate, iron carbonate, iron oxides, iron sulfides, and magnesium salts [943]. There are monographs (e.g., *Corrosion and Scale Handbook* [159]) and reviews [414] on scale depositions available in the literature.

Scale Inhibition

The problem is basically similar to preventing scale inhibition in washing machines; therefore similar chemicals are used to prevent scale deposition. Scale inhibition can be achieved either by adding substances that react with potential scale-forming substances so that from a thermodynamics standpoint the stable region is reached or by adding substances that suppress crystal growth.

Conventional scale inhibitors are hydrophilic, that is, they dissolve in water. In the case of down-hole squeezing, it is desirable that the scale inhibitor is adsorbed on the rock to avoid washing out the chemical before it can act as desired. However, adsorption on the rock may change the surface tension and the wettability of the system. To overcome these disadvantages, oil-soluble scale inhibitors have been developed. Coated inhibitors are also available. Often, scale inhibitors are not applied as such, but rather in combination with corrosion inhibitors.

Thermodynamic Inhibitors

Thermodynamic inhibitors are complexing and chelating agents, suitable for specific scales. For example, for scale inhibition of barium sulfate, common chemicals are ethylenediaminetetraacetic acid (EDTA) and nitrilotriacetic acid. The solubility of calcium carbonate can be influenced by varying the pH or the partial pressure of carbon dioxide (CO_2). The solubility increases with decreasing pH and increasing partial pressure of CO_2, and it decreases with temperature.

However, usually the solubility increases with higher temperature. The temperature coefficent of solubility is dependent on the enthalpy of dissolution. An exothermic enthalpy of dissolution causes a decrease in solubility with increased temperature, and vice versa.

Kinetic Inhibitors

Kinetic inhibitors for hydrate formation may also be effective in preventing scale deposition [1627]. This may be understood in terms of stereospecific and nonspecific mechanisms of scale inhibition.

Adherence Inhibitors

Another mechanism of scale inhibition is based on adherence inhibitors. Some chemicals simply suppress the adherence of crystals to the metal surfaces. These are surface-active agents.

Mathematical Models

Mathematical models have been developed [1144–1146, 1623]. The scale formation of iron carbonate and iron monosulfide has been simulated by thermodynamic and electrochemical models [49, 1144, 1154, 1893].

Optimal Dose

A method to estimate the optimal dose of a scale inhibitor has been described [1223]. The method starts with noting the chemical composition and temperature of the water. From these parameters a stability index is calculated, allowing for the prediction of the optimal dose of a scale inhibitor.

Chemicals in Detail

Both inorganic acids, such as hydrochloric acid and hydrofluoric acid, and organic acids, such as formic acid, can be used to increase the pH. Acids are used in combination with surfactants. Various phosphates, pyrophosphates, polyphosphates, and metaphosphates belong to the phosphate type of scale inhibitors. Of course, the classic ethylenediaminetetraacetic acid shows chelating properties, and there are various organic multifunctional acids and hydroxyacids.

Acids

Acids, when used as scale inhibitors, are extremely corrosive. Their effectiveness has been laboratory tested. Parameters include acid type, metallurgy, temperature, inhibitor type and concentration, duration of acid-metal contact, and the effect of other chemical additives [279]. Lead and zinc sulphide scale deposits can be removed by an acid treatment [922].

Encapsulated Scale Inhibitors

Encapsulated scale inhibitors have been developed. This type of scale inhibitor allows chemical release over an extended period [865, 1452]. Microencapsulated formulations may contain a gelatin coating with a multipurpose cocktail, such as [1006, 1007]:

- Scale inhibitor
- Corrosion inhibitor
- Biocide
- Hydrogen sulfide scavengers
- Demulsifier
- Clay stabilizer

Chelating Agents

Trace amounts of chelating agents, such as EDTA, citric acid, or gluconic acid, may lower the efficiency of scale inhibitors [150]. The concentration of calcium ions and magnesium ions affects the inhibition of barium sulfate [198]. Penta-phosphonate, hexa-phosphonate, phosphino-polycarboxylate, and polyvinyl sulfonate scale inhibitors were studied. String chelating agents, given in Table 7–1, also stabilize the coating of encapsulated formulations [1006].

Scale Inhibition at High Reservoir Temperatures

Conventional polymer and phosphonate scale inhibitors may not be appropriate for application in high-pressure and high-temperature reservoirs. Only a limited range of commercially available oil field scale inhibitor chemicals are sufficiently thermally stable at temperatures above 150° C. These chemicals are homopolymers of vinyl sulfonate and copolymers of acrylic acid

Table 7–1
Chelating Agents for the Stabilization of Coatings [1006]

Chelating agent	Acronym
N-(3-Hydroxypropyl)imino-N,N-diacetic acid	3-HPIDA
N-(2-Hydroxypropyl)imino-N,N-diacetic acid	2-HPIDA
N-Glycerylimino-N,N-diacetic acid	GLIDA
Dihydroxyisopropylimino-N,N-diacetic acid	DHPIDA
Methylimino-N,N-diacetic acid	MIDA
2-Methoxyethylimino-N,N-diacetic acid	MEIDA
Amidoiminodiacetic acid = sodium amidonitrilotriacetic acid	SAND
Acetamidoiminodiacetic acid	AIDA
3-Methoxypropylimino-N,N-diacetic acid	MEPIDA
Tris(hydroxymethyl)methylimino-N,N-diacetic acid	TRIDA

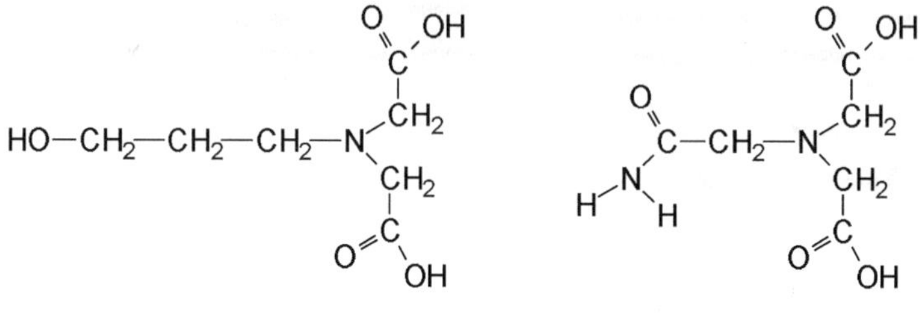

N-(3-Hydroxypropyl)imino-N-N-diacetic acid   Amidoiminodiacetic acid

Figure 7–1. N-(3-Hydroxypropyl)imino-N,N-diacetic acid, amidoiminodiacetic acid.

and vinyl sulfonate. Other polymers, such as polymaleic acid, polyitaconic acid, and maleic acid/acrylic acid copolymers, may offer similar thermal stability. However, they are not available (as of 1995) from oil field chemical suppliers [397]. Thermal stability tests, influence on pH, ionic strength, and oxygen on conventional polymer and phosphonate scale inhibitors, for example, on phosphinopolycarboxylate, polyvinyl sulfonate, penta-phosphonate, and hexa-phosphonate, have been presented [534, 731–733].

Characterization

Field desorption mass spectrometry [1606], ^{13}C nuclear magnetic resonance, and fourier-transform infrared spectroscopy [1337] have been used to characterize oil field chemicals, among them, scale inhibitors. Ion chromatography is suitable for the simultaneous determination of hydroxyethylsulfonate, sodium vinylsulfonate, chloride, and sulfate reaction byproducts [85, 1830].

Phase diagrams of a polyacrylate-phosphonate system with temperature and calcium ion concentration can be established with turbidimetric measurements [1830]. Conductometric titrations also are suitable to characterize the phase behavior of scale inhibitors [514] (Table 7–2).

Table 7–2
Scale Inhibitors

Chemical	References
1-Hydroxyethylidene-1,1-diphosphonic acid	[800]
Carbonic dihydrazide, $H_2NNHCONHNH_2$	[1269]
Polyaminealkylphosphonic acid and carboxymethylcellulose or polyacrylamide	[978]
Polyacrylic acid and chromium	[1866]
Polyacrylates[a]	[1828]
Amine methylene phosphonate[b]	[730]
Phosphonomethylated polyamine	[1633]
Oil soluble	[906, 1827]
Sulfonated polyacrylate copolymer	[373]
Tetrakis-hydroxymethyl-phosphonium sulfate	[1068]
Phosphonates	[848, 923]
Carboxymethylinulin	[1040]
Polycarboxylic acid salts	[496]
Phosphoric acid esters of rice bran extract	[1887]
Poly(phosphino maleic anhydride)	[1867]
N,N-Diallyl-N-alkyl-N-(sulfoalkyl) ammonium betaine copolymer (with N-vinyl pyrrolidone or acrylamide), diallylmethyltaurine hydrochloride ($CH_2 = CH - CH_2Cl \times CH_3 - NH - CH_2 - CH_2 - SO_3^- Na^+$)	[619]
Aminotri(methylenephosphonic acid)	[1007]
Polyaspartates	[571]
Polyacrolein	[1625]
Naphthylamine polycarboxylic acids	[327]
Phosphonic acid and hydrofluoric acid	[466]
Tertiary amines[c]	[1489]

a) In borate crosslinked fracturing fluids.
b) High temperature.
c) Oil soluble.

Chapter 8
Gelling Agents

Organic and inorganic gels are used to seal formations with high permeability. Sealing can be permanent or temporary. In this way, undesirable flows in the formation can be suppressed. Gelling agents are mainly used for water shutoff and limiting sand production; some agents can be used also for primary plugging of surface and flow strings, as well as for eliminating lost circulation [1539]. However, the high prices often limit wide applications.

Flow-deflecting technologies based on gel-forming agents are effective in highly water-invaded multizone reservoirs. Redistribution of the flow is achieved by equalizing the injectivity profile of the injection wells and reducing the content of produced water in the producing wells. Redistribution also reduces the consumption of electricity, demulsifiers, and fuels used for oil treating in the field. To broaden the use of gel technologies, it is necessary to solve certain problems of developing and manufacturing mobile units for preparing and injecting large volumes of gel compositions in wells and for producing nonfreezing agents for year-round stimulation treatments [754].

Basic Mechanisms of Gelling Agents

The subdivision of sealing agents into gelling agents and plugging agents is somewhat arbitrary. Gelling may be understood as a sealing that is less perfect than plugging. Therefore this chapter summarizes gelling agents that are formed by physical crosslinking, whereas Chapter 18 deals with gels that are formed by chemical reactions, for instance, in situ polymerization. Furthermore, if a chemical composition is exclusively used as a plugging formulation, it has been included in Chapter 18.

The formation of gels can be achieved by various chemical principles:

- Bringing polyanionic and polycationic substances together
- Physical crosslinking of carboxyl-functional polymers with multivalent ions
- Using chelating agents

Table 8–1
Examples of Polymers

Polymer
Polyacrylamide
Carboxymethylcellulose
Polyacrylonitrile, hydrolyzed

Table 8–2
Commonly Used Salts, Crosslinkers, and Chelating Agents

Salt
Aluminum citrate
Chromium sulfate
Ferrochromo-lignosulfonate
Manganese nitrate
Potassium bichromate
Sodium bichromate
Ferric acetylacetonate
Ammonium ferric oxalate

Polymer–Crosslinker–Retarder Systems

Typical compositions consist of water, at least one polymer capable of being gelled when contacted with a crosslinking agent, a crosslinking agent that is a polyvalent metal cation, and a reactivity-retarding chelating agent. The chelating agent is selected from water-soluble dicarboxylic acids, for example, hydroxy carboxylic acids or ketocarboxylic acids [1304]. Examples for polymers are given in Table 8–1; examples for crosslinkers are given in Table 8–2.

Carboxylic Acids as Retarders

Chelating agents retard the gelation of polymers with multiple carboxyl groups. The retardation results from a reaction of the multivalent cations with the chelating agent. Any water-soluble dicarboxylic acids, hydroxy carboxylic acids, ketocarboxylic acids, and the corresponding salts may serve as chelating agent [1304–1306].

Aluminum Trichloride

Aluminum trichloride, a cheap, abundant waste product of the chemical industry, forms a gel under certain conditions with carbonates and on mixing with alkalies. Laboratory and field tests showed that aluminum trichloride can be used as a gel-forming agent for reducing the permeability of water-conducting channels [673].

Aluminum Phosphate Ester Salts

Organic liquid gels are used for temporary plugging during fracturing operations. This type of gelling agent permits on-the-fly gelling of hydrocarbons, particularly those used in hydraulic fracturing of subterranean formations to enhance oil and gas production. A gel of an organic liquid, such as diesel or crude oil, can be formed using an aluminum phosphate diester in which all of the reagents are substantially free of water and pH-affecting substances [740,779,920]. The diester may be prepared by the reaction of a triester with phosphorous pentoxide to produce a polyphosphate, which is then reacted with an alcohol to produce a phosphate diester. The latter diester is then added to the organic liquid along with a nonaqueous source of aluminum, such as aluminum isopropoxide in diesel oil, to produce the metal phosphate diester. The conditions in the preceding two reaction steps are controlled to provide a gel with good viscosity versus temperature and time characteristics. The gel is useful in fracturing subterranean formations by entraining a solid particulate proppant therein and pumping the resultant mixture into the subterranean rock formation at sufficient pressure to fracture the formation.

A similar process allows reacting triethyl phosphate and phosphorous pentoxide to form a polyphosphate in an organic solvent [871]. An excess of 1.3 moles of triethyl phosphate with respect to phosphorous pentoxide is the most preferred ratio. In the second stage, a mixture of higher aliphatic alcohols from hexanol to decanol is added in an amount of 3 moles per 1 mole phosphorous pentoxide. Aluminum sulfate is used as a crosslinker. Hexanol results in a high-temperature viscosity of the gel, while maintaining at a pumpable viscosity at ambient temperatures [870].

Biopolymers

Curdlan

A process using a microbially gelled biopolymer was developed and used to modify the permeability in coreflood experiments [128]. Curdlan is a microbial carbohydrate with 1,3-β-linkages. Alkaline-soluble curdlan biopolymer was mixed with microbial nutrients and acid-producing alkaliphilic bacteria and injected into Berea Sandstone cores. Concurrent bottle tests with the polymer solution were incubated beside the core. In the bottle tests, the polymer formed rigid gel in 2 to 5 days at 27° C. After 7 days' incubation, 25- to 35-psi fluid pressure was required to begin flow through the cores. Permeability of the cores was decreased from 852 to 2.99 mD and from 904 to 4.86 mD, respectively, giving residual resistance factors of 334 and 186.

Poly-3-hydroxybutyrate

Alcaligenes eutrophus produces a massive amount of intracellular polyester (poly-3-hydroxybutyrate [PHB]), specifically as high as 70% of the cell weight. This bacterium was selected for porous-media plugging studies [1316].

To simulate the subsurface environment, both static drainage and pressurized pumping flow systems of *A. eutrophus* living cells and PHB suspensions through laboratory sand packs were investigated [1100]. A PHB water solution, a commercial product in powder form that disperses well but is not completely dissolved in water, showed plugging effects solely dependent on the concentration of PHB. These facts signify that *A. eutrophus* and its microbial product, PHB, are efficient plugging agents. They have potential applications in the MEOR process, such as selective plugging, because of their relative nonagglomerating cell size; their rod shape of 0.7 µ in diameter and 1.8 to 2.6 µ in length; and their lack of any exopolymer in culture solutions, especially the beneficial biopolymer (PHB) produced internally in cells.

Succinoglycan

Aqueous solutions of succinoglycan can be crosslinked by means of a polyvalent metal cation [449]. The gelation occurs after 3 to 24 hours but can be further delayed by an appropriate chelating agent. Chelating agents are multifunctional carboxylic acids or their metal salts, especially citrate, oxalate, and malate metal salts. The aqueous gels are useful in profile modification, or permeability reduction, of subterranean hydrocarbon-bearing formations in enhanced oil recovery. The application is simple. A gel is pumped down into a formation for a period of time sufficient to obtain the desired in-depth penetration and decrease in permeability of the high-permeability zones of the formation. Usually, an in-depth penetration of 75 to 900 feet from the well is sufficient.

Organic Polysilicate Ester

The permeability of subterranean oil-bearing formations can be controlled by injection of an organic polysilicate ester [851]. Polysilicate esters may be built up from simple, monohydroxylic alcohols such as methanol, ethanol, propanol, or butanol or from diols such as ethylene glycol or polyols such as glycerol or polyalkylene oxides. The polysilicates are injected into a formation through injection wells, suitably in an amount from 10% to 100% of the pore volume of the zone to be treated. In the formation, the polysilicate esters form gels that selectively decrease the permeability of the high-permeability regions of the formation.

Latex

The use of polyisoprene or butadiene-styrene latex with bentonite or chalk filler and polyoxypropylene as an additive has been used in a plugging solution for oil and gas wells [1042]. The solution can be pumped but coagulates within the formation at temperatures of 100° C within 2 hours. This causes a reduction in permeability. The formulation is particularly useful in deep oil deposits.

Carboxymethylcellulose

A mixture of lignosulfonate with modified carboxymethylcellulose and metal ions as crosslinkers has been suggested as a plugging agent [1375].

Carboxymethylcellulose, modified with polyoxyethylene glycol ethers of higher fatty alcohols, combines the properties of a surfactant and carboxymethylcellulose. Thus it reduces the viscosity of the composition and increases the strength of the produced gel. Sodium and potassium bichromates act as crosslinking agents. Ionic crosslinks are formed as a result of the reaction of Cr^{3+} and Ca^{2+} ions with molecules of modified carboxymethylcellulose. A gel-forming composition is obtained by mixing aqueous solutions of the respective components. Highly mineralized water also can be used and the gelation time can be controlled by changing the contents of $CaCl_2$ and bichromates.

Polydimethyl-diallyl Ammonium Chloride

Polydimethyl-diallyl ammonium chloride is a strongly basic cation-active polymer. A mixture of polydimethyl-diallyl ammonium chloride and the sodium salt of carboxymethylcellulose, which is an anion-active polymer, is applied in an equimolar ratio [497] in aqueous sodium chloride solution. The proposed plugging composition has high efficiency within a wide pH range.

Lignosulfonate and Carboxymethylcellulose

An aqueous solution of 3% to 6% lignosulfonate and 2% to 8% carboxymethylcellulose, modified with polyoxyethylene glycol ethers of higher fatty alcohols form the base of a plugging system [1375]. Lignosulfonate is a waste product from the cellulose-paper industry. Furthermore, as crosslinking agents, sodium or potassium bichromate and calcium chloride are added in amounts from 2% to 5%. The composition is obtained by mixing aqueous solutions of the components in a cement mixer.

A mixture of polymers that can serve as a plugging solution when taken in an equimolar ratio consists of polydimethyl-diallyl ammonium chloride, which is a strongly basic cation-active polymer, and the sodium salt of carboxymethylcellulose, which is an anion-active polymer. The aqueous solution contains 0.5% to 4% of each polymer. Gelling occurs because the macro ions link together from different molecules. The proposed plugging composition has high efficiency within a wide pH range [497].

Polyacrylamide-Based Formulations

Aqueous solutions of polyacrylamide may be used as plugging solutions for high-permeability formations. Partially hydrolyzed polyacrylamide polymer also has been used [1211] and completely hydrolyzed polyacrylonitrile has been proposed [1427].

The polymer solutions are pumpable as such, but in the presence of multivalent metal ions, gels are formed. The gel formation is caused by an intermolecular crosslinking, in which the metal ion forms bonds to the polymer.

The metal ions are often added as salts of organic compounds, which form chelates. This causes a delayed gelation. Likewise, the components of the gelling agent are pumped down in two stages. Some metal cations cannot be used with brines. On the other hand, brines are often produced from wells, and it is desirable to find uses for them to avoid disposal processes.

Delayed Gelation

Complexing Agents. Delayed gelation can be achieved by adding complexing agents to the mixture. The metal ions are initially complexed. Therefore all components of the gelling composition can be injected simultaneously. It is possible to dissolve the mixture in produced brines that have high salinity. The use of produced brines eliminates the need to treat or dispose of the brines.

Examples of the components of the gelling composition are a water-soluble polymer such as polyacrylamide, an iron compound such as ferric acetylacetonate or ammonium ferric oxalate, and a ketone such as 2,4-pentanedione [1247]. The composition forms a temporary gel that is useful for the temporary plugging of a formation. The temporary gels that are formed will disappear after 6 months.

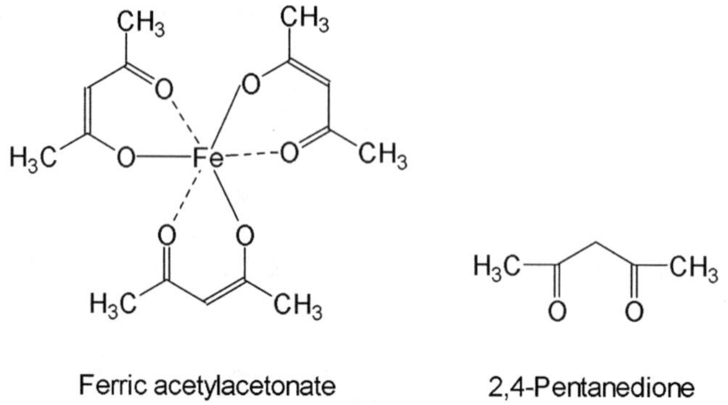

Figure 8–1. Ferric acetylacetonate, 2,4-pentanedione.

Figure 8–2. Hydrolysis of urea with water.

Adjustment of pH. Some organic reagents hydrolyze in aqueous solution at elevated temperature and release ammonia. Examples are urotropin and urea. The hydrolysis of urea is shown in Figure 8–2. Urotropin yields hydrolysis formaldehyde and ammonia. In this way, the pH is increased. The chemical reactions necessary for the formation of a gel with other components of the mixture can then take place.

Polyacrylamide and Urotropin-Based Mixture

In Table 8–3 a recipe for and the gel-forming properties of a polyacrylamide and urotropin-based mixture [1133] are shown in Table 8–4.

Reinforcement by Fibers

Fibers can be added to a gelation solution [1212, 1213]. Fibers that will not interfere with the gelation process and will provide adequate reinforcement must be chosen. In addition, they should not adversely affect the ability of the solution to be pumped and injected. In particular, glass fibers and cellulosic fibers meet the requirements as reinforcing fibers for plugging solutions.

Table 8–3
Gel-Forming Composition
Based on Polyacrylamide

Material	% by wt
Polyacrylamide	0.05–3
Urotropin	0.01–10
Sodium bichromate	0.01–1
Water	up to 100

Table 8–4
Gel-Forming Time at Various Temperatures

Temperature (° C)	Time (hr)
60	10–18
80	6–22
120	4.5–7

Metal Ions and Salts as Crosslinking Agents

Iron Salts. Iron, cations, and some divalent cations cannot be used in the brine environment.

Waste Material. Waste materials from other processes have been established to be active, such as waste from galvanizing processes [998]. In this case the components must be placed in two portions, however.

Iron and chromic salts from lignosulfonate are also a source for metal ions [1000]. Lignosulfonates are waste products from the paper industry.

Chromium (III) Propionate. A chromium (III) propionate–polymer system is suitable for gelation treatments in oil fields, where freshwater is not available [1303, 1307]. It produces good and stable gels in hard brines, as well as in freshwaters. The effectiveness of treatment is a function of gelling agent concentration (i.e., the higher the concentration of chromium propionate, the higher the residual resistance factors). The process can be used for in-depth treatment in the same way as with aluminum citrate and for nearwell treatment. Produced brines that contain iron and barium can be used. In comparison with aluminum citrate, the use of chromium (III) propionate gives more effective in-depth treatment at only half of the crosslinker concentration. This makes the chromium propionate process even more attractive for freshwater applications.

Gelation Process and Gel Breaking. The gelation process of polyacrylamide with chromium ions takes place through coordination bonding with the nitrogen moiety [1316]. Studies have indicated that better results can be expected in certain concentration ranges of the reactants. Gels formed at neutral pH have been observed to be comparatively stable and to sustain the reservoir temperature for 50 days. Solubility in chemicals like HCl, mud acid, and hydrogen peroxide indicates that these chemicals can be used as breakers.

Aluminum Citrate. Aluminum citrate can be used as a crosslinker for many polymers; the gels are made of low concentrations of polymer and aluminum citrate in water. This crosslinker provides a valuable tool, in particular, for in-depth blockage of high-permeability regions of rock in heterogeneous reservoirs. The formulations can be mixed as a homogeneous solution at the surface.

Partially hydrolyzed polyacrylamides, carboxymethylcellulose, polysaccharides, and acrylamido methylpropane sulfonate have been screened to investigate the performance of aluminum citrate as a chelate-type crosslinker. An overview of the performance of 18 different polymers has been presented in the literature [1646]. The performance of the colloidal dispersion gels depends strongly on the type and the quality of the polymer used. The gels were mixed with the polymers at two polymer concentrations, at three polymer-to-aluminum ratios, and in different concentrations of potassium chloride. The gels were quantitatively tested 1, 7, 14, and 28 days after preparation.

Interactions of Metal Salts with the Formation. Interactions of metal salts with the formation and distribution of the retained aluminum in a porous medium may significantly affect the location and strength of gels. This interaction was demonstrated with polyacrylamide–aluminum citrate gels [1514]. Solutions were displaced in silica sand. The major findings of this study are that as the aluminum-to-citrate ratio increases, the aluminum retention increases. Furthermore, the amount of aluminum retained by silica sand increases as the displacing rate decreases. The process is reversible, but the aluminum release rate is considerably slower than the retention rate. The amount of aluminum released is influenced by the type and the pH level of the flowing solution. The citrate ions are retained by silica sand primarily as a part of the aluminum citrate complex. Iron, cations, and some divalent cations cannot be used in the brine environment.

Bentonite Clay and Polyacrylamide

A water-expandable material based on bentonite clay and polyacrylamide is added to the circulating drilling solution [118]. The material expands in water to 30 to 40 times its initial volume. The swelling takes place within 2 to 3 hours. During the circulation of the drilling solution, the material enters the cracks and pit spaces of the natural stratal rock. When expanding slowly, the material converts within 30 to 40 minutes into the plugging material, which is strongly fixed to the rock. Tests showed that the additive effectively prevents the absorption of the drilling solution on the stratal rock as a result of the production of a strongly adhering and insulating film, which is not dislodged even after subjecting the material to an excess pressure of 3 atm for 2 hours. The time of expansion of the material is sufficiently slow to let it permeate into the slits and cracks of the stratal rock, yet is sufficiently quick to provide a compact insulation.

Polyacrylic Acid

Polymers of acrylic acid and methacrylic acid have been tested for their gel-formation ability [1396]. They are used with gel-forming additives similar to those described for polyacrylamides. Also, mixtures of latex with methacrylate–methacrylic acid copolymer as an additive have been described as plugging agents [1041].

Alkali-Silicate Aminoplast Compositions

A composition to be injected into the ground for plugging consists of an alkaline metal silicate and an aminoplastic resin [1662, 1663]. Urea-formaldehyde, urea-glyoxal, or urea-glyoxal-formaldehyde condensation products are suitable. The composition has been suggested to be useful for a reservoir rock that is subjected to enhanced recovery methods. It can also be used for consolidation of ground and in building tunnels, dams, and other underground structures of this type.

In Situ Polymerization and In Situ Polyaddition

Epoxide Resins

Epoxide resins have good adhesive properties. They can be cured at low temperatures with aminic hardeners and at elevated temperatures with organic anhydrides. Formulations can be adjusted to a long pot life and a low exotherm. The compositions are not miscible with well fluids. A disadvantage is the comparatively high price. Standard epoxide resins are based on bisphenol-A. A method for selectively plugging wells using a low-viscosity epoxide resin composition containing a single curing amine-based agent has been described [447,448]. The method proposed is applicable to plugging permeable zones in a gravel-packed well and may be used to repair leaks in well casing or production tubing and in cementing to prevent communication between subterranean regions.

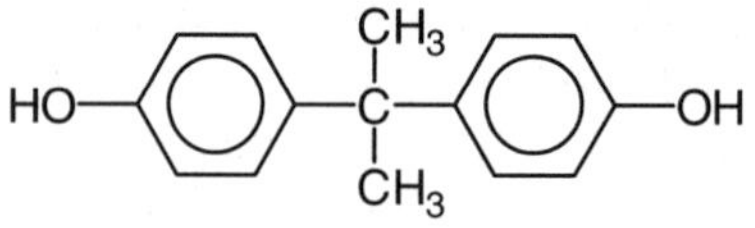

Bisphenol-A

Figure 8–3. Bisphenol-A.

Urea–Formaldehyde Resins

Urea-formaldehyde resins can be cured with isopropylbenzene production wastes containing 200 to 300 g/liter of $AlCl_3$ as an acid hardener [189]. Isopropylbenzene is formed as an intermediate in the Hock process by a Friedel-Crafts reaction from propene and benzene. The mixture hardens in 45 to 90 minutes and develops an adhesion to rock and metal of 0.19 to 0.28 MPa for 0.2% $AlCl_3$ and 0.01 to 0.07 MPa for 0.4% $AlCl_3$, respectively. A particular advantage is the increased pot life of the formulation.

A solution of a urea-formaldehyde condensate and/or a phenol-formaldehyde condensate with minor amounts of lignosulfonates can be used for isolation of absorption strata during drilling of oil and gas wells [1134]. The solution is prepared by mixing the resin component with lignosulfonate. Curing is achieved by thermosetting the solution at 80° to 120° C. The interaction between aldehyde and phenol groups of the lignosulfonates with methylol groups of urea-formaldehyde resins or phenol nuclei of phenol-formaldehyde resins results in the formation of bridges of lignosulfonate macromolecules, which act as hardeners of urea-formaldehyde or phenol-formaldehyde resins. The elasticity of the created fragments results in reduced shrinkage of the hardened compound. The composition has a low toxicity, in part because of the absence of acids and alkali. The shrinkage is only 0.1% to 0.6%. A wide range of curing rate at temperatures up to 120° C can be adjusted.

Vinyl Monomers

The use of vinyl monomers for gel formation requires a polymerization process in the formation. This technique is used to enable a solution to gel slowly even at high temperatures. An aqueous solution of a vinyl monomer is mixed with a radical-forming initiator, and if necessary, with a dispersant. The initiator decomposes at elevated temperatures and initiates the polymerization process. In this way, a gel is formed in place. The polymerization process is sensitive to molecular oxygen. To further delay curing, polymerization inhibitors may be added to the solution in small amounts. This technique is used in the treatment of subterranean formations, especially for plugging lost

N-Methylol acrylamide   N-Methylol methacrylamide

Figure 8–4. N-methylol acrylamide and N-methylol methacrylamide.

circulation in drilling operations, particularly at elevated temperatures. Acrylic acid, acrylamide, vinyl sulfonic acid, and N-vinylpyrrolidone may be used. The polymer formed may be crosslinked by a multifunctional vinyl monomer such as glycerol dimethacrylate or diacrylate [657, 658].

Monomers like N-methylol acrylamide or N-methylol methacrylamide [1081] also have been suggested. Inhibitors can be phenol derivatives [1079] such as N-nitrosophenylhydroxylamine salt [1179].

Chapter 9

Filter-Cake Removal

The drill-in fluids are typically composed of either starch or cellulose polymers, xanthan polymer, and sized calcium carbonate or salt particulates. Insufficient degradation of the filter-cakes resulting from even these clean drill-in fluids can significantly impede the flow capacity at the wellbore wall. Partially dehydrated, gelled drilling fluid and filter-cake must be displaced from the wellbore annulus to achieve a successful primary cement job.

Organic Acids

Citric Acid

A composition for dissolving filter-cake deposits left by drilling mud in wellbores is composed of an aqueous solution of citric acid and potassium chloride, alkali metal formate, acid tetraphosphate, alkaline earth chloride, and alkali metal thiophosphate [1012].

The composition is useful as an additive for clearing stuck pipe in wellbores and as a fixer spacer for cementing pipe in wellbores. Another use of the composition is as a well stimulation fluid in oil and gas production wells, in which the composition is effective to dissolve filter-cake that blocks pores in the production formation.

Horizontal Well Acid Breaker

Horizontal completions in unconsolidated formations are being enhanced by a hydrochloric acid (HCl) breaker system for well clean up. Typically, the use of HCl in open-hole environments is avoided because of wellbore stability concerns. However, HCl successfully removes salt fluid loss control materials in wells without noticeable hole collapse [33].

Acetic Acid

A case study was reported regarding the use of acetic acid [1323]. A large-scale acetic acid–based stimulation treatment was developed to remove drilling mud filter-cake in vertical wells in a carbonate reservoir in Saudi Arabia. The wellbore stability of this weakly consolidated carbonate formation can be easily reduced by contact with HCl-based acids. Laboratory testing had indicated that the formation was mechanically weak, became brittle upon contact with acids, and produced large amounts of fine particles that can cause severe damage. Furthermore, a chloride-free acid formula was required to minimize the interference with the pulsed-neutron logs. A special acid treatment was needed to reduce the damage while maintaining the integrity of the formation. Laboratory tests were done with both regular and emulsified acetic acid and an auxiliary chemical (mainly ethylenediaminetetraacetic acid [EDTA]). Pressure buildup tests after the treatment indicated that the acid was successful in removing the filter-cake in all cases investigated. Neither fine particles nor any type of emulsion was observed in the well flowback samples.

Bridging Agents

Magnesium oxide, manganese oxide, calcium oxide, lanthanum oxide, cupric oxide, and zinc oxide can be used in combination with a polymer (e.g., hydroxyethylcellulose as fluid loss agent and xanthan as suspension aid) as solid particle bridging agents. In addition to the bridging agent, the drilling or servicing fluid can also include an oxidizer, which is deposited in the filter-cake and is activated by the ammonium chloride in the cleaning solution to break up the polymer in the filter-cake. Magnesium peroxide in encapsulated form is the most suitable.

These compounds should be dissolvable in a clean-up solution containing a quaternary organic ammonium salt, or simply ammonium chloride [1749]. The solubilities of some selected particulate bridging agents are shown in Table 9–1. A chelating agent such as citric acid or its salts is also included in the clean-up solution.

However, production engineers have been reluctant to use particle bridging because of the possibility of particle transport into the formation, resulting in formation damage and/or costly and often ineffective stimulation treatments. A particle bridging fluid has been developed that quickly and effectively controls fluid loss in a wide range of permeabilities and pore diameters [916].

Water-soluble organic polymers, such as hydroxyethylcellulose, have been used to slow the leak-off rate of clear brines into permeable formations. Fluid loss or leak-off, however, can be effectively controlled only by bridging the pore openings with rigid or semirigid particles of sufficient size and number.

Table 9-1
Solubilities of Some Particulate Bridging Agents [1749]

Particulate bridging agent	Aqueous ammonium salt clean-up solution	Solubility [g/100 ml]
Magnesium oxide	4 M ammonium chloride	1.6
Magnesium oxide	8 M ammonium acetate	2.8
Magnesium oxide	1.3 M ammonium chloride plus 1 M sodium citrate	2.8
Magnesium carbonate	8 M ammonium acetate	2.2
Magnesium carbonate	4 M ammonium chloride plus 0.4 M trisodium salt of nitrilotriacetic acid (NTA)	2.9
Anhydrite ($CaSO_4$)	4 M ammonium chloride	1.7
Anhydrite ($CaSO_4$)	8 M ammonium acetate	2.9
Lime (CaOH)	1.3 M ammonium chloride	3.0
Zinc oxide	4 M ammonium chloride	3.0
Zinc oxide	1.3 M ammonium chloride plus 0.8 M sodium citrate	2.9
Zinc carbonate	4 M ammonium chloride	2.4
Lanthanum oxide	0.36 M diammonium salt of ethylenediaminetetraacetic acid (EDTA)	2.2 / 6.0
Manganese hydroxide	4 M ammonium chloride	1.5

The filter-cake formed in this process, however, is highly dispersible to the produced fluid and thus is effectively removed by placing the well on production. No acid treatment or other removal techniques are required. The primary bridging agent in this fluid is a sized calcium carbonate with particle sizes capable of initiating bridging in pore diameters in excess of 100 µ.

Enzymatic Breaker

Damaging materials such as filter-cakes and very viscous fluids within a subterranean formation of a wellbore can be removed by means of an enzyme treatment [1747]. The enzyme treatment degrades polysaccharide-containing filter-cakes and their damaging fluids as the treatment reduces their viscosity. The degraded filter-cake and damaging fluid can then be removed from the formation back to the well surface. The particular enzymes used are specific to a particular type of polysaccharide and are active at low to moderate temperatures. The enzymes attack only specific linkages in filter-cakes and damaging fluids and are active in the pH range of 2 to 10.

Enzymes to degrade crosslinked hydroxypropylated starch derivative and xanthan gum polymer systems are available [158, 1246]. Specific enzymes are efficient in reducing the near wellbore damage induced by the starch polymer to eventually return permeabilities to the range of 80% to 98% without the use of acid systems.

Peroxides

Hydrogen Peroxide

A reagent for removing clay deposits is based on an aqueous solution of H_2O_2 and Na_2CO_3 in a concentration range of 15 to 30 g/liter and 75 to 150 g/liter or a solution of sodium bicarbonate and HCl in a concentration range of 60 to 80 g/liter and 3.5 to 4.0 g/liter, respectively [274]. The injection is followed by a holding time of preferably 2 to 5 hours. Clay layer break-up products are washed out with a wash solution such as petroleum, circulating water of aqueous solution of a surfactant.

Metal Peroxides

A process that successfully removes filter-cake contains polysaccharide polymers and certain bridging particles and uses alkaline earth metal peroxides and zinc peroxide in an acidic aqueous solution [499, 1239]. On soaking the filter-cake, a loosely adherent mass is left behind on the walls of the borehole. Thereafter, a wash solution in which the bridging particles are soluble is used to remove the remaining filter-cake solids from the sides of the borehole.

Magnesium Peroxide in Filter-Cake

Magnesium peroxide is very stable in an alkaline environment and remains inactive when added to polymer-based drilling fluids, completion fluids, or workover fluids. Because the magnesium peroxide material is a powdered solid, it becomes an integral part of the deposited filter-cake [501]. The peroxide can be activated with a mild acid soak. This treatment produces hydrogen peroxide, which decomposes into oxygen and hydroxyl radicals (OH·) when catalyzed by a transition metal. These highly reactive OH· species attack positions on the polymers that are resistant to acid alone. Significant improvements in filter-cake removal can be realized by using magnesium peroxide as a breaker in alkaline water-based systems, especially in wells with a bottom-hole temperature of 150° F or less, in the following operations: drilling into a pay zone, underreaming, lost circulation pills, and fluid loss pills for gravel prepacks.

Oligosaccharide

A gelled and dehydrated drilling fluid or filter-cake can be removed from the walls of wellbores by injecting an aqueous sugar solution [1829]. The solution is kept in contact with the filter-cake for a period of time sufficient to cause the disintegration of the gelled drilling fluid and the filter-cake. The composition, with the disintegrated drilling fluid and filter-cake dispersed therein, is then displaced from the wellbore. Monosaccharide sugars, disaccharide sugars, and trisaccharide sugars can be used. Surface active agents, such as a blend of nonionic ethoxylated alcohols or a mixture of aromatic sulfonates, can be added.

Oscillatory Flow

To remove filter-cake, a physical method can be applied wherein a fluid is oscillated in the annulus prior to cementing [948, 949]. The direction of flow of the fluid in the annulus is changed at least twice. The oscillatory flow of the fluid removes the drilling mud and the filter-cake from the annulus. After this oscillatory flow treatment, the cement slurry is pumped into the annulus.

Chapter 10

Cement Additives

State-of-the-art cementing technology [1135, 1333, 1709] and recent developments are sketched out in the literature [426]. There are two basic kinds of activities in cementing, namely, primary and secondary cementing. Primary cementing fixes the steel casing to the surrounding formation. Secondary cementing is for filling formations, sealing, water shut-off, etc.

Primary Cementing

The main purposes of primary cementing are:

- Supporting vertical and radial loads to the casing
- Isolating porous formations
- Sealing subsurface fluid streams
- Protecting the casing from corrosion

Secondary Cementing

Secondary cementing refers to cementing operations that are intended to use cement in maintaining or improving the operation of the well. There are two general cementing operations: squeeze cementing and plug cementing.

Squeeze Cementing. Squeeze cementing is used for the following purposes:

- Repairing a faulty primary cementing operation
- Stopping intolerable loss of circulation fluid during drilling operations
- Sealing abandoned or depleted formations
- Repairing leaks of the casing
- Isolating a production zone by sealing adjacent unproductive zones

The slurry should be designed to allow the fluid loss of the formation to be squeezed into the respective formation. Low-permeability formations can have a formulation of the slurry with an American Petroleum Institute (API) fluid loss [68] of 100 to 200 ml/30 min, whereas high-permeability formations require a slurry with 50 to 100 ml/30 minutes water loss. A high-pressure squeeze operation with a short duration requires an accelerator.

Thick slurries will not fill a narrow channel well. Therefore squeeze cement slurries should be rather thin. Dispersants should be added for this reason. High compressive strength is not necessary for these types of slurries.

Plug Cementing. Plug cementing is used for plugging abandoned wells for environmental reasons. A kick-off plug is used to plug off a section of the borehole. The plug uses a hard surface to assist the kick-off procedure. Plug cementing is also used in drilling operations if extensive circulation loss is observed. The plug is set in the region of the thief zone and pierced again with the bit.

Often in open-hole completion operations and in production, it is necessary to shut off water flows. Additional cementation methods are used to provide an anchor for testing tools or for other maintenance operations.

Basic Composition of Portland Cement

Cement is made of calcareous and argillaceous rock materials, obtained from quarries. Thus from the chemical viewpoint, the main components are carbonates and silicates. The raw materials used for cement are given in Tables 10–1 and 10–2. The active components in cement formulations are listed in Table 10–3.

Manufacturing

Grinding and Mixing

There are two processes for manufacturing cements, the dry process and the wet process. The dry process is cheaper than the wet process, but in practice

Table 10–1
Raw Calcareous Materials

Material	Remarks
Limestone	$CaCO_3$, sedimentary rock formed by the accumulation of shells or corals
Cement rock	Sedimentary rock with composition similar to industrial cement
Chalk	Soft limestone
Marl	Loose deposit consisting mainly of $CaCO_3$
Alkali waste	Obtained from chemical plants; a lot of chemical processes require bases; $Ca(OH)_2$ is a preferred base. After the reaction, $CaCO_3$ is sometimes left behind.

Table 10–2
Raw Argillaceous Materials

Material	Remarks
Clay	Hydrous aluminum silicates
Shale	Consists of clay, mud, and silt, mainly aluminum silicates
Slate	Dense fine-grained rock containing mainly clay
Ash	Contains silicates; secondary product

Table 10–3
Active Components in Cement Formulations

Component	Remarks
Tricalcium aluminate $3CaO \times Al_2O_3$	Hydrates quickly, responsible for strength of cement in early stage; setting time can be controlled by addition of gypsum
Tricalcium silicate $3CaO \times SiO_2$	Responsible for strength in all stages
Dicalcium silicate $2CaO \times SiO_2$	Responsible for final strength
Tetracalcium aluminoferrite $4CaO \times Al_2O_3 \times Fe_2O_3$	Little effect on physical properties

more difficult to control. Limestone and clay materials are crushed, either dry or in water slurry, and stored in separate containers. The composition is analyzed and the contents are blended according to the result of the analysis and according to the desired properties. Blends obtained from the wet process must be dried to some extent. The blend is ground to a mesh size of 100 to 200 (i.e., 0.15 to 0.07 mm) [81, 83, 84].

Burning

The blends are heated in a long rotary kiln. In the first stage of heating, free water evaporates at temperatures exceeding 900° C. Calcium carbonate caustifies to calcium oxide (CaO). The CaO starts reacting with aluminum silicates and the materials liquify. Heating continues to a final temperature of 1500° C. When the material is cooled it forms irregular-shaped solids called *clinkers*. Small amounts of gypsum (1% to 3% by weight) are added to these clinkers. Gypsum prevents flash setting and controls the basicity, due to CaO.

The commercial product is actually a blend of different cements. This blending leads to a more constant quality. The chemical composition of Portland cement is typically 60% to 69% CaO, 18% to 24% SiO_2, 4% to 8%

$Al_2O_3 + TiO_2$, 1% to 8% Fe_2O_3, <5.0% MgO, <2.0% K_2O u. Na_2O, <3.0% SO_3, all in % by weight [569].

Active Components in Cements

There are four chemical compounds that are active components in cement formulations. See Table 10–3 for a list of the components.

Chemistry of Setting

The setting reactions are exemplified with tricalcium silicate ($3CaO \times SiO_2$) and dicalcium silicate ($2CaO \times SiO_2$). In contact with water, the calcium silicates react with water molecules to form calcium silicate hydrate ($3CaO \times 2SiO_2 \times 3H_2O$) and calcium hydroxide ($CaO \times H_2O$).

$$2(3CaO \times SiO_2) + 6(H_2O) = 1(3CaO \times 2SiO_2 \times 3H_2O) + 3(CaO \times H_2O)$$

$$2(2CaO \times SiO_2) + 4(H_2O) = 1(3CaO \times 2SiO_2 \times 3H_2O) + 1(CaO \times H_2O)$$

From these equations, the stoichiometric amounts of water required for setting can be calculated. The hydration reactions are exothermic. Then the hydration rests. The inhibition period can last several hours. This behavior is important; otherwise proper placement of the cement would not be possible. After the inhibition period, the hydration proceeds and the cement starts developing strength. The water uptake results in an amorphous gel with a variable stoichiometry. During setting, volume changes through molecular contraction, shrinkage, swelling, and so on occur.

Standardization of Cements

The API has nine classes of well cements [63]. The classification is similar to ASTM C 150, Type I [80]. The well cement classes are summarized in Table 10–4.

Mixing with Additives

The cement usually is mixed dry with the additives. Depending on the application of the cement, a wide variety of additives can be incorporated. These include accelerators, retarders, dispersants, extenders, weighting agents, gels, foamers, and fluid loss additives.

Table 10–4
Classes of Cements and Properties

Class	Depth [m]	Properties
A Portland	0–2000	General purpose
B Portland	0–2000	Sulfate resistant
C High early	0–2000	High early strength
D Retarded	2000–3000	General purpose or sulfate resistant
E Retarded	3000–4000	High temperature, high pressure; moderately or highly sulfate resistant
F Retarded	3000–5000	Extremely high temperature and high pressure; moderately or highly sulfate resistant
G Basic	0–2500	Covering a wide range of depth, temperatures and pressures; no additives beside $CaSO_4$; can be used with accelerators or retarders
H Basic	0–2500	Basic well cement; no additives beside $CaSO_4$; can be used with accelerators or retarders
J	3500–5000	Extremely high temperature and high pressure; no additives beside $CaSO_4$; can be used with accelerators or retarders

Important Properties of Cement Slurries and Set Cement

In cementing operations and applications, several properties are required or desired, respectively, for proper use. An overview is thus provided.

Specific Weight

Specific weight is one of the most important properties of a cement slurry. The specific weight of a certain dry cement regulates the minimum or maximum amount of water allowed to be added. The minimum amount of water, from the aspect of density, is greater than the stoichiometric quantity necessary for proper setting. If more than the maximum amount of water is used, pockets of free water will be formed in the set cement column. Typical amounts of water range from 38% to 46% by weight in the final mixture.

Thickening Time

The thickening time covers the time allowed for manipulation after mixing with water. It is similar to pot life time in thermoset resins. Viscosity increases with time, beginning from mixing with water, because of the chemical reaction of setting. When viscosity becomes too great, the slurry is no longer pumpable. Therefore it is necessary to place the cement within a certain time after mixing;

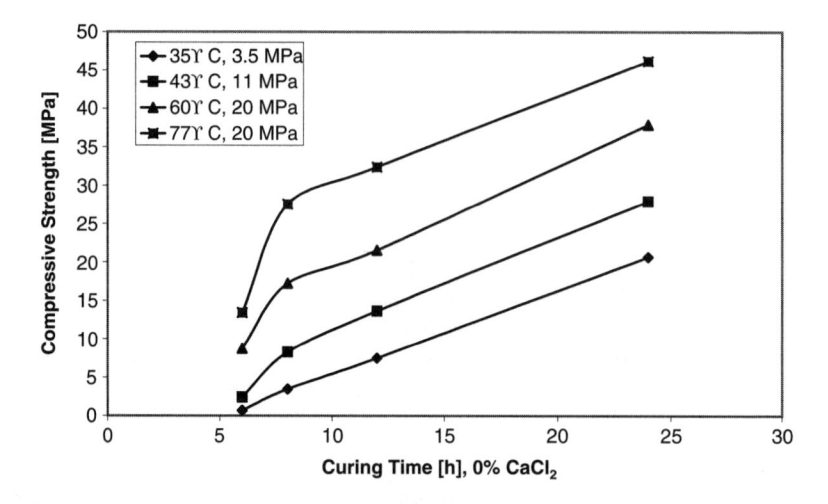

Figure 10–1. Influence of temperature and pressure on the comprehensive strength of an API Class H cement [1645].

otherwise, serious damage of the well could occur. An accurate knowledge of the time needed for the operation is necessary. The following factors have to be considered:

- Mixing time
- Displacing time (to bring down)
- Plug release time
- Safety time

In general, the thickening time decreases with increasing temperature. Therefore it is important that the temperature conditions in the well be known.

Strength of the Set Cement

The strength of a cement usually refers to the compressive strength. The development of strength in the course of setting is shown in Figures 10–1 and 10–2. In general, the rate of setting increases with increasing temperature and pressure.

Special Cement Types

Resin Cement

Resin cements or polymer cements have been reviewed by Chandra, Justnes, and Ohama [343]. Polymer cements are materials made by replacing the cement at least partly with polymers. Cements can be modified by latex, dispersions, polymer powders, water-soluble polymers, liquid resins, and monomers.

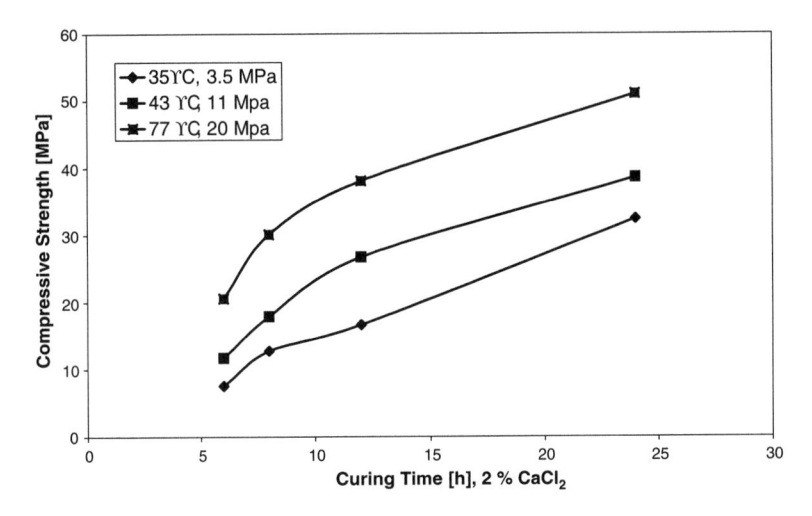

Figure 10-2. Influence of temperature and pressure on the comprehensive strength of an API Class H cement with 2% CaCl$_2$.

Table 10-5
Properties of Polymer Cement in Comparison with Conventional Cement

Property	Change	Remarks
Abrasion resistance	++	
Impact resistance	+	
Tensile strength	+	
Flexural strength	+	Depends on curing methods
Compressive strength	0	
Gas migration	−	Despite remaining porous structures, due to surface-tension modification
Chemical resistance	+	
Acid, alkali, salt	++	Pure epoxide, furan, and acrylic cements
Organics, alkali	−	Pure polyester
Pore volume	−	
Freeze-thaw resistance	+	Pure polymer cements absorb essentially no water [82]

+, better; −, bad; 0, no change in comparison with conventional cement.

When the polymer is used in small amounts, it modifies only the pore structure and does not behave like a binder.

The addition of polymer changes the properties of the set cement. The water to cement ratio is reduced with the increase of the polymer-to-cement ratio. This influences the mechanical properties. The polymer content results in the change of the properties as listed in Table 10-5.

Oil-Based Cement

To prepare an oil-based cement, the cement is suspended in hydrocarbons instead of water. In this organic environment, no setting takes place and the cement particles remain fine. Therefore they may penetrate into small pores. The setting starts when the cement particles come in contact with water. This takes place in the formation. Oil-based cements are mainly used as plugging cements and squeeze cements.

Small–particle-size cement has found a number of uses in production and injection well casing repair jobs [440]. Oil-based cement is particularly useful for water shutoff jobs, because the hydrocarbon slurry sets only in the presence of water, so the oil-producing sections of a reservoir remain relatively damage free after water shutoff. The selective water shutoff with oil-based cement also has been used with polymers crosslinked by metal crosslinkers [442, 1178].

High-Temperature Cement

High-temperature cement formulations are basically polymer concrete. Mainly unsaturated polyesters or vinyl ester resins with allylphthalate as vinyl monomer are in use [927]. Curing is achieved with peroxides, which decompose sufficiently fast at temperatures between 120° and 200° C. Polymer concrete requires additional materials, which compensate shrinkage.

Highly filled polymer composites, such as polymer concrete, suffer from setting stresses generated during the cure of the resin binder, when the polymerization shrinkage is hindered by the close packing of a filler and by aggregate particles. The setting stresses significantly decrease the strength of the cured composite. Producing zero-shrinkage with strength enhancement can be achieved by dispersing small amounts of the hydrated mineral montmorillonite into the resin. The addition of montmorillonite was found effective with three different resin binders (polyester, epoxies, and acrylics). A mineral-resin interaction mechanism has been suggested [774]. The organic molecules replace some of the ordered hydration water, released by the mineral at the temperatures generated by the exothermic polymerization reaction. A binder used for plugging hot drill holes contains mainly phosphoric slag, trisodium phosphate, and NaOH [151].

Silica Flour and Silica Fume

Silica powder has been studied as a stabilizer for oil well cement at high temperatures [1895]. Tests indicated that silica powder can improve the stability and pressure resistance strength of cement. The manufacture of very– high-strength concrete (after 28 days the compressive strength is greater than

80 MPa) often involves the addition of ultrafine particles together with large proportions of organic admixtures [465]. Silica fumes were found to be the most effective additive.

Substitution of silica flour with varying proportions of silica fume affects the strength and the permeability of hardened cements [727]. The positive effect on the strength regression at increased temperatures is due to a greater rate of carbonation of the set cement [1224]. Carbonation is controlled not by permeability but by the calcium hydroxide content at 150° C. The optimal silica addition for geothermal fluids, which contain high levels of CO_2, is ca. 15% to 20%. The coefficient of permeability was found to decrease with an increase in the degree of hydration. The use of silica fume was found to decrease the permeability only slightly [136].

Low-Temperature Cement

Special formulations have been developed for cementing operations in arctic regions or for deep water applications [206, 208, 256, 720, 739, 1792]. In low-temperature formations, wherein the cement is subjected to freeze–thaw cycling, freezing-point depressants must be added. Salts may serve as such, but traditional organic freezing-point depressants, such as ethylene glycol, also may be added [1022–1024].

In the case of epoxide cements at temperatures lower than 20° C, the viscosity increases so much that pumping becomes difficult [342]. Small amounts of aromatic solvents reduce the viscosity satisfactorily. Ethylene glycol butyl ether also changes the interfacial tension so that the polymer may penetrate into hairline cracks and fine capillaries. This is advantageous in blocking liquid or gas migration.

High-Alumina Cement

High-alumina cement is a rapid-hardening cement made from bauxite and limestone. It is comparatively resistant to chemical attack. Milling retards the setting of aluminous cement [1582]. On the other hand, setting accelerators such as lithium carbonate increase their effect by this treatment.

Compositions of high-alumina cement containing quartz or glass, calcium carbonate, microsilica, carbon black, iron oxide red mud or screened fly ash, and styrene-butadiene latex have been described [141, 1803, 1804].

Magnesian Cement

Innovations in cementing technology have made it possible to place a cement slurry across a given area, establish the desired seal, and subsequently remove the blockage by completely dissolving the cement using common oil field acids [393]. This type of cement is referred to as magnesian cement or magnesium oxychloride cement. Magnesian cement is completely soluble in hydrochloric acid. Retarded acid-soluble well cements contain magnesium oxide, magnesium chloride, additional borate, and a sugar (e.g., sucrose) as a set retarder. Lower-density compositions are foamed with appropriate amounts of gas [968, 1754–1757].

Fiber Cement

Fiber-containing cement was initially developed as a high-strength material that could be used to line a borehole [1789]. Several relatively simple and cheap spin-off applications of fiber cement were identified, such as the use of fiber cement in cement plugs for borehole stabilization and as a lost circulation material. Several companies are already applying or offering fiber cement for these purposes in the field, in both organic fibers and metal fibers [372, 1077, 1682–1684].

On the other hand, the addition of fibers may cause undesired properties of the cement [1759]. Fibers can actually increase pore and fracture systems in latex cements. The amount of fibers in a fiber seal cement influences the porosity and permeability while affecting compressive strength. During acid treatment of the formation, the fibers in the cement can be easily dislodged and extracted from the cracks, leaving pore spaces behind.

The benefits of carbon-fiber technology have been demonstrated in the building industry, but have not been investigated for improving set cements in gas and oil wells. A great potential exists for a technologic breakthrough in difficult cementing operations. A fourfold increase in toughness and a 50% increase in compressive strength can be obtained through carbon fiber addition. However, carbon fibers are very expensive, so the benefits may be jeopardized by the high price. Carbon fiber oxidized by a hot NaOH solution appear to have potential for use as reinforcement in high-temperature cementitious material systems [1434, 1691]. It has been determined that active carboxylic acid and sodium carboxylate functional groups introduced on the fiber surfaces by extensive oxidation react preferentially with Ca^{2+} ions released from cement in a hydrothermal environment at 300° C. This interfacial interaction leads to a linkage between the fiber and the cement matrix, thereby enhancing the bond strength.

Fibers may also have disadvantages. An increase in the number of fibers in the cement leads to an increase in the porosity and permeability of the cement, resulting in the decrease of compressive strength. During the acid treatment of the formation, the fibers in the cement can be easily dislodged and extracted from cracks as smooth flakes leaving pore spaces [1759].

Table 10–6
Classification of Cement Additives

Additives for cementing
Basic cements
Accelerators and salts
Extenders and density-reducing additives
Silica to reduce or prevent high-temperature strength retrogression
Dispersants
Bond improving and expanding agents
Retarders
Fluid loss additives
Antigas migration agents
Antifoam and defoaming agents
Additives and mixtures to reduce or prevent lost circulation
Density-increasing or weighting agents
Free water control and solid suspending agents
Spacers and chemical washes or preflushes
Specialty cement blends

Classification of Cement Additives

Cement additives have been classified into six categories [221]. In fact, this classification is crude. The World Oil Cementing Supplements regularly review commercially available cementing additives in certain *World Oil* issues [62, 66, 69, 70]. Here comprehensive listings of cementing products and additives available from major suppliers can be found. The products are grouped into functional categories as shown in Table 10–6. Products commercially available also can be found readily on the World Wide Web. Individual compounds may emerge in more than one of the categories listed in Table 10–6. For example, rubber particles reduce the density of the slurry and are also suitable as lost circulation additives. Also, there is a difference between fluid loss and lost circulation. Fluid loss refers to filtration of certain components of the fluid into the formation. Lost circulation is the total material lost into high-permeability thief zones. Table 10–7 gives a summary of additives to control special problems [1135].

Additives in Detail

Lightweight Cement

Adding low-density materials reduces the density of a cement composition. These additives are referred to as *extenders*, because they reduce the demand of

Table 10–7
Summary of Additives to Control Special Problems

Purpose	Action or agents
Gel-strength additives	Preparation of spacers
Permeability control	Silica flour, gas bubble–producing additives
Corrosion control	Various nitrogen compounds, polyoxylated amines, amides, and imidazolines
Radioactive traces	Helpful in finding the region of actual placement of the cement
Bactericides	Paraformaldehyde, sodium chromate
Strength increasers	Nylon, metal fibers
Defoamers	Controlled inclusion of air during mixing
Encapsulation	Controlled release of various additives [1486]

Table 10–8
Lightweight Cement Additives

Compound class	Remarks	References
Bentonite	Increases viscosity	
Furnace slag cement	High service temperatures	[1128]
Porous glass particles		[898, 1896]
Gilsonite	Soluble in organic solvents	
Pozzolan	Inexpensive	[803, 804]
Silica fume	Substitute for natural pozzolan	[1278, 1668]
Rubber		[1078]
Polyacrylonitrile		[1124]
Hydroxyethylcellulose	Free-fluid inhibitor	[1677, 1678]
Expanded polystyrene		[204]
Perlite	Much additional water	[1314]
Coal	No additional water	[542]
Diatomaceous earth	No increase of viscosity	
Hollow aluminosilicate microspheres		[1017, 1018]

the cement itself, at the cost of other desirable properties, such as set strength. Extenders, such as silicates, generally are not chemically inert, but may be reactive. Lightweight additives are summarized in Table 10–8.

Bentonite

Bentonite is used in secondary cementing operations [1634]. High percentages of bentonite significantly reduce the specific weight of the slurry.

Bentonite is dry blended to the cement. The addition of bentonite requires more water. Bentonite has been used up to 25% by weight. The additive also increases the viscosity of the slurry. High amounts of bentonite increase the permeability and reduce the final strength, however. Therefore high concentrations of bentonite are not recommended.

Furnace Slag

A furnace slag cement slurry can have a density of 1500 to 1600 kg/m^3. A combination of silica flour and furnace slag may be used to achieve service temperatures exceeding 200° C [670]. A gas may be suitable as a foaming agent [358].

Hollow Glass Microspheres

Glass beads with diameters of 2 to 200 μ and densities of 600 to 700 kg/m^3 can be mixed with a cement slurry in certain proportions to form low-density glass bead slurry. The glass beads are used to reduce weight. Laboratory experiments showed that the slurry is effective in improving cementing quality for low-pressure and low-permeability formations. Low-density cement with hollow microspheres is often not stable. This results from the comparatively low plastic viscosity [1139]. The plastic viscosity of the cement slurry must exceed 40 MPa to improve the cement quality and eliminate fluid channeling. When porous glass particles are used, they are filled with water. The cement slurry will absorb extra water from the pores of the porous particles, thus counteracting the shrinkage during setting [898].

Hollow Aluminosilicate Microspheres

Materials formed by acid–base reactions between calcium aluminate compounds and phosphate-containing solutions yield high-strength, low-permeability, CO_2-resistant cements when cured in hydrothermal environments. The addition of hollow aluminosilicate microspheres to the uncured matrix constituents yields slurries with densities as low as approximately 1200 kg/m^3, which cure to produce materials with properties meeting the criteria for well cementing. These formulations also exhibit low rates of carbonation. The cementing formulations are pumpable at temperatures up to 150° C.

Gilsonite

Gilsonite can be used to reduce the density of cement. Gilsonite (Uintaite) is an asphalt with a density of 1050 to 1150 kg/m^3, a melting point of 140° to 160° C, and an ultimate composition of approximately 85% C, 10% H,

2.5% N, 1.5% O. It is soluble in organic solvents and occurs naturally in Utah and Colorado. More details are given on p. 28.

Pozzolan

Pozzolan is a very finely ground pumice or fly ash. The specific gravity of pozzolan is only slightly less than cement. Therefore only a slight reduction of the specific weight can be achieved. On the other hand, pozzolan is inexpensive. Pozzolan has been proposed in several formulations. Silica fume has been proposed as a substitute for natural pozzolan. Silica fume is a pozzolanic material composed of extremely fine, amorphous spheres produced as a by-product in the manufacture of silicon metals. It has a high water demand and it is more reactive than natural pozzolan or fly ash. It increases the compressive strength significantly.

Rubber

Addition of rubber particles of 30% to 100% by weight to cement with a grain size of approximately 40 to 60 mesh (0.4 to 0.25 mm) will produce a lightweight cement. The addition of rubber particles also creates a low permeability. The compositions are advantageous for cementing zones subjected to extreme dynamic stresses such as perforation zones and the junctions of branches in a multi-sidetrack well. Recycled, expanded polystyrene lowers the density of a hydraulic cement formulation and is an environmentally friendly solution for downcycling waste materials.

Coal

Coal has been used as a very–low-gravity additive. Coal does not require a significant amount of additional water when added to the cement.

Diatomaceous Earth

Diatomaceous earth has a lower specific gravity than bentonite. In contrast to bentonite, it will not increase the viscosity of the slurry. Diatomaceous earth concentrations of up to 40% have been used.

Perlite

Expanded perlite requires a large amount of water when added to the slurry. It is often used in a blend with volcanic glass fines, or with pozzolan, along with bentonite. Without additional bentonite, perlite tends to separate and float to the upper part of the slurry.

Table 10–9
Surfactants for Foam Cement

Surfactant	References
Alkyl sulfates and alkyl ether sulfates	[1561]
Alkylpolyoxyalkylene sulfonates	[220]
Polyoxyethylene	[1365]

Table 10–10
Weighting Agents for Cement

Compound class	Remarks
Ilmenite	No additional water
Hematite	Some water
Barite	Still more water
Manganese compounds	

Foam Cement

Foam cement is a special class of lightweight cement. The gas content of foamed cement can be up to 75% by volume. The stability of the foam is achieved by the addition of surfactants, as shown in Table 10–9. A typical foamed cement composition is made from a hydraulic cement, an aqueous rubber latex in an amount up to 45% by weight of the hydraulic cement, a latex stabilizer, a defoaming agent, a gas, a foaming agent, and a foam stabilizer [359, 362]. Foamed high-temperature applications are based on calcium phosphate cement [257].

Density-Increasing or Weighting Agents

Weighting agents (Table 10–10) are added to increase the density of the cement. They are typically used to combat high bottom-hole pressures. Common additives are powdered iron, ferromat, powdered magnetite, and barite. Hematite can be used to increase the density of a mixture up to 2200 kg/m^3 (19 lb/gal). Hematite requires the addition of some water.

Ilmenite has a specific gravity of 4700 kg/m^3. It requires no addition of water when added to the slurry. Ilmenite has a minimal effect on the thickening time and compressive strength. Barite requires more water then hematite when added to the cement. This results in a decrease of the compressive strength of the set cement.

Thickening and Setting Time Control

Often it is necessary to influence the setting time, either by accelerators or by retarders. If a cement is to be placed into a shallow depth, then acceleration of setting will be desirable to avoid unnecessary waiting times. On the other hand, in a deep formation more open time is required, which may require the addition of retarders.

Cement Retarders

Examples of retarders are shown in Table 10–11. They are added to prevent cement from setting too rapidly. These additives are also referred to as *set retarders*. Cements with retarders to prevent rapid setting may be used at the high-temperature and high-pressure environments of deep wells. Common retarders are lignosulfonate and certain carbohydrate derivatives, such as welan gum, xanthan gum, cellulose, and polyanionic cellulose.

Table 10–11
Cement Retarders

Compound class	References
Scleroglucan	[326]
Copolymer of isobutene and maleic anhydride	[1118]
Amino-N-([alkylidene] phosphonic acid) derivatives	[1332]
Alkanolamine-hydroxy carboxy acid salts (e.g., tartaric acid and ethanolamine)	[361, 455]
Phosphonocarboxylic acids	[456, 457]
Dicyclopentadiene bis(methylamine) methylenephosphonate	[430]
Lignosulfonate derivatives	[480, 481]
Carbohydrates grafted with vinyl polymers	[557]
Carboxymethyl hydroxyethylcellulose	
Wellan gum[a]	[1475]
Borax based	[328]
Carrageenan	[1470]
Polyethylene amine derivatives and amides	[1517–1520, 1522]
Copolymers from maleic acid, of 2-acrylamido-2-methylpropane sulfonic acid and others[b]	[1521]
Ethylenediamine-tetramethylene phosphonic acid, polyoxyethylene phosphonic acid, or citric acid[c]	[1221]
Polyacrylic acid phosphinate[c]	[412]

a) Coadditive for retarded formulations.
b) When bentonite is used, calcium–sodium–lignosulfonate is the best retarder for such cement slurries.
c) High-alumina cement.

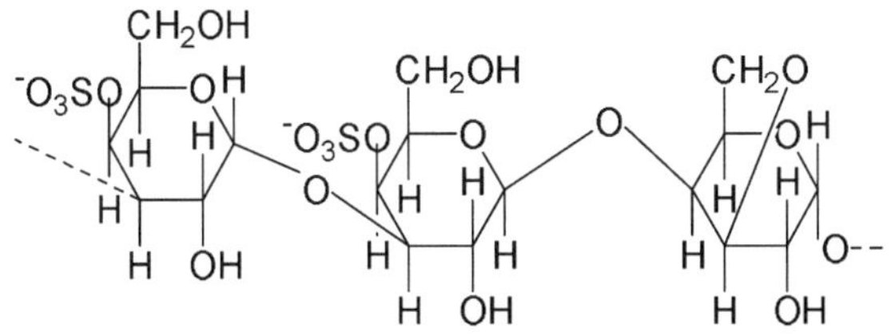

κ-Carrageenan

Figure 10–3. κ-Carrageenan.

Carrageenan, a high–molecular-weight polysaccharide derived from seaweed, produces an exceptionally stable storable liquid that is superior to those typically employed in storable cement slurries. Carrageenan gums are ionic linear polysaccharides composed of repeating galactose units that individually may be sulfated or unsulfated. Specific carrageenan types include κ-, ι-, and λ-carrageenans. Typically, ι-carrageenan is employed. Mixtures of carrageenan types are also possible. The properties of the individual carrageenan types primarily depend on the number and position of sulfate groups on the repeating galactose units. In the presence of excess cations, κ and ι carrageenans form gels [1470]. The structure of carrageenan is shown in Figure 10–3. Carrageenan is also used as a thickener in food applications, such as ice cream.

Cement Accelerators

Cement accelerators are shown in Table 10–12. The most common accelerators are calcium chloride and sodium chloride. Calcium chloride may be used in concentrations up to 4% by weight in wells with bottom-hole temperatures less than 50° C. Calcium chloride tends to increase the final strength under pressure conditions.

Sodium chloride can be used as an accelerator in formulations that are bentonite free. The maximal bottom-hole temperature is 70° C. In concentrations above 5%, the effectiveness is reduced. Saturated sodium chloride solutions act as retarders.

Special grades of gypsum hemihydrate are blended with Portland cement for types with reduced thickening time and setting time. Gypsum requires significantly more water addition. The maximal application temperatures are 70° to 80° C. Sodium silicate is used for cement slurries with diatomaceous earth. It can be used up to 7% by weight.

The addition of sodium sulfate to Portland cement accelerates the cement hardening and increases mobility of the solution. Chloride-free accelerators

Table 10–12
Cement Accelerators

Compound	References
Propylene carbonate[a]	[1111]
Sodium and calcium chlorides[b]	[961]
Aluminum oxide and aluminum sulfate	[19]
Sodium sulfate	[1688]
Calcium chloride	[299]
2,4,6-Trihydroxybenzoic acid and disodium 4,5-dihydroxy-m-benzenedisulfonate[c]	[651]
Esters of formic acid and formamide	[190, 191, 323]
Monoethanolamine, diethanolamine triethanol amine	[322]

a) Also has thixotropic properties.
b) Waste of isopropylbenzene production by the alkylation method.
c) Chloride-free set accelerator.

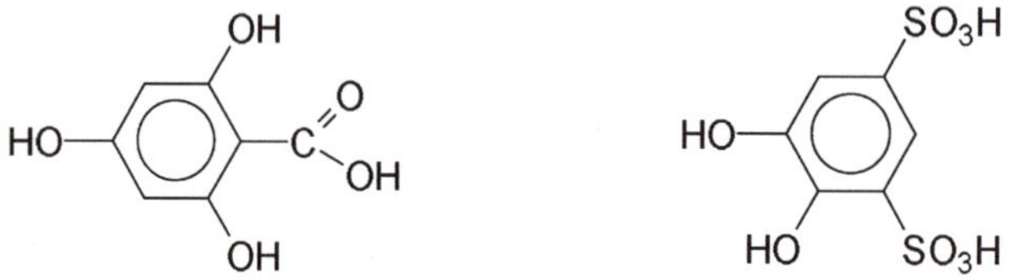

2,4,6-Trihydroxybenzoic acid   4,5-Dihydroxy-m-benzenedisulfonic acid

Figure 10–4. 2,4,6-Trihydroxybenzoic acid, 4,5-dihydroxy-m-benzenedisulfonic acid.

have been developed. Trihydroxybenzoic acid is a weak accelerator; it eliminates the dormant period in the curing process, as does as 4,5-dihydroxy-m-benzenedisulfonate. The recommended amount of dry cement is 0.1% to 5.0% by weight [651].

Viscosity Control

The viscosity of a cement affects the pumping properties. The viscosity must be kept low enough to ensure pumpability of the slurry during the entire operation period. In deep wells, because of the increased temperature, the viscosity becomes increasingly lower, which leads to undesirable flow characteristics of the slurry. This effect can be serious, because the viscosity follows the Arrhenius law. Some of the additives used for viscosity control also

Table 10–13
Additives for Viscosity Control

Compound	References
Latex	[255]
Scleroglucan[a]	[324–326]
Calcium lignosulfonate[b]	
Phenol-formaldehyde resin modified with furfuryl alcohol[c]	[1352]
Hectorite clay[d]	[321]
Sulfonic acid copolymer, castor oil[e]	[183, 184]

a) High-temperature viscosifying additive.
b) Also retarder.
c) Polymer cement.
d) Thixotropic cement.
e) Multifunctional additive.

may act as accelerators. Additives for viscosity control are shown in Table 10–13.

Thermal Thinning. To combat the effect of thermal thinning, latex is added to a cement slurry without a latex-stabilizing surfactant, resulting in a slurry having low mixing viscosity and good solid-suspension properties at downhole temperatures. The latex emulsion inverts or breaks down-hole, thereby providing the necessary viscosity and gel strength to compensate for thermal thinning and to keep the solids suspended in the cement slurry. The inverted latex emulsion coagulates, forming rubberlike particles that increase the viscosity and gel strength of the cement slurry. No additional viscosifying agents are required to prevent the solids from settling.

Surface Active Agents as Dispersants

The use of surface active agents in cements has primarily two main goals. They act as retarding agents and dispersing agents. Dispersants improve the rheologic properties. In particular, dispersants are added to cement slurries to facilitate the blending at high densities without the demand for excess water. They also enhance the flow behavior of the cement slurry and allow the slurry to be pumped in turbulent flow, thereby effecting a better bonding between the well casing and the rock formation. Furthermore, dispersants may exhibit side effects: They enhance the action of fluid loss additives and enhance the retarder effectivity. Dispersants are widely used as oil field chemicals, besides in cementing activities. The applications in the field of cementing technology are summarized in Table 10–14.

Table 10–14
Dispersants

Dispersant	References
Polyoxyethylene sulfonate[a]	[374, 802]
Acetone formaldehyde cyanide resins	[408, 558]
Polyoxethylated octylphenol[b]	[717–719]
Copolymers of maleic anhydride and 2-hydroxypropyl acrylate[c]	
Allyloxybenzene sulfonate or allyloxybenzene phosphonate[d]	[319]
Ferrous lignosulfonate, ferrous sulfate, and tannic acid	[1108]
Alkali lignosulfonate[e]	[356, 360]
Acetone, formaldehyde polycondensate[f]	[1363, 1364]
Sulfonated napththalene formaldehyde condensate	[242, 421, 423, 489, 1220]
Sulfonated indene and indene–cumarone resins	
Melamine sulfonate polymer, vinyl sulfonate polymer, styrene sulfonate polymer	[1250]
Polyethyleneimine phosphonate	[422]
Casein with polysaccharides	[1801, 1802]

a) For squeeze cementing.
b) Nonionic surfactant, 1–3 kiloDalton (kD).
c) 1–20 kD.
d) As copolymer with various vinyl monomers.
e) Phenolic hydroxy group blocked; biodegradable.
f) For dispersing silica fume.

The set retarding characteristics of lignosulfonates can be substantially eliminated by blocking the phenolic hydroxy group content of the lignosulfonate. To minimize the presence of free phenolic hydroxy groups in the lignin, such groups are blocked by reacting the lignosulfonate with, for example, propylene oxide.

Expansion Additives

During setting, cement normally undergoes shrinkage. By adding expandable swelling additives into the matrix, the shrinkage can be combated. Expanding cement is used in water shutoff technology and plugging.

A series of test methods and procedures have been developed to measure these phenomena [67]. Cracks should be avoided because these increase the permeability of the cement. The expansion of the cement, without the formation of macro-fissures, depends on the time at which the expanding

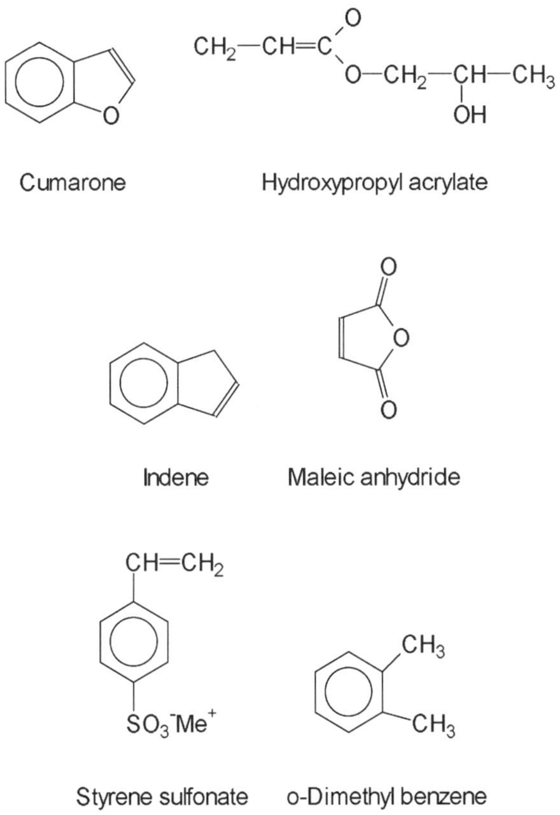

Figure 10–5. Cumarone, hydroxypropyl acrylate, indene, maleic anhydride, styrene sulfonate, O-dimethyl benzene.

additives (e.g., CaO or MgO) crystallize out [443, 697, 698]. The cement expansion should occur predominantly in the setting phase of plastic deformability of the cement matrix to fill in existing cavities. A certain controlled amount of expansion should also occur in a phase of higher strength of the matrix [699]. The expansion characteristics of CaO and MgO [700] depend on the thermal history of burning, which can be optimized for specific requirements.

Dead burned magnesium oxide is suitable as an expanding additive [368–370]. The expansion occurs by a hydration mechanism. The additive is particularly effective when used at setting temperatures greater than approximately 150° C. Enhanced adhesion of expanded cements can be achieved by the addition of urea–formaldehyde resins [1720].

Set Strength Enhancement

Dialkanol aminoalkyl phenols as admixtures enhance the strength [675]. The additives are useful in very small amounts and do not affect the initial properties of the fluid. The strength additive does not cause set acceleration or early set strength enhancement but provides enhanced compressive strength of the cement in later stages. Addition of small amounts of potassium ferricyanide and nitrile-trimethyl phosphonic acid promotes the formation of complex compounds and thus increases the strength of cement rock [1771].

Fibers

A mixture of a reactive aluminum silicate and a fibrous mineral, such as wollastonite [1271], improves the compressive strength, flexural strength, and tensile strength over conventional cement compositions. Wollastonite is typically available as very fine fibers or microfibers having diameters similar to those of cement particles (typically approximately 25 to 40 µm) and a fiber length typically between 0.4 and 0.6 mm. Alternatively, the cement composition may include consolidating fibers, such as nylon or polypropylene fibers. These reduce the potential for cement debris formed under high-stress conditions. Consolidating fibers are typically added to a cement composition in an amount between 0.25 and 5.0 pounds per sack.

Adhesion Improvement

A cement slurry additive consisting of methylcellulose, melamine-formaldehyde resin, and trioxane has been proposed for better bonding of cement to the casing string [20]. Bisphenol-A epoxy resins, with amine-based curing agents, sand filler, and a mixture of n-butanol and dimethyl benzene as a diluent, have been proposed as additives to increase adhesion properties of cement [572].

Quaternary ammonium salts (C_{10} to C_{18} alkyl benzyl dimethyl ammonium chloride) added in 40 ppm and 2% of sodium chloride effect an increase in the strength of the cement rock and the adhesion properties by 50% to 80% [1769].

Filtration Control

Filtration control additives are added to cements for the same reason they are used in drilling fluids. Untreated cement slurries, however, have much greater filtration rates than do untreated drilling muds. Therefore it is very important to limit the loss of water from a slurry into a permeable formation. This is necessary for several reasons:

- To minimize hydration of water-sensitive formations
- To allow sufficient water to be available for cement hydration
- To avoid a modification of the slurry properties (e.g., rheology, density, thickening time)
- To avoid a bridging of the annular gap

However, the mechanism of action of filtration control additives is not yet completely understood. Examples are bentonite, latex, various organic polymers, and copolymers. Many additives for fluid loss are water-soluble polymers. Vinyl sulfonate fluid loss additives based on the 2-acrylamido-2-methylpropane sulfonic acid (AMPS) monomer are in common use in field cementing operations [363]. The copolymerization of AMPS with conjugate monomers yields a fluid loss agent whose properties include minimal retardation, salt tolerance, high efficiency, thermal stability, and excellent solids support.

Gilsonite is active as a fluid loss additive because the permeability of cement is reduced. Latex additives also act as fluid loss additives. They also act as bonding aids, gas migration preventers, and matrix intensifiers. They improve the elasticity of the cement and the resistance to corrosive fluids [921]. A styrene-butadiene latex in combination with nonionic and anionic surfactants shows less fluid loss. The styrene-butadiene latex is added in an amount up to 30% by weight of the dry cement. The ratio of styrene to butadiene in the latex is typically 2:1. In addition, a nonionic surfactant (octylphenol ethoxylate and polyethylene oxide) or an anionic surfactant, a copolymer of maleic anhydride, and 2-hydroxypropyl acrylate [719] can be added in amounts up to 2%.

Furthermore, anionic aromatic polymers (AAPs) have been identified that simultaneously impart to salt-rich cement slurries improved fluid loss control and adequate rheologic properties. An advantage offered by these AAPs is their use even at low bottom-hole circulating temperatures. Commonly, the inclusion of fluid loss control additives, or friction reducers, is contraindicated in salt-rich systems when circulating temperatures are below 65° C, because the conventional additives cause an excessive retardation of the setting time. AAP systems still display acceptably short thickening times (<6 hours) and good early strength development when circulating temperatures are as low as 50° C.

Clay Control Additives

The cementing technology can be improved in wells with zones containing clays or shales that are sensitive to freshwater cement filtrate. Sodium chloride (i.e., natural salt) and potassium chloride (KCl) have been the primary materials of choice to yield a filtrate that damages these zones less. However, the unfavorable effects of salt on cement have been extensively documented, in particular, the negative impact of KCl on cements. Instead of inorganic salts, quaternary ammonium salts of aliphatic tertiary amines have been tested [487, 488, 1403]. The material shows superior clay controlling properties without the undesirable side effects of either sodium or potassium chloride.

Anti–Gas-Migration Agents

Gas channeling can occur during the setting of a cement slurry. The formation of channels is dependent on the setting characteristics, and additives can influence this. During the setting period of a cement, two time cycles of cement expansion and contraction are observed [1710, 1711]. This is due to the individual contribution of each component in the cement mixture. To obtain the optimal tightness of the cement, the final contraction is crucial for blocking gas migration. Ironite sponge, synthetic rubber powder, and anchorage clay were tested [1711, 1712] as permeability reducers. These materials are environmentally safe and inexpensive. Proper amounts of these additives can be used to optimize the compressive strength and to eliminate both microfractures and the formation of a microannulus. A copolymer of AMPS, N-vinylacylamide, and acrylamide reduces gas channeling [668].

Carbon black may serve as a low-cost additive for controlling the gas migration in cement slurries [303]. It is intended as a suitable substitute for polymer latex and silica fume and has been tested in field applications [304, 1256]. The concentration of carbon black varies from 2 to 20 parts, based on the weight of the dry cement [1220]. The particle size varies from 10 to 200 nm. A surfactant is necessary for dispersion, for example, formaldehyde-condensed naphthalene sulfonate or sulfonated cumarone or indene resins.

A mixture of lignosulfonates, alkali-treated brown coal, and minor amounts of organic silicon compounds (e.g., ethyl silicone) reduces the permeability of cements [1019]. The additives may interact with the crystallization centers of the cement slurry and form a gel system in its pores and capillaries, thus reducing the permeability of the cement and increasing its isolating capability. Furthermore, it is claimed that the additive retards the setting rate of cement up to 200° C and increases the resistance to corrosive media.

Phosphorated aluminum powder reacts with calcium hydroxide in the cement slurry during setting and produces hydrogen. The gas swells the cement slurry. Thus the additive prevents the channeling of oil, gas, and water [214, 215, 1824].

Corrosion Inhibitors

Ground water with high salinity, high CO_2 content, and sulfate-reducing bacteria corrodes the cement by dissolution, chemical, and expansion actions. By improving the permeability of the cement and reducing the content of $Ca(OH)_2$ in the cement, anticorrosion effects can be promoted. Addition of silica flour changes the proportion of calcium and silica and the composition of the hydrates. Homogeneity; tight, small pore channels; and low permeability improve the cement strength and anticorrosion effects [1104]. Hydrazine chloride has been proposed for inhibiting the corrosion of the casing [1767].

Chemical Attack

Portland cement is susceptible to corrosion by CO_2 and H_2S. The chemical attack by CO_2 is called *carbonation*. A microsample technique has been developed to study the CO_2 corrosion in cements, because the corrosion is difficult to monitor with common test procedures [264]. This technique is also advantageous as an accelerated testing method. A polymer-modified cement has been tested in field studies [694]. The addition of silica also improves chemical resistance [146], in particular brine corrosion.

Uses from Waste of Other Industrial Branches

Proper waste management becomes increasingly important with increasing amounts of waste created by modern civilization. In cementing technology, wastes from various sources can be used in various ways. The activities may be subdivided into two main classes:

- Use of wastes as raw material or secondary fuel in cement manufacture
- Direct use of wastes as additives for oil well cements

We discuss several issues here some of which were mentioned in previous sections.

Cement Manufacture

In the manufacture of Portland cement, many otherwise-waste materials can be used either as a substitute for the traditional raw material, or as a secondary fuel (e.g., used tires) [334, 1577]. In particular, drilling wastes can be introduced in the clinker burning process [878]. For both waste disposal and cement manufacturers, a mutual benefit will emerge. The cement manufacturing companies reduce their demand for traditional raw materials and save the limited capacity of landfills and other waste-treatment industries.

Waste water-based drilling fluids can be solidified by adding cement mixtures [1729], in particular, those with low-quality blast furnace slags [168, 410, 413, 1544]. Such mixtures have already been applied in wells at temperatures from approximately 4° to 315° C. The disposal of rock cuttings is achieved by combining the cuttings with water and blast furnace slag; injecting the slurry into the annulus surrounding a wellbore casing; and solidifying the cuttings, water, and slag. Solidification in blast furnace slag cement is

Table 10–15
Fluid Loss Additives for Cements

Compound	References
Water-soluble polymers[a]	[1469]
Gilsonite[b]	
AMPS-based fluid loss additives[c]	[1277]
Styrene-butadiene latex[d]	[718]
Anionic aromatic polymers[e]	[258]
Polynaphthalene sulfonate and acrylic terpolymer	[963]
Polyvinylacetate[f]	[1250–1252]
Copolymers of acrylic acid and long side chain acrylic esters and several similar materials (e.g., methacrylics)	[107]
Hydrophobically modified hydroxypropyl guar	[99]

a) General purpose.
b) Also density reducer.
c) Various other advantages.
d) Also has thixotropic properties.
e) Low bottom-hole temperatures.
f) With a dispersing sulfonated polymer and surfactant.

inexpensive [1183]. The slag is compatible with both oil- and water-based drilling muds. Drilling fluids, therefore, do not need to be removed from the drilling cuttings prior to solidification in the wellbore annulus [757].

Disposal of Oil Sludge

The proper disposal of oil sludge is one of the fundamental problems of petroleum production. Increasingly stringent environmental control regulations, lack of final disposal sites, and high costs involved in disposal have resulted in limited oil sludge disposal options. Two options for disposing the oil sludge have been compared [930]:

- The use of solid–liquid centrifugal separation
- The use of oil sludge as a cementing material

If the initial oil concentration in the sludge is high, high-temperature centrifugation with biodegradable surfactant is recommended to lower the concentration to a reasonable value. The resulting solid extract can be mixed with cement to obtain cement of a quality suitable for masonry.

The major potential uses of wastes in cementing technology are summarized in Table 10–16. There are many Russian patents dealing with the use of wastes from the production of organic chemicals as cement additives.

Table 10–16
Use of Waste in Cementing Technology

Waste type	References
Water-based drilling fluids[a]	[168, 410, 413]
Oil-based drilling fluids containing water[b]	[320, 930]
Rock cuttings[c]	[757]
Slags from nonferrous metal industries[d]	[1474]
Slags from nonferrous metal industries[e]	[720, 1031, 1032, 1331, 1886]
Waste from silicon production	[1044]
Sludge waste from nonferrous metal industries[f]	[976]
Waste from titanium and magnesium industries[g]	[1534]
Waste from semiconductor production	[56]
Waste from soda production	[907, 1288, 1445, 1457–1459, 1465, 1471, 1628, 1661, 1662]
Production wastes from organic chemicals[h]	[631, 1265, 1427, 1428, 1902]
Polymer wastes[i]	[259, 960, 1033]

a) With low-quality blast furnace slags.
b) Moderate-strength cement.
c) No removal of drilling fluids; less drilling fluids to be disposed.
d) Nickel slag.
e) Aluminum slag.
f) Magnesium sludge.
g) Contains chlorides, as accelerator.
h) Maleic anhydride, sebacic acid.
i) Rubber.

Chapter 11

Transport

Pipelines have a long history. In ancient times, pipelines were used for water transport. Examples are still visible in archeologic areas. However, it is clear that these early constructions could not bear large pressures. The advent of gas pipelines started between 1820 and 1830 with the distribution of town gas. Nowadays pipelines are indispensable in petroleum industries for the transport of various materials, including natural gas, crude oil of various types, and refined products.

The construction and operation of pipelines are described in the literature [221, 1008, 1234] and are not a subject of oil field chemicals. Here the additives and chemicals that facilitate the transport of fossil fuel products are discussed.

Pretreatment of the Products

Certain requirements concerning the purity of the product to be transported must be fulfilled. For natural gas, the water content should be kept below a certain level to reduce hydrate formation. In addition, oxygen and sulfur must be controlled effectively. Some issues are presented in Table 11–1.

Pretreatment for Corrosion Prevention

Methods used to control presumptive corrosion include deaeration and dehydration. Carbon dioxide and hydrogen sulfide are the main corrosives in pipelines for natural gas, but they are only aggressive in the presence of water. Therefore sweetening and drying the gas are useful to prevent corrosion. In oil pipelines, water emulsified in crude oil can cause corrosion problems [251]. Emulsified crude oil in separated produced water is also an environmental and disposal problem.

Table 11–1
Main Classes of Additives and Chemicals Used for Transport

Additive	Remarks
Drag reducers	For both liquid and gas transport
Pour-point depressants	Reduce pour points of waxy crudes
Odorizing additives	For safety
Gas hydrate inhibitors	For gas and multi-phase transport to prevent hydrate deposits
Surfactants	For multiphase transport of heavy crudes
Corrosion inhibitors	Both chemical inhibitors and biocides
Paraffin inhibitors	Prevent paraffin depositions

Table 11–2
Composition of the Natural Gas Transmitted from Alberta, Canada to Ontario, Canada [1261]

Component	% per Volume
Nitrogen	0.01270
Carbon dioxide	0.00550
Methane	0.95400
Ethane	0.01970
Propane	0.00510
i-Butane	0.00170
n-Butane	0.00080
i-Pentane	0.00020
n-Pentane	0.00010
n-Hexane	0.00020

Natural Gas

Natural gas consists mainly of methane although there are trace amounts of higher hydrocarbons, nitrogen, and even helium. It is typical in the gas transportation and storage industry to try to strip out higher hydrocarbons such as ethane, propane, butane, and unsaturated hydrocarbons from natural gas if the gas is to be transmitted through pipelines. This leaves mostly methane (with some traces of nitrogen and carbon dioxide) to be transported by the gas pipeline. The materials that are stripped out are then transported or stored separately, often as liquids. A typical composition of natural gas transmitted through pipelines is shown in Table 11–2.

The compressibility factor z of methane is always less than 1.0 in normal temperature ranges (i.e., between $-40°$ and $50°$ C). Furthermore, the compressibility factor decreases as the pressure rises or the temperature falls. Therefore, less energy is needed to pump a given volume of methane (measured at standard volume) at any given normal temperature than would be expected at that temperature if the methane were an ideal gas. This effect is more marked at higher pressures. Similarly, as the pressure is increased at a constant temperature, more methane (measured at standard volume) can be stored in a given volume than would be predicted from the ideal gas equation.

Below 7 MPa, the dominant variable for the compressibility factor in the PVT equation is the molecular weight of the gas. At this pressure level, the addition of ethane or propane increases the molecular weight of the gas more rapidly than the z factor decreases. Thus there is an advantage to removing ethane, propane, etc. from the gas.

At pressures greater than 7 MPa it is advantageous to add substances such as C_2 or C_3 hydrocarbon compound, carbon monoxide, hydrogen fluoride, ammonia, or a mixture of these with the natural gas. Ammonia without other additives is useful as an additive for gas storage at pressures down to about 5.5 MPa. Above a lower limit, which varies with the additive being added and the pressure, this results in a smaller $M_w \times z$ product, and therefore a decrease in the amount of power needed to compress the mixture for storage and to keep it compressed. It is also advantageous to add ammonia to natural gas to be transmitted through pipelines at pressures above 5.5 MPa. Depending on the cost, it also can be advantageous to add carbon monoxide. Hydrogen fluoride is also effective, but is prohibitive because of the toxicity and corrosive properties. The precise amount of each additive that can be added at any pressure for beneficial results can be found by calculating the product of the molecular weight times the z factor of the resulting mixture $M_w \times z$, and comparing it with the product of the molecular weight times the z factor of the original natural gas. If the product $M_w \times z$ is smaller for the mixture than for the natural gas, energy can be saved in pumping and compression [1262].

The use of two or more additives has a synergistic effect in many cases, so that an even smaller amount of each is needed than would be if only one were present to produce the z factor over that of an equivalent standard volume of natural gas at the pressure and temperature involved.

When the mixture is pumped through a pipeline, however, an additional effect with ammonia emerges. In a pipeline, there are pumping stations at intervals along the pipeline. At each pumping station, the gas is compressed. As the gas moves toward the next pumping station, it gradually loses pressure and expands. The compression during passage through the compressor station heats the gas; the gas cools while passing through the pipeline, transferring some of its heat to the surrounding soil through the pipeline wall. Ammonia has the property of being a refrigerant, which absorbs heat as it expands.

Thus when an ammonia–natural gas mixture is compressed and then is subsequently allowed to flow through a gas pipeline, the ammonia cools the mixture as it expands. This is regarded as an additional advantage [1261].

Sulfur Contamination of Refined Products

If refined products, such as gasoline, diesel, jet fuel, or kerosene, are transported in a pipeline, where otherwise sour hydrocarbon fluids are transported, there may be an undesired enrichment of sulfur in the refined products. This can be avoided if the oxygen level of the transportant is maintained at below 20 ppm [570]. The dissolved oxygen level in the hydrocarbon product is controlled by reducing the amount of air injection employed in mercaptan or disulfide reduction or by the use of oxygen scavengers prior to the introduction of the refined hydrocarbon product into the pipeline.

Demulsifiers

A gas containing entrained asphaltene-containing hydrocarbons is conditioned for pipeline transportation by injecting a surface active compound that is composed of a mixture of demulsifying agent, antifoaming agent, dispersant, aromatic solvent, and alcohol solvent, whereby the formation of emulsion from asphaltene-containing hydrocarbons is prevented [1206]. A surface-active composition is shown in Table 11–3.

Heavy Crudes

The most relevant parameters in pipeline transportation of heavy crude oil are velocity, viscosity, temperature, density, and pour point [691]. Heavy crude

Table 11–3
Surface Active Composition for Conditioning a Gas Containing Entrained Asphaltenes

Action	Compound	% by wt
Demulsifying agent	Solution of a sulfonic acid, a phenolic resin, and alcohol	5 to 15
Antifoaming agent	Silicone	1 to 3
Dispersant	Imidazoline	10 to 40
Aromatic solvent	Benzene, toluene, xylene, residues of BTX distillation	10 to 60
Alcohol solvent	From methanol to hexanol	20 to 60

can be transported on trunk systems in a variety of modes, including segregation, blending, and batching. Segregation requires separate pipelines, blending consists of mixing crudes, and batching refers to shipping crude in discrete batches. There are a number of methods for enhancing the transportability of heavy crude. These include oil-in-water emulsion formation, droplet suspension, dilution, the use of drag-reducing additives, and heating.

Emulsions for Heavy Crudes

Oil-in-water emulsions provide a cost-effective alternative to the methods mentioned previously, namely, heating or diluting. A typical transport emulsion is composed of 70% crude oil, 30% aqueous phase, and 500 to 2000 ppm of a stabilizing surfactant formulation [1497]. Nonionic surfactants are relatively insensitive to the salt content of the aqueous phase; ethoxylated alkylphenols have been used successfully for the formation of stable emulsions that resist inversion.

Activation of Natural Surfactants

With chemical treatment, the natural surfactants in crude oil can be activated [1384]. This method has been shown to be effective for highly viscous crude oil from the *Orinoco Belt* that has been traditionally transported either by heating or diluting. The precursors to the surfactants are preferably the carboxylic acids that occur in the crude oil. The activation occurs by adding an aqueous buffer solution [1382, 1383]. The buffer additive is either sodium hydroxide in combination with sodium bicarbonate or sodium silicate. Water-soluble amines also have been found to be suitable [1506].

Additional stabilizers for the emulsion can be multivalent inorganic salts, such as aluminum nitrate [1523], in small quantities of 30 ppm.

Low Temperature Transportation

In addition to the surfactant, a freezing-point depressant can be added for low-temperature transportation. Possible depressants include salts, sugars, and alcohols such as glycerol [736].

Corrosion Control

Coatings, cathodic protection, and chemical additives are used extensively to prevent internal and external pipeline corrosion. The excessive use of incompatible chemical additives has caused severe problems in gas-transporting systems. Costs arising from these problems often exceed the costs of the chemicals themselves. The careful evaluation and selection of chemical additives can minimize these problems and reduce operating costs [1860].

Crude Oil Treatment

Some crude oils contain certain organic compounds that are corrosive. In particular, these include naphthenic acids. Such crude oils cause problems in transportation, refining, and processing. The naphthenic acid content can be reduced simply with alcohol treatments, such as methanol, to form the corresponding ester. Hence, treatment temperatures will preferably be around 350° C. Pressures from about 100 to 300 kPa are typical and generally result from the system itself [1556].

Chemical Inhibition

Types of Inhibitors

Inhibitors may be classified according to their solution properties as either oil-soluble inhibitors, water-soluble inhibitors, or dispersible inhibitors. Chemical inhibitors act as film formers to protect the surface of the pipeline. Corrosion inhibitors, used for the protection of oil pipelines, are often complex mixtures. The majority of inhibitors used in oil production systems are nitrogenous and have been classified into the broad groupings given in Table 11-4. Typical corrosion inhibitors are shown in Table 11-5. For details, see also Chapter 6.

Synergism with Drag Reducers

Copolymers of acrylamide and acrylic acid that are added as drag reducers also enhance the activity of corrosion inhibitors in brine solutions by decreasing turbulence so that the corrosion inhibitor can more thoroughly contact the internal surface of the pipeline. Therefore the copolymer acts synergistically with corrosion inhibitors to increase their efficiencies [914]. The effect of a chemical drag reducer on oxygen corrosion of carbon steel has been investigated in a large-scale flow loop simulating a seawater injection line.

Table 11-4
Groupings of Corrosion Inhibitors Used in Transportation [1194]

Grouping
Amides or imidazolines
Salts of nitrogenous molecules with carboxylic acids (fatty acids, naphthenic acids)
Nitrogen quaternaries
Polyoxylated amines, amides, and imidazolines
Nitrogen heterocyclics

Table 11–5
Corrosion Inhibitors

Inhibitor	References
Glutaraldehyde[a]	[543]
2-Phenylbenzimidazole 2-methylbenzothiazole[b]	[995]
Benzotriazole 1,3-dimethyl-2-thiourea, thiourea, sodium-hexa-metaphosphate	
Tribasic sodium orthophosphate with polyurethane foam[c]	[200]

a) Bacterial corrosion.
b) Chemical corrosion.
c) For external protection.

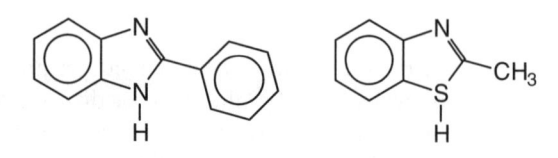

2-Phenylbenzimidazole   2-Methylbenzothiazole

Benzotriazole   1,3-Dimethyl-2-thiourea

Figure 11–1. Corrosion inhibitors: 2-phenylbenzimidazole, 2-methylbenzotriazole, benzotriazole, and 1,3-dimethyl-2-thiourea.

A drag reduction of up to 48% was achieved. In addition, a reduction of corrosion occurred [1197].

Coatings

Coatings protect the wall material by preventing aggressive substances from coming in contact with the wall. A coating is actually not an additive, but this section is included for completeness. Polyethylene coating with special antioxidant stabilizers has an excellent resistance to thermal oxidation at 80° C. A polypropylene coating is suitable for external corrosion protection of steel line pipes at temperatures between −30° C and 120° C. Coatings are summarized in Table 11–6.

Table 11-6
Coatings for Pipelines

Coating Material	Remarks	References
Bitumen	Oxidized bitumen with hexamethylenetetramine	[939]
Concrete	Styrene-butadiene copolymer latex additions on centrifugally cast concrete	[271]
Epoxide resins	Glass fiber reinforced	[1371]
Polyethylene		[1230]
Polypropylene	External corrosion protection	[72]
Polyurethane foams	Insulating systems for high-temperature marine pipelines	[1389]
Polyurea		[953]

Alternative Plastic Materials

Fiber-reinforced epoxide pipes have many advantages for the petroleum industry. These pipes offer corrosion resistance, high strength-to-weight ratio, light weight, desirable electrical properties, dimensional stability, pressure and temperature stability in given ranges, and low maintenance costs.

Paraffin Inhibitors

The techniques of paraffin removal and paraffin prevention have been reviewed [810]. In particular, inhibitors for paraffin deposits are copolymers of ethylene with vinylacetate [525–527, 1597] or polymers from p-nonylphenyl methacrylate and p-dodecylphenyl methacrylate [773]. These materials lower the pour point of the oil. It has been shown that for oils which differ in the content of n-paraffins and asphalt-resinous substances, it is necessary to use blends of copolymers of different compositions and molecular weights to obtain optimal efficiency. Polyacrylamide and wastes from the production of glycerol with a concentration of 400 mg/liter of oil have also been claimed to be effective as paraffin inhibitors [536].

Pour Point Depressants

Some crude oils are so waxy that their transportation by cold pipelines is very difficult, especially in winter. This is because of the high pour points of such crudes, which adversely affect the transportation process.

The crystallization of waxes at lower temperatures causes reduced liquidity of waxy crude oils, which considerably hampers the transportation of crude oils through long distance pipelines. Taking into consideration all of the economic aspects, additive treatment, which depresses the pour point and improves the

Table 11–7
Pour Point Depressants

Chemicals	References
Copolymer of acrylic esters with allyl ethers[a]	[1013, 1855]
Urea and derivatives	[970]
Homopolymer of acrylic esters	[1217]
Graft polymers[b]	[135, 582]
Substituted fullerenes[c]	[1578]

a) In amounts of several hundred ppm.
b) Ethylene-vinylacetate copolymer as backbone and graft components: unsaturated dicarboxylic acid imides, dicarboxylic acid amides, dicarboxylic acid half-amides, or ammonium salts of the half-amides.
c) Fullerene–aniline, fullerene–phenol adducts.

flow characteristics of the crude at lower temperatures, was found to be the most suitable method for transporting waxy crude oil [1698]. Typical pour-point depressants are shown in Table 11–7.

Alternatively, the pour point is reduced by modifying the crude oil itself, for example, by cracking [655].

Drag Reducers

Pipeline-flow improvers, or drag-reducing agents, have been utilized in the petroleum industry for many years [41]. The first application of drag reducers in the petroleum industry was to reduce the down-hole pressure loss during the pumping of fluids down-hole to fracture-tight formations. One of the first large-scale pipeline applications was to increase the throughput of crude oil on the Trans-Alaskan pipeline in 1979. Because of the reduction of the apparent viscosity, drag reducers are useful for saving energy required for pumping. Drag reducers are discussed in more detail in Chapter 12. Some drag reducers are shown in Table 11–8.

Effect of Flow Improver Additives on Paraffin Deposition

In waxy crudes, the wax has a tendency to become deposited during storage of the crude oil in tanks or while flowing through pipelines. The deposition in the pipeline severely affects the pipeline throughput. The deposits have to be removed periodically from the storage tanks and the pipelines by pigging operations.

The wax deposition behaviors of Bombay high crude oil have been studied under different conditions using a cold disk-type assembly [769]. It is observed that the deposition occurs much less with additive-treated crude than with

Table 11–8
Drag Reducers

Chemicals	References
Linear low-density polyethylene[a]	[88]
Copolymer of a linear α-olefin with crosslinkers[b]	[692]
Polyacrylamides, polyalkylene oxide polymers and their copolymers[c]	[726, 1087, 1150]
Fluorocarbons[d]	[938]
Polyalkylmethacrylates[e]	[1157, 1158]
	[1499, 1500, 1503]
Terpolymer of styrene, alkyl acrylate, and acrylic acid or methacrylic acid[f]	[1315]

a) Olefin up to 10 mole-%.
b) α-Olefins are 1-hexene, 1-octene, 1-decene, and 1-dodecene; crosslinkers are divinylbenzene or organo-siloxanes with pendent vinyl groups.
c) Water-soluble drag reducers for emulsions.
d) For asphaltenic crude oils.
e) Esters with C_{10} to C_{18} and ionic monomers; reduces friction in the flow of hydrocarbons by a factor of 5 at concentrations of 25 ppm.
f) Styrene also includes tertiary-butylstyrene (drag reducer for hydrocarbon fluids).

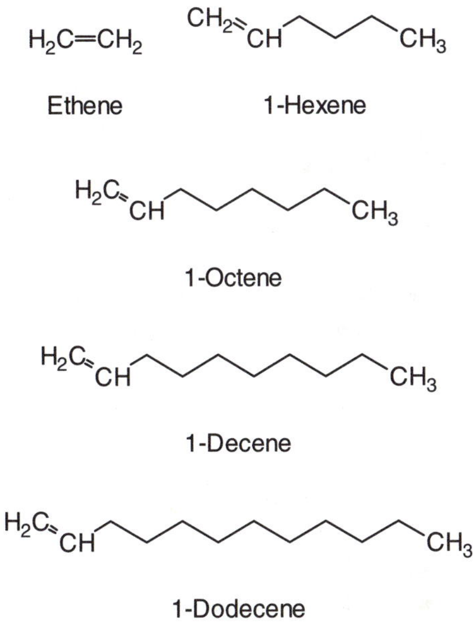

Figure 11–2. Ethene, 1-hexene, 1-octene, 1-decene, 1-dodecene.

Table 11–9
Classes of Additives for Hydrate Control [715]

Class	Remark
Thermodynamic inhibitors	Methanol or glycol modify stability range of hydrates.
Antinucleants	Prevent nucleation of hydrate crystals.
Growth modifiers	Control the growth of hydrate crystals.
Slurry additives	Limit the droplet size available for hydrate formation.
Anti-agglomerates	Dispersants that remove hydrates.

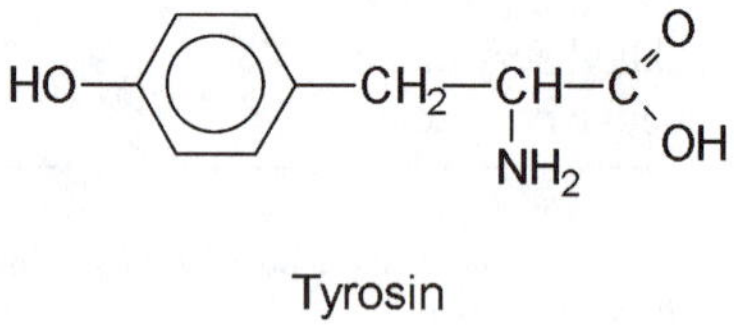

Figure 11–3. Tyrosin.

untreated crude under otherwise identical conditions. However, the deposits obtained from treated crude have slightly higher melting points and less oil content. The wax separation temperature of crude oils can be determined from their viscosity behavior [954].

Drag Reduction in Gas Transmission Lines

Drag reduction in gas transmission lines can be achieved by applying a thin chemical coating on the pipe's inner surface to reduce friction between the flowing gas and pipe wall. An important criterion for gas drag reduction is that the additive can bond strongly onto metal surfaces and form a smooth film to mitigate the turbulence at the gas-solid interface [1102]. Effective gas drag reducers have properties similar to corrosion inhibitors, lubricants, and surfactants. Fatty acid amines or amides exhibit these properties.

Hydrate Control

The usual practice for avoiding the plugging of production facilities by hydrates is to add thermodynamic inhibitors, such as methanol or glycol. A newer concept is the injection of low-dosage additives: either kinetic inhibitors, which delay nucleation or prevent the growth of hydrate crystals, or hydrate dispersants, which prevent the agglomeration of hydrate particles and allow them to be transported within the flow [880, 1387]. Hydrate control is discussed extensively in Chapter 13. Classes of hydrate control agents are shown in Table 11–9, and additives are shown in Table 11–10.

Table 11–10
Additives for Hydrate Control for Pipelines

Additive	References
Methanol, glycol[a]	
Poly(N-vinyl-2-pyrrolidone)[b]	[1638]
Alkyl glycosides[c]	[1493]
L-tyrosine and the methyl ester of L-tyrosine[d]	[524]

a) Thermodynamic inhibitor.
b) Growth inhibitor.
c) C_8–C_{18} alkyl glycoside with glucose, fructose, etc. For example, 1-O-octyl-β-D-glucopyranoside, dodecyl-β-maltoside.
d) Amino acid.

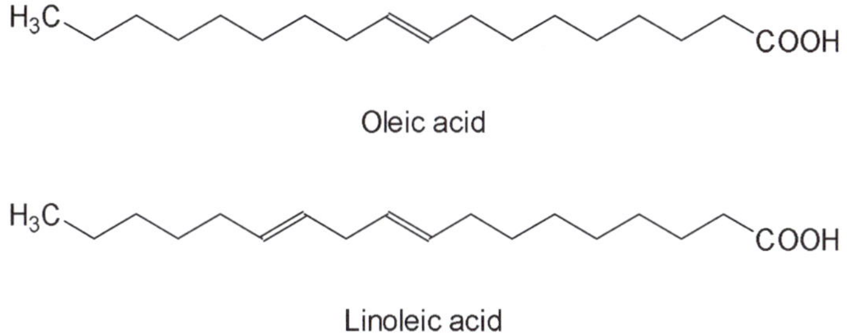

Oleic acid

Linoleic acid

Figure 11–4. Components for additives for slurry transport: oleic acid, linoleic acid.

Additives for Slurry Transport

Examples of additive uses are the slurry pipeline transport of minerals, the removal of the solids produced during the drilling of wells, and the removal of solids formed during the polishing and grinding of metals. Anionic, cationic, or nonionic surfactants may be used to alter the viscosity (e.g., cetyl-trimethyl ammonium chloride, dodecyl diphenyl oxide disulfonate, and soya bis[2-hydroxyethyl]amine) [1531]. Fatty amines are prepared from the corresponding fats by both conversion into the nitrile and hydrogenation. Soybean oil contains a mixture of various long-chain acids; the major acids are oleic acid (9-octadecenoic acid) and linoleic acid (9,12-octadecadienoic acid), that is, C_{18} single and double unsaturated acids, respectively. Fatty amines are highly active surfactants.

Table 11-11
Additives for Odorization [568]

Additive
Ethylmercaptan or mixture of ethyl-, propyl-, and butylmercaptans dimethyl disulfide Diethyl disulfide

$$CH_3-CH_2-SH \quad CH_3-CH_2-CH_2-CH_2-SH$$

Ethylmercaptan   Butylmercaptan

$$CH_3-CH_2-S-CH_2-CH_3$$

Diethylsulfide

Figure 11-5. Additives for odorization: ethylmercaptan, butylmercaptan, diethylsulfide.

Additives for Odorization

Odorization serves as a warning in the detection of natural gas in air before it reaches combustible levels. The most important compounds are given in Table 11-11 and Figure 11-5. The topic is detailed in Chapter 15.

Cleaning

Even carefully built pipelines have been found to contain up to 10 tons of waste materials, mostly iron rust, sand, mud, and welding rods. Even shoes, gloves, and bottles have been removed from pipes. Cleaning can generally be achieved by pumping suitable solvents in pigging operations.

Gelled Pigs

Gelled fluid pigs will perform most of the functions of conventional pigs, but they have additional chemical capabilities. In addition, they can be injected into a pipeline through a valve. However, for displacement by a gas, gel pigs must be propelled by a mechanical pig. Most pipeline gels are water-based, but a variety of chemicals, solvents, and acids can be gelled. Gelled diesel [1459], an organic gel, was first patented for pipeline use in 1973. The gels can be used for waste material removal, separation of products, placement of biocides and inhibitors, and removal of trapped mechanical pigs [951, 1215].

An ablating gelatin pig has been described for use in pipelines. Because of the properties of gelatin, the pig will ablate, thereby depositing a protective layer onto the wall of the pipe [1127]. The pig can be molded outside the pipe or it can be formed in situ. The pig is formed by mixing gelatin with a heated liquid and then allowing the mixture to cool to ambient temperature. Preferably, the liquid contains a corrosion inhibitor or a drag reducer. In some applications, a slug of the treating solution is also passed through the pipeline between two ablating gelatin pigs. For high-temperature applications, a hardener may be added to increase the melting temperature of the pig.

Chapter 12

Drag Reducers

This chapter deals almost exclusively with the state of drag reduction in pipelines for liquid transportation. Drag-reducing additives are important in oil drilling applications and the maintenance of pumping equipment in pipelines. Flow drag in pipes can be reduced by adding a small amount of high-molecular polymer. Pipeline flow improvers, or drag-reducing agents (DRAs), have been utilized in the petroleum industry for many years [41]. The first application of drag reducers was the use of guar in oil well fracturing, presently a routine practice. The quantities of DRA used in this application were very large. One of the first large-scale pipeline applications was to increase the throughput of crude oil on the Trans-Alaskan pipeline in 1979 with oil-soluble polymers. This agent was highly successful in applying a drag-reducing phenomenon. Since then, DRA use has increased in refined products pipelines. Part of the reason for the increased use of DRA is an economic factor, namely, to offset power costs. The DRA cost for moving an additional barrel of product through a pipeline system can be less than $0.05/bbl. This cost level has been made possible by improved performance in commercially available DRAs and a nearly unchanging price structure.

Operating Costs

Pipeline operators of liquid hydrocarbon products can decrease operating costs by using a drag-reducing flow improver to eliminate the need for underutilized intermediate or booster pump stations [725, 1309]. Product lines operating below their capacity or that only use boosters intermittently can also realize cost savings. The overall benefits are likely to be most significant in 6- to 8-inch lines operating between 67% and 92% of their rated throughput capacity. Using computer modeling techniques, engineers have demonstrated potential power savings of up to 22% (from lower demand charges and reduced energy use) for systems using booster stations 85% of the operating time. When stations operate only 70% of the time, total energy cost savings can approach 35%, depending on the diameter of the line and electricity costs.

Mechanism of Drag Reducers

A review on drag-reducing polymers is given in the literature [1359]. It has been suggested that drag reduction occurs by the interactions between elastic macromolecules and turbulent-flow macrostructures. In turbulent pipe flow, the region near the wall, composed of a viscous sublayer and a buffer layer, plays a major role in drag reduction.

The most serious problem in the effectiveness of drag reducers is the chain degradation of polymers by shear strains in turbulent flow. Ultra-high–molecular-weight polymers are more susceptible to shear-induced degradation [667]. Polymers with linear-chain structures are more vulnerable than branched polymers [347] and natural gums with semi-rigid structures [478]. The mechanism of shear degradation is assumed to be associated with chain elongation. The chain degradation is often observed when the shear rate is increased to a critical point, after which drag reduction sharply decreases.

The friction drag and heat–transfer-reduction phenomena associated with turbulent flows of so-called drag-reducing fluids are not well understood [997]. It is believed that elastic fluid properties are strongly related to these phenomena. However, not all drag-reducing fluids are viscoelastic, nor are all viscoelastic fluids drag reducing, suggesting that drag reduction and viscoelasticity are probably incidentally-accompanying phenomena. It is argued that turbulence suppression (i.e., flow laminarization) is a determining factor for the reduction phenomena—not the fluid elasticity—because of the flow-induced anisotropic fluid structure and the associated properties, whereas the fluid elasticity may be a major cause for the laminar heat-transfer augmentation.

Damping of Transmission of Eddies

One of the mechanisms of drag reduction is that transmission of eddies can be damped by the viscoelastic properties of fluids. The transfer process of an isolated eddy in Maxwell fluids with viscoelastic properties was studied, and the expressions describing such phenomena were obtained [1103]. The results of the study showed that eddy transmission was damped significantly with an increase of the viscoelastic properties of the fluids.

Viscoelastic Fluid Thread

In the extensive literature on polymer drag reduction, it has occasionally been reported that a continuous thread of a high-concentration polymer solution injected into the axis of a pipe produces a drag-reduction effect on the water flow in the pipe [856]. The thread seems to persist through the length of the pipe and little, if any, diffusion of polymer to the walls of the pipe is apparent.

A polymer of the polyacrylamide type was injected as a 0.5% solution from an axially-placed nozzle at the bellmouth entrance. The experiments showed that the central thread provided drag reduction almost equivalent to premixed solutions of the same total polymer concentration flowing in the pipe. Overall concentrations of 1, 2, 4, and 20 ppm were used. Moreover, the effects were additive: 2 ppm thread overall concentration plus 2 ppm premixed gave drag reductions equivalent to 4 ppm of either type. Reynolds numbers of up to 300,000 were investigated. In other experiments, a number of different polymer fluids were injected on the centerline of a water pipe-flow facility [857]. Two distinct flow regions were identified:

- Reynolds numbers above 25,000, at which centerline injection acted as a rather efficient mixing device for water-soluble polymer and no drag-reduction, resulted from materials insoluble in water
- Reynolds numbers from 10,000 to 25,000, at which strong evidence exists that under certain conditions, a viscoelastic fluid thread can interact with turbulence eddies and reduce the overall flow friction in the pipe

Polymer Degradation in Turbulent Flow

Drag reduction in turbulent flow is of great potential benefit to many industrial processes, including long-distance transportation of liquids, oil well operations, and transportation of suspensions and slurries, but it is complicated by the problem of polymer degradation. A capillary rheometer was used to investigate the effect of various parameters on polymer degradation in turbulent flow [1270]. These parameters included polymer concentration, contraction ratio, pipe length, pipe diameter, number of passes, solvent weight, and molecular weight of polymer. A commercial organic drag reducer, two grades of polyacrylamide, and a high-molecular weight polyisobutylene were used. In turbulent flow, the polymer degraded more in a poor solvent at low Reynolds numbers, whereas an opposite effect was observed at high Reynolds numbers. The critical Reynolds number (Re_c) or critical apparent shear/extensional rate (V/d) was found to increase with polymer concentration and molecular weight as represented by the dimensionless concentration $c(\eta)$.

Drag Reduction in Two-Phase Flow

The drag-reducing properties of a polyacrylamide type were tested in two-phase air/water flow, using a horizontal pipe of 31-mm diameter [1545]. The properties of the polymer were tested in single-phase water flow, and the results were found to comply with the reduction in pressure drop found by other workers. Positive effects in two-phase flow were found to depend on the Reynolds number of the liquid flow. Thus the drag reduction in stratified flow was small or negative. In slug flow, the drag reduction seems to occur in the liquid slug, not in the layer below the bubble. The flow regime seems unaffected by the polymer. It has been established that in multiphase flow, drag reducers act significantly as corrosion inhibitors because they smooth the flow profile near the walls [935].

Drag Reduction in Gas Flow

For storage or pipeline transportation of natural gas at pressures over 5.5 MPa (800 psi), it is advantageous to add ammonia to the natural gas. The ammonia should not create a liquid phase at the temperature and pressure used. Such an ammonia–natural gas mixture can be compressed or pumped with a lower energy expenditure than would be needed for an equivalent volume of natural gas alone. When more than 4% by volume of ammonia is present, pumping through pipelines is also aided by the refrigerant effect of the ammonia, which reduces the temperature of the gas being transported [1261].

Microfibrils

Friction loss in liquids can be reduced with maximum stability by adding a predetermined amount of selected organo-polymeric microfibrils to a liquid [1615]; these microfibrils are insoluble but highly dispersible in the liquid. Organo-polymeric microfibril designates a solid, organic polymer in the form of microfibrils having an average diameter in the range of 100 to 1000 angstroms (Å), an average length in the range of 1 to 500 μ, and an aspect ratio (length/diameter) of 10 to 1,000,000. Polymeric materials to be processed into microfibrils should be insoluble but highly dispersible in a given liquid.

Drag-Reducing Surfactant Solutions

The behavior of two types of drag-reducing surfactant solutions was studied in turbulent flows in pipes of different diameters [174]. The surfactant systems contained rod-like micelles consisting of equimolar mixtures of n-tetradecyltrimethylammonium bromide, n-hexadecyltrimethylammonium bromide, and sodium salicylate. The structure of the turbulence was studied using a laser-Doppler anemometer in a 50-mm pipe. In the turbulent-flow regimen, both surfactant solutions exhibited characteristic flow regimes. In the turbulent-flow regimen at low Reynolds numbers, velocity profiles similar to those observed for dilute polymer solutions are found, whereas at maximal drag reduction conditions, more S-shaped profiles that show deviations from a logarithmic profile occur.

Soapy Industrial Cleaner

Experiments have been conducted to investigate the effect of a soapy industrial cleaner on reducing the skin friction of a Jordanian crude oil flowing turbulently in pilot-scale pipes of different sizes. Experiments showed that a concentration of only 2 ppm of the chemical additive injected into the crude oil line caused an appreciable amount of drag reduction [1165]. The effects of additive concentration and pipe diameter on drag reduction have been investigated.

Lyophobic Performance of the Lining Material

An experimental study was conducted on the characteristics of frictional drag for a lyophobic surface, with the surface consisting of polytetrafluoroethylene and the working media being water and machine oil [1545]. The test results indicate that, depending on the lyophobic performance of the lining material, the pipes lined with polytetrafluoroethylene have a better drag-reducing effect than conventional steel pipes. A drag reduction of approximately 12% or 6% is achieved with the working medium being either water or machine oil, respectively. In other words, polytetrafluoroethylene has a higher lyophobic performance against water than against machine oil. The theoretical analysis of the flow mechanism on the lyophobic surface shows that surface lyophobic treatment, which can lower the surface energy level to such a degree that the attraction of the solid wall on liquid molecules becomes weaker than the liquid molecular absorption, causes a gliding flow adjacent to the pipe wall, thus reducing the drag.

Interpolymer Complexes

It has been shown that hydrogen bonding–mediated interpolymer complexes can be powerful drag reducers. The drag-reduction levels in such polymer systems increase dramatically by a factor of 2 to 6 when compared with their nonassociating polymeric precursors. Their shear stability is also shown to be significantly enhanced [1156].

Hydrocarbon-soluble polymers containing small percentages of polar associating groups are used to determine the effects of polymer associations on solution drag reduction. Experimental data suggest that intrapolymer associations generally decrease the dilute solution drag-reduction activities of single associating polymers with like polar groups [1005]. Interpolymer complexes formed by one polymer with anionic groups and one polymer with cationic groups can overcome this limitation and provide an enhanced dilute solution drag-reduction activity as a result of favorable interpolymer associations, which build larger structures of higher apparent molecular weight. The latter associations may also increase the polymers' resistance to degradation in turbulent flows.

Drag Reducers in Detail

Ultra-High Molecular Weight Polyethylene

The flow of liquid hydrocarbons can be enhanced by introducing into the stream a nonagglomerating suspension of ultra-high molecular weight polyethylene [490, 1647] in water with small amounts of surfactant. The finely divided polyethylene is prepared by polymerization and then cryogrinded below the glass-transition temperature.

Copolymers of α-Olefins

Several copolymers of α-olefins are used as drag reducers. Suggested recipes are summarized in Table 12–1. Linear low-density polyethylene is a copolymer of ethylene and α-olefins. It is obtained by copolymerization utilizing Ziegler–Natta catalysts or metallocene catalysts. Concentrates may be prepared by

Table 12–1
Copolymers from α-Olefins and Others Used as Drag Reducers

Copolymer	References
Divinylbenzene /1-hexene, 1-octene, 1-decene, and 1-dodecene[a]	[88, 250, 538, 692]
Styrene/N-vinylpyridine	[1004]
Ethene/α-olefins[b]	[852]
Homo- or co-polymers that α-olefins[c]	[1533]
Polyisobutene[d]	[1169]
(Meth) acrylic acid esters	[1500–1505]
C_{12} to C_{18} acrylate or methacrylate/ionic monomer[e]	[1157, 1158]
tert-Butylstyrene/alkyl acrylate, acrylic acid or methacrylic acid	[1315]
Acrylamide-acrylate	[1579, 1580]
Ultrahigh-molecular-weight polyolefin	[1142]
Styrene/methyl styrene sulfonate/N-vinylpyridine (NVP)[f]	[1417]

a) Synthesis by a Ziegler-Natta process.
b) Up to C_{30}, Ziegler-Natta.
c) Molecular weight of up to 15,000 and an isotacticity of 75% or greater.
d) Oil-soluble polymer.
e) Reduce friction in the flow of a hydrocarbon fluid by a factor of 5 at concentrations as low as 1 to 25 ppm.
f) Polyampholytes.

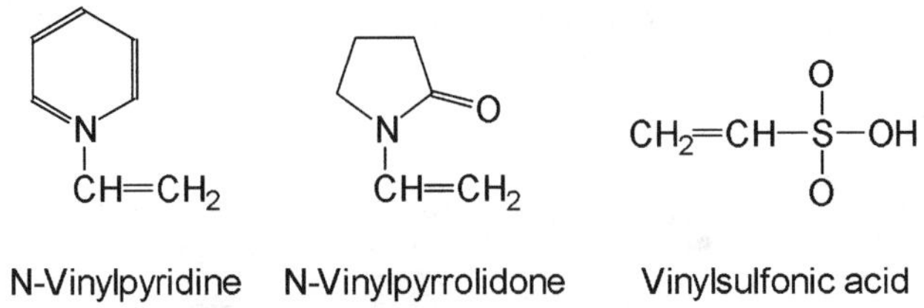

Figure 12–1. N-Vinylpyridine, N-vinylpyrrolidone, vinylsulfonic acid.

precipitating the polymer from a kerosene solution with isopropanol [565]. The resulting slurry concentrate dissolves rapidly in flowing hydrocarbon streams.

By coating poly-α-olefins with a fatty acid wax as a partitioning agent and dispersing it in a long-chain alcohol, a nonagglomerating, nonaqueous suspension can be obtained [918].

Polyether Compounds for Oil-Based Well Drilling Fluids

A liquid oil, an emulsifier, and a friction modifier, which includes certain polyether compounds, can be added to a drilling fluid consisting of a water-in-oil emulsion formed from a brine [1155]. The friction modifier serves to decrease the coefficient of friction of the well drilling fluid. Decreasing the coefficient of friction lowers the force required to turn the drill bit in the hole. Gravitational forces increase the coefficient of friction in deviated-, horizontal-, and extended-reach wells.

Tylose

Tylose is not as effective in drag reduction as other substances described in literature. Detailed mean velocity, normal Reynolds stress, and pressure drop measurements were performed with 0.4% to 0.6% by weight aqueous solutions of tylose, a methylhydroxylcellulose (molecular weight 6000 Dalton), after a selection process from a set of low–molecular-weight fluids [174]. The viscosity measurements of the tylose solutions showed shear-thinning behavior, and the oscillatory and creep tests measured elastic components of the stress in the order of the minimal detectable values by the rheometer. These low–molecular-weight polymer solutions delayed transition from the laminar to the turbulent regime and showed drag reductions of half that reportedly occur with other low-elasticity, shear-thinning, high-molecular, aqueous polymer solutions.

Microencapsulated Polymers

Highly concentrated drag-reducing agents may be prepared by microencapsulating a polymer or a monomer. The microencapsulation may be performed before, during, or after the polymerization of a monomer into an effective drag-reducing polymer. If the encapsulation is done before or during polymerization, a catalyst may be present, but little or no solvent is required. The result is very small-scale bulk polymerization within the microcapsule. The inert capsule or shell may be removed before, during, or after introduction of the microencapsulated drag reducer into a flowing liquid. No injection probes or other special equipment should be required to introduce the drag-reducing slurry into the liquid stream, nor is grinding (cryogenic or otherwise) of the polymer necessary to form a suitable drag-reducing agent [989, 990].

Chapter 13

Gas Hydrate Control

The Relevance of Gas Hydrates

The main problem of hydrate formation will arise in pipelines transporting natural gas, because gas hydrates are solids and will leave deposits. The solid deposits reduce the effective diameter of the pipeline and can therefore restrict or even clog the flow properties. Furthermore, the formation of condensates, hydrates, or ice may occur in the course of decompression of natural gas stored in natural reservoirs (e.g., in salt caverns). The operation of oil and gas pipelines in the deep sea is significantly complicated by the formation of gas hydrates [1204]. Experience indicates that large gas hydrate plugs in gas and oil pipelines form most actively during the period of an unforeseen long shut-down. In static conditions, three types of hydrate crystals can be formed [1153]:

1. Surface-contact films and massive hydrates, which form by sorption of gas and water molecules on the surfaces of growing crystals
2. Bulk diffusional whiskerlike hydrate, which forms both in the volume of gas and in the bulk of liquid water through sorption of molecules on the growing crystal surface and by tunnel sorption of molecules at the base of the crystal
3. Gel-like soft crystals, which form in the bulk of liquid water at a deficiency of dissolved gas in water

Under the appropriate pressure and temperature conditions for hydrate formation, there may be a simultaneous formation of some crystals and a decomposition of other crystals.

Inclusion Compounds, Clathrates

Clathrates are crystalline-addition compounds of at least two species of molecules. They are bound mainly by van der Waals forces. One compound, the host, makes up the structure. The other partner, the guest, is placed in free

Channels   Layers

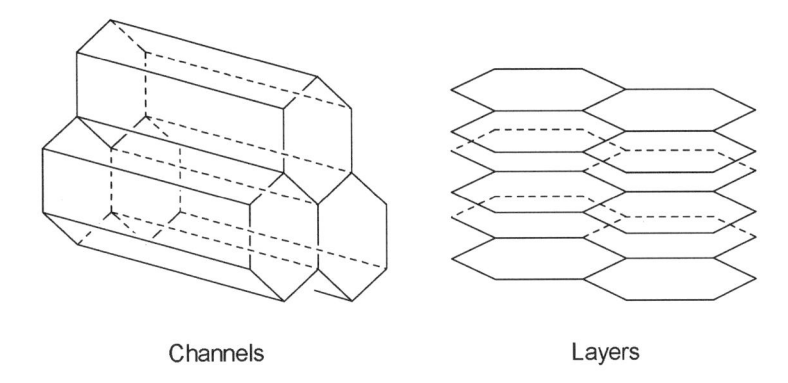

Figure 13-1. Structure of clathrates.

spaces of the crystal lattice. If the free space is closed in every direction then these compounds are considered clathrates in the closer sense. The structure of clathrates is illustrated in Figure 13–1. Other types of inclusion compounds are channel-inclusion compounds and layer-inclusion compounds. There are monographs on this topic [1152, 1830].

The size of the free space varies slightly as a result of the size and the shape of the molecule to be included. This fact is used in the separation of molecules. A relevant example in petroleum refinement is the separation of paraffins from other compounds with urea. In this case, a channel-like lattice is formed by urea. In the free space linear alkanes (n-octane) find space, whereas branched alkanes (i-octane) cannot be included.

An example for a host molecule with a layerlike structure is graphite. Various types of both organic and inorganic inclusion compounds, as well as stoichiometric and nonstoichiometric compounds, are known.

Gas Hydrates

Gas hydrates are a special form of clathrates. Here water is the host molecule. The first gas hydrate (with chlorine) was described in 1818 by Sir Humphrey Davy. Naturally-occurring gas hydrates in Siberia are methane hydrates.

Inspection of Table 13–1 shows that the number of water molecules required to form the hydrate increases with the size of the guest molecule to be captured. Hydrates are classified into simple and mixed hydrates depending on whether one or more host or guest molecules compose the compound. They also exhibit different structures.

Heavy Hydrates, Structure H–Hydrates

Higher molecular compounds, such as benzene, cyclopentane, methylcyclopentane, cyclohexane, methylcyclohexane, isopentane, and 2,3-dimethylbutane

Table 13-1
Composition of Gas Hydrates

Hydrate	Chemical composition
Methane	$CH_4 \times 6H_2O$
Ethane	$C_2H_6 \times 8H_2O$
Propane	$C_3H_8 \times 17H_2O$
Isobutane	$C_4H_{10} \times 17H_2O$
Nitrogen	$N_2 \times 6H_2O$
Carbon dioxide	$CO_2 \times 6H_2O$
Hydrogen sulfide	$H_2S \times 6H_2O$

Adamantane   Hexamethylene tetramine   Urea

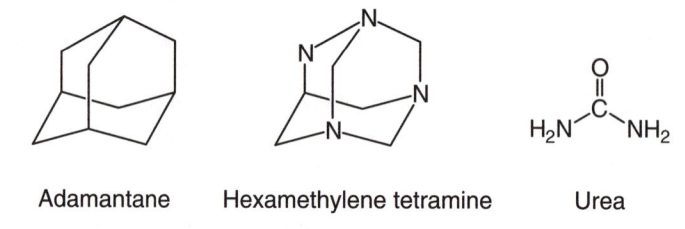

Figure 13-2. Adamantane, hexamethylene tetramine, and urea.

may also form gas hydrates. Recent studies [1752] show that in certain circumstances, the effect of the heavy hydrate formers cannot be ignored and, in some cases, they significantly reduce the hydrate-free zone.

In 1987 a new type of hydrates was discovered. Structure H formation requires the presence of two different types of guest molecules: a smaller component, such as methane or hydrogen sulfide, and a large molecule, such as i-pentane or heavier hydrocarbons, which may occur naturally in petroleum. No aromatic molecules have been found to be possible guests for this hydrate structure [1731]. Structure H–hydrates are somewhat more stable than methane hydrates at higher temperatures. A variety of systems have been investigated with respect to stability-covering systems of methane-2-methylbutane-water, methane-2,2-dimethylbutane-water, methane-tert-butyl methyl ether-water, and methane-adamantane-water [877, 1151]. A substantial inhibiting effect is observed in the presence of the NaCl.

Thermodynamic data suggest the possibility of forming structure H–hydrates in gas and oil reservoirs and industrial installations. Hydrates formed with methane and adamantane have been found in situ in Mobile Bay in the Gulf of Mexico [1202].

Therefore structure H–hydrates are now recognized as a potential problem in the petroleum industry. The stable occurrence of structure H-hydrates calls into question existing hydrate-prediction programs and suggests that the hydrate phase itself should be measured, in contrast to previous experimental practices [1203].

Conditions for Formation

Water Content

The water content of a natural gas is a crucial parameter for the formation of gas hydrates, at least in transporting systems. Clearly a water-free gas may not be able to form gas hydrates. Natural gas contains water in the range of 8000 to 10,000 ppm (by volume). Specifications for pipelining restrict the water content to 120 to 160 ppm. The reasons for this limitation are not only the risk of hydrate formation, but also potential corrosion problems. Therefore drying of a wet gas is essential.

The water content is measured most conveniently via the dew point, but electrical and optical methods are also available. The dew point is connected most directly to hydrate formation, because it is believed that condensate formation is essential before the formation of hydrates.

Stability Diagram

Hydrates are stable below a certain temperature and above a certain pressure. A p–T stability diagram is shown in Figure 13–3.

For gases with contents of compounds with higher molecular weight, the curve is shifted significantly to lower pressures. For gas mixtures without hy-

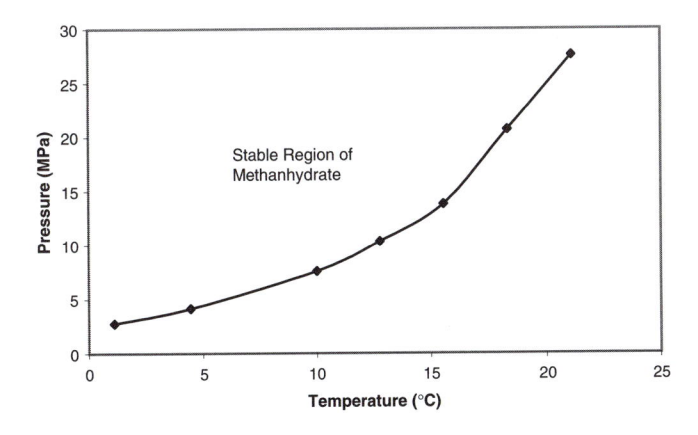

Figure 13–3. p–T Stability diagram of methane hydrate.

drogen sulfide as a first approximation, the gas density may serve as parameter to estimate the conditions of stability of hydrates. Similar p–T diagrams, such as that from methane, are given with the relative density of the gas as parameter in the literature.

Analytical Equations

In some regions, the equilibrium pressure of coexistence of gas hydrate with the corresponding gaseous state follows a Clausius–Clapeyron relation:

$$\ln p = A + \frac{B}{T} \tag{13-1}$$

At elevated temperatures the relationship changes into

$$\ln p = A + BT + CT^2 \tag{13-2}$$

The parameters A, B, and C are dependent on the particular nature of the gas. Katz developed a simple method for gas mixtures that takes the composition of the gas into account [942, 1086]. Furthermore, a graphic method is available that permits the estimation of the hydrate-forming temperatures at pressures for natural gas containing up to 50% hydrogen sulfide [129].

Decomposition

Clathrates and, in particular, gas hydrates can be decomposed very easily by dissolving or melting the crystal lattice of the host molecule.

Formation and Properties of Gas Hydrates

Knowledge concerning the mechanism of hydrates formation is important in designing inhibitor systems for hydrates. The process of formation is believed to occur in two steps. The first step is a nucleation step and the second step is a growth reaction of the nucleus. Experimental results of nucleation are difficult to reproduce. Therefore, it is assumed that stochastic models would be useful in the mechanism of formation. Hydrate nucleation is an intrinsically stochastic process that involves the formation and growth of gas–water clusters to critical-sized, stable hydrate nuclei. The hydrate growth process involves the growth of stable hydrate nuclei as solid hydrates [129].

Two-Step Mechanism of Formation

Surfactants like sodium dodecyl sulfate reduce the surface tension at the liquid–gas interface considerably. In fact, the addition of surfactants in small quantities has a substantial effect on the kinetic characteristics of hydrate formation without changing the equilibrium parameters. Concentrations of surfactant that increase the induction period of hydrate formation of propane hydrates also increase the rate of subsequent growth of the hydrate phase [1039]. The effect of stirring on the kinetics of formation of hydrate formation does not seem to be pronounced [383]. This is in contrast to other results in which it was found that the induction time is strongly dependent on the stirring rate and the driving force [1636]. Experiments conducted at the same stirring rate and with a high driving force seem to show that the induction time varies exponentially with the size of the driving force. Surface effects are reported on the nucleation. For example, diatomaceous earth and synthetic amorphous silica were found to nucleate hydrate formation [1639].

Nucleation Particle Sizes

Nucleation and growth of gas hydrate crystals have been investigated with optical methods under different pressures and temperatures. The particle sizes measured during gas hydrate nucleation ranged from 2 to 80 nm [1334, 1335]. The nucleation process is nondeterministic, because of a probabilistic element within the nucleation mechanism [1393].

Clustering Before Nucleation

A controversy exists regarding the early stages of formation of gas hydrates. The mechanism proposed by Sloan and Fleyfel [384, 613, 1637] for the kinetics of hydrate formation is composed of

1. The formation of clusters of hydrogen-bonded water molecules around different sizes of apolar molecules
2. The joining of these clusters to create a hydrate nucleus

The hypothesis was extended to nucleation of hydrates from liquid water. An alternative hypothesis was proposed by Rodger [1516]. The main difference between these two sets of theories is that Rodger's hypothesis relates the initial formation process to the surface of the water, whereas the theory of Sloan and coworkers considers clusters related to soluted hydrate formers in liquid water as the primary start for joining, agglomeration, and crystal growth. The theories of Sloan and coworkers have been discussed and related to elements of the hypothesis proposed by Rodger [1043].

Experimental Methods and Models

One of the major goals is to find additives that can inhibit crystallization. This goal is not restricted to gas hydrates, but applies to all scales that could be formed. Inhibitors of crystallization may be generally effective for a variety of mineral scales or, in some cases, may be selected for a narrower range of scales. Such generality and specificity of action may be understood in terms of stereospecific and nonspecific mechanisms of scale inhibition. New techniques for comparing the effectiveness and activity of new hydrate inhibitors and laboratory results on various chemicals have been described.

Modern experimental methods are used to obtain information about interactions between potential crystallization inhibitors and the substrates themselves. By combining atomic force microscopy, scanning electron microscopy, and optical microscopy, both stereospecific and nonspecific interactions of inhibitors with various crystal species (for example, calcite, calcium oxalate monohydrate, and ice) have been examined. The crystals were chosen as representatives of strong ionic, hydrated ionic, and hydrogen-bonded lattices. Both stereospecific and nonspecific interactions were observed in each case [1627]. The strongest interactions of the adsorbate with the crystal surfaces were elucidated at the angstrom level with good agreement between experiment and theory. Such definition of the weaker interactions requires more work, and in fact may be beyond the reach of current methodology. However, reasonable models of each of the interactions have been proposed in that research.

Modeling the Formation of Gas Hydrates

Several program packages are available to predict conditions with respect to pressure, temperature, and some special inhibitors. The models are based on thermodynamic considerations rather than on kinetic arguments. Therefore kinetic inhibitors cannot be treated with these programs. On the basis of purely thermodynamic correlations, an algorithm has been developed to predict the formation of hydrates in systems containing oil or gas in equilibrium with water [120]. For a specified temperature and feed composition, the program computes the equilibrium pressure at the hydrate point. The authors report that the predictions are satisfactory for gas–water hydrate forming mixtures, but not for an oil–water system. A practical model for the effect of alcohol and salinity on gas hydrate formation has been implemented as a stand-alone computer program that either accesses the model via a spreadsheet or uses the model as an object code. A critical comparison of the various available packages has been given in literature [450]. Four methods of hydrate prediction were evaluated, namely, GPA/CSM, EQUI–PHASE, GPA/AQUA*SIM, and API/HYDRATE.

Inhibition of Gas Hydrate Formation

Both thermodynamic and kinetic factors affect the inhibition of hydrate deposits.

Drying

Hydrate formation can be prevented by drying a gas to such an extent that no condensate can be formed. This method is the preferable one, but inhibition of hydrate formation from the liquid phase can be achieved.

Lowering the Hydrate-Formation Temperature with Additives

The hydrate-formation temperature can be reduced by the addition of antifreeze agents such as methanol, glycols [1430], or brines. The depression of the freezing point is given by

$$\Delta T = K \frac{I}{100 - I} \tag{13-3}$$

where K is a specific parameter dependent on the nature of the additive and I is the amount of antifreeze in percent by weight with respect to water. Note that in the case of volatile additives, the additive added will be only partially present in the water phase, and to some extent in the gas phase.

Brines have inherent corrosive properties and therefore are not suitable. Ethylene glycol is preferred because of its low cost and low solubility in hydrocarbons.

Kinetic Inhibition of Hydrates

As mentioned previously, the classic additive to prevent hydrate formation is alcohol. Traditional hydrate inhibitors such as methanol and glycols have been in use for many years, but demand for cheaper methods of inhibition is great. Therefore the development of alternative, cost-effective, and environmentally acceptable hydrate inhibitors is a technologic challenge for the oil and gas production industry [947].

Crystallization Inhibitors

Certain alkylated ammonium, phosphonium, or sulfonium compounds are effective, in relatively low concentrations, in interfering with the growth of gas hydrate crystals [972] and therefore are useful in inhibiting plugging by gas hydrates in conduits containing low-boiling hydrocarbons and water. For example, tetrabutylammonium bromide will be active. Gas hydrate or ice formation is further inhibited in lines by adding amino acids or amino alcohols [523].

The tendency of hydrates to agglomerate can be reduced by adding a condensation product from polyalkenylsuccinic acid and a polyethyleneglycolmonoether [530]. The product is nonionic and has amphiphilic properties. In general, a concentration between 0.1% and 5% by weight, based on the water present, is sufficient to prevent agglomeration. The following copolymers, as well as several other polymers (e.g., copolymers from acrylamides, acrylates, methacrylamides, methacrylates, N-vinyl heterocyclics, vinyl ethers, and N-vinyl amides), are useful for inhibiting the formation of clathrate hydrates in fluids [54, 55, 396, 401, 973, 974, 1713]:

- Acrylamide-maleimide
- N-Vinyl amide-maleimide
- Vinyl lactam-maleimide
- Alkenyl cyclic imino ether-maleimide
- Acryloylamide-maleimide

Modified amino acids such as N-acyl-dehydroalanine polymers and copolymers with N-vinyl-N-methyl acetamide seem to be particularly effective [396]. The crystallization kinetics in the presence of polyvinylpyrrolidone and tyrosine have been tested by time-resolved experiments [981]. An influence is evident on the particle size distribution of the hydrate [1433].

Anti-Agglomerants

Dispersing hydrates into a condensate phase by anti-agglomerants is an alternative to kinetic or thermodynamic inhibitors to prevent hydrate-plug formation in a gas production pipeline [649, 880]. This method has been taken into consideration for pipeline transport. In laboratory experiments it was shown that several commercial dispersants were successful. At low-water concentrations, dodecyl-2-(2-caprolactamyl) ethanamide was shown to be superior [880]. Anti-agglomerants have been used in several field applications in deepwater systems—both in subsea wells and dry tree wells—under both flowing and shut-in conditions [649]. Potential advantages of the anti-agglomerants over methanol include smaller umbilicals, smaller pumps, smaller storage facilities, and less frequent supply trips.

Hydrate Inhibitors for Drilling Fluids

Low-density gas hydrate–suppressive drilling fluids have been developed for deep-water applications. These fluids are glycol based [764, 766].

Chapter 14

Antifreeze Agents

An antifreeze is defined as an additive that, when added to a water-based fluid, will reduce the freezing point of the mixture [1671]. Antifreezes are used in mechanical equipment in environments below the freezing point to prevent the freezing of heat-transfer fluids. Another field of application is in cementing jobs to allow operation below the freezing point.

Hydrate control is not included in this chapter, but is discussed in Chapter 13 because of the relative importance and difference in chemical mechanism. Many chemicals added to water will result in a depression of the freezing point. The practical application is restricted, however, because of some other unwanted effects, such as corrosion, destruction of rubber sealings in engine parts, or economic aspects.

Theory of Action-Colligative Laws

Freezing point depression follows the colligative laws of thermodynamics at low concentrations added to water. At the same time the boiling point generally will be increased. The freezing point depression can be readily explained from the theory of phase equilibria in thermodynamics.

In equilibrium the chemical potential must be equal in coexisting phases. The assumption is that the solid phase must consist of one component, water, whereas the liquid phase will be a mixture of water and salt. So the chemical potential for water in the solid phase μ_s is the chemical potential of the pure substance. However, in the liquid phase the water is diluted with the salt. Therefore the chemical potential of the water in liquid state must be corrected. x refers to the mole fraction of the solute, that is, salt or an organic substance. The equation is valid for small amounts of salt or additives in general:

$$\mu_s = \mu_l + RT \ln(1 - x) \qquad (14\text{--}1)$$

The equation is best expressed in the following form:

$$\frac{\mu_s - \mu_l}{RT} = \ln(1 - x) \cong -x \qquad (14\text{--}2)$$

184   Oil Field Chemicals

The derivative with respect to temperature will give the dependence of equilibrium concentration on temperature itself:

$$\frac{d\frac{\mu_s - \mu_l}{RT}}{dT} = -\frac{dx}{dT} = \frac{\Delta H}{RT^2} \qquad (14\text{--}3)$$

ΔH is the heat of melting of water. Because the heat of melting is always positive, an increase of solute will result in a depression of the freezing point. For small freezing point depressions, the temperature on the right-hand side of the equation is treated as a constant. Furthermore, it is seen that additives with small molecular weight will be more effective in depressing the freezing point. Once more it should be noted that the preceding equation is valid only for small amounts of additive. Higher amounts of additive require modifications of the equation; in particular, the concept of activity coefficient has to be introduced. The phase diagram over a broader range of concentration can be explained by this concept.

Overview of Antifreeze Chemicals

Some data concerning the activity of antifreeze chemicals are presented in Table 14–1. Inspection of Table 14–1 shows that there are two different types of antifreeze chemicals, that is, liquids that are miscible over the full range of concentration with water and salts, often salts which are soluble only to a certain amount. In the case of liquids, a mixture of 50% by weight with water is given. In the case of solids, the ethylene glycol forms with water an

Table 14–1
Anti-Freeze Chemicals

Component	Concentration in water (% by wt)	Depression of freezing point (° C)
Calcium chloride	32	−50.0
Ethanol	50	−38.0
Ethylene glycol	50	−36.0
Glycerol	50	−22.0
Methanol	50	−50.0
Potassium chloride	13	−6.5
Propylene glycol	50	−32.0
Seawater	100 (6% salt)	−3.0
Sodium chloride	23	−21.0
Sucrose	42	−5.0
Urea	44	−18.0

The depression of freezing point is dependent on concentration. A mixture of glycol in water will lower the freezing point with an increasing amount of ethylene glycol, as given in Table 14–2.

Table 14-2
Depression of the Freezing Point in a Mixture of Ethylene glycol–Water

Amount of ethyleneglycol (% by wt)	Depression of freezing point (° C)
10	−4
20	−9
30	−15
40	−24
50	−36

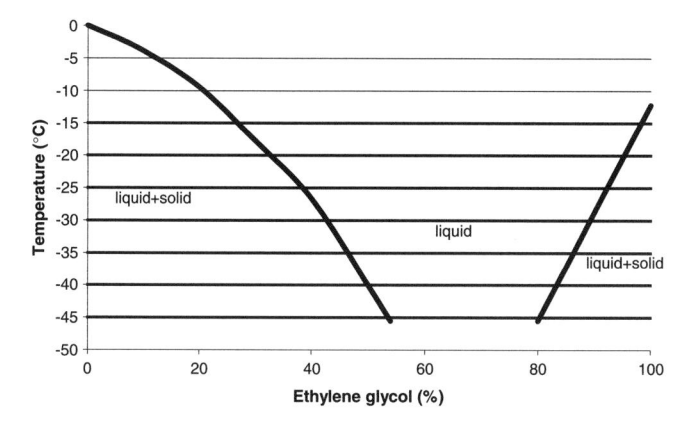

Figure 14-1. Phase diagram of the binary system for ethylene glycol–water.

eutectic point between 65% and 80% at around −70° C. Pure ethylene glycol will solidify at −14° C, however. Mixtures of propylene glycol with water can supercool at higher concentrations of propylene glycol. The equilibrium freezing points cannot be measured.

The depression of the freezing point in a mixture of ethylene glycol and water is shown in Table 14–2. The phase diagram of the binary system ethylene glycol–water is plotted in Figure 14–1. Some organic anti-freeze agents are depicted in Figure 14–2.

Heat-Transfer Liquids

The classic antifreeze agents in heat-transfer liquids are brine solutions and alcohols.

Brines

Of the commonly used antifreeze agents, brines are the most corrosive to metals of the engines and exhibit scale deposition characteristics that are

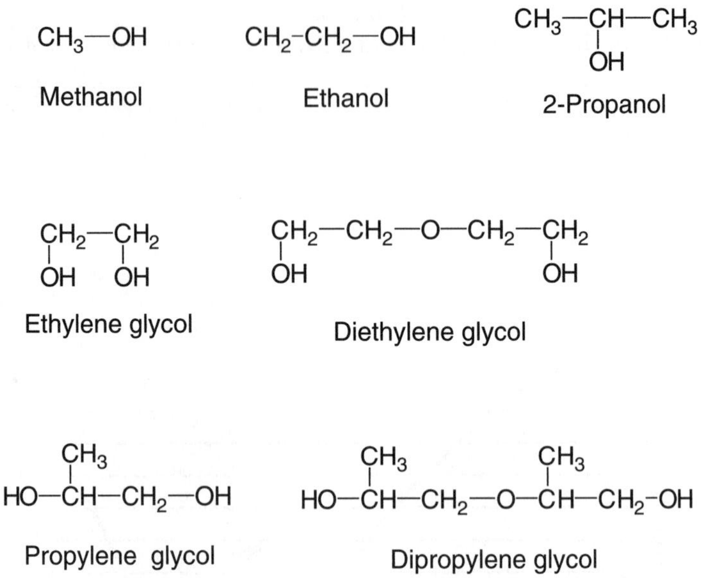

186 Oil Field Chemicals

$$CH_3-OH \qquad CH_2\text{-}CH_2-OH \qquad \underset{\underset{OH}{|}}{CH_3-CH-CH_3}$$

Methanol Ethanol 2-Propanol

$$\underset{\underset{OH}{|}\quad\underset{OH}{|}}{CH_2-CH_2} \qquad \underset{\underset{OH}{|}\qquad\qquad\underset{OH}{|}}{CH_2-CH_2-O-CH_2-CH_2}$$

Ethylene glycol Diethylene glycol

$$\underset{}{\overset{CH_3}{\underset{|}{HO-CH-CH_2-OH}}} \qquad \overset{CH_3\qquad\qquad CH_3}{\underset{|\qquad\qquad\qquad|}{HO-CH-CH_2-O-CH-CH_2-OH}}$$

Propylene glycol Dipropylene glycol

Figure 14–2. Some organic anti-freeze agents: methanol, ethanol, 2-propanol, ethylene glycol, diethylene glycol, propylene glycol, dipropylene glycol.

highly restrictive to heat transfer. Today brines (seawater) still find applications in offshore uses because they are cheap.

Mono-alcohols

Alcohols such as methanol and ethanol are readily available and are occasionally used despite significant disadvantages, such as low boiling points. During summer months significant amounts of alcohol can be lost due to evaporation. Such losses lead to costly replacement of the additive. Furthermore, alcohols have low flash points, which may cause safety problems. Moreover, methanol is highly poisonous. Therefore, the use of alcohols has ceased almost completely in recent years.

Glycols

Ethylene glycol is not as active in depression of the freezing point as methanol, but it has a very low vapor pressure; evaporation loss in a coolant system is due more to the evaporation of water than to the evaporazation of ethylene glycol. Furthermore, the flammability problem is literally eliminated. 1:1 mixtures of ethylene glycol and water do not exhibit a flash point at all.

Ethylene glycol-based antifreeze formulations may contain small amounts of other glycols such as diethylene glycol or triethylene glycol. Propylene-based glycols such as propylene glycol and propylene glycol ethers have limited use, especially in areas in which regulations about human toxicity apply. Ethylene glycol proves most effective in depression of the freezing point and heat transfer activities.

Properties of Glycol-Based Antifreeze Formulations

Pour Point

The desired concentration of an antifreeze agent will be governed by several features. The freezing point of a mixture is the point at which the first ice crystal can be observed. This does not mean, however, that this temperature would be the lowest allowable temperature in the application. In the case of heat transfer agents, the fluid will not function efficiently, but because the fluid will not freeze completely to a solid state it may still be operational. Pure water will expand by complete freezing at about 9%. The addition of antifreeze, such as ethylene glycol, will significantly lower the amount of expansion, thus protecting the system from damage. At the freezing temperature the crystals are mainly water themselves; therefore, the concentration of the antifreeze agent still in the solution will be increased. This causes a further depression of the freezing point of the residual liquid. At higher glycol concentrations the fluid never solidifies completely. The fluid becomes thick and taffy-like. The point at which the fluid ceases to flow is referred to as the pour point. The pour point is significantly lower than the freezing point. However, the use of such a system down to the pour point will significantly increase the energy required for pumping. Furthermore, because of the decreased ability for heat transfer, it is generally not recommended to regularly use systems beyond the freezing point of the mixture.

Corrosion

Alcohols may be corrosive to some aluminum alloys. In an aqueous mixture corrosion may still occur because of dissolved ions from residual salts. At high temperatures and in the presence of residual oxygen from the air, glycols are oxidized slowly to the corresponding acids. These acids can corrode metals.

Inhibition of acid corrosion can be achieved by adding buffer systems that essentially keep the pH constant and neutralize the acids. For example, a formulation of 100 kg ethylene glycol with 400 g KH_2PO_4, 475 g Na_2HPO_4, and 4 liters water is used as an antifreezing agent, which can be diluted accordingly with water (approximately 50:50). This formulation will be highly anticorrosive. Also, borax can be used to protect metal surfaces from corrosion.

188 Oil Field Chemicals

HS—CH₂-CH₂-SH

Dithioglycol

Triazol

H₂N—CH₂-CH₂—OH

Aminoethanol

Urotropin

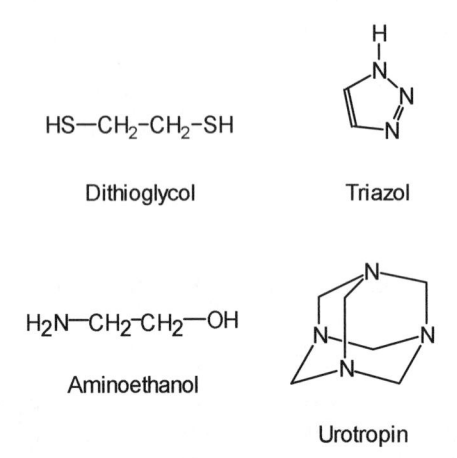

Figure 14–3. Corrosion inhibitors: dithioglycol, triazol, aminoethanol, urotropin.

Besides pure chemical corrosion, solid products of corrosion in the system will give rise to erosive corrosion, in which the particles moving with the fluid will impact onto the surfaces and can remove protective surface layers. Such corrosion effects are most pronounced in regions of high fluid-stream velocity.

The most common corrosion inhibitors, which may form protective films on the metal surfaces, are borates, molybdates, nitrates, nitrites, phosphates, silicates, amines, triazoles, and thiazioles (e.g., monoethanolamine, urotropin, thiodiglycol, and mercaptobenzothiazole). The addition of such inhibitors does not effectively protect against corrosion [137]. Some corrosion inhibitors are shown in Figure 14–3.

Dibasic salts of dicyclopentadiene dicarboxylic acid are claimed to be active as corrosion inhibitors [444]. Certain salts of fatty acids (metal soaps), together with benzotriazole, are claimed to give synergistic effects for corrosion in antifreeze-agent formulations [446].

The choice of a corrosion inhibitor as an additive in antifreezing agents is also dependent on the mode of operation. For instance, cars are operated intermittently. Here the corrosion inhibitors must also protect the system when it is idle. Film-forming silicates can protect the system while idle. This is especially true of aluminum parts, which are introduced in cars for the sake of weight reduction. But silicones can react with ethylene glycol to form crosslinked polymers. These gels may clog lines.

The engines in the oil industry are usually heavy stationary diesels that run continuously. Also, aluminum is normally not used in this type of engine. For these types of engines, corrosion inhibitors for glycol systems based on

Table 14-3
Compatibility of Ethylene Glycol with Some Elastomers

Material	25° C	80° C	160° C
Polyurethane	good	poor	poor
Acrylonitrile-butadiene-copolymer	good	good	
Styrene-butadiene-copolymer	good	fair	poor
Ethylene-propylene-diene-copolymer	good	good	good
Natural rubber	good	poor	poor
Silicone rubber polydimethylsiloxane	good	good	
Vinylidene fluoride-hexafluoropropene rubber	good	good	poor

silicate-forming films are not recommended, because of gel formation. Appropriate blends of corrosion inhibitors added to the glycol–water mixture to minimize corrosion problems in applications for coolants have been developed [843].

Therefore, coolant formulations for engines involved in applications such as natural gas transmissions consist of phosphate for ferrous metal protection and a triazole for the protection of brass parts. Corrosion is discussed in detail in Chapter 6.

Foam Inhibitors

Although glycol–water formulations are not prone to foaming, mechanical and chemical factors may cause foaming in the system. The use of corrosion inhibitors and the presence of contaminants may enhance the tendency to form foams. For these reasons, antifoaming agents, such as silicones, polyglycols, or oils, are sometimes added.

Damage of Elastomers

Some elastomer sealings that are in contact with the antifreeze mixture may not be stable in such a medium because of consequences such as swelling. The compatibility of ethylene-glycol with certain plastics is shown in Table 14-3.

Toxicity and Environmental Aspects

Human toxicity and aquatic toxicity have been measured typically with fresh formulations only. Spent fluids may contain various contaminants and degradation products that may change the toxic effects.

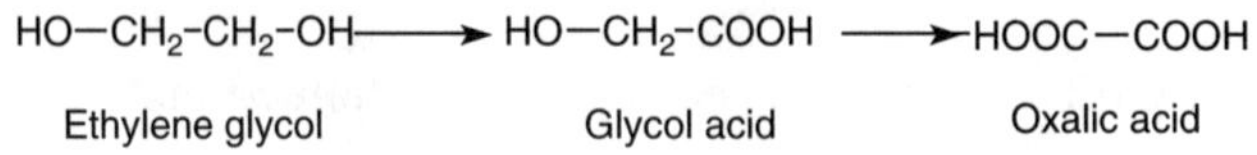

$$HO-CH_2-CH_2-OH \longrightarrow HO-CH_2-COOH \longrightarrow HOOC-COOH$$

Ethylene glycol　　　　　Glycol acid　　　　　Oxalic acid

Figure 14-4. Oxidation of ethylene glycol to glycolic acid and oxalic acid.

Human Toxicity

The toxicity of antifreeze agents is mainly due to the main ingredient, ethylene glycol. It is often believed that glycols are healthy to the skin, because the compounds are related to glycerol. This is completely wrong, because the degradation metabolism (that is, catabolism) is completely different due to a difference of a single carbon atom.

Ethylene glycol is acutely toxic to humans and animals if ingested. Ethylene glycol is defined as an animal teratogenic. Propylene glycol has not shown teratogenic effects, and the oral toxicity is lower. On the other hand, propylene glycol is more irritating to the skin than ethylene glycol.

Aquatic Toxicity

The aquatic toxicity of antifreeze agents is not strictly a function of the main component. Aquatic toxicity may come also from minor components in the formulation. Both ethylene glycol and propylene glycol are believed to be essentially nontoxic for aquatic life.

Biodegradation

Laboratory tests of ethylene glycol containing formulations have shown a complete bio-oxidation within 20 days. The rate of bio-oxidation is stationary over the full period. On the other hand, propylene glycol initially degrades more rapidly during the first 5 days of the test to an extent of 62%, slowing to 79% conversion after 20 days.

Recycling

Recycling is achieved either by simply filtering or by redistillation. In the case of filtering, only deposits are removed. There is essentially no information concerning the activity of other additives as corrosion inhibitors. Redistillation is more effective because this process recovers the glycol in high quality, although it is more complicated and cost intensive than filtering. The refined glycol must be re-inhibited before use.

Spent antifreeze formulations can be purified before recycling. If antifreeze is kept separate from waste oils, it can be easily treated and recycled into a new product.

Hydraulic Cement Additives

Antifreezing agents for cement consist mainly of salts such as calcium chloride, magnesium chloride, sodium chloride, and soda. Calcium chloride is highly corrosive and very restricted in use. Some salts, especially potassium chloride, will affect the curing time of cement. The latter chemical is in fact used to increase the pot life of cement. Likewise, alcohol freezing-point depressants, such as ethylene glycol, can be also included in the composition [1022].

Pipeline Transportation of Aqueous Emulsions of Oil

Mixtures of aqueous emulsions of oil can be more effectively transported through pipelines if certain antifreeze formulations are added to the system. Stable oil-in-water emulsions for pipeline transmission by using 0.05% to 4% ethoxylated alkylphenol as an emulgator and a freezing-point depressant for water enable pipeline transmission at temperatures below the freezing point of water [736].

Highly viscous petroleum oil containing 30% to 80% water can be transported through pipes more efficiently when a 1:1 mixture of washing liquid and antifreeze (i.e., ethylene glycol with borax) is added to the oil in amounts of 0.002% to 0.2% by weight. In addition to increased efficiency of transport, reduced corrosion of pipes can be achieved [893].

Low-Temperature Drilling Fluids

Antifreeze agents are occassionally added to reduce the freezing point of the drilling fluid itself [756, 758]. Such a water-based drilling fluid is composed of water, clay or polymer, and a poly-glycerol. The drilling fluid is useful in low-temperature drilling.

Chapter 15

Odorization

The primary objective of gas odorization is safety. Odorization serves as a warning in the detection of natural gas in air before it reaches combustible levels. Certain federal pipeline safety regulations require that combustible gases in pipelines be detectable at one-fifth of the lower explosive limit by a person with a normal sense of smell, either by the natural odor of the gas or by means of artificial odorization [574]. Therefore the proper odorization and odorants are integral parts of safety [813, 1753].

Odorization is a primary concern for any gas transmission company [813, 1379]. Accurate injection of the odorant, proper monitoring techniques, and complete record maintenance are important factors in developing and sustaining a successful odorization program.

A review has been presented concerning the aspects of odorization. Important points to consider are which pipelines require odorization, the detectable limits of gas odor, odorants and odorizing considerations, and monitoring a pipeline system to ensure that the odorization program is meeting the regulatory requirements [574].

Additives for Odorization

Because of their inherent penetrating smell, certain organic sulfur compounds are used for odorization. Repellents from the skunk contain compounds such as trans-2-butene-1-thiol and 3-methyl-1-butanethiol. Ethylmercaptan, because of its extremely low odor threshold, is the favorite compound used as an odorant in natural gas and liquid propane for leak detection. Tetrahydrothiophene is also often used. Common odorization reagents are summarized in Table 15–1 and Figures 15–1 to 15–3.

Measurement and Odor Monitoring

The methods of odor monitoring are reviewed by Klusman [975] and by Wetteman and Wilson [1838].

Table 15–1
Additives for Odorization

Additive	References
Ethylmercaptan or mixture of ethyl-, propyl-, and butylmercaptans dimethyl disulphide 15%–45%, diethyl disulphide 10%–30%, and balance methylethyl disulphide[a]	[567, 568, 890]
Tetrahydrothiophene, thiophene mercaptans with additional pyridine and picoline	[1868]
Mixture of ethyl-, propyl-, butyl-, and amylmercaptans	[566]

a) From wastes.

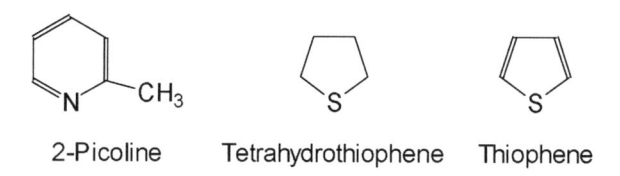

Figure 15–1. Picoline, tetrahydrothiophene, thiophene.

Chemical and Physical Methods

The concentration of odorants in gases can be measured by the absorbance in the ultraviolet region [1612, 1616]. The absorbance of odorized gas is much higher than the absorbance of untreated gas. Gaschromatography with an electrochemical detector is also suitable [1822] for analysis of mercaptans.

Olfactoric Response

The olfactoric response of humans, males and females between the ages of 16 and 82 years, was tested with various odorants, including tert-butylmercaptan, thiophene, ethyl mercaptan, dimethyl sulfide, isopropyl mercaptan, and mixtures of these odorants. The goal was to establish the warning levels below the explosion limit in the event of a gas leak [1498]. The study suggests that ethyl mercaptan is the most suitable odorant. Trained dogs can detect odorizing agents in concentrations as small as 10 to 18 ppb [185, 1463, 1464].

Uses and Properties
Leak Detection

Leaks in pipelines can be detected by means of a test-fluid. The test-fluid, a mixture of dimethyl sulfide in solvent, is injected into a pipeline. In the case of a leak, the test-fluid escapes through the leak, and the odorant is released from the closed compartments [1465, 1466].

Odor-fading

One specific problem of odorization is odor-fading. Odor-fading from organosulfur chemical-odorized liquefied petroleum gas stored in carbon steel containers can occur by catalytic effects of the containers. To postpone this effect, the respective steel surfaces can be deactivated by treating the surface with a deactivating agent [1336] before exposing the walls to the liquefied petroleum gas containing the odorizer. Examples of such deactivating agents are benzotriazole, tolyl triazole, mercaptobenzothiazole, benzothiazyl disulfide, or mixtures of these compounds. It has been suggested that a mathematical model and adequate software should be developed to predict the odorant fade [47].

Thermodynamic Properties of Odorants

The effectiveness of an odorant depends on the partition coefficients and the solubility. Vapor-liquid equilibria data for sulfur compounds in liquified natural gas are available [745, 944].

Environmental Problems

If natural gas for storage in natural reservoirs is odorized with sulfur compounds, then a possible environmental impact can result. Some of the odorant is lost in the formation [1557]. If the loss occurs in a reservoir adjacent to an aquifer, it could contaminate the water and cause environmental problems. When gas is drawn off, water is also often injected into the reservoir. A case was described in which the respective water had a strong characteristic odor [707]. A stripping column has been recommended to overcome this problem. Contaminated groundwater can also be decontaminated by the reaction with iron [867]. This technique was proposed to remedy groundwater contaminated with ethylmercaptan in situ. Studies suggest chemical reactions with iron rather than an irreversible surface adsorption. Gas odorizers can be

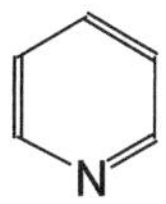

Pyridine

Figure 15–2. Pyridine.

CH$_3$—CH$_2$—SH CH$_3$—CH$_2$—CH$_2$—CH$_2$—SH
Ethylmercaptan Butylmercaptan

CH$_3$—CH$_2$—S—CH$_2$—CH$_3$
Diethylsulfide

Figure 15–3. Additives for odorization: ethylmercaptan, butylmercaptan, diethylsulfide.

removed by extraction [925, 1524], similar to the usual glycol dehydration and desulphurization process.

Another cleaning process for the removal of tetrahydrothiophene process uses an advanced oxidation technique, consisting of water treatment by UV–radiation in combination with a dosage of hydrogen peroxide [1392]. It is possible to keep the concentrations of odorant and condensate in the effluent below 0.1 ppb.

Chapter 16

Enhanced Oil Recovery

Approximately 60% to 70% of the oil in place cannot be produced by conventional methods [22]. Enhanced oil-recovery methods gain importance in particular with respect to the limited worldwide resources of crude oil. The estimated worldwide production from enhanced oil-recovery projects and heavy-oil projects at the beginning of 1996 was approximately 2.2 million barrels per day (bpd). This is approximately 3.6% of the world's oil production. At the beginning of 1994, the production had been 1.9 million bpd [1254].

Enhanced oil-recovery processes include chemical and gas floods, steam, combustion, and electric heating. Gas floods, including immiscible and miscible processes, are usually defined by injected fluids (carbon dioxide, flue gas, nitrogen, or hydrocarbon). Steam projects involve cyclic steam (huff and puff) or steam drive. Combustion technologies can be subdivided into those that autoignite and those that require a heat source at injectors [521].

Chemical floods are identified by the chemical type that is injected. The most common processes are polymers, surfactants, and alkalis, but chemicals are often combined. For example, polymer slugs usually follow surfactant or alkaline slugs to improve sweep efficiency. Injection of materials that plug permeable channels may be required for injection profile control and to prevent or mitigate premature water or gas breakthrough. Crosslinked or gelled polymers are pumped into injectors or producers for water shutoff or fluid diversion. Cement squeezes often can effectively fix near-wellbore water channeling problems. The design of chemical injection–enhanced oil recovery projects can be more complicated than that of waterflood projects. Down-hole conditions are more severe than are those for primary or secondary recovery production. Well injectivity is complicated by chemicals in injected waters, so in addition to precautions used in waterfloods, chemical interactions, reduced injectivity, deleterious mixtures at producers, potential for accelerated corrosion, and possible well stimulations to counter reduced injectivity must be considered [522]. Monographs on enhanced oil-recovery technologies have been presented by Green and Willhite [735], Sorbie [1661], and Littmann [1114].

Table 16–1
Estimated Production by Enhanced Oil Recovery and Heavy Oil

Region	BPD[a] (1996) [1254]	BPD[a] (1998) [1255]
United States	724,000	760,000
Canada	515,000	400,000
China	166,000	280,000
Former Soviet Union	200,000	200,000
Others	593,000	700,000
Total	2,198,000	2,340,000

a) Barrels of oil per day; includes in situ thermal heavy oil projects and primary heavy recovery oil projects.

Waterflooding

The surfactants described or characterized for waterflooding are summarized in Table 16–2. Commercial alkene sulfonates are a mixture of alkene sulfonate, hydroxyalkane sulfonate, and olefin disulfonate [211].

Caustic Waterflooding

Injection Strategies

To develop improved alkali-surfactant flooding methods, several different injection strategies were tested for recovering heavy oils. Oil recovery was compared for four different injection strategies [641]:

- Surfactant followed by polymer
- Surfactant followed by alkaline polymer
- Alkaline surfactant followed by polymer
- Alkali, surfactant, and polymer mixed in a single formulation

The effect of alkaline preflush was also studied under two different conditions. All of the oil-recovery experiments were conducted under optimal conditions with a viscous, nonacidic oil and with Berea sandstone cores.

Foam-Enhanced Caustic Waterflooding

An alkaline waterflooding process is enhanced by the injection of aqueous solutions of foam-forming surfactant and gases, or preformed foams, either ahead of or behind conventional alkaline slugs. A slug of an aqueous solution containing an alkaline agent, followed by a driving fluid, is injected into the formation and displaces oil through the relatively high-permeability

Table 16–2
Surfactants for Waterflooding

Surfactant	References
Ethoxylated methylcarboxylates	[1687]
Propoxyethoxy glyceryl sulfonate	[903]
Alkylpropoxyethoxy sulfate as surfactant, xanthan, and a copolymer of acrylamide and sodium 2-acrylamido-2-methylpropane sulfonate	[111]
Carboxymethylated ethoxylated surfactants (CME)	[663]
Polyethylene oxide (PEG) as a sacrificial adsorbate	[113]
Polyethylene glycols, propoxylated/ethoxylated alkyl sulfates	[1374]
Mixtures of sulfonates and nonionic alcohols	[110]
Combination of lignosulfonates and fatty amines	[467]
Alkyl xylene sulfonates, polyethoxylated alkyl phenols, octaethylene glycol mono n-decyl ether, and tetradecyl trimethyl ammonium chloride	[1655]
Anionic sodium dodecyl sulfate (SDS), cationic tetradecyl trimethyl ammonium chloride (TTAC), nonionic pentadecylethoxylated nonylphenol (NP–15), and nonionic octaethylene glycol N-dodecyl ether	[1656]
Dimethylalkylamine oxides as cosurfactants and viscosifiers	[1362]
(N-Dodecyl)trimethylammonium bromide	[112]
Petrochemical sulfonate and propane sulfonate of an ethoxylated alcohol or phenol	[79]
Petrochemical sulfonate and α-olefin sulfonate	[79, 1555]

zones of the formation, and oil is recovered via the production well. Thereafter, a slug of an aqueous solution consisting of a foaming agent is coinjected into the formation with a gas and creates a foam upon mixing with the gas. The foam created from the aqueous foaming agent and gas will go preferentially into the formation zones of relatively high-permeability and low oil saturation, substantially plugging these zones. Then a slug of an aqueous alkaline agent is injected, followed by a driving fluid that displaces the alkaline solution and oil through the less permeable zones toward the production well [881].

Polymers

Polymers can be used for mobility control. The interaction between polymers and surfactants is shown to be affected by pH, ionic strength, crude oil type, and the properties of the polymers and surfactants [642].

Interphase Properties

Alkaline agents can reduce surfactant losses and permit the use of low concentrations of surfactants. Laboratory tests show that alkali and synthetic surfactants produce interfacial properties that are more favorable for increased oil mobilization than either alkali or surfactant alone [639, 640].

Clay Dissolution

During caustic waterflooding the alkali can be consumed by the dissolution of clays and is lost in this way. The amount lost depends on the kinetics of the particular reaction. Several studies have been performed with kaolinite, using quartz as a yardstick, because the kinetic data are documented in the literature. The initial reaction rate has been found pH independent in the pH range of 11 to 13 [517]. The kinetics of silica dissolution could be quantitatively described in terms of pH, salinity, ion-exchange properties, temperature, and contact time [1549].

Acid Flooding

Acid flooding can be successful in formations that are dissolvable in the particular acid mixture, thus opening the pores. Hydrochloric acid is common, in a concentration of 6% to 30%, sometimes also with hydrofluoric acid and surfactants added (e.g., isononylphenol) [130, 723]. The acidic environment has still another effect on surfactants. It converts the sulfonates into sulfonic acid, which has a lower interfacial tension with oil. Therefore a higher oil forcing-out efficiency than from neutral aqueous solution of sulfonates is obtained. Cyclic injection can be applied [4, 494], and sulfuric acid has been described for acid treatment [25, 26, 1535]. Injecting additional aqueous lignosulfonate increases the efficiency of a sulfuric acid treatment [1798].

Hydrochloric acid in combination with chlorine dioxide can be used as a treatment fluid in water-injection wells that get impaired by the deposition of solid residues [332, 333]. The treatment seems to be more effective than the conventional acidizing system when the plugging material contains iron sulfide and bacterial agents because of the strongly oxidative power of chlorine dioxide. Mixtures of chlorine dioxide, lactic acid, and other organic acids [1172, 1173] also have been described.

Iron control chemicals are used during acid stimulation to prevent the precipitation of iron-containing compounds. The precipitation of these compounds in the critical near-wellbore area can decrease well productivity or injectivity. Acetic acid, citric acid, NTA, EDTA, and erythorbic acid are applied [1726, 1727]. A time dependence of iron (III) hydroxide precipitation was observed. Acetic acid can prevent the precipitation of iron (III) at high acetic acid concentrations at low temperatures.

If the injected acid itself contains iron (III), a precipitation of the asphaltic products can occur when it comes in contact with certain crude oils. This leads to practically irreversible damage of the zone treated. The amount of precipitate generally increases with the strength and concentration of the acid. Certain organic sulfur compounds, such as ammonium thioglycolate, mercaptoethanol, cysteamine, thioglycerol, cysteine, and thiolactic acid [581], can reduce the iron (III).

Emulsion Flooding

Optimizing the formulation of micellar surfactant solutions used for enhanced oil recovery consists of obtaining interfacial tensions as low as possible in multiphase systems, which can be achieved by mixing the injected solution with formation fluids. The solubilization of hydrocarbons by the micellar phases of such systems is linked directly to the interfacial efficiency of surfactants. Numerous research projects have shown that the amount of hydrocarbons solubilized by the surfactant is generally as great as the interfacial tension between the micellar phase and the hydrocarbons. The solubilization of crude oils depends strongly on their chemical composition [155].

Micellar flooding is a promising tertiary oil-recovery method, perhaps the only method that has been shown to be successful in the field for depleted light oil reservoirs. As a tertiary recovery method, the micellar flooding process has desirable features of several chemical methods (e.g., miscible-type displacement) and is less susceptible to some of the drawbacks of chemical methods, such as adsorption. It has been shown that a suitable preflush can considerably curtail the surfactant loss to the rock matrix. In addition, the use of multiple micellar solutions, selected on the basis of phase behavior, can increase oil recovery with respect to the amount of surfactant, in comparison with a single solution. Laboratory tests showed that oil recovery–to–slug volume ratios as high as 15 can be achieved [439].

A solids-stabilized water-in-oil emulsion may be used either as a drive fluid for displacing hydrocarbons from the formation or to produce a barrier for diverting the flow of fluids in the formation. The solid particles may be formation solid particles or nonformation solid particles, obtained from outside the formation (e.g., clays, quartz, feldspar, gypsum, coal dust, asphaltenes, polymers) [228, 229].

The problems of enhanced oil-recovery methods have been summarized by Bragg [230]:

> Oil recovery is usually inefficient in subterranean formations (hereafter simply referred to as formations) where the mobility of the in situ oil being recovered is significantly less than that of the drive fluid used to

displace the oil. Mobility of a fluid phase in a formation is defined by the ratio of the fluid's relative permeability to its viscosity. For example, when waterflooding is applied to displace very viscous heavy oil from the formation, the process is very inefficient because the oil mobility is much less than the water mobility. The water quickly channels through the formation to the producing well, bypassing most of the oil and leaving it unrecovered. In Saskatchewan, Canada, primary production crude has been reported to be about 2 to 8% of the oil in place, with waterflooding yielding only another 2 to 5% of that oil in place. Consequently, there is a need to either make the water more viscous, or use another drive fluid that will not channel through the oil. Because of the large volumes of drive fluid needed, it must be inexpensive and stable under formation flow conditions. Oil displacement is most efficient when the mobility of the drive fluid is significantly less than the mobility of the oil, so the greatest need is for a method of generating a low-mobility drive fluid in a cost-effective manner.

Oil recovery can also be affected by extreme variations in rock permeability, such as when high-permeability thief zones between injectors and producers allow most of the injected drive fluid to channel quickly to producers, leaving oil in other zones relatively unrecovered. A need exists for a low-cost fluid that can be injected into such thief zones (from either injectors or producers) to reduce fluid mobility, thus diverting pressure energy into displacing oil from adjacent lower-permeability zones.

In certain formations, oil recovery can be reduced by coning of either gas downward or water upward to the interval where oil is being produced. Therefore, a need exists for a low-cost injectant that can be used to establish a horizontal pad of low mobility fluid to serve as a vertical barrier between the oil producing zone and the zone where coning is originating. Such low mobility fluid would retard vertical coning of gas or water, thereby improving oil production.

For modestly viscous oils—those having viscosities of approximately 20–100 centipoise (cP)–water-soluble polymers such as polyacrylamides or xanthan gum have been used to increase the viscosity of the water injected to displace oil from the formation. For example, polyacrylamide was added to water used to waterflood a 24 cP oil in the Sleepy Hollow Field in Nebraska. Polyacrylamide was also used to viscosify water used to flood a 40 cP oil in the Chateaurenard Field, France. With this process, the polymer is dissolved in the water, increasing its viscosity.

While water-soluble polymers can be used to achieve a favorable mobility waterflood for low to modestly viscous oils, usually the process cannot economically be applied to achieving a favorable mobility displacement of more viscous oils—those having viscosities of from approximately 100 cP or higher. These oils are so viscous that the amount of polymer needed to achieve a favorable mobility ratio would usually be uneconomic. Further, as known to those skilled in the art, polymer dissolved in water often is desorbed from the drive water onto surfaces of the formation rock, entrapping it and rendering it ineffective for viscosifying the water. This leads to loss of mobility control, poor oil recovery, and high polymer costs. For these reasons, use of polymer floods to recover oils in excess of 100 cP is not usually technically or economically feasible. Also, performance of many polymers is adversely affected by levels of dissolved ions typically found in formations, placing limitations on their use and/or effectiveness.

Water-in-oil macroemulsions have been proposed as a method for producing viscous drive fluids that can maintain effective mobility control while displacing moderately viscous oils. For example, the use of water-in-oil and oil-in-water macroemulsions have been evaluated as drive fluids to improve oil recovery of viscous oils. Such emulsions have been created by addition of sodium hydroxide to acidic crude oils from Canada and Venezuela. In this study, the emulsions were stabilized by soap films created by saponification of acidic hydrocarbon components in the crude oil by sodium hydroxide. These soap films reduced the oil/water interfacial tension, acting as surfactants to stabilize the water-in-oil emulsion. It is well known, therefore, that the stability of such emulsions substantially depends on the use of sodium hydroxide (i.e., caustic) for producing a soap film to reduce the oil/water interfacial tension.

Various studies on the use of caustic for producing such emulsions have demonstrated technical feasibility. However, the practical application of this process for recovering oil has been limited by the high cost of the caustic, likely adsorption of the soap films onto the formation rock leading to gradual breakdown of the emulsion, and the sensitivity of the emulsion viscosity to minor changes in water salinity and water content. For example, because most formations contain water with many dissolved solids, emulsions requiring fresh or distilled water often fail to achieve design potential because such low-salinity conditions are difficult to achieve and maintain within the actual formation. Ionic species can be dissolved from the rock and the injected fresh water can mix with higher-salinity resident water, causing breakdown of the low-tension stabilized emulsion.

Various methods have been used to selectively reduce the permeability of high-permeability thief zones in a process generally referred to as profile modification. Typical agents that have been injected into the reservoir to accomplish a reduction in permeability of contacted zones include polymer gels or crosslinked aldehydes. Polymer gels are formed by crosslinking polymers such as polyacrylamide, xanthan, vinyl polymers, or lignosulfonates. Such gels are injected into the formation where crosslinking reactions cause the gels to become relatively rigid, thus reducing permeability to flow through the treated zones.

In most applications of these processes, the region of the formation that is affected by the treatment is restricted to near the wellbore because of cost and the reaction time of the gelling agents. Once the treatments are in place, the gels are relatively immobile. This can be a disadvantage because the injected fluid (for instance, water in a waterflood) eventually finds a path around the immobile gel, reducing its effectiveness. Better performance should be expected if the profile modification agent could slowly move through the formation to plug off newly created thief zones, penetrating significant distances from injection or production wells.

Chemical Injection

The state of the art in chemical oil recovery has been reviewed [1732]. More than two thirds of the original oil remains unrecovered in an oil reservoir after primary and secondary recovery methods have been exhausted. Many chemically based oil-recovery methods have been proposed and tested in the laboratory and field. Indeed, chemical oil-recovery methods offer a real challenge in view of their success in the laboratory and lack of success in the field. The problem lies in the inadequacy of laboratory experiments and the limited knowledge of reservoir characteristics. Field test performances of polymer, alkaline, and micellar flooding methods have been examined for nearly 50 field tests. The oil-recovery performance of micellar floods is the highest, followed by polymer floods. Alkaline floods have been largely unsuccessful. The reasons underlying success or failure are examined in the literature [1732].

Ammonium Carbonate

Ammonium carbonate decomposes in acid medium into ammonium salts and carbon dioxide. It is thus valuable for the in situ generation of carbon dioxide [495, 1674, 1675].

Hydrogen Peroxide

The physical properties of hydrogen peroxide indicate that hydrogen peroxide injection has the potential of combining the more favorable aspects of many enhanced oil-recovery processes, namely:

1. Steam
2. Combustion
3. Oxygen-water combustion
4. Carbon dioxide injection

Hydrogen peroxide decomposes to form water and oxygen. Both products are environmentally desirable and effective in recovering oil. Heat is generated in the oil reservoir when the decomposition reaction occurs. The available heat from the chemical reactions supports steam and hot waterflooding operations, among others. Continued injection of liquid hydrogen peroxide advances the heat bank, steam zone, hot-water zone, oxygen-burning front, and CO_2 bank through the formation, effectively displacing oil [1268].

Combinations of hydrogen peroxide, sulfuric acid, and urea have been proposed [1]. The temperature influences the urea decomposition into ammonia and carbon dioxide that provokes pressure buildup in a formation model and a 19% increase of oil-displacement efficiency in comparison with water.

Reactions of hydrogen peroxide with near-wellbore formation and liquids creates high temperatures for lowering the oil viscosity and removing formation damage. The application of this chemical technique for heat-bank–type flooding is noted as a technically superior method, but it is probably not economically viable [156]. There is a wide potential field of applications of hydrogen peroxide, including pressure generation, hydrate melting in subsea equipment, and metal cutting for offshore structure decommissioning [157].

Alcohol Waterflooding

Butanol

n-Butanol and other C_4 alcohols are suitable for hot waterflooding in medium to heavy oil reservoirs at depths greater than 1500 m [1494].

Residue from the Production of Glycerol or Ethylene Glycol

Waste water–soluble alcohols are useful for miscible waterflooding [886].

Chemical Injection of Waste Gases

Waste gas from produced hydrocarbons can be safely disposed by reinjecting into a formation. The waste gas is mixed with a surfactant to form a foam that, in turn, is placed within a disposal zone of a subterranean formation. The waste gas is trapped within the foam, thereby reducing the mobility of the gas in the formation, which, in turn, restricts the ability of the waste gas to readily flow out of the disposal zone and into the producing zone of the formation. The waste gas foam can be placed into the formation by coinjecting the surfactant and the waste gas, or it can be formed in situ by first injecting the surfactant and then injecting the waste gas [1356].

Thermal conversion of organic waste material, such as plastics, or of biomass under the influence of oxygen into crude synthesis gas yields a hydrogen gas.

The crude mixture of synthesis gas can be injected into a depleted crude oil well. In such a well, high-molecular organic material, which will not flow readily, is inside. Hydrogen itself will crack the long chains of the sticking organics in situ and will make it more prone to flow. In this way, improved oil recovery and plastics waste disposal by oxidative pyrolysis can be achieved, followed by in situ degradative hydrogenation of geopolymers. Thus more organic material can be recovered than was initially put into the well [598].

Polymer Waterflooding

The polymer in a polymer waterflooding process acts primarily as a thickener. It decreases the permeability of the reservoir and thus improves the vertical and lateral sweep efficiency.

Associative copolymers of acrylamide with N-alkylacrylamides, terpolymers of acrylamide, N-decylacrylamide, and sodium-2-acrylamido-2-methylpropane sulfonate (NaAMPS), sodium acrylate (NaA), or sodium-3-acrylamido-3-methylbutanoate (NaAMB) have been shown to possess the required rheologic behavior to be suitable for enhanced oil-recovery processes [1184].

Other copolymers of acrylamide with the zwitterionic 3-(2-acrylamido-2-methylpropyldimethyl ammonio)-1-propane sulfonate (AMPDAPS) monomer also have been examined.

Low-Tension Polymer Flood Technique

The low-tension polymer flood technique consists of combining low levels of polymer-compatible surfactants and a polymer with a waterflood. This affects mobility control and reduces front-end and total costs. [929].

Table 16–3
Polymers Used in Polymer Waterflooding

Polymer	References
Partially hydrolyzed polymer polyacrylamide	[366]
Polyacrylamide, bentonite clay	[722]
Polyacrylamide	[1137, 1138, 1460, 1490]
Polydimethyldiallyl ammonium chloride, biopolymers	[1160]
Exopolysaccharide produced by *Acinetobacter*	[1670]
Xanthan	[772, 1320]
Wellan	[850]

The synergism of surfactant-polymer complex formation has been studied by gel permeation chromatography [114].

Influence of Viscosity on Ionic Strength

The viscosity and non-Newtonian flooding characteristics of the polymer solutions decrease significantly in the presence of inorganic salts, alkali silicates, and multivalent cations. The effect can be traced back to the repression of the dissociation of polyelectrolytes, to the formation of a badly dissociating polyelectrolyte metal complex, and to the separation of such a complex from the polymer solution [1054].

Modified Acrylics

A modified acrylamide polymer that is hydrophobically associating has remarkably improved the properties of salt resistance and temperature resistance, compared with high–molecular-weight polyacrylamide [1351].

Biopolymers

Pseudozan

Pseudozan is an exopolysaccharide produced by a *Pseudomonas* species. It has high viscosities at low concentrations in formation brines, forms stable solutions over a wide pH range, and is relatively stable at temperatures up to 65° C. The polymer is not shear degradable and has pseudoplastic behavior. The polymer has been proposed for enhanced oil-recovery processes for mobility control [1075].

Xanthan

Xanthan exhibits an interaction with anionic surfactants (petroleum sulfate), which is a beneficial synergistic effect for mobility control in chemical-enhanced oil-recovery processes [1115].

Combination Flooding

Combination flooding comprises the combination of at least two basic techniques from gas flooding, caustic flooding, surfactant flooding, polymer flooding, and foam flooding. There may be synergisms between the various chemical reagents used. There are specific terms that clarify the individual combination of the basic methods, such as surfactant-enhanced alkaline flooding, alkaline-assisted thermal oil recovery, and others.

Table 16–4
Methods Summarized Under Combined Flooding

Type
Alkali/polymer flooding
Alkali/surfactant/polymer flooding
Alkaline-assisted thermal oil recovery
Alkaline steamflooding
Polymer-assisted surfactant flooding
Water-alternating gas technology

Low-Tension Polymer Flood

Coinjection of a low-concentration surfactant and a biopolymer, followed by a polymer buffer for mobility control, leads to reduced chemical consumption and high oil recovery. There may be synergistic effects between the surfactant and the polymer in a dynamic flood situation. The chromatographic separation of surfactant and polymer is important to obtain good oil recovery and low surfactant retention [1721].

In buffered surfactant-enhanced alkaline flooding, it was found that the minimum in interfacial tension and the region of spontaneous emulsification correspond to a particular pH range, so by buffering the aqueous pH against changes in alkali concentration, a low interfacial tension can be maintained when the amount of alkali decreases because of acids, rock consumption, and dispersion [1826].

Effect of Alkaline Agents on the Retention of Enhanced Oil-Recovery Chemicals

The effectiveness of alkaline additives tends to increase with increasing pH. However, for most reservoirs, the reaction of the alkaline additives with minerals is a serious problem for strong alkalis, and a flood needs to be operated at the lowest effective pH, approximately 10. The ideal process by which alkaline agents reduce losses of surfactants and polymers in oil recovery by chemical injection has been detailed in the literature [1126].

Alkaline Steamflooding

The performance of steamflooding often suffers from channeling and gravity segregation. Alkaline additives may be used with steam for certain types of crude oils to improve the steamflood performance. Experimental results show that sodium orthosilicate outperforms sodium hydroxide and sodium metasilicate [1233].

Aluminum Trichloride and Trisodium Phosphate as Sediment-Forming Material

Aluminum trichloride and trisodium phosphate can be injected as sediment-forming material [721].

Water-Alternating Gas Technology

The oil production from thin under–gas cap zones with an active aquifer is not efficient because of the rapid breakthrough of gas or water. The water-alternating gas technology based on the injection of water solution with oil- and water-soluble polymers seems to be promising to stimulate such wells. For heavy oils, this technology can be considered as an alternative to thermal-enhanced oil recovery [1673].

Hydrocarbon-Assisted Steam Injection

In steam injection, the mobility of the hydrocarbons is greater if a C_1 to C_{25} hydrocarbon is added than if steam is used alone under substantially similar formation conditions [647, 1321, 1322].

Foam Flooding

Basic Principles of Foam Flooding

A process for enhancing the recovery of oil in a subterranean formation comprises injecting a foam having oil-imbibing and transporting properties. A foam having such properties is selected either by determination of the lamella number or by micro-visualization techniques. The method for selecting a surfactant capable of forming a foam functional to both imbibe and transport an oil phase in a subterranean formation comprises

1. Determining the surface tension of the foaming solution
2. Measuring the radius of a foam lamella plateau border where it initially contacts the oil or of an emulsified drop
3. Determining the interfacial tension between the foaming solution and the oil
4. Correlating these measurements with a mathematical model to obtain a value indicative of the oil imbibing properties of the foam

The foam, having a viscosity greater than the displacing medium, will preferentially accumulate in the well-swept and/or higher permeability zones of the formation. The displacing medium is thus forced to move into the unswept or underswept areas of the formation. It is from these latter areas that the additional oil is recovered. However, when a foam is used to fill a low oil content area of the reservoir, the oil contained therein is, for all practical purposes, lost. This is because the foam functions to divert the displacement fluid from such areas [1574–1576].

Foam stability in the presence of oil can be described from thermodynamics in terms of the spreading and entering coefficients S and E respectively. These coefficients are defined as follows:

$$S = Y_F^o - Y_{OF} - Y_O^o \qquad (16\text{--}1)$$

wherein

Y_F^o is the foaming solution surface tension;
Y_{OF} is the foaming solution-oil interfacial tension; and
Y_O^o is the surface tension of the oil.

$$E = Y_F^o + Y_{OF} - Y_O^o \qquad (16\text{--}2)$$

Based on these coefficients, one can predict that three types of oil-foam interactions could take place.

First, (Type A) an oil will neither spread over nor enter the surface of foam lamellae when E and S are less than zero.

Secondly, (Type B) oil will enter but not spread over the surface of foam lamellae when E is greater than zero but S is less than zero.

Thirdly, (Type C) oil will enter the surface of foam lamellae and then spread over the lamellae surfaces if both E and S are greater than zero. This latter behavior, typically, will destabilize the foam.

However, experimental results have not borne out these predictions. Furthermore, the theory was developed assuming that the oil droplets are readily imbibed into the foam lamellae. Again however, experimental results show that some foams, particularly those of type A supra do not readily imbibe oil.

There exists, therefore, a need to distinguish between foams which are stable to oil but do not significantly imbibe oil, as in type A supra, foams which are stable to oil and do imbibe oil as in the second type above and finally, foams that are unstable to oil as in the third predicted type [1823].

A foam drive method comprises the following steps:

1. Injecting into the reservoir an aqueous polymer solution as preceding slug

2. Periodically injecting simultaneously or alternately a noncondensable gas and a foaming composition solution containing alkalis, surfactants and polymers to form combined foam or periodically injecting the gas and the foam previously formed from the solution
3. Injecting a polymer solution as a protecting slug and then continuing with waterflooding [1823]

Ambient-Pressure Foam Tests

Several surfactants were studied in ambient-pressure foam tests, including alcohol ethoxylates, alcohol ethoxysulfates, alcohol ethoxyethylsulfonates, and alcohol ethoxyglycerylsulfonates [210]. Surfactants that performed well in the 1-atm foaming experiment were also good foaming agents in site cell and core flood experiments performed in the presence of CO_2 and reservoir fluids under realistic reservoir temperature and pressure conditions.

Laboratory studies of foam flow in porous media suggest that the relative foam mobility is approximately inversely proportional to the permeability. This means that foam has potential as a flow-diverting agent, in principle sweeping low-permeability regions as effectively as high-permeability regions [716].

Sandpack Model

A one-dimensional sandpack model has been used to investigate the behavior of four anionic sulfonate surfactants of varying chemical structure with steam. The study was performed with a crude oil at a residual oil saturation of approximately 12% of the pore volume. The observed pressure drops across the various sections of the pack were used to study the behavior of the surfactant. The tested surfactants varied in chain length, aromatic structure, and number of ionic charges. A linear toluene sulfonate produced the highest strength foam in the presence of oil at residual saturations, in comparison with α-olefin sulfonates. This is in contrast to the behavior of the surfactants in the absence of oil, where the α-olefin sulfonates perform better. The reason for this change in behavior is the relative propagation rate of the foams produced by the surfactants [1484, 1485].

Foaming Agents

When an oil reservoir is subjected to steam injection, steam tends to move up in the formation, whereas condensate and oil tend to move down due to the density difference between the fluids. Gradually, a steam override condition develops, in which the injected steam sweeps the upper portion of the formation but leaves the lower portion untouched. Injected steam will tend to follow the path of least resistance from the injection well to a production well. Thus areas of high permeability will receive more and more of the injected steam which further raises the permeability of such areas. This phenomenon exists to an even larger degree with low injection rates and thick formations. The steam override problem worsens at greater radial distances from the injection well because steam flux decreases with increasing steam zone radius.

Although residual oil saturation in the steam swept region can be as low as 10%, the average residual oil saturation in the formation remains much higher due to poor vertical conformance. Thus it is because of the creation of steam override zones that vertical conformance in steamfloods is usually poor.

A similar conformance problem exists with carbon dioxide flooding. Carbon dioxide has a large tendency to channel through oil in place since carbon dioxide viscosity may be 10 to 50 times lower than the viscosity of the oil in place. This problem of channeling through oil is exacerbated by the inherent tendency of a highly mobile fluid such as carbon dioxide to preferentially flow through more permeable rock sections. These two factors, unfavorable mobility ratios between carbon dioxide and the oil in place and the tendency of carbon dioxide to take advantage of permeability variations, often make carbon dioxide flooding uneconomical. Conformance problems increase as the miscibility of the carbon dioxide with the oil in place decreases.

Although not much attention has been devoted to carbon dioxide conformance, it has long been the concern of the oil industry to improve the conformance of a steamflood by reducing the permeability of the steam swept zone by various means. The injection of numerous chemicals such as foams, foaming solutions, gelling solutions or plugging or precipitating solutions have been tried. Because of the danger of damaging the reservoir, it is considered important to have a non-permanent means of lowering permeability in the steam override zones. For this reason, certain plugging agents are deemed not acceptable. In order to successfully divert steam and improve vertical conformance, the injected chemical should be

1. Stable at high steam temperatures of about 150° C to about 315° C
2. Effective in reducing permeability in steam swept zones
3. Non-damaging to the oil reservoir
4. Economical

The literature is replete with references to various foaming agents which are employed to lower permeability in steam swept zones. The vast majority of the foaming agents require the injection of a non-condensable gas to generate the foam in conjunction with the injection of steam and the foaming agent [1372].

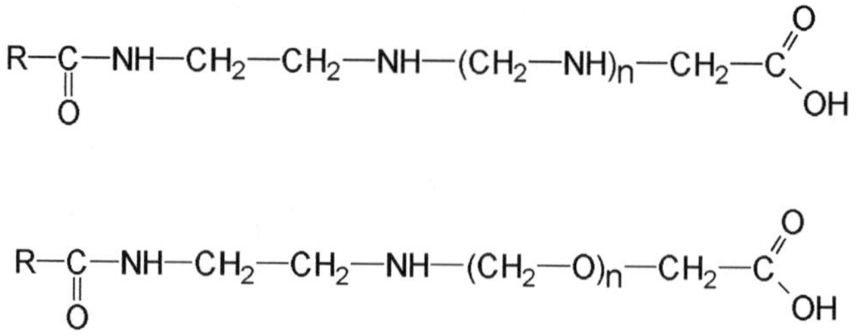

212 Oil Field Chemicals

$$R-\underset{\underset{O}{\|}}{C}-NH-CH_2-CH_2-NH-(CH_2-NH)_n-CH_2-C\overset{O}{\underset{OH}{\nwarrow}}$$

$$R-\underset{\underset{O}{\|}}{C}-NH-CH_2-CH_2-NH-(CH_2-O)_n-CH_2-C\overset{O}{\underset{OH}{\nwarrow}}$$

Figure 16–1. Examples of amphoteric tensides.

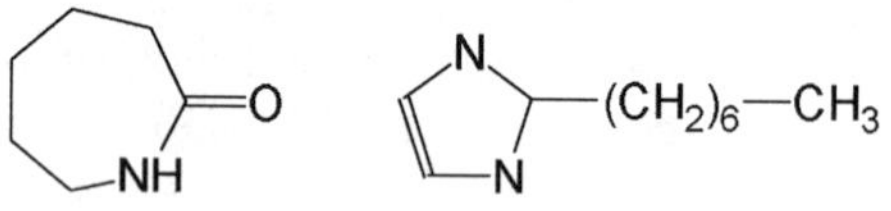

Caprolactam   2-Heptylimidazoline

Figure 16–2. Caprolactam, 2-heptylimidazoline.

C_{12} to C_{15} alcohols and α-olefin sulfonate are highly effective with steam foaming agents or carbon dioxide foaming agents in reducing the permeability of flood-swept zones [1372].

The sodium salt of tall oil acid is suitable as a foam surfactant. Experimental results show that sodium tallates are effective foaming agents that can produce pressure gradients of hundreds of pounds per square inch (psi) per foot in a sandpack [1373].

The foam-holding characteristics of foam from surfactants in oil field jobs can be tailored by adding an imidazoline-based amphoacetate surfactant. Amphoacetates are a special class of amphoteric tensides (Figure 16–1). Imidazoles, such as 2-heptylimidazoline, are reacted with fatty acids under the ring opening. For alkylation, the imidazoline is reacted with, for example, chloroacetate [493].

Residues from caprolactam productions have been proposed as surfactants [1772].

Fluorocarbon Surfactant

A foam can be generated by using an inert gas and a fluorocarbon surfactant solution in admixture with an amphoteric or anionic hydrocarbon surfactant solution. A relatively small amount of the fluorocarbon surfactant is operative when mixed with the hydrocarbon surfactant and foamed. The foam has better stability than a foam made with hydrocarbon surfactant alone when in contact with oil [1491].

Polymer-Enhanced Foams

Some of the parameters affecting the foam performance are polymer concentration, the chemical nature of the surfactants and their concentration, aqueous phase salinity and pH, and shear rate [1901]. The performance of polymer-enhanced foams was shown to be much better in comparison with conventional foams. Polyacrylamide polymers were used as an additive. Higher foam resistance and longer foam persistence were achieved by using relatively low concentrations of polymers. The studies also showed that the foam performance was significantly improved over a broad range of polymer concentrations. Foams are severely affected in the presence of oil, but polymer-enhanced foam reduced the negative impact of oils on foam mobility. Polymer-enhanced foams are suitable for plugging fracture reservoirs [1705].

Carbon Dioxide Flooding

In the 1990s certain research groups focused on the development of CO_2-soluble polymers usable as direct thickeners, in particular, ionomers [1003].

Sandstone rock surfaces are normally highly water-wet. These surfaces can be altered by treatment with solutions of chemical surfactants or by asphaltenes. Increasing the pH of the chemical treating solution decreases the water wettability of the sandstone surface and, in some cases, makes the surface medium oil-wet [1644]. Thus the chemical treatment of sandstone cores can increase the oil production when flooded with carbon dioxide.

A cosolvent used as a miscible additive to CO_2 changed the properties of the supercritical gas phase. The addition of a cosolvent resulted in increased viscosity and density of the gas mixture and enhanced extraction of the oil compounds into the CO_2-rich phase. Gas phase properties were measured in an equilibrium cell with a capillary viscometer and a high-pressure densitometer. Cosolvent miscibility with CO_2, brine solubility, cosolvent volatility, and relative quantity of the cosolvent partitioning into the oil phase are factors that must be considered for the successful application of cosolvents. The results indicate that lower–molecular-weight additives, such as propane, are the most effective cosolvents to increase oil recovery [1472].

By adding common solvents as chemical modifiers, the flooding fluid shows marked improvement in solvency for heavy components of crudes because of its increased density and polarity [884]. Miscible or immiscible carbon dioxide injection is considered to be one of the most effective technologies to improve oil recovery from complicated formations and hard-to-recover oil reserves. Application of this technology can increase ultimate oil recovery by 10% to 15%. One of the main advantages of this technology is that it can be applied in a wide range of geologic conditions for producing both light and heavy oils.

The main factors restraining wide application of CO_2 flooding are the dependence on natural CO_2 sources, transportation of CO_2, safety and environmental problems, breakthrough of CO_2 to the production wells, and corrosion of well and field equipment. This restrains the wide implementation of CO_2 injection technology, not only in many European countries, but also worldwide, where this technology could significantly improve oil production and ultimate oil recovery from depleted oil formations. Another technology for in situ CO_2 generation has been developed and described. It is based on an exothermic chemical reaction between gas-forming water solution and low-concentration active acids [535].

Hydrocarbons and other fluids are recovered at a production well by a mixture of CO_2 and 0.1% to 20% by weight trichloroethane at a temperature and pressure above the bubble point of the mixture, which ensures that the mixture will be in a single phase [866].

Steamflooding

Carbon Dioxide

When the temperature of a carbonate reservoir that is saturated with high-viscosity oil and water increases to 200° C or more, chemical reactions occur in the formation, resulting in the formation of considerable amounts of CO_2. The generation of CO_2 during thermal stimulation of a carbonate reservoir results from the dealkylation of aromatic hydrocarbons in the presence of water vapor, catalytic conversion of hydrocarbons by water vapor, and oxidation of organic materials. Clay material and metals of variable valence (e.g., nickel, cobalt, iron) in the carbonate rock can serve as the catalyst. An optimal amount of CO_2 exists for which maximal oil recovery is achieved [1538]. The performance of a steamflooding process can be improved by the addition of CO_2 or methane [1216].

Air Injection

Air as a steam additive results in an increased rate at which oil is recovered because of low-temperature oxidation reactions [894].

Chemical Reactions

The reactivity of steam can be reduced via pH control. The injection or addition of a buffer such as ammonium chloride inhibits the dissolution of certain mineral groups, controls the migration of fines, inhibits the swelling of clays, controls chemical reactions in which new clay minerals are formed, and helps prevent the precipitation of asphaltenes and the formation of emulsions as a result of steam injection [1862].

The reaction of sulfate with sulfide is strongly pH dependent, and the oxidation potential of sulfate in the neutral pH region is very low. At atmospheric pressure and temperatures up to the boiling points of the inorganic and organic media (e.g., water, alkanes, alkyl-substituted arenes), no reaction takes place within 100 hours. However, the reaction may proceed very slowly over geochemical time periods. During the steamflooding processes, huge amounts of H_2S are produced together with CO_2 and small amounts of elemental hydrogen. In the producing zones; the temperatures lie in the range of 250° to 270° C, which is significantly below the conditions described in the literature. The H_2S production rises from 50 ppm to up to 300,000 ppm, causing enormous corrosion and health safety risks [840].

Addition of 2% to 5% urea with respect to water is claimed to reduce the viscosity of the heavy hydrocarbon by at least 50% [308].

In Situ Combustion

A significant increase in light oil production can be achieved with air injection. A total consumption of 5% to 10% of the remaining oil in place can be expected to maintain a propagation of the in situ oxidation process. The flue gas and steam generated at the combustion front strip, swell, and heat the contacted oil. The light oil is displaced at near-miscible conditions with complete utilization of injected oxygen [1700].

Special Techniques

Viscous Oil Recovery

Special techniques have been developed for the recovery of viscous oils. In particular, viscous oils are candidates for thermal methods.

Low-Temperature Oxidation

Cap Gas. Both crude and asphaltene-free oil were used to determine the consequences of low-temperature oxidation. It was found that the oxygen content in an artificial gas cap was completely consumed by chemical reactions (i.e., oxidation, condensation, and water formation) before the asphaltene content had reached equilibrium.

The application of a pillow (cap) gas containing air and oxygen, aimed at improving the gravitational segregation in offshore production technology, may offer an appropriate alternative to increase the recovery factor in heavy oil–bearing reservoirs [1052].

Special Surfactant Formulations. An alkaline polyacrylamide solution in liquid hydrocarbons has been suggested for enhanced oil recovery [40]. Special surfactant formulations have been tried to recover heavy crude oils. Ternary surfactant formulations, so called "mixed-surfactant–enhanced alkaline systems," were successful in reversing the trend of increasing interfacial tension with time that is typical in additive-free alkaline crude oil systems. On the other hand, the initial interfacial tension values were higher. However, at higher temperatures (65° C), these ternary surfactant formulations were capable of generating very low interfacial tension values against the crude oil, which suggests that they could be suitable candidates for commercial exploitation of heavy oil–recovery processes [376].

Visbreaking

In situ visbreaking with steam and a catalyst can produce crude oils with reduced viscosity [821]. A special variety of visbreaking that involves a partial steam reforming, which produces smaller hydrocarbon components and additional hydrogen free radicals and carbon dioxide, has been described.

Low-Permeability Flooding

Oil recovery from diatomaceous formations is usually quite limited because a significant portion of oil saturation may be bypassed using conventional production techniques such as primary, waterflooding, cyclic or drive steaming. Significant improvement of oil recovery would require that a method of displacing oil from the interior of the diatoms into the flow channels between the diatoms be provided. Furthermore, it would be necessary to improve permeability in the natural flow channels so that the oil can be recovered. A combination of chemical additives is used to increase water wetness of a rock thereby increasing the capillary pressure which forces oil and water from the diatomaceous formation. These chemical additives include wetting agents (e.g., mono-, di-, tribasic forms of sodium or potassium phosphate, and sodium silicate) and surfactants. These surfactants can be either sulfonates, ammonium salts of linear alcohol, ethoxy sulfates, or calcium phenol ethoxylated alkyl sulfonates. These surfactants lower the interfacial tension between oil and water thereby allowing oil to flow more freely through the diatomaceous matrix. Imbibition experiments with up to 3 wt % of active surfactant concentration indicate a 31% improvement in oil recovery over that obtainable with brine alone [277, 1357].

It has also been proposed to inject a solvent [454], for example, jet fuel, petroleum naphtha, aromatic hydrocarbons, or naphthenic hydrocarbons, before injecting the surfactant solution.

Microbial-Enhanced Oil-Recovery Techniques

Microbial-enhanced oil recovery (MEOR) was first proposed in 1926 by A. Beckman [1780]. Between 1943 and 1953, C. E. Zobell [1903, 1904] laid the foundations of MEOR techniques. The results were largely dismissed in the United States because there was little interest in finding methods to enhance the recovery of oil at this time. However, in some European countries, the interest for MEOR increased and several field trials were conducted. The first MEOR water flood field project in the United States was initiated in 1986. The site selected was in the Mink Unit of Delaware–Childers Field in Nowata County, Oklahoma [268].

Microbiologists laid the foundations of MEOR. After the petroleum crisis in 1973, the interest in MEOR generally increased [1074]. Monographs about the underlying ideas and the practice of MEOR are available [1236, 1884].

Basic Principles and Methods

The very methods, ranging from a single well treatment to fieldwide treatments, can be subsumed as MEOR techniques. The injection of microbes into the formation is a common practice in all of these techniques. This should stimulate the in situ microflora, resulting in the production of certain compounds that increase the oil recovery of exhausted reservoirs. The following basic effects can be achieved by microbes [1780]:

- In situ production of gels for selective water shutoff
- In situ production of biosurfactants for surfactant flooding
- In situ production of acids for dissolving carbonate rocks
- In situ production of CO_2
- In situ degradation of long chain molecules to reduce viscosity and paraffin content
- Displacement of oil by metabolites of inoculated bacteria grown in situ
- Huff and puff technique: (1) Huff: Migration of cells and synthesis of metabolic products following inoculation and closing of injection well; (2) Puff: Production and recovery of oil after incubation period

It is often stressed that the technology is environmentally friendly. The stimulation of oil production by in situ bacterial fermentation is thought to be initialized by one or a combination of the following mechanisms:

1. Improvement of the relative mobility of oil to water by biosurfactants and biopolymers
2. Partial repressurization of the reservoir by methane and CO_2 gases
3. Reduction of oil viscosity through the dissolution of organic solvents in the oil phase

218   Oil Field Chemicals

4. Increase of reservoir permeability and widening of the fissures and channels through the etching of carbonaceous rocks in limestone reservoirs by organic acids produced by anaerobic bacteria
5. Cleaning of the wellbore region through the acids and gas from in situ fermentation: The gas serves to push oil from dead space and dislodge debris that plugs the pores; the average pore size is increased and, as a result, the capillary pressure near the wellbore is made more favorable for the flow of oil
6. Selective plugging of highly permeable zones by injecting slime-forming bacteria followed by sucrose solution that turns on the production of extracellular slimes

Successful microbial MEOR requires

1. The selection, injection, dispersion, metabolism, and persistence of organisms with properties that facilitate the release of residual oil
2. The coinjection of growth-effective nutrients into the extreme environments that characterize petroleum reservoirs [1604]

Economics

The most widely practiced technique for applying MEOR involves cyclic stimulation treatments of producing wells. Improvements in oil production can result from removal of paraffinic or asphaltic deposits from the near-wellbore region or from mobilization of residual oil in the limited volume of the reservoir that is treated. An alternate method involves applying microbes in an ongoing waterflood to improve oil recovery [267]. In the laboratory, microorganisms have been shown to produce chemicals such as surfactants, acids, solvents (alcohols and ketones), and gases (primarily CO_2) that can be effective in mobilizing crude oil under reservoir conditions. Microbial growth and polymer production in porous media have been shown to improve the sweep efficiency by permeability modification. In general, cost-effective MEOR methods are best applied in shallow, sandstone reservoirs in mature producing fields.

The function of aerobic MEOR is based on the ability of oil-degrading bacteria to reduce the interfacial tension between oil and water. This process implies pumping water containing oxygen and mineral nutrients into the oil reservoir to stimulate growth of aerobic oil-degrading bacteria. Based on core flood experiments, the amount of bacterial biomass responsible for dislodging the oil can be calculated. The process is limited by the amount of oxygen available to the bacteria to degrade the oil. The bacterial biomass is more efficient than synthetic surfactants in dislodging the oil [1694].

Experiments have shown that bacterial cells may penetrate a solid porous medium with at least 140-mD permeability and that a bacterial population may be established in such a medium if suitable substrates are supplied. Enhanced oil-recovery organism suitability is governed by parameters such as capacity to produce a surfactant/cosurfactant, cell morphology and relationship of bacterial size to pore size, and pore size distribution of the porous rock. The activity of the organism is directly affected by conditions in the reservoir, such as oxygen availability, temperature, pressure, and substrate availability [269].

A physical model to predict the large-scale application for MEOR has been developed. This model simulates both the radial flow of fluids toward the wellbore and bacteria transport through porous media [1235].

Field studies of MEOR processes require routine monitoring to determine the effects that microorganisms exert in the release of oil from petroleum-bearing formations. Careful monitoring of oil production, flow rates, oil/water ratios, temperature, pH, viscosity, ionic strength, and other factors allows observation of real changes that occur as a result of microbial activities after selected microbes are injected. Simple techniques to determine microbial counts can be used to determine viability and transport of injected microbes. The effect of injected energy sources, such as molasses, on indigenous microbes inhabiting a reservoir can also be detected [431].

For example, a process for recovering hydrocarbons from a subterranean, hydrocarbon-bearing formation comprises the following:

1. Introducing microbes into the formation, the microbes being effective to render at least a portion of the hydrocarbons in the formation more easily recoverable
2. Passing electrical energy through at least a portion of the formation to increase the mobility of the microbes in the formation
3. Recovering hydrocarbons from the formation

The specific microbes used depends on many factors, for example, the particular formation involved, the specific hydrocarbons in the formation, and the desired microbial action on these formation hydrocarbons. The microbes may be aerobic or anaerobic and may or may not require one or more additional nutrients (e.g., naturally ocurring or injected) to be included in the formation. Highly mobile microbes, such as flagellated or ciliated bacilli, are useful. The microbes are sized so that they are mobile in the connate water of the formation [966].

Bacillus licheniformis produces a water-insoluble levan that has potential application as a selective plugging agent in MEOR. The microorganisms grow on sucrose, glucose, and fructose but produce levan only on sucrose. Thus plugging may be selectively controlled in the reservoir by substrate manipulation. Oil reservoirs that have a temperature of less than 55° C, a pH between 6 and 9, a pressure less than 500 atm, and a salt concentration of 4% or less are potentially suitable [1480].

A possible approach to MEOR consists of the additional aeration of the water injected into the formation, together with the addition of mineral salts of nitrogen and phosphorus. The result is the activation of the vital activities of aerobic microorganisms and the oxidation of the residual oil. The metabolic products of the petroleum-oxidizing bacteria are CO_2 and water-soluble organic compounds. These compounds enter the nonoxygenated zone of the formation and can act as oil-recovery agents. The compounds also may serve as additional substrates for anaerobic bacteria, particularly for methanogens. The methane formed can be easily recovered. It increases the mobility of the oil in place [892, 1664].

Potential Health Hazard of Bacteria

Practically all life forms may be infected by one or more kinds of microorganisms. Some of these have mutual advantage, such as in symbiosis, whereas some result in a disease of the host. The use of bacteria in MEOR operations necessitates a consideration of possible untoward effects against man and other living creatures. Because large numbers of bacteria are going to be placed into the ground and possibly come into direct contact with the oil field workers who know little about them, it is necessary to closely examine possible hazards that may be associated with the routine use of bacteria in oil field endeavors relating to enhanced oil recovery [744].

Metabolism

MEOR methods mainly utilize the metabolites (biosurfactant, biopolymer, organic acid, and biogas) generated in situ or ex situ by bacteria to improve the oil phase mobility. In situ MEOR is mainly targeted toward the residual oil left after primary production or secondary production by waterflooding, and its success depends strongly on the penetration and the stability of recovering agents. To contact trapped oil with bacteria that have favorable oil-displacement properties, the microbes must be transported from a wellbore to locations deep within the reservoir [899].

Salt-tolerant clostridium was shown to increase oil production by in situ production of gas and solvents [743].

When microbial activity develops in a subsurface geologic environment, the geologic, mineralogic, hydrologic, and geochemical aspects of the environment will have a profound effect on the microorganisms and, in turn, the microbial population will have some effect on the rocks and fluids. The most significant geologic changes are:

1. The precipitation of dissolved minerals, especially carbonates
2. The change of permeability caused by precipitation in pore throats
3. A change of porosity, either an increase or decrease, depending on the equilibria of dissolved salts and products of organic acids [270]

Various bacterial species have proven useful in MEOR. The principle is based on the species' biochemical byproducts produced, such as gases, surfactants, solvents, acids, swelling agents, and cosurfactants, which facilitate the displacement of oil. In field experiments, in situ fermentation is often desirable for producing a great quantity of gases. *Clostridium hydrosulfuricum 39E* was found to have surface-active properties during simulated enhanced oil recovery experiments [1874].

Key mechanisms important for improved oil mobilization by microbial formulations have been identified, including wettability alteration, emulsification, oil solubilization, alteration in interfacial forces, lowering of mobility ratio, and permeability modification. Aggregation of the bacteria at the oil-water-rock interface may produce localized high concentrations of metabolic chemical products that result in oil mobilization. A decrease in relative permeability to water and an increase in relative permeability to oil was usually observed in microbial-flooded cores, causing an apparent curve shift toward a more water-wet condition. Cores preflushed with sodium bicarbonate showed increased oil-recovery efficiency [355].

Microorganisms inhabiting petroleum-bearing formations or introduced into subterranean environments are subject to extremes of redox potential, pH, salinity, temperature, pressure, ecologic pressure, geochemistry, and energy and nutrient availability. Successful MEOR requires the selection, injection, dispersion, metabolism, and persistence of organisms with properties that facilitate the release of residual oil and the coinjection of growth-effective nutrients into the hostile environments that characterize petroleum reservoirs [1605].

Microbial Control of the Production of Sulfide

A microbial process was developed for controlling the production of hydrogen sulfide by sulfate-reducing bacteria, using mutant strains of *Thiobacillus denitrificans*. *T. denitrificans* oxidizes sulfide to sulfate, using oxygen or nitrate as the electron acceptor, but is inhibited by sulfide concentrations above 100 to 200 µM. A mutant of *T. denitrificans* resistant to glutaraldehyde (40 mg/liter) and sulfide (1500 µM) was obtained by repeated subculturing at increasing concentrations of the inhibitors. This strain prevented the accumulation of sulfide by *Desulfovibrio desulfuricans* when both organisms were grown in liquid medium or in Berea sandstone cores. The wild-type strain of *T. denitrificans* did not prevent sulfide accumulation by *T. desulfuricans*. The mutant also prevented the accumulation of sulfide by a mixed population of sulfate-reducing bacteria enriched from an oil field brine. Fermentation balances showed that this strain stoichiometrically oxidized the biogenically produced sulfide to sulfate. The mutant grew at temperatures up to 40° C, in salinities up to 2%, and at pressures up to 120 atm [1193].

C. hydrosulfuricum 39E was found to have surface-active properties during simulated enhanced oil-recovery experiments [1875].

Microbial Ecology of Corrosion

Among the bacteria that can inhabit an oil reservoir are the sulfur bacteria that use sulfur compounds in their metabolism. These bacteria produce hydrogen sulfide, which has been responsible for extensive corrosion in the oil field. Thus exclusion of these bacteria from MEOR is highly desirable. The net effect of souring a reservoir is a decrease in the economic value of the reservoir [1835].

B. licheniformis JF–2 and *Clostridium acetogutylicum* were investigated under simulated reservoir conditions. Sandstone cores were equilibrated to the desired simulated reservoir conditions, saturated with oil and brine, and flooded to residual oil saturation. The waterflood brine was displaced with a nutrient solution. The MEOR efficiency was directly related to the dissolved gas/oil ratio. The principal MEOR mechanism observed in this work was solution gas drive [505].

Bacillus Licheniformis

The *B. licheniformis JF-2* strain produces a very effective surfactant under conditions typical of oil reservoirs. The partially purified biosurfactant from JF-2 was shown to be the most active microbial surfactant found, and it gave an interfacial tension against decane of 0.016 mN/m. An optimal production of the surfactant was obtained in cultures grown in the presence of 5% NaCl at a temperature of 45° C and pH of 7. The major endproducts of fermentation were lactic acid and acetic acid, with smaller amounts of formic acid and acetoin. The growth and biosurfactant formation were also observed in anaerobic cultures supplemented with a suitable electron acceptor, such as NaNO$_3$ [1106].

Strict Anaerobic Bacteria

Several strict anaerobic bacteria belonging to different phylogenetic taxons were isolated from the Tatar and Siberian oil fields under different physico-chemical conditions. All isolated strains are capable of producing oil-releasing compounds: biopolymers, organic acids, or gases. Methanogenic bacteria were shown to produce polysaccharides. The polysaccharide of *Methanococcoides euhalobius* was partly purified and characterized. Some acetogenic strains capable of producing volatile fatty acids from CO_2 and H_2 were isolated from stratal waters of the Tatar and Siberian oil fields [167].

In highly saline brines, which were collected from the Vassar Vertz Sand Unit, Payne County, Oklahoma, diverse populations of anaerobic, heterotrophic bacteria were present. All strains grew in a mineral salts medium containing glucose, yeast extract, and casamino acids in the presence of NaCl concentrations of up to 20% by weight [177].

Methanohalophilus

A methanogenic bacterium was isolated from oil reservoir brines by enrichment with trimethylamine. Methane production occurred only with trimethylamine compounds or methanol as substrates. Sodium ions, magnesium ions, and potassium ions were all required for growth. This organism appears to be a member of the genus *Methanohalophilus* based on substrate utilization and general growth characteristics [695].

Sulfate-Reducing Desulfovibrio

A sulfate-reducing bacterium was isolated by enrichment with a lactate-sulfate medium containing 3% NaCl. This isolate utilized lactate as an electron donor for sulfate reduction and contained desulfoviridin, typical of the genus *Desulfovibrio* [695].

Ultramicrobacteria

Selective plugging of high-permeability areas in a reservoir rock will increase the oil recovery during waterflooding. The injected water flows through the low-permeability, oil-bearing zones, pushing oil along its path. Many plugging agents that have been developed do not deeply penetrate the reservoir and often wash out when the injection pressure is reduced. On the other hand, ultramicrobacteria penetrate deep into formations and grow exponentially with nutrient stimulation to preferentially plug high-permeability zones [434].

The microorganisms reduce the nitrate and produce sulfuric acid, which eventually dissolves the rock formation, thus releasing oil. The microorganisms can be denitrifying thiobacilli, such as *T. denitrificans* [1667].

Lactic Acid Bacteria

Particularly preferred bacteria are those of the genera *Lactobacillus* or *Pediococcus*. Lactic acid produced by bacteria may also be used for removal of carbonate or iron scale in oil field equipment [395].

Scale Inhibitors as a Microbial Nutrient

Organic phosphates and organic phosphonates are known as *scale inhibitors*. At the same time, substances in this class can be nutrients for certain bacteria. Therefore a phosphorous nutrient injection system can both prevent scales and act as a nutrient in favorite cases [903, 904].

Interfacial Properties

Interfacial Tension

The interfacial tension plays an important role in the success of enhanced oil-recovery methods. An additional complication arises when the components undergo a chemical reaction or change.

The dynamic interfacial tension behavior of reacting acidic oil–alkaline solutions has been studied for both an artificially acidified synthetic oil and a real crude oil at various concentrations [131, 132] with either a drop volume tensiometer or a spinning drop tensiometer.

The spinning drop technique measures the shape of the oil drop in the flooding solution in a capillary tube. An automatic measuring system has been developed by combining a video-image analysis, an automatic recording system, and a computer for calculation of the interfacial tension [1865].

Interfacial Rheologic Properties

The interfacial rheologic properties are extremely sensitive parameters toward the chemical composition of immiscible formation liquids [1053]. Therefore comparison and interpretation of the interfacial rheologic properties may contribute significantly to extension of the spectrum of the reservoir characterization, better understanding of the displacement mechanism, development of more profitable enhanced and improved oil-recovery methods, intensification of the surface technologies, optimization of the pipe line transportation, and improvement of the refinery operations [1056].

Interfacial rheologic properties of different crude oil–water systems were determined in wide temperature and shear rate ranges and in the presence of inorganic electrolytes, surfactants, alkaline materials, and polymers [1056].

Caustic Waterflooding. In caustic waterflooding, the interfacial rheologic properties of a model crude oil–water system were studied in the presence of sodium hydroxide. The interfacial viscosity, the non-Newtonian flow behavior, and the activation energy of viscous flow were determined as a function of shear rate, alkali concentration, and aging time. The interfacial viscosity drastically decreases in the presence of sodium hydroxide. Under favorable conditions, the change may exceed three or four orders of magnitude. Simultaneously, the sodium hydroxide effectively suppresses the non-Newtonian flow behavior of the interfacial layer [1055].

Tracers

The addition of tracer chemicals to an injection fluid provides information on the permeability of a reservoir. Small amounts of a tracer are added to the injected fluid and the distribution of the tracer at the production well is monitored with respect to time. Radioactive tracers and nonradioactive tracers exist.

Isotopic-labeled tracers behave like the components in the fluid of interest. For example, tritium water behaves like water. If less similar chemicals are used as tracers, selective adsorption, chemical reaction, and liquid-liquid distribution must be considered. The tracer must be chosen so that the analytic method is sufficiently sensitive to detect the tracer in the desired amounts.

Tritiated or ^{14}C-tagged hydrocarbons (including tritium gas) can be measured by using a liquid scintillation counter or a gas proportional counter [1717, 1718].

Isotopic tracers are not exclusively radioactive. For instance, ^{13}C is a nonradioactive element that is suitable for a nonradioactive labeling technique.

Application of Tracers

Sensitive analytic procedures enable detection and measurement of very low tracer levels. In tracer studies, an identifiable tracer material is injected through one or more injection wells into the reservoir being studied. Water or other fluid is then injected to push the tracer to one or more recovery wells in the reservoir. The output of the recovery wells is monitored to determine tracer breakthrough and flow through the recovery wells. Analysis of the breakthrough times and the flows yields important information regarding how to perform the secondary or enhanced recovery processes.

A sharp breakthrough of tracers in a two-well tracer test is achieved by the following method [1672]:

1. Injecting a solution consisting of a water-soluble tracer and a partitioning tracer that distributes between the formation oil and water into the formation through a temporary injection well
2. Discontinuing injection into the temporary injection well after a slug of the tracer solution has been injected
3. Producing formation fluids from the production well

4. Monitoring the concentration of each tracer and the volumes of fluids produced from the producing well borehole
5. Determining the formation of residual oil saturation from the chromatographic separation of the water-soluble tracer and the partitionable tracer

Retention of the Tracer

To improve the evaluation of a water and gas pilot, tracers were injected in the gas phase at the beginning of the first two-gas injection periods. Perfluoromethylcyclopentane and perfluoromethylcyclohexane were used. In laboratory studies, these compounds were shown to have a higher partitioning to the oil phase than did tritiated methane. This caused a minor retention of the tracer [518, 1119].

Radioactive Tracers

Water tracers were tested for use in carbonate reservoirs. Seven substances were tested: tritiated water (HTO) and the ions $S^{14}CN^-$, $^{36}Cl^-$, $^{131}I^-$, $^{35}SO_4^{2-}$, $H^{14}CO^{3-}$, and $^{22}Na^+$. HTO is the ideal water tracer, although $S^{14}CN^-$ and $^{36}Cl^-$ may be considered as near-ideal tracers for water flow in chalk. However, $^{131}I^-$ and $^{35}SO_4^{2-}$ show a more complicated behavior because of ion exclusion, adsorption, desorption, and chemical reactions [186].

Nonradioactive Tracers

Analysis of the halohydrocarbons, halocarbons, and sulfur hexafluoride is usually achieved by gas chromatography that is equipped with an electron capture detector. Complex metal anions, such as cobalt hexacyanide, are used as nonradioactive tracers in reservoir studies. The cobalt in the tracer compound must be in the complex anion portion of the molecule, because cationic cobalt tends to react with materials in the reservoir, leading to inaccurate analytic information [1226].

> In most production reservoirs, the produced brines are injected into the formation for purposes of maintaining reservoir pressure and avoiding subsidence and environmental pollution. In the case of geothermal fields, the brines are also injected to recharge the formation. However, the injected brines can adversely affect the fluids produced from the reservoir. For example, in geothermal fields, the injected brine can lower the temperature of the produced fluids by mixing with the hotter formation fluids. In order to mitigate this problem, the subsurface paths of the injected fluids must be known.

Tracers have been used to label fluids in order to track fluid movement and monitor chemical changes of the injected fluid. Radioactive materials are one class of commonly used tracers. These tracers have several drawbacks. One drawback is that they require special handling because of the danger posed to personnel and the environment. Another drawback is the alteration by the radioactive materials of the natural isotope ratio indigenous to the reservoir—thereby interfering with scientific analysis of the reservoir fluid characteristics. In addition, the half life of radioactive tracers tends to be either too long or too short for practical use.

A number of organic compounds are suitable for use as tracers in a process for monitoring the flow of subterranean fluids. The following traces have been proposed: benzene tetracarboxylic acid, methylbenzoic acid, naphthalenesulfonic acid, naphthalenedisulfonic acid, naphthalenetrisulfonic acid, alkyl benzene sulfonic acid, alkyl toluene sulfonic acid, alkyl xylene sulfonic acid, α-olefin sulfonic acid, salts of the foregoing acids, naphthalenediol, aniline, substituted aniline, pyridine, substituted pyridines [883].

Thermal Stability of Alkylbenzene Sulfonate

The thermal degradation of alkylbenzene sulfonates in alkaline media is important because of the application at elevated temperatures. The half-lives, with respect to thermal degradation, of several commercially available sulfonates were estimated at hundreds to thousands of years at 204° C. The degradation mechanism was predominately a clipping of the alkyl chain to yield an alkylbenzene sulfonate with the phenyl group attached to the α-carbon; however, desulfonation also occurred [1624].

Asphaltene Deposition

Asphaltenes are components of crude oils. They contain numerous individual compounds, especially high–molecular-weight condensed aromatic components with heteroatoms. Because of the complexity of their chemistry, asphaltenes are summarized as the oil fraction that is soluble in benzene, but not in n-pentane. In crude oil, asphaltenes are normally present as a colloidal dispersion. Such a dispersion is stabilized by oleoresins. During the production, refining, transportation, and storage of crude oil, asphaltenes may precipitate. The precipitation of asphalts, asphaltene, and other organics in the porous media near the wellbore will reduce the permeability. Commonly encountered causes of such a precipitation are a temperature drop or a composition change (e.g., vaporization of highly volatile components). Asphaltenes may also precipitate during flow through porous media. In particular, CO_2 flooding during the production process may bring about flocculation or precipitation of asphaltenes. Partial esters of phosphoric acid with carboxylic acids are dispersants for asphaltenes [1225].

Furthermore, the injection of organic aromatic solvents and soaking is a feasible method to remove the precipitates [924]. The precipitation of asphalt from crude oil can be reduced by adding an N,N-dialkylamide of a fatty acid [1525, 1527]. When asphaltenes are precipitated out, they can be removed from the walls of a well, pipeline, and so forth by washing with a hydrocarbon solvent. However, it has been shown that isopropyl benzoate is exceptionally useful as a solvent for asphaltene removal [1583].

Stabilizer Dispersant

The addition of hydrogenated castor oil to a copolymer of acrylamide and sodium acrylate formulation will suspend the copolymer and retard the settling process [1657].

Reservoir Properties

Reservoir Models

There are several simulators for modeling the processes in enhanced oil recovery; refer to Table 16–5.

Table 16–5
Simulators and Models for Modeling Enhanced Oil Recovery

Simulator	References
Second-order Godunov-type finite difference for 2-dimensional, 3-component incompressible polymer floods	[845]
Alkaline/surfactant/polymer compositional reservoir simulator, 3-dimensional compositional reservoir simulator, for high-pH chemical flooding processes	[178]
New flux correcting (NFC)	[1659]
PC-GEL 3-dimensional, 3-phase (oil, water, and gas) permeability modification simulator	[348]
Fully implicit total-variation-diminishing (TVD) high-order algorithm for compositional simulation and chemical flooding simulator	[1116]
Compositional chemical flooding simulator (UTCHEM)	[441, 1477]
Front-tracking model for in situ combustion oil recovery	[1515]
Caustic waterflooding	[179, 889, 1550]

Profile Control

Profile control occurs by artificially changing the permeability as is done in water shutoff; for more information, see Chapter 18.

Acrylamide Polymers

The permeability of a high-permeability zone in a high-temperature oil can be reduced by placing a gellable polymeric aqueous solution and gelling this solution in situ [1122, 1123]. The gel is made of a copolymer or a mixture of a synthetic polymer and a biopolymer (e.g., xanthan gum) [1121]. Crosslinking agents are trivalent chromium ions [1706] or aldehydes. The pH is adjusted to a value within the range of 1.5 to 5.5, depending on the desired gelling time. Antisyneresis properties can be obtained with various organic acids and their alkali metal or ammonium salts. A delay in gelation also occurs when malonic acid is used [24]. Acrylamides also can be crosslinked in situ by a hydroxyphenylalkanol (e.g., o-hydroxyphenylmethanol [1249] or furfuryl alcohol, formaldehyde [1248]).

Melamine and Phenol–Formaldehyde Resins

Melamine resins [1617] and phenol–formaldehyde resins [1620] can be gelled in situ to reduce the permeability. Various classes of polymers can be gelled by similar principles [882].

Latex

Latex particles may flocculate when injected in a reservoir with high formation temperatures. When the particles flocculate, shrink, and harden, they form a more effective blocking agent than the dispersed, expanded, and softer particles [1654].

In Situ Carbonate Precipitation

Carbon dioxide flooding is the most promising enhanced oil-recovery method. To overcome the tendency of CO_2 to bypass the smaller pores containing residual oil, one approach is to plug the larger pores by chemical precipitation. Several relatively inexpensive water-soluble salts of the earth alkali group react with CO_2 to form a precipitate.

Laboratory experiments have indicated that carbonate precipitation can alter the permeability of the core samples under reservoir conditions. The precipitation reduces the gas permeability in favor of the liquid permeability. This indicates that precipitation occurs preferentially in the larger pores.

Once the precipitate is formed in the larger pores, the subsequent fluid flow will be diverted to relatively smaller pores, thereby increasing the sweep efficiency. Additional experimental work with a series of connected cores suggested that the permeability profile can be successfully modified. However, pH control plays a critical role in the propagation of the chemical precipitation reaction [48].

In Situ Silica Cementation

The permeability profile of a formation in which temperatures higher than 90° C are encountered is modified by the following process: Initially, an aqueous solution of an alkali metal hydroxide, ammonium hydroxide, or organoammonium hydroxide is injected into a zone of greater permeability in a formation. After this, a spacer volume of a water-miscible organic solvent is injected into the zone. Then a water-miscible organic solvent containing an alkylpolysilicate is injected into the greater permeability zone. A silica cement is formed in situ, thereby substantially closing off the higher permeability zone to fluid flow. Finally, a steamflooding, waterflooding, carbon dioxide stimulation, or fireflooding enhanced oil-recovery operation is commenced in a lower permeability zone [1618, 1619].

Hydratable Clay

Hydratable clays may be used as plugging agents for profile control. The swelling of the clay is desirable, unlike when formation damage occurs [1899]. First, an aqueous solution containing the salt of certain cations that inhibit clay swelling is prepared. The cations K^+, Ca^{2+}, and Mg^{2+}, among others, are inhibitive of clay swelling. The clay slurry is introduced into the formation, where it enters the channels of high permeability. There the slurry contacts the NaCl brine solution already present in natural or injected drive fluids, and the inhibitive cations bound to the clay particles are replaced by Na^+ ions, which attract water molecules and promote clay swelling. The Na^+ clay swells up to 10 times its original volume, causing the slurry to acquire a gel-like consistency. The clay gel formed is capable of blocking the flow of water and can resist a differential pressure gradient of up to 500 kPa/m.

Silicate Gel

Silicate gel enhances the sweep efficiency of a waterflood, gasflood, or steamflood operation by reducing the permeability of the high-permeability zones. Weak acids may be added to control gel generation rate [377].

Formation Damage

The in situ release of fine particles in a porous medium resulting from changes in the colloidal character of the fines has been studied. Changes in the electrolytic condition of the permeating fluid induce damage. The results showed that high pH and low salinity cause the fines to be released. The release causes a drastic decline in the permeability of the medium. These findings establish the interplay between salinity changes, cation exchange, and pH during a water shock and elucidate the vital role of the ion-exchange process in formation damage [1781].

Formation damage caused by clay migration may be observed when the injected brine replaces the connate water during operations such as waterflooding, chemical flooding including alkaline, and surfactant and polymer processes. These effects can be predicted by a physicochemical flow model based on cationic exchange reactions when the salinity decreases [1665]. Other models have also been presented [345, 1245].

The pH variation of the flowing fluid suggests that chemical reactions are occurring in the formation. A high pH promotes formation damage by particle deposition within porous media. The permeability reduction is minimized by using brines and high oil recoveries. Suspended solid particles are released and moved with the injection water when the salt concentration drops below the critical salt concentration, causing a reduction of permeability and eventually formation damage [124].

Wettability

The wettability of the rock is responsible for the behavior of a reservoir subjected to any oil-recovery process. Because the chemical composition of the mineral surface is mostly responsible for its wetting behavior, the relationship between wettability and chemical composition of the surface is key information.

Most suitable for the examination of the surface is x-ray photoelectron spectroscopy, whereas the wettability determination can be established by a detailed interpretation of core flooding experiments and wettability index measurements. The results of such studies show that the organic carbon content in the surface is well correlated with the wetting behavior of the material characterized by petrophysical measurements [1467, 1468].

Flooding of Oil in Chalk

Chalk reservoirs encounter some specific problems during secondary recovery of oil by waterflooding. Displacement experiments in several formations indicated that the shape of the leaching front depends not only on the nature of the fluids but also on the morphology of the formations. The following must be distinguished from each other:

- Voids filled with unrecoverable oil
- Easily accessible pores
- Preferential paths

The injection of water in oil-saturated chalk induces a regular front of leaching in some formations (and an irregular front in others), with a high percentage of preferential paths, especially when chalk contains a high proportion of rounded grains [1241].

Soil Remediation

A chemical-enhanced oil-recovery technology can be used to remove oily contaminants from soil. Laboratory studies demonstrated that a variety of alkaline-surfactant combinations can be used with a polymer to reduce the residual oil saturation in waterflooding [1435].

Polyaromatic hydrocarbons absorb strongly to humus and other soil components, rendering these contaminants difficult to remove by thermal, physical, or chemical means, and unavailable for biodegradation. To desorb polyaromatic hydrocarbons from soil, surfactant flooding processes and soil-washing processes or treatments to enhance the biodegradation of polyaromatic hydrocarbons have been considered.

However, surfactant flooding may contaminate groundwater. Soil washing requires excavation and biodegradation of polyaromatic hydrocarbons and is incomplete even with surfactants. Biodegradable surfactants that can form reasonably stable foams in the presence of up to 50% ethanol have been developed. These ethanol-based foams can readily desorb the polyaromatic hydrocarbons from gas plant soils and are moving well through soils at pressures of 1.5 psi/ft or less [964, 965]. A partially hydrolyzed copolymer of acrylamide and n-octylacrylamide together with sodium alkyl sulfates has also been described for in situ decontamination by flooding operations [1790].

Chapter 17

Hydraulic Fracturing Fluids

Hydraulic fracturing is a technique to stimulate the productivity of a well. A hydraulic fracture is a superimposed structure that remains undisturbed outside the fracture. Thus the effective permeability of a reservoir remains unchanged by this process. The increased productivity results from increased wellbore radius, because in the course of hydraulic fracturing, a large contact surface between the well and the reservoir is created.

Stresses and Fractures

Hydraulic fracturing belongs to the newer techniques in petroleum sciences, not older than approximately 40 years. The classic treatment [869] of hydraulic fracturing states that the fractures are approximately perpendicular to the axis of the least stress. For most deep reservoirs, the minimal stresses are horizontal; therefore in fracturing, vertical stresses will occur. The actual stress can be calculated by balancing the (vertical) geostatic stress and the horizontal stress by the common tools of the theory of elasticity. For example, the geostatic stress must be corrected in a porous medium filled with a liquid having a poroelastic constant and hydrostatic pressure. The horizontal stress can be calculated from the corrected vertical stress with the *Poisson ratio*. Therefore, under some circumstances, in particular in shallow reservoirs, horizontal stresses can also be created rather than vertical stresses. The possible modes of stresses are summarized in Table 17–1.

Fracture Initialization Pressure

Knowledge of the stresses in a reservoir is essential to get information about the pressure at which initialization of a fracture can take place. The upper bound of the fracture initialization pressure can be estimated using a formula given

Table 17–1
Modes of Stesses in Fractures

p_b	Fracture initialization pressure
$3s_{H,min}$	Minimal horizontal stress
$s_{H,max}$	Maximal horizontal stress (= minimal horizontal stress + tectonic stress)
T	Tensile strength of rock material
p	Pore pressure

by Terzaghi [1807], which states that

$$p_b = 3s_{H,min} - s_{H,max} + T - p \qquad (17\text{–}1)$$

Fracture Closure Pressure

The closure pressure indicates the pressure at which the width of the fracture becomes zero. This is normally the minimal horizontal stress.

Pressure Decline Analysis

The pressure response during a fracturing job provides important information about the success of the operation. The fluid efficiency can be estimated from the closure time.

Comparison of Stimulation Techniques

In addition to hydraulic fracturing, there are other stimulation techniques such as acid fracturing or matrix stimulation. Hydraulic fracturing finds use not only in the stimulation of oil and gas reservoirs, but also in coal seams to stimulate the flow of methane from there.

Fracturing fluids are often classified into water-based fluids, oil-based fluids, alcohol-based fluids, emulsion fluids, and foam-based fluids.

Several reviews have been given in the literature [540, 551, 1091] dealing with the basic principles of hydraulic fracturing and the guidelines to select a particular formulation for a specific job. Polymer hydration, crosslinking, and degradation are the key processes influencing their application. Technologic improvements over the years have focused primarily on improved rheologic performance, thermal stability, and cleanup of crosslinked gels.

Action of a Fracturing Fluid

Fracturing fluids are injected into a subterranean formation for the following purposes:

- To create a conductive path from the wellbore extending into the formation
- To carry proppant material into the fracture to create a conductive path for produced fluids

Stages in a Fracturing Job

A fracturing job has several stages that involve injecting a prepad, a pad, a proppant containing fracturing fluid, and finally, a treatment with flush fluids. A prepad is a low-viscosity fluid used to condition the formation. The prepad may contain fluid loss additives, surfactants, and a defined salinity to prevent formation damage. The generation of the fractures takes place by injecting the pad, a viscous fluid, but without proppants. After the development of the fractures, a proppant must be injected to keep the fractures open. The job ends eventually with a cleanup stage in which flush fluids and other cleanup agents are applied. The actual detailed time schedule depends on the particular system used.

After the completion of the fracturing treatment, the fluid viscosity should decrease to allow the placement of the proppant and a rapid fluid return through the fracture. It is important to control the time at which the viscosity break occurs. In addition, the degraded polymer should produce little residue to restrict the flow of fluids through the fracture.

Basic Constituents

A general review of commercially available additives for fracturing fluids is given in the literature [71]. Possible components in a fracturing fluid are listed in Table 17–2.

In particular, Table 17–2 reflects the complexity of a fracturing fluid formulation. Some additives may not be used together reasonably, such as oil-gelling additives in a water-based system. More than 90% of the fluids are water based. Aqueous fluids are economical and can provide control of a broad range of physical properties as a result of additives developed over the years.

Additives for fracturing fluids serve two purposes:

1. To enhance fracture creation and proppant-carrying capability
2. To minimize formation damage [780]

Table 17-2
Components in Fracturing Fluids

Component/category	Function/remark
Water-based polymers	Thickener, to transport proppant; reduces leak-off in formation
Friction reducers	Reduce drag in tubing
Fluid-loss additives	Form filter-cake; reduce leak-off in formation if thickener is not sufficient
Breakers	Degrade thickener after job or disable crosslinker (wide variety of different chemical mechanisms)
Emulsifiers	For diesel premixed gels
Clay stabilizers	For clay-bearing formations
Surfactants	Prevent water-wetting of formation
Nonemulsifiers	
pH-control additives	Increase stability of fluid (e.g., for elevated-temperature applications)
Crosslinkers	Increase viscosity of thickener
Foamers	For foam-based fracturing fluids
Gel stabilizers	Keeps gels active longer
Defoamers	
Oil-gelling additives	Same as crosslinkers for oil-based fracturing fluids
Biocides	Prevent microbial degradation
Water-based gel systems	Common
Crosslinked gel systems	Increase viscosity
Alcohol-water systems	
Oil-based systems	Used in water-sensitive formation
Polymer plugs	Used also for other tasks
Continuous mix gel concentrates	Premixed gel on diesel base
Resin-coated proppants	Proppant material
Intermediate- to high-strength ceramics	Proppant material

Additives that assist the creation of a fracture include viscosifiers, such as polymers and crosslinking agents; temperature stabilizers; pH control agents; and fluid loss control materials. Formation damage is reduced by such additives as gel breakers, biocides, surfactants, clay stabilizers, and gases.

Types of Hydraulic Fracturing Fluids

Table 17-3 summarizes the various types of fluids and techniques used in hydraulic fracturing.

Table 17–3
Various Types of Hydraulic Fracturing Fluids

Type	Remarks
Water-based fluids	Predominant
Oil-based fluids	Water sensitive; increased fire hazard
Alcohol-based fluids	Rare
Emulsion fluids	High pressure, low temperature
Foam-based fluids	Low pressure, low temperature
Noncomplex gelled water fracture	Simple technology
Nitrogen-foam fracture	Rapid cleanup
Complexed gelled water fracture	Often the best solution
Premixed gel concentrates	Improve process logistics
In situ precipitation technique	Reduces the concentration of the scale-forming ingredients [858, 859]

Comparison of Different Techniques

Certainly the optimal techniques depend on the type of reservoir. Reports that compare the techniques in a related environment are available. In the Kansas Hugoton field (Mesa Limited Partnership), several hydraulic fracturing methods were tested [403]. A method in which a complexed gelled water fracture was applied was the most successful when compared with a foam technique and with older and simpler techniques. The study covers some 56 wells where such techniques were applied.

Expert Systems

A PC-based interactive computer model has been developed to help engineers choose the best fluid and additives and the most suitable propping agent for a given set of reservoir properties [844, 1863]. The computer model also optimizes the treatment volume based on reservoir performance and economics. To select the fluids, additives, and propping agents, the expert system surveys stimulation experts from different companies, reviews the literature, and then incorporates the knowledge into rules using an expert system shell.

Modeling Fluid Leak-Off

The fluid leak-off during hydraulic fracturing can be modeled, calculated, and measured experimentally. Procedures for converting laboratory data to an estimate of the leak-off under field conditions have been given in the literature [1426].

Characterization of Fracturing Fluids

Historically, viscosity measurements have been the single most important method to characterize fluids in petroleum-producing applications. Whereas the ability to measure a fluid's resistance to flow has been available in the laboratory for a long time, a need to measure the fluid properties at the well site has prompted the development of more portable and less sophisticated viscosity-measuring devices [1395]. These instruments must be durable and simple enough to be used by persons with a wide range of technical skills. As a result, the Marsh funnel and the Fann concentric cylinder, both variable-speed viscometers, have found wide use. In some instances, the Brookfield viscometer has also been used.

However, it has been established that an intense control of certain variables may improve the execution of a hydraulic fracturing job and the success of a stimulation. Therefore an intense quality control is recommended [552, 553]. Such a program includes monitoring the breaker performance at low temperatures and measuring the sensitivity of fracturing fluids to variations in crosslinker loading, temperature stabilizers, and other additives at higher temperatures.

Rheologic Characterization

To design a successful hydraulic fracturing treatment employing crosslinked gels, accurate measurements of rheologic properties of these fluids are required. Rheologic characterization of borate-crosslinked gels turned out to be difficult with a rotational viscometer. In a laboratory apparatus, field pumping conditions (i.e., crosslinking the fluid on the fly) and fluid flow down tubing or casing and in the fracture could be simulated [1589]. The effects of the pH and temperature of the fluid and the type and concentration of the gelling agent on the rheologic properties of fluids have been measured.

These parameters have significant effect on the final viscosity of gel in the fracture. Correlations to estimate friction pressures in field size tubulars have been developed from laboratory test data. In conjunction with field calibrations, these correlations can aid in accurate prediction of friction pressure of borate-crosslinked fluids.

Characterization Methods

Zirconium-Based Crosslinking Agent

The concentration of a crosslinking agent containing zirconium in a gel is determined by first adding an acid to break the gel and converting the zirconium into the ionic (noncomplexed) form [340]. This is followed by the addition of

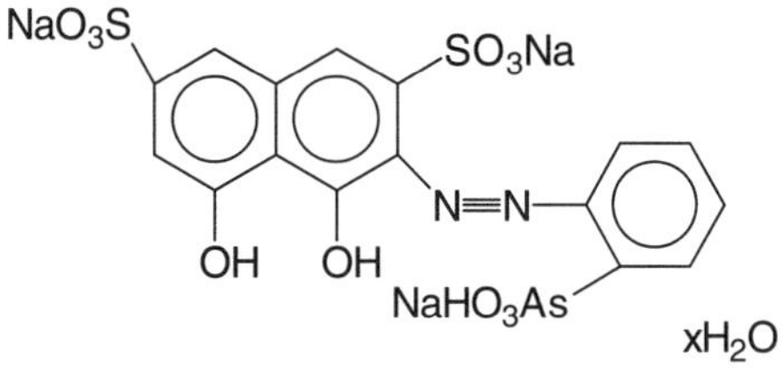

Figure 17-1. Colorimetric reagent to measure zirconium in gels.

Arsenazo (III) to produce a colored complex, which can be determined with standard colorimetric methods. Arsenic compounds are highly toxic.

Oxidative Gel Breaker

The concentration of an oxidative gel breaker can be measured by colorimetric methods, by periodically or continuously sampling the gel [341]. The colorimetric reagent is sensitive to oxidizing agents. It contains iron ions and thiocyanate. Thus the quantity of breaker added to the fracturing fluid can be controlled.

The method is based on the oxidation of ferrous ions to ferric ions, which form a deep red complex with thiocyanate.

$$Fe^{2+} \rightarrow Fe^{3+} + e^- \tag{17-2}$$

Size Exclusion Chromatography

Size exclusion chromatography [240, 665] has been used to monitor the degradation of the thickeners initiated by various oxidative and enzymatic breakers.

Gel Breaking of Guar

Maximal well production can be achieved only when the solution viscosity and the molecular weight of the gelling agent are significantly reduced after the treatment, that is, the fluid is degraded. However, the reduction of the fracturing fluid viscosity, the traditional method of evaluating these materials, does not necessarily indicate that the gelling agent has been thoroughly degraded also. The reaction between hydroxypropylguar and the oxidizing agent (ammonium peroxydisulfate) in an aqueous potassium chloride solution was studied [799] under controlled conditions to determine changes in solution viscosity and the weight average of the molecular mass of hydroxypropylguar.

Formation Damage in Gas Wells

Studies [664] on formation damage using artificially fractured, low-permeability sandstone cores indicated that viscosified fracturing fluids can severely restrict the gas flow through narrow fractures. Polysaccharide polymers such as hydroxypropylguar, hydroxyethylcellulose, and xanthan gum caused a significant reduction—up to 95% of the gas flow through the cracked cores. In contrast, polyacrylamide gels caused little or no reduction in the gas flow through cracked cores after a liquid cleanup. Other components of fracturing fluids (e.g., surfactants, breakers) caused less damage to gas flows.

Water-Based Systems

Thickeners

Naturally occurring polysaccharides and their derivatives form the predominant group of water-soluble species generally used as thickeners to impart viscosity to treating fluids [1092]. Other synthetic polymers and biopolymers have found ancillary applications. Polymers increase the viscosity of the fracturing fluid in comparatively small amounts. The increase in fluid viscosity of hydraulic fracturing fluids serves for improved proppant placement and fluid loss control. Table 17–4 summarizes polymers suitable for fracturing fluids.

Table 17–4
Summary of Polymers Suitable for Fracturing Fluids

Thickener	References
Hydroxypropylguar[a]	
Galactomannans[b]	[1237]
Hydroxyethylcellulose-modified vinyl phosphonic acid	[847]
Carboxymethylcellulose	
Polymer from N-vinyl lactam monomers, vinyl sulfonates[c]	[175]
Reticulated bacterial cellulose[d]	[1836]
Bacterial xanthan[e]	[839]

a) General purpose eightfold power of thickening in comparison to starch.
b) Increased temperature stability, used with boron-based crosslinkers.
c) High-temperature stability.
d) Superior fluid performance.
e) Imparts high viscosity.

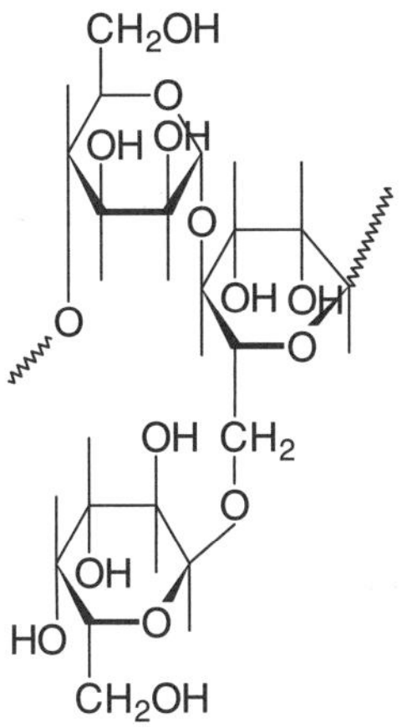

Figure 17-2. Structural unit of guar. In hydroxypropylguar, some of the hydroxyl groups are etherified with oxopropyl units.

Guar

Fracturing fluids have traditionally been viscosified with guar and guar derivatives. Actually, guar is a branched polysaccharide from the guar plant *Cyamopsis tetragonolobus*, originally from India, now also found in the southern United States, with a molar mass of approximately 220,000 Dalton. It consists of mannose in the main chain and galactose in the side chain. The ratio of mannose to galactose is 2:1. Polysaccharides with a mannose backbone and side chains unlike mannose are referred to as *heteromannans* according to the nomenclature of polysaccharides, in particular as *galactomannans*. Derivatives of guar therefore are sometimes called *galactomannans*.

Guar-based gelling agents, typically hydroxypropylguar, are widely used to viscosify fracturing fluids because of their desirable rheologic properties, economics, and ease in hydration. Nonacetylated xanthan is a variant of xanthan gum, which develops a synergistic interaction with guar that exhibits a superior viscosity and particle transport at lower polymer concentrations.

Anionic galactomannans, which are derived from guar gum, in which the hydroxyl groups are partially esterified with sulfonate groups that result from 2-acrylamido-2-methylpropane sulfonic acid and 1-allyloxy-2-hydroxypropyl sulfonic acid [1872], have been claimed to be suitable as thickeners. The composition is capable of producing enhanced viscosities, when used either alone or in combination with a cationic polymer and distributed in a solvent.

Boron-crosslinked galactomannan fracturing fluids have an increased temperature stability. The temperature stability of fracturing fluids containing galactomannan polymers is increased by adding a sparingly soluble borate with

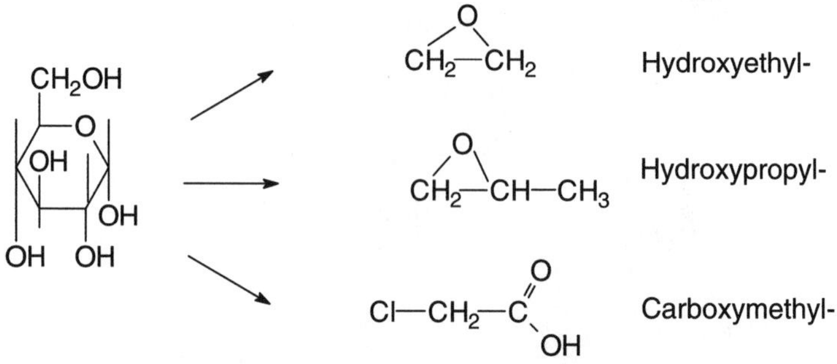

Figure 17-3. Modifiers for polyhydroxy compounds, shown for dextrose as the model compound.

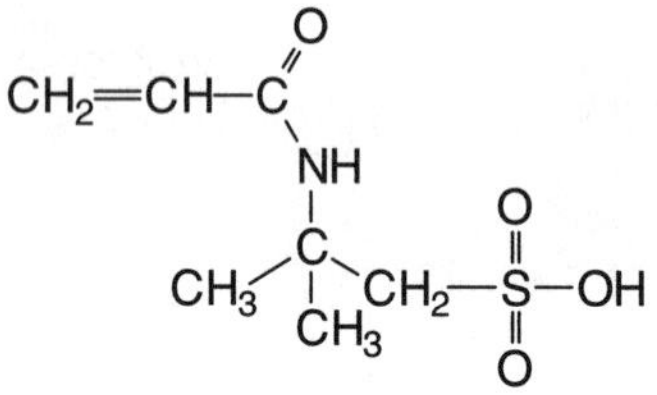

2-Acrylamido-2-methylpropane sulfonic acid

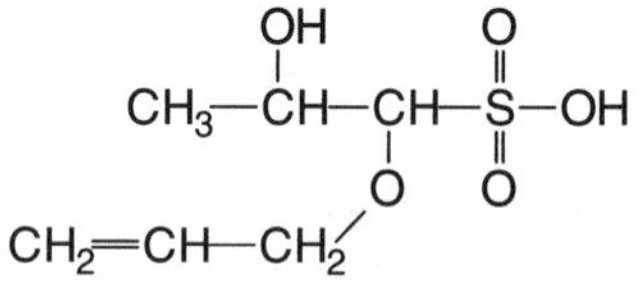

1-Allyloxy-2-hydroxypropyl sulfonic acid

Figure 17-4. Vinyl modifiers for guar gum: 2-acrylamido-2-methylpropane sulfonic acid, 1-allyloxy-2-hydroxypropyl sulfonic acid.

a slow solubility rate to the fracturing fluid. This provides a source of boron for solubilizing at elevated temperatures, thus enhancing the crosslinking of the galactomannan polymer. The polymer also improves the leak-off properties of the fracturing fluid.

Hydroxyethylcellulose

Hydroxyethylcellulose can be chemically modified by the reaction with vinyl phosphonic acid in the presence of the reaction product of hydrogen

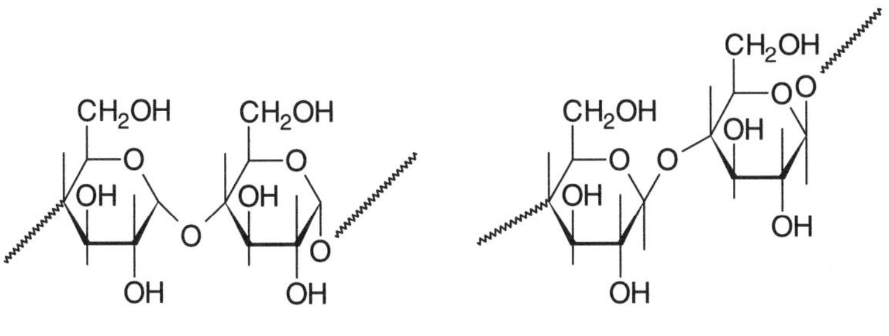

Figure 17–5. Amylose, cellulose. Amylose consists of a water-soluble portion, a linear polymer of glucose, the amylose; and a water-insoluble portion, the amylopectin. The difference between amylose and cellulose is the way in which the glucose units are linked. In amylose, α-linkages are present, whereas in cellulose, β-linkages are present. Because of this difference, amylose is soluble in water and cellulose is not. Chemical modification allows cellulose to become water soluble.

peroxide and a ferrous salt. The hydroxyethylcellulose forms a graft copolymer with the vinyl phosphonic acid. This type of modified hydroxyethylcellulose has been proposed as a thickener for hydraulic fracturing fluids [847]. Polyvalent metal cations may be employed to crosslink the polymer molecules to further increase the viscosity of the aqueous fluid.

Biotechnologic Products

Gellan Gum and Wellan Gum. Gellan gum is the generic name of an extracellular polysaccharide produced by the bacterium *Pseudomonas elodea*. Gellan gum is a linear anionic polysaccharide with a molecular mass of 500,000 Dalton. It is composed of 1,3-β-D-glucose; 1,4-β-D-glucuronic acid; 1,4-β-D-glucose; and 1,4-α-L-rhamnose.

Wellan gum is produced by aerobic fermentation. The backbone of wellan gum is identical to gellan gum, but it has a side chain consisting of L-mannose or L-rhamnose. It is used in fluid loss additives and is extremely compatible with calcium ions in alkaline solutions.

Reticulated Bacterial Cellulose. A cellulose with an intertwined reticulated structure, produced from bacteria, has unique properties and functionalities unlike other conventional celluloses. When added to aqueous systems, reticulated bacterial cellulose improves the fluid rheology and the particle suspension over a wide range of conditions [1836]. Test results showed advantages in fluid performance and significant economic benefits by the addition of reticulated bacterial cellulose.

Table 17–5
Variant Xanthan Gums

Number	Repeating units	Ratio
Pentamer	D-glucose: D-mannose: D-glucuronic acid	2:2:1
Tetramer	D-glucose: D-mannose: D-glucuronic acid	2:1:1

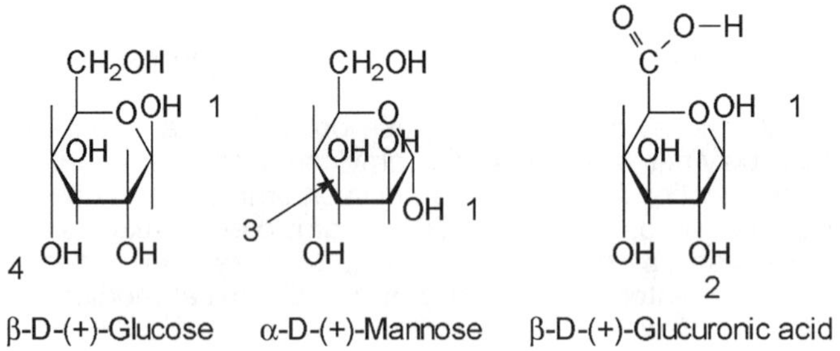

Figure 17–6. D-Glucose, D-mannose, and D-glucuronic acid.

Xanthan Gum. Xanthan gum is produced by the bacterium *Xanthomonas campestris*. Commercial productions started in 1964. Xanthans are water-soluble polysaccharide polymers with the following repeating units [502], as given in Table 17–5 and Figure 17–6.

The D-glucose moieties are linked in a β-(1,4) configuration. The inner D-mannose moieties are linked in an α-(1,3) configuration, generally to alternate glucose moieties. The D-glucuronic acid moieties are linked in a β-(1,2) configuration to the inner mannose moieties. The outer mannose moieties are linked to the glucuronic acid moieties in a β-(1,4) configuration.

Most of the xanthan gum used in oil field applications is in the form of a fermentation broth containing 8% to 15% polymer. The viscosity is less dependent on the temperature in comparison with other polysaccharides.

Miscellaneous Polymers

Acrylics. A copolymer of 2-ethylhexylacrylate and acrylic acid is not soluble either in water or in hydrocarbons. The ester units are hydrophobic and the acid units are hydrophilic. An aqueous suspension with a particle size smaller than 10 μ can be useful in preparing aqueous hydraulic fracturing fluids [776].

Polymer from N-Vinyl Lactams and Vinyl Sulfonates. A water-soluble polymer from N-vinyl lactam monomers or vinyl-containing sulfonate monomers

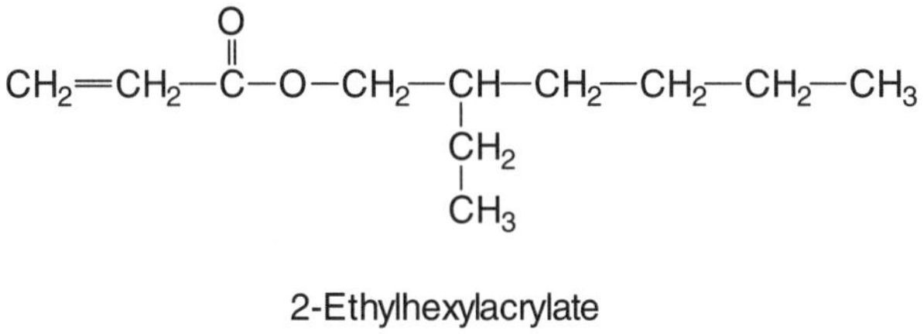

$$CH_2{=}CH_2{-}\overset{\overset{O}{\|}}{C}{-}O{-}CH_2{-}\underset{\underset{CH_3}{|}}{\underset{CH_2}{|}}{CH}{-}CH_2{-}CH_2{-}CH_2{-}CH_3$$

2-Ethylhexylacrylate

Figure 17–7. 2-Ethylhexylacrylate.

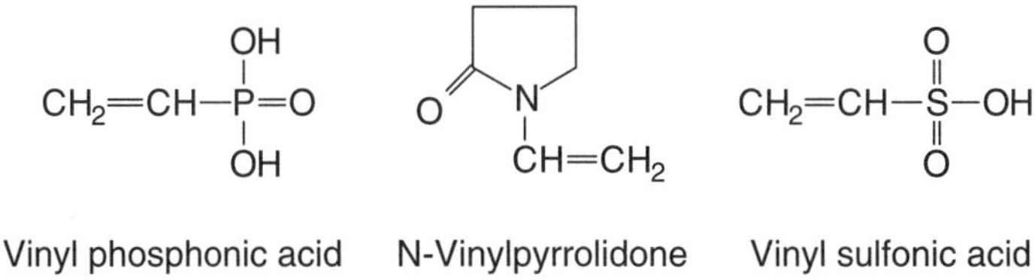

Vinyl phosphonic acid   N-Vinylpyrrolidone   Vinyl sulfonic acid

Figure 17–8. Monomers for synthetic thickeners: vinyl phosphonic acid, N-vinyl-2-pyrrolidone, vinyl sulfonic acid.

reduces the water loss and enhances other properties of well-treating fluids in high-temperature subterranean environments [175]. Lignites, tannins, and asphaltic materials are added as dispersants.

Concentrates

Historically, fracture-stimulation treatments have been performed by using conventional batch mix techniques. This involves premixing chemicals into tanks and circulating the fluids until a desired gelled fluid rheology can be obtained. This method is time-consuming and burdens the oil company with disposal of the chemically laden fluid if the treatment ends prematurely. Environmental concerns, such as when spillage occurs or disposal is involved, can be avoided if the fluid is capable of being gelled as needed. Thus a newer technology involves a gelling-as-needed technology with water, methanol, and oil [737]. This procedure eliminates batch mixing and minimizes handling of chemicals and base fluid. The customer is charged only for products used, and environmental concerns regarding disposal are virtually eliminated. Computerized chemical addition and monitoring, combined with on-site procedures, ensure quality control throughout a treatment. Fluid rheologies can be accurately varied during the treatment by varying polymer loading.

The use of a diesel-based concentrate with hydroxypropylguar gum has been evolved from the batch-mixed dry powder types [778]. The application of such a concentrate reduces system requirements. Companies can benefit

Table 17–6
Components of a Slurry Concentrate

Component	Example
Hydrophobic solvent base	Diesel
Suspension agent	Organophilic clay
Surfactant	Ethoxylated nonylphenol
Hydratable polymer	Hydroxypropylguar gum

Sorbitan monooleate   Ethoxylated nonylphenol

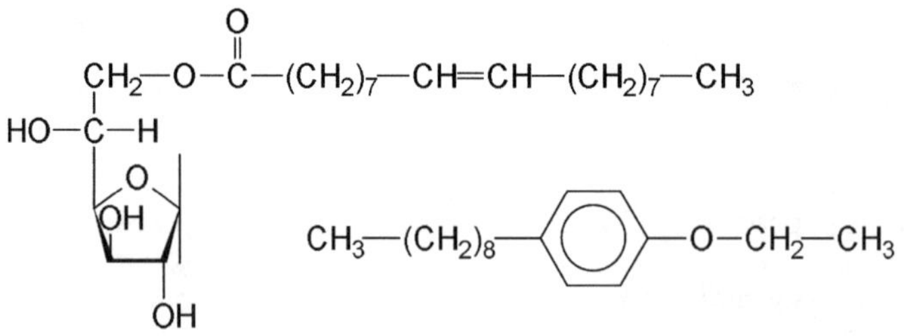

Figure 17–9. Surfactants in a slurry concentrate: sorbitan monooleate, ethoxylated nonylphenol.

from the convenience of the reduced logistic burden that comes from using the diesel hydroxypropylguar gum concentrate.

A fracturing fluid slurry concentrate has been proposed [238] that consists of the components shown in Table 17–6. Such a polymer slurry concentrate will readily disperse and hydrate when admixed with water at the proper pH, thus producing a high-viscosity aqueous fracturing fluid. The fracturing fluid slurry concentrate is useful in producing large volumes of high-viscosity treating fluids at the well site on a continuous basis.

Aqueous Concentrate Suspensions

Fluidized aqueous suspensions of 15% by weight or more of hydroxyethylcellulose, hydrophobically modified cellulose ether, hydrophobically modified hydroxyethylcellulose, methylcellulose, hydroxypropylmethylcellulose, and polyethylene oxide are prepared by adding the polymer to a concentrated sodium formate solution containing xanthan gum as a stabilizer [278]. The xanthan gum is dissolved in water before sodium formate is added. Then the polymer is added to the solution to form a fluid suspension of the polymers. The polymer suspension can serve as an aqueous concentrate for further use.

Friction Reducers

Low pumping friction pressures are achieved by delaying the crosslinking, but there are also specific additives available to reduce the drag in the tubings. The first application of drag reducers was using guar in oil well fracturing, now a routine practice.

Relatively small quantities of a bacterial cellulose (0.60 to 1.8 g/liter) in hydraulic fracturing fluids enhance their rheologic properties [1425]. Proppant suspension is enhanced and friction loss through well casings is reduced.

Fluid Loss Additives

Fluid loss additives are used widely as additives for drilling fluids. When fracturing zones have high permeability, concern exists about damage to the matrix from deeply penetrating fluid leak-off along the fracture or caused by the materials in the fluid that minimize the amount of leak-off.

Several fracturing treatments in high-permeability formations, which are characterized by short lengths, and often by disproportionate widths, exhibit positive posttreatment skin effects. This is the result of fracture face-damage [12]. If the invasion of the fracturing fluid is minimized, the degree of damage (e.g., permeability impairment resulting from filter-cake or polymer invasion) is of secondary importance. Thus if the fluid leak-off penetration is small, even severe permeability impairments can be tolerated without exhibiting positive skin effects. The first priority in designing fracture treatments should be maximizing the conductivity of the fracture. In high-permeability fracturing, using high concentrations of polymer-crosslinked fracturing fluids with fluid loss additives and breakers is recommended.

Materials used to minimize leak-off also have the potential to damage the conductivity of the proppant pack. High shear rates at the tip of the fracture may prevent the formation of external filter-cakes, increasing the magnitude of spurt losses in highly permeable formations. Therefore, particularly for fracturing tasks, nondamaging additives are needed. Enzymatically degradable fluid loss additives are available. Table 17–7 summarizes some fluid loss additives suitable for hydraulic fracturing fluids.

Degradation of Fluid Loss Additives

A fluid loss additive for fracturing fluids, which is a mixture of natural starch (cornstarch) and chemically modified starches (carboxymethyl and hydroxypropyl derivatives) plus an enzyme, has been described [1848, 1850]. The enzyme degrades the α-linkage of starch but does not degrade the β-linkage of guar and modified guar gums when used as a thickener. The starches can be

Table 17-7
Fluid Loss Additives for Hydraulic Fracturing Fluids

Chemical	References
Calcium carbonate and lignosulfonate[a]	[915, 917]
Natural starch and carboxymethyl and hydroxypropyl derivatives[b]	
Hydroxyethylcellulose with crosslinked guar gums[c]	[544, 1324, 1325]
Granular starch and particulate mica	[338]

a) Wellan or xanthan gum polymer can be added to keep the calcium carbonate and lignosulfonate in suspension.
b) Show synergistic effect, see text.
c) 500-mD permeability.

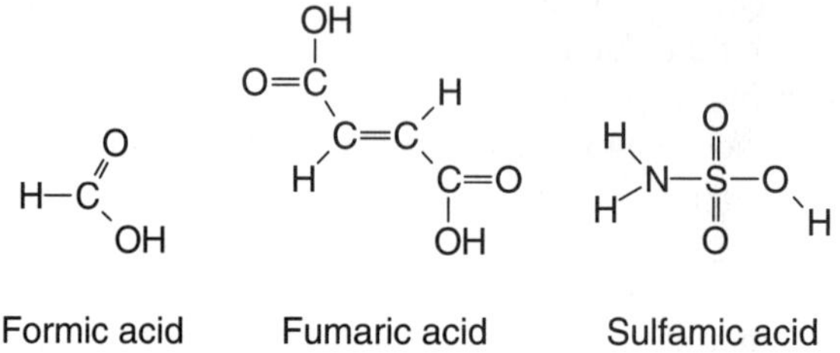

Formic acid   Fumaric acid   Sulfamic acid

Figure 17-10. Weak organic acids: formic acid, fumaric acid, and sulfamic acid.

coated with a surfactant, such as sorbitan monooleate, ethoxylated butanol, or ethoxylated nonylphenol, to facilitate the dispersion into the fracturing fluid. Modified starches or blends of modified and natural starches with a broad particulate size distribution have been found to maintain the injected fluid within the created fracture more effectively than natural starches [1849]. The starches can be degraded by oxidation or by bacterial attack.

Fluid Loss of Crosslinked Guars

Static leak-off experiments with borate-crosslinked and zirconate-crosslinked hydroxypropylguar fluids showed practically the same leak-off coefficients [1883]. An investigation of the stress-sensitive properties showed that zirconate filter-cakes have viscoelastic properties, but borate filter-cakes are merely elastic. Noncrosslinked fluids show no filter-cake–type behavior for a large range of core permeabilities, but rather a viscous flow dependent on porous medium characteristics.

Table 17–8
Common Buffer Solutions

Buffer	pK_A
Sulfamic acid/sulfamate	1.0
Formic acid/formate	3.8
Acetic acid/acetate	4.7
Dihydrogenphosphate/hydrogenphosphate	7.1
Ammonium/ammonia	9.3
Bicarbonate/carbonate	10.4
Fumaric acid/hydrogen fumarate	3.0
Benzoic acid/benzoate	4.2

Table 17–9
Composition of a Defoamer for Hydraulic Fracturing Fluids

Compound	Amount (% by wt)
C_6–C_{12} high-boiling mixture of ketones, ethers, alcohols, esters, and aldehydes	50 to 90
Sorbitan monooleate	10 to 50
Polyglycol M = 3800 Dalton	10

pH Control Additives

Buffers are necessary to adjust and maintain the pH. Buffering agents can be salts of a weak acid and a weak base. Examples are ammonium, potassium, sodium carbonates (caustic soda), bicarbonates, and hydrogen phosphates [1345]. Weak acids such as formic acid, fumaric acid, and sulfamic acid also are recommended. Common aqueous buffer ingredients are shown in Table 17–8.

For example, an increased temperature stability of various gums can be achieved by adding sodium bicarbonate to the fracturing fluid and thus raising the pH of the fracturing fluid to 9.2 to 10.4.

Defoamers

A defoamer and an antifoamer composition are described for defoaming aqueous fluid systems [1908]. The composition of a typical defoamer for hydraulic fracturing fluids is shown in Table 17–9.

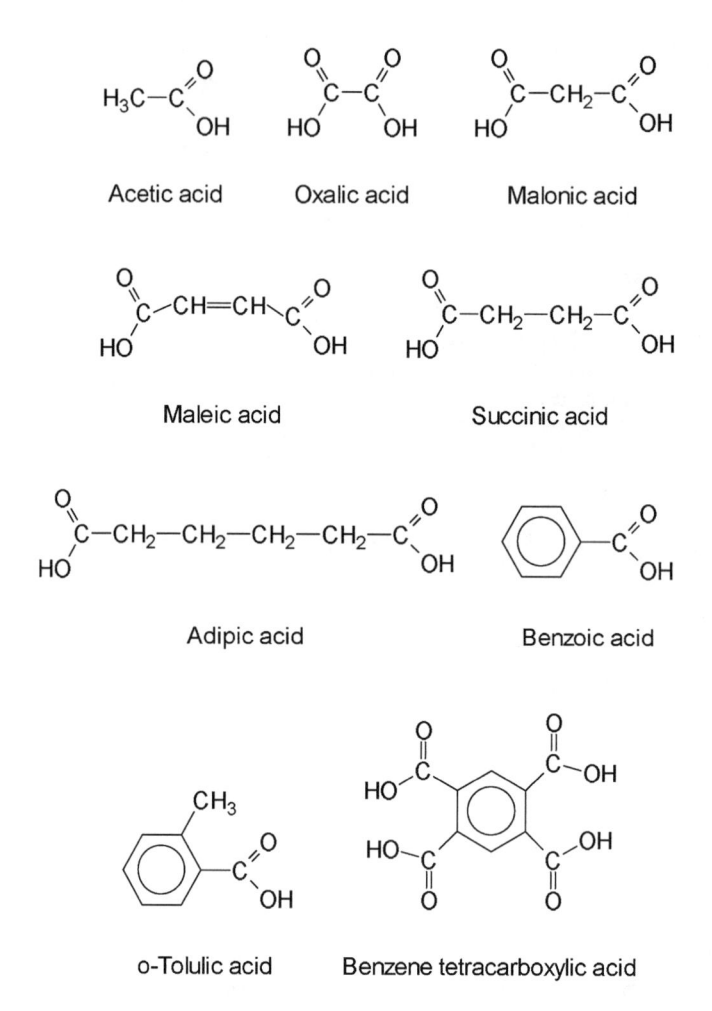

Figure 17–11. Carbonic acids: acetic acid, oxalic acid, malonic acid, maleic acid, succinic acid, adipic acid, benzoic acid, o-tolulic acid, and benzene tetracarboxylic acid.

Clay Stabilizers

Advances in clay-bearing formation treatment have led to the development of numerous clay-stabilizing treatments and additives. Most additives used are high–molecular-weight cationic organic polymers. However, it has been shown that these stabilizers are less effective in low-permeability formations [834].

The use of salts, such as potassium chloride and sodium chloride, as temporary clay stabilizers during oil well drilling, completion, and servicing, has been

Table 17–10
Clay Stabilizers

Compound	References
Ammonium chloride	
Potassium chloride[a]	[1870]
Dimethyl diallyl ammonium salt[b]	[1733]
N-alkylpyridinium halides	
N,N,N-trialkylphenylammonium halides	
N,N,N-trialkylbenzylammonium halides	
N,N-dialkylmorpholinium halides[c]	[829, 831]
Reaction product of a homopolymer of maleic anhydride and an alkyl diamine[d]	[1566]
Tetramethylammonium chloride and methyl chloride quaternary salt of ethylene-ammonia condensation polymer[d]	[9]
Quaternary ammonium compounds[e]	[763]

a) Added to a gel concentrate with a diesel base.
b) Minimum 0.05% by weight to prevent swelling of clays.
c) Alkyl equals methyl, ethyl, propyl, and butyl.
d) Synergistically retards water absorption by the clay formation.
e) Hydroxyl-substituted alkyl radials.

in practice for many years. Because of the bulk and potential environmental hazards associated with the salts, many operators have looked for alternatives to replace their use. Recent research has developed a relationship between physical properties of various cations (e.g., K^+, Na^+) and their efficiency as temporary clay stabilizers.

These properties were used to synthesize an organic cation (Table 17–10) with a higher efficiency as a clay stabilizer than the typical salts used in the oil industry to this point. These additives provide additional benefits when used in conjunction with acidizing and fracturing treatments. A much lower salt concentration can be used to obtain the same clay-stabilizing effectiveness [830, 833]. The liquid product has been proven to be much easier to handle and transport. It is environmentally compatible and biodegradable in its diluted form.

Recent developments have led to the synthesis of a new class of clay-stabilizing chemical additives capable of successfully stabilizing clays in very–low-permeability sandstones, that is, low molecular salts.

Biocides

A hydraulic fracturing fluid containing guar gum or other natural polymers can be stabilized against bacterial attack by adding heterocyclic sulfur

Table 17–11
Biocides

Compound	References
Mercaptobenzimidazole[a]	[933]
1,3,4-Thiadiazole-2,5-dithiol[b]	[931, 932, 934]
(= 2,5-dimercapto–1,3,4-thiadiazole),	
2-mercaptobenzothiazole,	
2-mercaptothiazoline,	
2-mercaptobenzoxazole,	
2-mercaptothiazoline,	
2-Thioimidazolidone	
(= 2-imidazolidinethion),	
2-thioimidazoline,	
4-ketothiazolidine-2-thiol,	
N-pyridineoxide-2-thiol	

a) Guar gum.
b) Xanthan gum and guar gum.

compounds. This method of stabilization prevents any undesired degradation of the fracturing fluid, such as reduction of its rheologic properties (which are necessary for conducting the hydraulic fracturing operation) at high temperatures. Biocides suitable for fracturing fluids are shown in Table 17–11 and Figures 17–12 and 17–13.

Surfactants

Surface-active agents are included in most aqueous treating fluids to improve the compatibility of aqueous fluids with the hydrocarbon-containing reservoir. To achieve maximal conductivity of hydrocarbons from subterranean formations after fracture or other stimulation, it is the practice to cause the formation surfaces to be water-wet. Alkylamino phosphonic acids and fluorinated alkylamino phosphonic acids adsorb onto solid surfaces, particularly onto surfaces of carbonate materials in subterranean hydrocarbon-containing formations, in a very thin layer. The layer is only one molecule thick and thus significantly thinner than a layer of water or a water-surfactant mixture on water-wetted surfaces [1421–1423]. These compounds so adsorbed resist or substantially reduce the wetting of the surfaces by water and hydrocarbons and provide high interfacial tensions between the surfaces and water and hydrocarbons. The hydrocarbons displace injected water, leaving a lower water saturation and an increased flow of hydrocarbons through capillaries and flow channels in the formation.

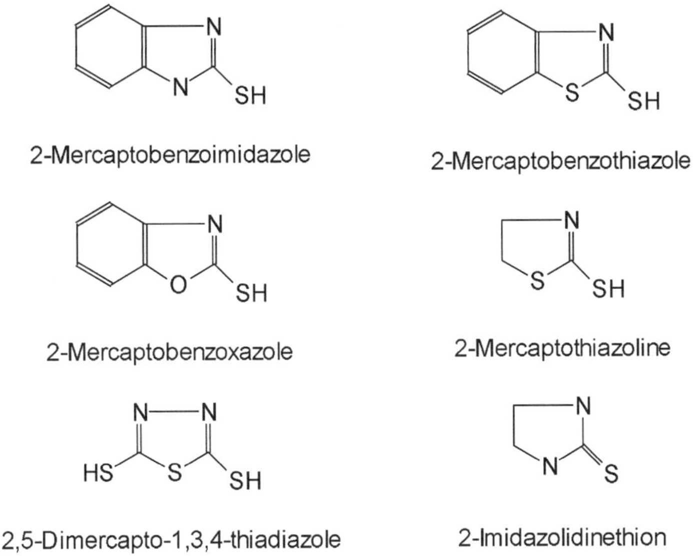

2-Mercaptobenzoimidazole 2-Mercaptobenzothiazole

2-Mercaptobenzoxazole 2-Mercaptothiazoline

2,5-Dimercapto-1,3,4-thiadiazole 2-Imidazolidinethion

Figure 17–12. Biocides for hydraulic fracturing fluids: 2-mercaptobenzoimidazole, 2-mercaptobenzothiazole, 2-mercaptobenzoxazole, 2-mercaptothiazoline, 2,5-dimercapto-1,3,4-thiadiazole, and 2-imidazolidinethion.

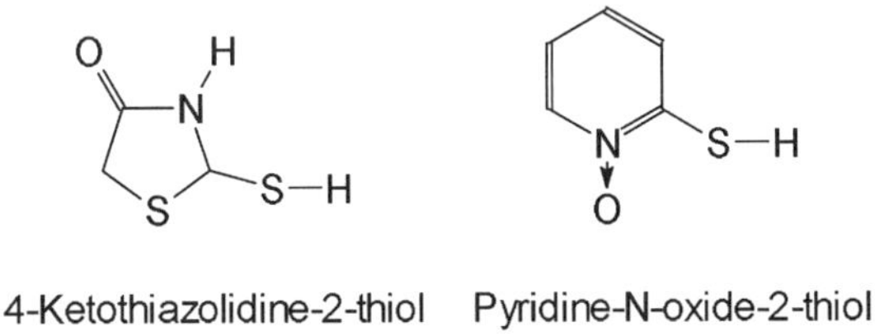

4-Ketothiazolidine-2-thiol Pyridine-N-oxide-2-thiol

Figure 17–13. 4-Ketothiazolidine-2-thiol, Pyridine-N-oxide-2-thiol.

A methyl quaternized erucyl amine [660] is useful for aqueous viscoelastic surfactant-based fracturing fluids in high-temperature and high-permeability formations.

Crosslinkers

Kinetics of Crosslinking

The rheology of hydroxypropylguar is greatly complicated by the crosslinking reactions with titanium ions. A study to better understand the rheology of the reaction of hydroxypropylguar with titanium chelates and how the rheology depends on the residence time, shear history, and chemical composition has been performed [145]. Rheologic experiments were performed to obtain information about the kinetics of crosslinking in hydroxypropylguar. Continuous flow and dynamic data suggest a crosslinking reaction order of approximately 4/3 and 2/3, respectively, with respect to the crosslinker and hydroxypropylguar concentration. Dynamic tests have shown that the shearing time is important in determining the final gel properties. Continued steady shear and dynamic tests show that high shear irreversibly destroys the gel structure, and the extent of the crosslinking reaction decreases with increasing shear. Studies at shear rates below $100\ s^{-1}$ suggest a shear-induced structural change in the polymer that affects the chemistry of the reaction and the nature of the product molecule.

Delayed Crosslinking

Delayed crosslinking is desirable because the fluid can be pumped down more easily. A delay is a retarded reaction rate of crosslinking. This can be achieved with the methods explained in the following section.

Borate Systems

Boric acid can form complexes with hydroxyl compounds. The control of the delay time requires control of the pH, the availability of borate ions, or both. Control of pH can be effective in freshwater systems [17]. However,

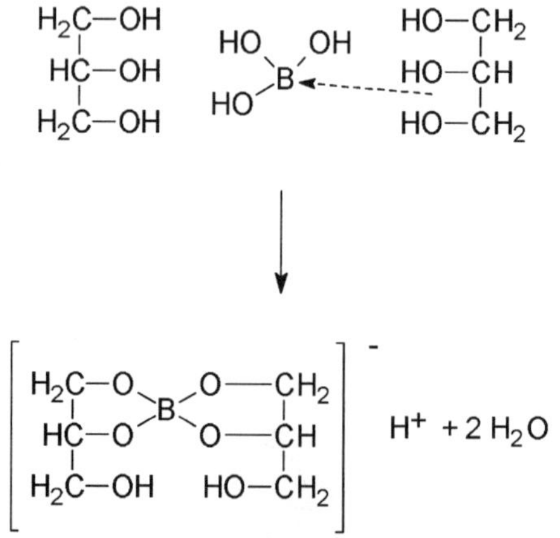

Figure 17–14. Formation of complexes of boric acid with glycerol. Three hydroxyl units form an ester and one unit forms a complex bond. Here a proton will be released that lowers the pH. The scheme is valid also for polyhydroxy compounds. In this case, two polymer chains are connected via such a link.

the control of borate is effective in both freshwater and seawater. This may be accomplished by using low-solubility borate species or by complexing the borate with a variety of organic species.

Borate-crosslinked fracturing fluids have been successfully used in fracturing operations. These fluids provide excellent rheologic, fluid loss, and fracture conductivity properties over fluid temperatures up to 105° C. The mechanism of borate crosslinking is an equilibrium process that can produce very high fluid viscosities under conditions of low shear [336].

Preparation of Fracturing Fluids.   A fracturing fluid is prepared in the following way:

1. Introducing a polysaccharide polymer into (sea) water to produce a gel
2. Adding an alkaline agent to the gel to obtain a pH of at least 9.5
3. Adding a borate crosslinking agent to the gel to crosslink the polymer [783]

A dry granular composition [781] can be prepared in the following way:

1. Dissolving from 0.2% to 1.0% by weight of a water-soluble polysaccharide in an aqueous solution
2. Admixing a borate source with the aqueous gel formed in step 1
3. Drying the borate-crosslinked polysaccharide formed in step 2
4. Granulating the product of step 3

A borate crosslinking agent can be boric acid, borax, an alkaline earth metal borate, or an alkali metal alkaline earth metal borate. The borate source, calculated as boric oxide, must be present in an amount of 5% to 30% by weight.

Borated starch compositions are useful for controlling the rate of crosslinking of hydratable polymers in aqueous media for use in fracturing fluids. The borated starch compositions are prepared by reacting, in an aqueous medium, starch and a borate source to form a borated starch complex. This complex provides a source of borate ions, which cause crosslinking of hydratable polymers in aqueous media [1552]. Delayed crosslinking takes place at low temperatures.

Delayed Crosslinking Additives.   Glyoxal [458, 460, 461] is effective as a delay additive within a certain pH range. It bonds chemically with both boric acid and the borate ions to limit the number of borate ions initially available in solution for subsequent crosslinking of a hydratable polysaccharide (e.g., galactomannan). The subsequent rate of crosslinking of the polysaccharide can be controlled by adjusting the pH of the solution.

Other dialdehydes, keto aldehydes, hydroxyl aldehydes, ortho-substituted aromatic dialdehydes, and ortho-substituted aromatic hydroxyl aldehydes have been claimed to be active in a similar way [459].

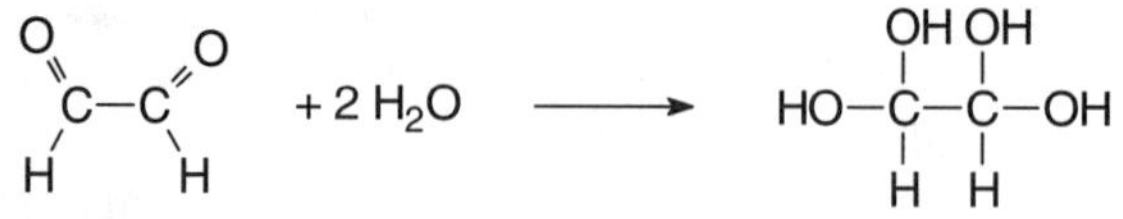

$$\underset{H}{\overset{O}{\underset{\|}{C}}} - \underset{H}{\overset{O}{\underset{\|}{C}}} + 2\,H_2O \longrightarrow HO-\underset{\underset{H}{|}}{\overset{\overset{OH}{|}}{C}}-\underset{\underset{H}{|}}{\overset{\overset{OH}{|}}{C}}-OH$$

Figure 17–15. Glyoxal. Low molecular aldehydes may form hydrates that act as complex forming agents.

Elevated Temperatures. Borate-crosslinked guar fracturing fluids have been reformulated to allow use at higher temperatures in both freshwater and seawater.

The temporary temperature range is extended for the use of magnesium oxide–delayed borate crosslinking of a galactomannan gum fracturing fluid by adding fluoride ions that precipitate insoluble magnesium fluoride [1347]. Alternatively, a chelating agent for the magnesium ion may be added. With the precipitation of magnesium fluoride or the chelation of the magnesium ion, insoluble magnesium hydroxide cannot form at elevated temperatures, which would otherwise lower the pH and reverse the borate crosslinking reaction. The addition effectively extends the use of such fracturing fluids to temperatures of 135° to 150° C.

Polyols [16], such as glycols or glycerol, can delay the crosslinking of borate in hydraulic fracturing fluids based on galactomannan gum. This is suitable for high-temperature applications up to 150° C. In this case, low–molecular-weight borate complexes initially are formed but exchange slowly with the hydroxyl groups of the gum.

Titanium Compounds

Organic titanium compounds are useful as crosslinkers [1462]. Aqueous titanium compositions often consist of mixtures of titanium compounds.

Zirconium Compounds

Various zirconium compounds are used as delayed crosslinkers, (see Table 17–12). The initially formed complexes with low–molecular-weight compounds are exchanged with intermolecular polysaccharide complexes, which cause delayed crosslinking.

Gel Breaking in Water-Based Systems

Gel Breaking

After the fracturing job, the properties of the formation should be restored. Maximal well production can be achieved only when the solution viscosity

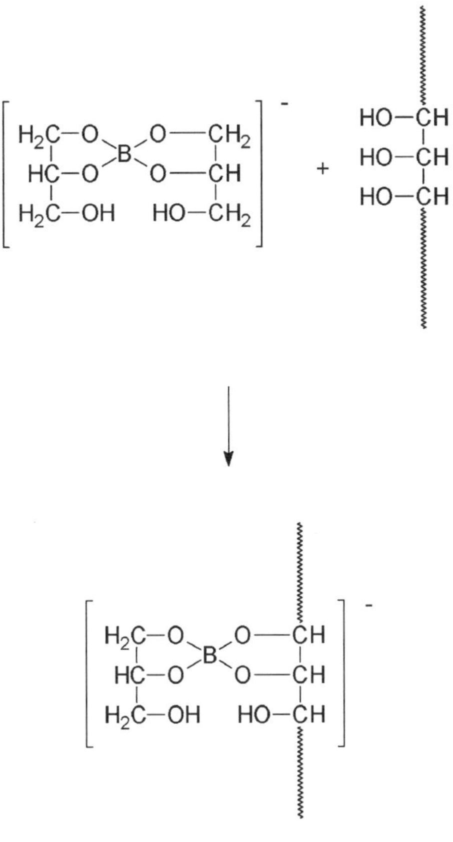

Figure 17-16. Delayed crosslinking by exchange of hydroxyl groups (idealized). If two low–molecular-weight hydroxyl compounds are exchanged with high–molecular-weight compounds, the hydroxyl units belonging to different molecules, then a crosslink is formed.

and the molecular weight of the gelling agent are significantly reduced after the treatment, that is, the fluid is degraded.

Basic Studies

Comprehensive research on the degradation kinetics of a hydroxypropylguar fracturing fluid by enzyme, oxidative, and catalyzed oxidative breakers was performed [415,416,418]. Changes in viscosity were measured as a function of time. The studies revealed that enzyme breakers are effective only in acid media at temperatures of 60° C or below. In an alkaline medium and at temperatures below 50° C, a catalyzed oxidative breaker system was the most effective breaker. At temperatures of 50° C or higher, hydroxypropylguar fracturing fluids can be degraded by an oxidative breaker without a catalyst.

Table 17–12
Zirconium Compounds Suitable as Delayed Crosslinkers

Zirconium crosslinker/chelate	References
Hydroxyethyl-tris-(hydroxypropyl) ethylene diamine (for hydroxypropylguar gum and carboxymethylcellulose)[a]	[1461]
Reaction products of a zirconium halide with sorbitol and citric, malic, or lactic acids	[1495]
Boron zirconium chelates from ammonium hydroxide; water-soluble amines; sodium or potassium zirconium; and organic acid salts such as lactates, citrates, tartrates, glycolates, malates, gluconates, glycerates, and mandelates; with polyols such as glycerol, erythritol, arabitol, xylitol, sorbitol, dulcitol, mannitol, inositol, monosaccharides, and disaccharides[b]	[463, 464, 1592, 1593]

a) Good high-temperature stability.
b) High-temperature application, enhanced stability.

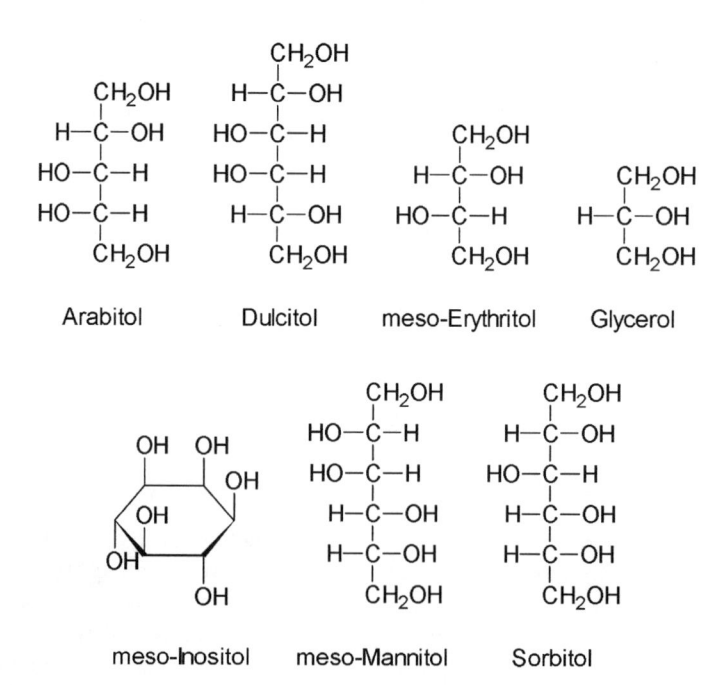

Figure 17–17. Arabitol, dulcitol, erythritol, glycerol, inositol, mannitol, and sorbitol.

Figure 17–18. Hydroxyethyl-tris-(hydroxypropyl) ethylene diamine.

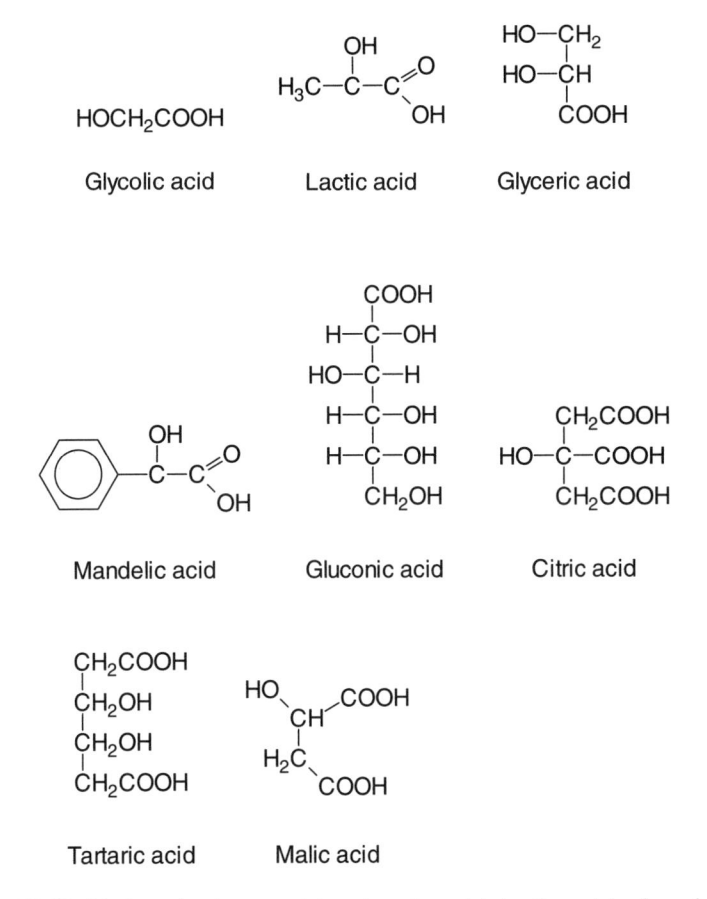

Figure 17–19. Various hydroxy acids: glycolic acid, lactic acid, glyceric acid, mandelic acid, gluconic acid, citric acid, tartaric acid, and malic acid.

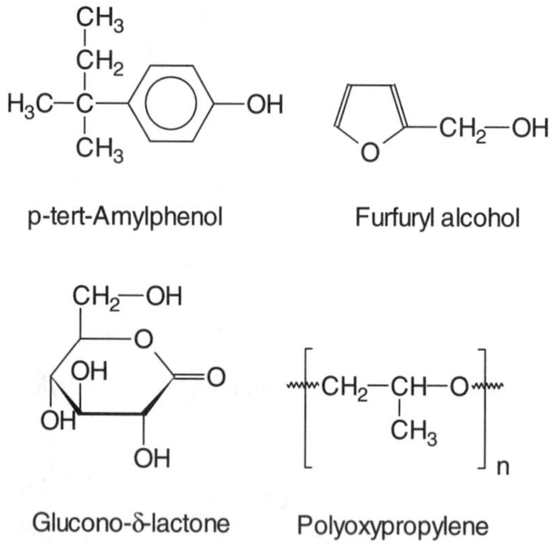

Figure 17-20. p-tert-Amylphenol, furfuryl alcohol, glucono-δ-lactone, and polyoxypropylene.

Oxidative Breakers

Among the oxidative breakers, alkali, metal hypochlorites, and inorganic and organic peroxides have been described in literature. These materials degrade the polymer chains by oxidative mechanisms. Carboxymethylcellulose, guar gum, or partially hydrolyzed polyacrylamides were used for testing a series of oxidative gel breakers in a laboratory study [180].

Hypochlorite Salts. Hypochlorites are powerful oxidants and therefore may degrade polymeric chains. They are often used in combination with tertiary amines [1846]. The combination of the salt and the tertiary amine increases the reaction rate more than the application of a hypochlorite alone. A tertiary amino galactomannan may serve as an amine source [1062]. This also serves as a thickener before breaking. Hypochlorites are also effective for breaking stabilized fluids [1817]. Sodium thiosulfate has been proposed as a stabilizer for high-temperature applications.

Peroxide Breakers. Alkaline earth metal peroxides have been described as delayed gel breakers in alkaline aqueous fluids containing hydroxypropylguar [1238]. The peroxides are activated by increasing the temperature of the fluid.

Perphosphate esters (e.g., methyl or ethyl esters) or amides can be used for oxidative gel breaking [1066]. Whereas the salts of the perphosphate ion interfere with the action of the crosslinkers, the esters and amides of perphosphate do not. Fracturing fluids that contain these breakers are useful for fracturing deeper wells operating at temperatures between 90° and 120° C and using metal ion crosslinkers, such as titanium and zirconium. Breaker systems based on persulfates have also been described [775].

Organic peroxides also are suitable for gel breaking [462]. The peroxides need not be completely soluble in water. The time needed to break is controlled in the range of 4 to 24 hours by adjusting the amount of breaker added to the fluid.

Redox Gel Breakers

Gel breakers may also act according to a redox reaction. Copper (II) ions and amines can degrade various polysaccharides [1621].

Delayed Release of Acid

Regained permeability studies with hydroxyethylcellulose polymer in high-permeability cores revealed that persulfate-type oxidizing breakers and enzyme breakers do not adequately degrade the polymer. Sodium persulfate breakers were found to be thermally decomposed, and the decomposition was accelerated by minerals present in the formation. The enzyme breaker adsorbed onto the formation but still partly functioned as a breaker. Dynamic fluid loss tests with reduced-pH, borate-crosslinked gels suggest that accelerated leak-off away from the wellbore could be obtained through the use of a delayed-release acid.

Rheologic measurements confirmed that a soluble delayed-release acid could be used to convert a borate-crosslinked fluid into a linear gel [1353].

Hydroxyacetic Acid Condensates. A condensation product of hydroxyacetic acid can be used as a fluid loss material in a fracturing fluid in which another hydrolyzable aqueous gel is used [313–315, 317]. The hydroxyacetic acid condensation product degrades at formation conditions to set free hydroxyacetic acid, which breaks the aqueous gel. This mechanism may be used for delayed gel breaking. Here permeability is restored without the need for separate addition of a gel breaker, and the condensation product acts a fluid loss additive.

Enzyme Gel Breakers

Enzymes specifically cleave the backbone structure of the thickeners and eventually of the fluid loss additive. They offer several advantages to other

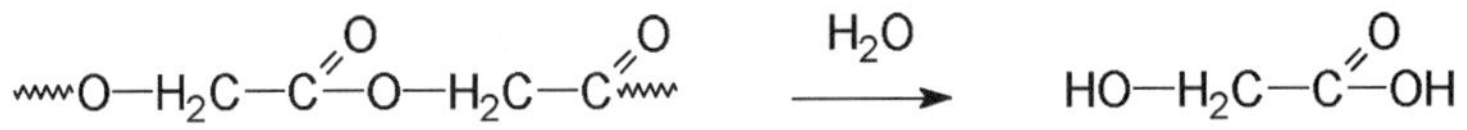

Figure 17–21. Hydrolysis of polyhydroxyacetic acid (polyglycol acid).

Table 17-13
Polymer Enzyme Systems

Polymer	References
Xanthan[a]	[14]
Mannan-containing hemicellulose[b]	[616]

a) Elevated temperatures and salt concentrations.
b) High alkalinity and elevated temperature.

breaker systems because of their inherent specificity and the infinite polymer-degrading activity. Initially the application of enzymes has been limited to low-temperature fracturing treatments because of the perceived pH and temperature constraints. Only recently, extreme temperature–stable and polymer-specific enzymes have been developed [241].

Basic Studies. Basic studies have been performed to investigate the performance of enzymes. The products of degradation [1642]; the kinetics of degradation; and limits of application [417], such as temperature and pH, have been analyzed. Because enzymes degrade chemical linkages highly selectively, no general-purpose enzyme exists, but for each thickener, a selected enzyme must be applied to guarantee success. Enzymes suitable for particulate systems are shown in Table 17–13.

Enzymes are suitable to break the chains of the thickener directly. Other systems also have been described that enzymatically degrade polymers, which degrade into organic acid molecules. These molecules are actually active in the degradation of the thickener [784].

Interactions Between Fracturing Fluid Additives and Enzyme Breakers.
Despite their advantages over conventional oxidative breakers, enzyme breakers have limitations because of interferences and incompatibilities with other additives. Interactions between enzyme breakers and fracturing fluid additives including biocides, clay stabilizers, and certain types of resin-coated proppants have been reported [1455].

Encapsulated Gel Breakers

The breaker chemical in encapsulated gel breakers is encapsulated in a membrane that is not permeable or is only slightly permeable to the breaker. Therefore the breaker may not come in contact initially with the polymer to be degraded. Only with time can the breaker diffuse out from the capsulation, or the capsulation is destroyed so that the breaker can act successfully.

Table 17-14
Use of Encapsulation in Delayed Gel Breaking

Breaker system	References
Ammonium persulfate[a]	[747, 748, 748, 748, 749, 969]
Enzyme breaker[b]	[753]
Complexing materials: fluoride, phosphate, sulfate anions[c]	[205]

a) Guar or cellulose derivatives.
b) Open cellular coating.
c) For titanium and zirconium; wood resin encapsulated.

Table 17-15
Membranes for Encapsulated Breakers

Membrane material	References
Polyamide[a]	[1558, 1559]
Crosslinked elastomer	[1161]
Partially hydrolyzed acrylics crosslinked with aziridine prepolymer or carbodiimide[b]	[879, 1354, 1355]
7% Asphalt and 93% neutralized sulfonated ionomer	[1701]

a) For peroxide particle sizes 50 to 420 μ.
b) Enzyme coated on cellulose derivative.

Encapsulated gel breakers find a wide field of application for delayed gel breaking. The breaker is prepared by encapsulating it with a water-resistant coating. The coating shields the fluid from the breaker so that a high concentration of breaker can be added to the fluid without causing premature loss of fluid properties, such as viscosity or fluid loss control. Critical factors in the design of encapsulated breakers are the barrier properties of the coating, release mechanisms, and the properties of the reactive chemicals. For example, a hydrolytically degradable polymer can be used as the membrane [1273].

This method of delayed gel breaking has been reported for oxidative breaking and for enzyme gel breaking. Formulations of encapsulated gel breakers are shown in Table 17-14. Membranes for encapsulators are shown in Table 17-15.

Granules

Granules may also be helpful in delayed breaking. Granules with 40% to 90% of sodium or ammonium persulfate breaker and 10% to 60% of an inorganic powdered binder, such as clay, have been described [1190]. The granules exhibit a delayed release of the breaker.

Other chemicals acting as delayed gel breakers are also addressed as controlled solubility compounds or cleanup additives that slowly release certain salts. Polyphosphates have been described as such [1229]. Granules composed of a particulate breaker chemical dispersed in a wax matrix are used in fracturing operations to break hydrocarbon liquids gelled with salts of alkyl phosphate esters. The wax granules are solid at surface temperature and melt or disperse in the hydrocarbon liquid at formation temperature, releasing the breaker chemical to react with the gelling agent [7].

Scale Inhibitors

The formation of calcium carbonate ($CaCO_3$), calcium sulfate, and barium sulfate scales in brine may create problems with permeability. Therefore it is advantageous that newly made fractures have a scale inhibitor in place in the fracture to help prevent the formation of scale. Formulations of hydraulic fracturing fluids containing a scale inhibitor have been described in the literature [1828].

Interference of Chelate Formers

Trace amounts of metal chelate–forming additives, which are used in fracture fluids, have been shown to have a debilitating effect on the performance of widely used barium sulfate scale inhibitors. Ethylenediaminetetraacetic acid, citric acid, and gluconic acid render some scale inhibitors, such as phosphonates, polycarboxylates, and phosphate esters, completely ineffective at concentrations as low as 0.1 mg/liter. Such low concentrations may be expected to return from formation stimulation treatments for many months and would appear to jeopardize any scale inhibitor program in place. This conclusion follows from experiments with a simulated North Sea scaling system at pH 4 and 6. The scale inhibitor concentrations studied were 50 and 100 mg/liter. The large negative effect of the organic chelating agents was observed at pH 4 and 6. The only scale inhibitors studied that remained unaffected by these interferences were polymeric vinyl sulfonates [150].

Encapsulated Scale Inhibitors

A solid, encapsulated scale inhibitor (calcium-magnesium polyphosphate) has been developed and extensively tested for use in fracturing treatments [1451–1453]. The inhibitor is compatible with borate- and zirconium-crosslinked fracturing fluids and foamed fluids because of coating. The coating exhibits a short-term effect on the release-rate profile. The composition of the solid derivative has the greatest effect on its long-term release-rate profile.

Oil-Based Systems

One advantage of fracturing with hydrocarbon gels compared with water-based gels is that some formations may tend to imbibe large quantities of water, whereas others are water-sensitive and will swell if water is introduced.

Organic Gel Aluminum Phosphate Ester

A gel of diesel or crude oil can be produced using a phosphate diester or an aluminum compound with phosphate diester [740]. The metal phosphate diester may be prepared by reacting a triester with phosphorous pentoxide to produce a polyphosphate, which is then reacted with an alcohol (usually hexanol) to produce a phosphate diester [870]. The latter diester is then added to the organic liquid along with a nonaqueous source of aluminum, such as aluminum isopropoxide (aluminum-triisopropylate) in diesel oil, to produce the metal phosphate diester. The conditions in the previous reaction steps are controlled to provide a gel with good viscosity versus temperature and time characteristics. All the reagents are substantially free of water and will not affect the pH.

Enhancers for phosphate esters are amino compounds [687]. The 2-ethylhexanoic acid trialuminum salt has been suggested with fatty acids as an activator [1690]. Another method to produce oil-based hydrocarbon gels is to use ferric salts [1648] rather than aluminum compounds for combination with orthophosphate esters. The ferric salt has the advantage of being usable in the presence of large amounts of water, up to 20%. Ferric salts can be applied in wide ranges of pH. The linkages that are formed can still be broken with gel breaking additives conventionally used for that purpose.

Increasing the Viscosity of Diesel

A copolymer of N,N-dimethylacrylamide and N,N-dimethylaminopropyl methacrylamide, a monocarboxylic acid, and ethanolamine may serve to increase the viscosity of diesel or kerosene [846].

Gel Breakers

Gel breakers used in nonaqueous systems have a completely different chemistry than those used in aqueous systems. A mixture of hydrated lime and

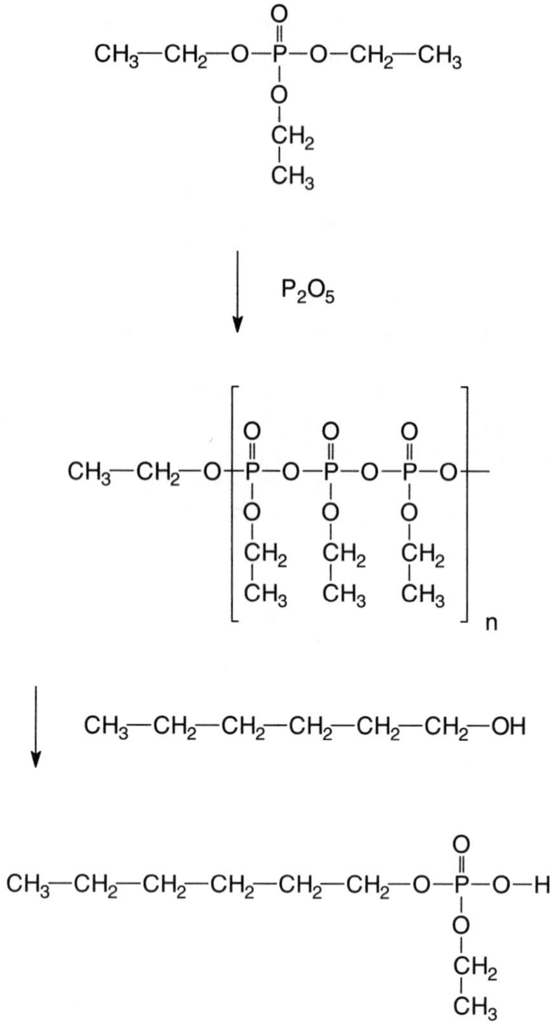

$$CH_3-CH_2-O-\underset{\underset{\underset{CH_3}{CH_2}}{O}}{\overset{\overset{O}{\|}}{P}}-O-CH_2-CH_3$$

$$\downarrow P_2O_5$$

$$CH_3-CH_2-O\left[\underset{\underset{\underset{CH_3}{CH_2}}{O}}{\overset{\overset{O}{\|}}{P}}-O-\underset{\underset{\underset{CH_3}{CH_2}}{O}}{\overset{\overset{O}{\|}}{P}}-O-\underset{\underset{\underset{CH_3}{CH_2}}{O}}{\overset{\overset{O}{\|}}{P}}-O\right]_n$$

$$\downarrow CH_3-CH_2-CH_2-CH_2-CH_2-CH_2-OH$$

$$CH_3-CH_2-CH_2-CH_2-CH_2-CH_2-O-\underset{\underset{\underset{CH_3}{CH_2}}{O}}{\overset{\overset{O}{\|}}{P}}-O-H$$

Phosphoric acid ethyl hexylester

Figure 17-22. Polyadduct from triethyl phosphate and phosphorous pentoxide and reaction with hexanol to produce a diester.

sodium bicarbonate is useful in breaking nonaqueous gels [1707]. Sodium bicarbonate used by itself is totally ineffective for breaking the fracturing fluid for aluminum phosphate– or aluminum phosphate ester–based gellants. An alternative is to use sodium acetate as a gel breaker for nonaqueous gels.

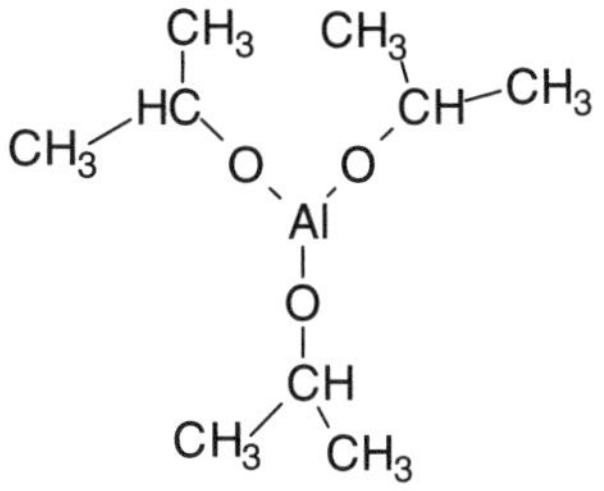

Figure 17-23. Aluminum isopropoxide.

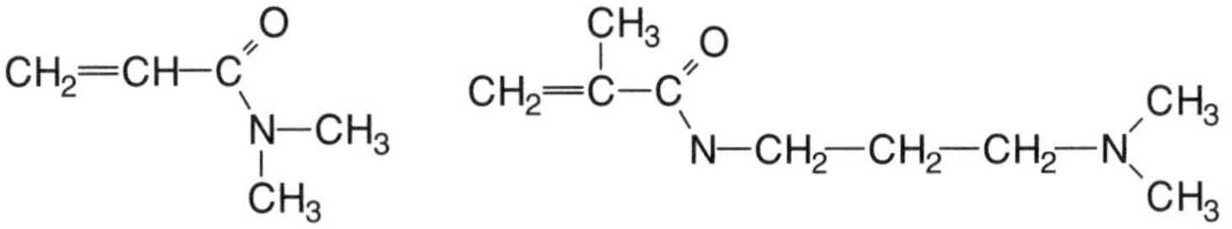

N,N-Dimethylacrylamide   Dimethylaminopropyl methacrylamide

Figure 17-24. Monomers in a copolymer for viscosifying diesel: N,N-dimethylacrylamide and N,N-dimethylaminopropyl methacrylamide.

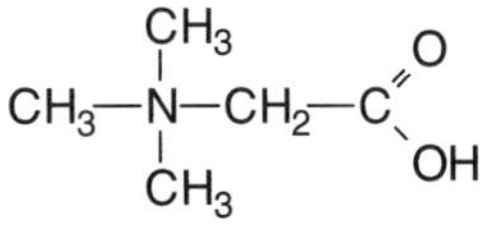

Betaine

Figure 17-25. Betaine.

Foam-Based Fracturing Fluids

Foam fluids can be used in many fracturing jobs, especially when environmental sensitivity is a concern [1669]. Foam-fluid formulations are reusable, are shear stable, and form stable foams over a wide temperature range. They exhibit high viscosities even at relatively high temperatures [209].

In addition to the normal additives, foam-fluid formulations contain surfactants, nitrogen, and carbon dioxide as essential components. Cocobetaine and α-olefin sulfonate have been proposed as foamers [1386].

The content of the gas is called *quality;* therefore a 70 quality contains 70% gas. Recently, foams with 95% gas have been examined. For such foam types, only foam prepared from 2% of an anionic surfactant with plain water had uniform, fine-bubble structure [782].

Fracturing in Coal-Beds

The production of natural gas from coal typically requires a stimulation with hydraulic fracturing. Basic studies on the effectiveness of various treatment methods for coal-beds have been presented in the literature [398, 1424].

Treating a coal seam with a well treatment fluid containing a dewatering agent will enhance methane production through a well. This additive enhances the permeability of the formation to water production and binds tenaciously to the coal surface so that the permeability-enhancement benefits are realized over a long production term.

Dewatering surfactants can be polyoxyethylene, polyoxypropylene, and polyethylene carbonates [1348] or p-tert-amylphenol condensed with formaldehyde, or they can be composed of a copolymer from 80% to 100% alkyl methacrylate monomers and hydrophilic monomers [777]. Such a well treatment fluid may be used in both fracturing and competition operations to enhance and maintain fracture conductivity over an extended period of production.

Propping Agents

For worthwhile oil or gas well stimulation, the best proppant and fluids have to be combined with a good design plan and the right equipment. The selection of a proppant is an important factor in determining how successful the stimulation treatment can be. To select the best proppant for each well, a general understanding of available proppants is imperative.

The propping agents should have high permeability at the respective formation pressures, high resistance to compression, low density, and good resistance to acids. Some propping agents are listed in Table 17–16.

Sand

Sand is the simplest proppant material. Sand is cheap, but at higher stresses it shows a comparatively strong reduction in permeability.

Table 17–16
Basic Propping Materials

Material	Description/property	References
Bauxite	Standard	[52, 608]
Bauxite + ZrO_2	Stress corrosion resistant	[957, 958]
Sand	Low permeability at higher pressures	
Light weights	Specific gravity control	[181]
Ceramic		[702, 703]
Clay		[609, 610, 956]

Ceramic Particles

Fired ceramic spheroids have been described for use as a well proppant [1051]. Each spheroid has a core made from raw materials comprising mineral particulates, silicium carbide, and a binder. The mixture includes a mineral with chemically bound water or sulfur, which blows the mixture during firing. Therefore the core has a number of closed air cells. Each spheroid has an outer shell surrounding the core, comprising a metal oxide selected from aluminum oxide and magnesium oxide. The fired ceramic spheroids have a fired density less than 2.2 g/cm^3.

Bauxite

Sintered bauxite spheres containing silica are standard proppant materials. The particles have a size range from 0.02 to 0.3 μ. They are enhanced to resist stress corrosion by inclusion of 2% zirconia in the mix before firing. A process for manufacturing a material suitable for use as a proppant is characterized by the following steps [53]: A fine fraction is separated from a naturally occurring bauxite. The fine fraction is an uncalcinated natural bauxite fraction composed largely of monomineralic particles of gibbsite, boehmite, and kaolinite. The kaolinite represents no more than 25% of the total. The separated fine fraction is pelletized in the presence of water. The pellets produced are treated to remove water.

Lightweight Proppants

Lightweight propping agents have a specific gravity of less than 2.60 g/cm^3. They are made from kaolin clay and eventually with a light-weight aggregate. Special conditions of calcination are necessary [1093–1096]. The alumina content is between 25% and 40% [1703]. A high-strength proppant has been described [182] with a specific gravity of less than 1.3 g/cm^3.

Porous Pack with Fibers

It is possible to build within the formation a porous pack that is a mixture of fibers and the proppant. The fibrous material may be any suitable material (e.g., natural or synthetic organic fibers, glass fibers, ceramic fibers, carbon fibers).

A porous pack filters out unwanted particles, proppant, and fines, while still allowing production of oil. Using fibers to make a porous pack of fibers and a proppant within the formation reduces the energy consumption of equipment. Pumping the fibers together with the proppant provides significant reductions in the frictional forces that otherwise limit the pumping of fluids containing a proppant [318].

Table 17–17
Polymer Coatings for Propping Agents

Material	References
Phenolic/furan resin or furan resin[a]	[77]
Novolak epoxide resin[a]	[701]
Pyrolytic carbon coating[a]	[875, 876]
Bisphenolic resin[b]	
Phenolic resin coated	[910, 911]
Furfuryl alcohol resin[b]	[549]
Bisphenol-A resin (curable)[b]	[909]
Epoxide resin with N-β-(aminoethyl)-δ-aminopropyltrimethoxysilane crosslinker	[1340]
Polyamide and others	[1339]

a) Chemically resistant.
b) Flowback prevention.

Coated Propping Agents

The propping particles can be individually coated with a curable thermoset coating. The coating enhances the chemical resistance of the proppants. This modification is necessary if a proppant is not stable against the additives in the fracturing fluid, such as an acid gel breaker. Resole-type phenolic resins are recommended as coating materials in the presence of oxidative gel breakers [482]. Multiple coating of particulate material results in a final coated product that has a smooth, uniform surface. Polymers suitable as coatings for propping agents are shown in Table 17–17.

Proppant Flowback

The flowback of a proppant following fracture stimulation treatment is a major concern because of the damage to equipment and loss in well production. The mechanisms of flowback and the methods to control flowback have been recently discussed in the literature [1343]. To reduce proppant flowback, a curable resin-coated proppant can be applied [1349]. The agent must be placed across the producing interval to prevent or reduce the proppant flowback.

Thermoplastic Films. Recently, thermoplastic film [1342, 1343] materials have been developed to reduce the proppant flowback that can occur after fracturing treatments. A heat-shrinkable film cut into thin slivers provides flowback reduction over broad temperature ranges and closure stress ranges and was found to cause little impairment to fracture conductivity with some dependency on use concentration, temperature, and closure stress.

Adhesive-Coated Material. The addition of an adhesive-coated material [335] to proppants decreases the flowback of the particulates. Such adhesive-coated materials can be inorganic or organic fibers, flakes, and the like. The adhesive-coated material interacts mechanically with the proppant particles to prevent the flowback of particulates to the wellbore. The consolidation of a proppant also may occur via a polyurethane coating, which will slowly polymerize after the fracturing treatment because of a polyaddition process [1856].

Magnetized Material. A magnetized material in the form of beads, fibers, strips, or particles can be placed with a proppant. The magnetized material moves to voids or channels located within the proppant bed and forms clusters, which are held together by magnetic attraction in the voids or channels, which in turn facilitate the formation of permeable proppant bridges therein. The magnetized material–proppant bridges retard and ultimately prevent the flowback of proppant and formation solids, but still allow production of oil and gas through the fracture at sufficiently high rates [389]. In a similar way, fibrous bundles placed with the proppant may act in a fracture as a flowback preventer [1338].

Acid Fracturing

A difference exists between acid fracturing and matrix acidizing. Acid fracturing is used for low-permeability, acid-soluble rocks. Matrix acidizing is a technique used for high-permeability reservoirs. Candidates for acid fracturing are formations such as limestones ($CaCO_3$) or dolomites ($CaCO_3 \times MgCO_3$). These materials react easily with hydrochloric acid to form chlorides and carbon dioxide. In comparison with the fracturing technique with proppants, acid fracturing has the advantage that no problem with proppant clean-out will appear. The acid etches the fracture faces unevenly, which on closure retain a highly conductive channel for the reservoir fluid to flow into the wellbore [1274].

On the other hand, the length of the fracture is shorter, because the acid reacts with the formation and therefore is spent. If traces of fluoride are in the hydrochloric acid, then insoluble calcium fluoride is precipitated out. Therefore plugging by the precipitate can jeopardize the desired effect of stimulation.

Fluid Loss

Fluid loss limits the effectiveness of acid fracturing treatments. Therefore formulations to control fluid loss have been developed and characterized

[1551, 1839]. It was discovered that viscosifying the acid showed a remarkable improvement in acid fluid loss control. The enhancement was most pronounced in very–low-permeability limestone cores. The nature of the viscosifying agent also influenced the success. Polymeric materials were more effective than surfactant-type viscosifiers [682].

A viscosity controlled acid contains gels that break back to the original viscosity 1 day after being pumped. These acids have been used both for matrix acidizing and for fracture acidizing to obtain longer fractures. The pH of the fluid controls the gel formation and breaking. The gels are limited to formation temperatures of 50° to 135° C [1871].

Encapsulated Acids

Acids, in particular, and etching agents, in general, may be mixed with a gelling agent and encapsulated with oils and polymers [713, 714].

Gel Breaker for Acid Fracturing

A particulate gel breaker for acid fracturing for gels crosslinked with titanium or zirconium compounds is composed of complexing materials such as fluoride, phosphate, sulfate anions, and multicarboxylated compounds. The particles are coated with a water-insoluble resin coating, which reduces the rate of release of the breaker materials of the particles so that the viscosity of the gel is reduced at a retarded rate [205].

Special Problems

Corrosion Inhibitors

Water-soluble 1,2-dithiol-3-thiones for fracturing fluids and other workover fluids have been described as corrosion inhibitors [1378] for aqueous environments. These compounds are prepared by reacting a polyoxylated starting material (polyethyleneoxide capped with isopropylphenol) with elemental sulfur. These compounds perform better in aqueous systems than their nonoxylated analogs. The concentration range is usually in the 10 to 500 ppm range, based on the weight of the water in the system.

The Problem of Iron Control in Fracturing

Results from laboratory tests and field jobs show that iron presents a significant and complex problem in stimulation operations [1653]. The problem presented by an acidizing fluid differs from that presented by a nonacidic or weakly acidic fracturing fluid. In general, acid dissolves iron compounds from the equipment and the flow lines as it is mixed and pumped into the formation. The acid may dissolve additional iron as it reacts with the formation. If the fluid does not contain an effective iron control system, the dissolved iron could precipitate. This precipitate may then accumulate as it is carried toward the wellbore during flowback. This accumulation of solids may decrease the natural and the created permeability and can have a detrimental effect on the recovery of the treating fluid and production.

Iron can be controlled with certain complexing agents, in particular glucono-δ-lactone, citric acid, ethylenediaminetetraacetic acid, nitrilotriacetic acid, hydroxyethylethylene diaminetriacetic acid, hydroxyethyliminodiacetic acid, and the salts from the aforementioned compounds. These compounds must be added together with nitrogen-containing compounds such as hydroxylamine salts or hydrazine salts [486, 643, 1815].

In general, chelating agents possess some unique chemical characteristics. The most significant attribute of these chemicals is the high solubility of the free acids in aqueous solutions. Linear core flood tests were used to study the formation of wormholes. Both hydroxyethylethylene diaminetriacetic acid and hydroxyethyliminodiacetic acid produced wormholes in limestone cores when tested at 150° F. However, the efficiency and capacities differ. Because these chemicals have high solubility in the acidic pH range, it was possible to test acidic (pH less than 3.5) formulations [644].

To control the iron in an aqueous fracturing fluid having a pH below 7.5, a thioalkyl acid may be added [243]. This is a reducing agent for the ferric ion, contrary to the complexants described in the previous paragraph.

Enhanced Temperature Stability

During the initial fracturing process, a degradation, which results in a decrease of viscosity, is undesirable. The polymer in fracturing fluids will degrade at elevated temperatures.

One method to prevent degradation too early is to cool down the formation with large volumes of pad solution before the fracturing job. Furthermore, the temperature stability of the fracturing fluid is extended through the addition of quantities of unhydrated, particulate guar or guar-derivative polymers before pumping the fracturing fluid into the formation [1346]. Finally, adjustment of the pH to moderate alkaline conditions can improve the stability.

The preferred crosslinkers for high-temperature applications are zirconium compounds. A formulation for a high-temperature guar-based fracturing fluid is given in Table 17–18. The fracturing fluid exhibits good viscosity and is stable at moderate to high temperatures, that is, 80° to 120° C.

Table 17-18
Formulation for a High-Temperature Guar-Based Fracturing Fluid [239]

Component	Action
Guar gum	Thickener
Zirconium or hafnium compound	Crosslinking agent
Bicarbonate salt	Buffer

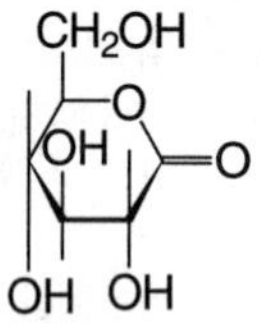

Glucono-δ-lactone

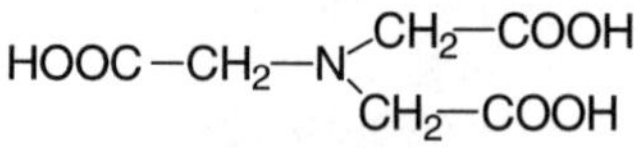

Nitrilotriacetic acid

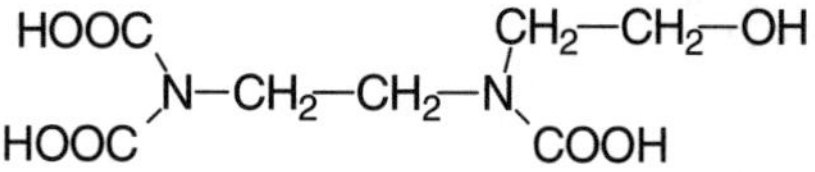

Hydroxyethylethylene diaminetriaceticacid

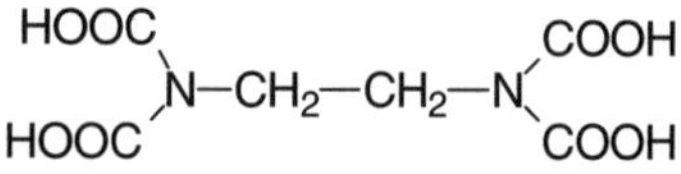

Ethylenediaminetetraacetic acid

Figure 17-26. Complexing agents for iron control: glucono-δ-lactone, nitrilotriacetic acid, hydroxyethylene diaminetetraacetic acid, and ethylenediaminetetraacetic acid.

Chemical Blowing

The efficiency of a fracturing fluid produced back from a formation can be increased by adding blowing agents [5, 905]. After placing the blowing agent (agglomerated particles and granules containing the blowing agent [e.g., dinitrosopentamethylenetetramine, sodium hydrogen carbonate and p-toluene sulfonyl hydrazide, azodicarbonamide, and p,p′-oxybis(benzenesulfonyl hydrazide)]) and fracturing the formation, the blowing agent decomposes, thereby causing the filter-cake to become more porous or providing a driving force for

Table 17–19
Frost-Resistant Formulation for Hydraulic Fracturing Fluids [147]

Component	% by wt
Hydrocarbon phase[a]	2–20
Surfactant	
Mineralized water	
Sludge from production of sulfonate additives (hydrocarbons 10% to 30%, calcium sulfonate 20% to 30%, calcium carbonate and hydroxide 18% to 40%)[b]	10–35
Emultal[c]	0.5–2.0

a) Gas condensate, oil, or benzene.
b) Slows down filtration and increases sand-holding capability, frost resistance, and stability.
c) Surfactant-emulsifier.

fluid load removal from the matrix. Increased porosity enhances the communication between the formation and the fracture, thus increasing the efficiency of the production of the fracturing fluid. The gas liberation within the matrix establishes the communication pathways for subsequent fracture and the well.

Frost-Resistant Formulation

A frost-resistant formulation from a Russian patent [147] is given in Table 17–19. The composition has a frost resistance down to $-35°$ to $-45°$ C.

Chapter 18

Water Shutoff

This chapter is closely related to the subject of gelling agents (refer to Chapter 8, p. 108, for more information). The action of both gelling and plugging agents have a similar performance: to minimize the permeability of a formation.

Basic Principles

The formation of plugs can be achieved by various chemical principles:

- Polymerization of vinyl monomers
- Polyaddition of epoxies
- Polycondensation of aminoplasts and phenoplasts
- Crosslinking of polymers
- Cementing

Chemicals for Water Shutoff

In Situ Polymerization, Polyaddition, and Condensation

In Situ Polymerization of Vinyl Monomers

Vinyl monomers are used as such and are allowed to polymerize in place to form a gel. This technique is used to enable a solution to gel slowly, even at high temperatures. An aqueous solution of a vinyl monomer is mixed with a radical-forming initiator and optionally with a dispersant. The initiator decomposes at elevated temperatures and initiates the polymerization process. In this way, a gel is formed in place. Radical polymerization is sensitive to molecular oxygen. For a more pronounced delay, polymerization inhibitors may be added to the solution in small amounts. The technique is used in the treatment of subterranean formations, especially for plugging lost circulation in drilling operations, particularly at elevated temperatures.

N-Methylol acrylamide   **N-Methylol methacrylamide**

Figure 18–1. N-methylol acrylamide and N-methylol methacrylamide.

As monomers, N-methylol acrylamide and N-methylol methacrylamide [1081] have been suggested. Inhibitors can be phenol derivatives [1079], such as an N-nitrosophenylhydroxylamine salt [1179].

Epoxide Resins

Epoxide resins have good adhesive properties. They can be cured at low temperatures with amine hardeners and at elevated temperatures with organic anhydrides. Formulations can be adjusted to a long pot life and a low exothermal reaction in the course of curing. The compositions are not miscible with well fluids. A disadvantage is the comparatively high price. Standard epoxide resins are based on bisphenol-A. A method for selectively plugging wells using a low-viscosity epoxide resin composition containing a single curing amine-based agent has been described [447, 448].

Wells can be treated selectively by using a low-viscosity epoxide resin formulation [447, 448]. A liquid bisphenol-A–based epoxide material and an amine hardener are used for curing at ambient temperature. The epoxide material has a very low viscosity at well surface temperatures and is immiscible with well fluids. The polyamine-curing agent used is an amber-colored mobile liquid having a low viscosity at ambient temperature. It has a long pot life and low exothermal. The method is applicable to plugging permeable zones in a gravel-packed well and may be used to repair leaks in well casing or production tubing and in cementing to prevent communication between subterranean regions. Aliphatic epoxide resins are compatible with water, which is an advantage over more commonly used aromatic epoxide resins that cannot tolerate water contamination [560].

Urea–Formaldehyde Resins

Urea–formaldehyde resins can be cured with isopropylbenzene production wastes containing 200 to 300 g/liter of aluminum trichloride ($AlCl_3$) as an acid hardener [189]. Isopropylbenzene is formed as an intermediate in the Hock process by a Friedel-Crafts reaction from propene and benzene. The mixture

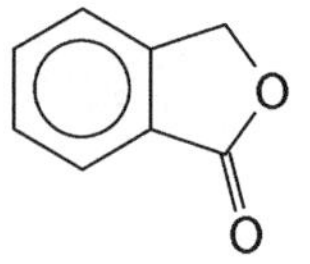

Phthalide

Figure 18–2. Phthalide.

hardens in 45 to 90 minutes and develops an adhesion to rock and metal of 0.19 to 0.28 and 0.01 to 0.07 MPa, respectively. A particular advantage is the increased pot life of the formulation.

N-methylol acrylamide or N-methylol methacrylamide can be polymerized with peroxides to obtain plugging [1079, 1080]. Suitable inhibitors may be used as retarders of the polymerization process to ensure sufficient pot life time. The components are in the form of a preemulsion.

The gelling of the vinyl monomers in an aqueous medium in the presence of an organic peroxide polymerization initiator is delayed by using an inhibitor consisting of an alkali metal or ammonium salt of the N-nitrosophenylhydroxylamine combined with an aminocarboxylic acid [1179].

Curing of Urea-Formaldehyde and Phenol-Formaldehyde. Urea-formaldehyde resins and phenol-formaldehyde resins can be cured by various mechanisms.

Acid Curing. Urea-formaldehyde resins and resol-phenol-formaldehyde resins can be acid-cured by wastes from the production of maleic anhydride [1902]. The waste from the production of maleic anhydride contains up to 50% maleic anhydride, in addition to phthalic anhydride, citraconic anhydride, benzoic acid, o-tolulic acid, and phthalide. The plugging solution is prepared by mixing a urea-formaldehyde resin with a phenol-formaldehyde resin, adding the waste from production of maleic anhydride, and mixing thoroughly.

Aluminum Trichloride. In an analogous way, an $AlCl_3$ containing waste of isopropylbenzene production [189] can be used as an acid hardener for urea-formaldehyde resins. The waste from the production of isopropylbenzene (via a Friedel-Crafts reaction) contains approximately 200 to 300 g/liter of $AlCl_3$.

Alkaline Curing. A plugging solution based on phenol-formaldehyde resin, formaldehyde, and water was described. To improve plugging efficiency under conditions of low (but above zero) temperatures, the solution also contains a bituminous emulsion [662]. Curing is achieved with free formaldehyde under alkaline conditions. The plugging solution is prepared by mixing an aqueous solution of a phenol-formaldehyde resin with a bituminous emulsion and adding the hardener (formaldehyde) directly before pumping solution into

Table 18-1
Plugging Material with 2-Furaldehyde–Acetone Monomer and Silicone Oligomers

Components	% by wt
2-Furaldehyde–acetone monomer	70–96
Silicoorganic compound	1–10
Acidic or alkali hardener	3–20

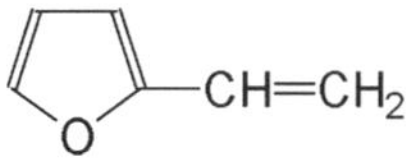

2-Furfurylidene

Figure 18-3. 2-Furfurylidene.

the stratum. Tests showed that the formulation has a low viscosity and a reduced hardening time. The solution has an improved plugging capability at low (but above zero) temperatures.

2-Furaldehyde–Acetone Monomer and Silicone

A plugging material with 2-furaldehyde–acetone monomer and silicone oligomers has been described [1099]. The components for this material are shown in Table 18-1. The 2-furaldehyde–acetone monomer can contain monofurfurylidene-acetone and difurfurylidene-acetone. The hardener can be iron chloride, benzene-sulfonic acid, hexamethylene diamine, or polyethylene polyamine. The plugging stone has improved strength, elastic-deformation, and anticorrosion and adhesion properties.

Cement with Additives

Polymethylmethacrylate Modified with Monoethanolamine

Polymethylmethacrylate can be modified with monoethanolamine to form a water-soluble polymer (Deman). Deman is used as a cement additive to increase the strength in amounts smaller than 0.5% of the total weight of the composition [1595]. The produced plugging stone has improved strength characteristics within a temperature range from $-30°$ to $+300°$ C.

Aluminum oxychloride [1596] and modified polymethylmethacrylate as cement additives exhibit a reduction of the permeability of the stratum up to 99.69%.

Crude Light Pyridine Bases

Small amounts of pyridine bases increase the corrosion resistance of cement stone without any associated loss of strength [1016]. The use of nitrilo-trimethyl phosphonic acid and an adduct between hexamethylene tetramine and chlorinated propene or butene improves the adhesion to the metal, hardening times, mobility, and strength [1770]. The latter adducts are further claimed to be useful as additives in cementing oil and gas wells in salt-bearing strata [1768].

Granulated Fly Ash

Granulated fly ash [6] can substitute for Portland cement to an extent of 40% to 60%. Fly ash is used in granulated form and has a moisture content around 10% to 20%. The formulation can be used for cementing oil and gas wells within a temperature range of 20° to 250° C. The solution has reduced water absorption and increased sedimentation stability. A formulation [1388] is shown in Table 18–2. Hydrosil (Aerosil) is used to increase the adhesion of the produced cement rock to the casing string. It also reduces the density and water absorption.

Nitrilo-Trimethyl Phosphonic Acid Hydrazine Hydrochloride

A corrosion-resistant formulation [1767] is achieved by adding phosphonic acid and hydrazine hydrochloride, as shown in Table 18–3. Hydrazine hydrochloride, combined with nitrilo-trimethyl phosphonic acid, provides a synergetic effect of increased inhibition of the plugging solution.

The hydrazine hydrochloride can bind the free oxygen present in the plugging solution. It also reduces the amount of sulfur oxides in the cement rock formed after hardening, thus preventing corrosion. The plugging rock has an increased corrosion stability in hydrogen sulfide–containing media. The

Table 18–2
Portland Cement with Fly Ash as Active Constituent

Components	% by wt
Portland cement	50–65
Fly ash	20–32
Sodium and/or calcium sulfate	2–8
Hydrosil	9–14

Table 18–3
Corrosion-Resistant Formulation

Component	% by wt
Plugging Portland cement	73.20–73.38
Nitrilo-trimethyl phosphonic acid	0.02–0.04
Hydrazine hydrochloride	0.18–0.32
Water	up to 100

Table 18–4
Plugging Solution with Portland Cement and Phosphonium Complexone

Components	% by wt
Portland cement	69–70
Calcium chloride	2.0–3.5
Phosphonium complexone	0.020–0.035
Water	up to 100

final product has a high adhesion to metal. The recipe can be modified with polyoxyethylene and a water-soluble cationic polyelectrolyte to increase the sedimentation stability [1361].

Phosphonium Complexone

Small amounts of phosphonium complexone [1560] are sufficient to increase adhesion to the stratal rock. Table 18–4 illustrates an example for plugging solution with Portland cement and phosphonium complexone. Calcium chloride acts as a regulator of the setting time in the suggested composition. More precisely, phosphonium complexone stands for certain chelating phosphorous compounds (e.g., oxyethylidene diphosphonic acid, nitrilo-trimethyl phosphonic acid, sodium tripolyphosphate, or amiphol) [1540]. The mixture is applicable at low temperatures from 20° to 75° C.

Aerated Plugging Solution

An aerated plugging solution has been proposed that uses certain waste products from chemical industry [520], as shown in Table 18–5. The water-glycol mixture contains monoethylene glycol, ethylcellosolve, diethylene glycol, triethylene glycol, and ethyl carbitol. It is obtained as a waste from production of oligomers (e.g., oligoester acrylates, oligoesters) as a result of washing the equipment with water.

Table 18–5
Composition for Preparation of Aerated Plugging Solution

Composition	% by wt
Portland cement	100
Oxethylated monoalkylphenols of propylene trimers	0.2–0.6
Water-glycol mixture	0.1–1.4
Air	0.01–0.02
Water	45–80

The composition has an improved thermal stability and, when combined with a water-glycol mixture, it provides a high foam-forming ability and an increased degree of aeration of the plugging solution. It can be used as part of an aerated plugging solution, which is suitable for wells under normal geologic conditions and wells having zones of abnormally low pressure.

Compressed Foam Mixture

Thread leaks, small holes, and leaks around packers in well casings and production tubing can be repaired by applying a compressed foam mixture [770]. The mixture contains discrete solid particles of various sizes for forcing the suspended particles into the opening to achieve a high friction seal. The foam mixture is moved along the inside of the conduit sandwiched between fluid bodies to keep the foam mixture intact. A backpressure is applied to the foam mixture to force the mixture through the openings.

Antifoaming with Sulfite-Waste Liquor as Plasticizer for Cements

A widespread additive for plasticizing plugging agents is sulfite-waste liquor (SWL), but its use is limited because of pronounced foaming of the agents when SWL is added to the cement in amounts greater than 0.5%. To prevent foaming when large amounts of SWL are added, a drilling mud antifoam agent should be added at the same time [1905]. With the use of this antifoaming agent, consisting of polymethylsiloxane and tributyl phosphate, a synergistic effect is observed—polymethylsiloxane simultaneously displaces the adsorbed molecules of the foam stabilizer (i.e., lignosulfonate) from the foam film, and the tributyl phosphate reduces the surface viscosity of the film.

Furfuramide

The addition of furfuramide to a plugging cement in portions of 1% to 10% produces a plugging rock of increased corrosion resistance and reduced water permeability [1098].

Table 18–6
Acid Resistant Cement Formulation

Component	% by wt
Slag-sand cement	100
Iron chloride	0.5–1.0
Polyacrylamide	0.05–0.10
Ethyl silicate	1–3
Water	48–51

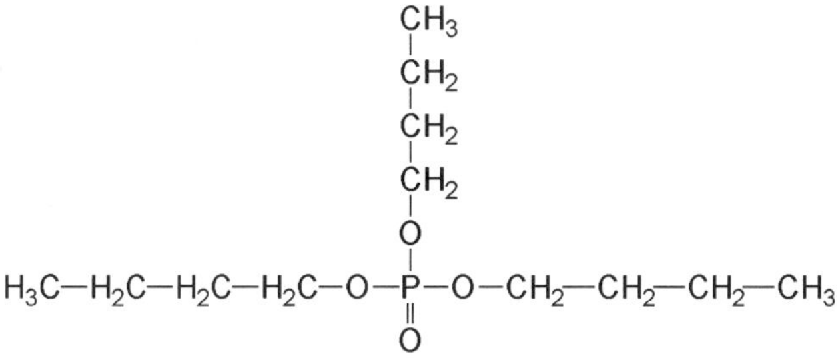

Tributyl phosphate

Figure 18–4. Tributyl phosphate.

Modified Methylcellulose with Mono-Substituted Sodium Phosphate

Small amounts of modified methylcellulose and mono-substituted sodium phosphate (both 0.1%), may be added to plugging cement [1766]. The mono-substituted sodium phosphate weakens the effect of calcium ions on modified methylcellulose and prevents its coagulation. Stabilization of the plugging solution and increased strength of the cement rock result.

Polyacrylamide

A cement formulation, as indicated in Table 18–6, will be useful as a plugging solution in deep and super-deep wells [34] at temperatures of 100° to 160° C and in the presence of hydrogen sulfide. The solution has a high stability against hydrogen sulfide aggression.

The trivalent iron chloride is used in hexahydrate form. The solution is prepared by first making a 1.5% aqueous solution of polyacrylamide (by stepwise dissolving of portions of dry polyacrylamide powder in water at temperatures of 40° to 60° C), preparing a solution of iron chloride in water separately and adding to it the required amount of ethyl silicate, then mixing the produced solution with slag-sand cement, and finally, adding the prepared 1.5% polyacrylamide solution.

Pentaerythrite

Pentaerythrite acts as an accelerator in Portland cement [1603].

Smectite Clays

Smectite clays, such as hectorites, have been proposed as an additive to Portland cement to obtain a thixotropic formulation. Good results have been obtained with the synthetic clay known as Laponite [1385]. The material is capable of gelling reversibly in less than 60 seconds. The thixotropic material generally finds use in oil well applications, for example to plug lost circulation zones, as grouts to repair damaged or corroded casing, and to limit annular gas migration. The material also finds particular application in techniques for the completion of horizontal wells to be completed with slotted or predrilled liners.

Plasticizers

When drilling deep wells, successfully filling the annulus with cement slurry can be achieved by treating the plugging slurry with plasticizers to increase its mobility. It is desirable that the plasticizer be effective in liquefying the cement slurry. Of course, it should not adversely affect other parameters of the slurry and hardened cement, and it should be readily available and inexpensive.

In favorite cases, these requirements have been met by caprolactam production wastes [166]. The alkaline waste waters of caprolactam production (AWCP) represents an aqueous solution of the sodium salts of mono- and dicarboxylic acids of 23% to 34% concentration, with a solution density of 1.14 to 1.16 g/cm^3 and a pH of 10.4. The AWCP agent increases the mobility of plugging agents prepared from Portland cement and gel-cement slurries with various bentonite contents. The AWCP is added to the mixing water of the cement slurry at a rate of 0.08% to 0.41%, with respect to the dry product, to the mass of mixing water. A valuable property of AWCP is that it plasticizes the plugging slurries; it does not change the setting time of the cement slurries or the strength of the hardened cement.

Water Glass

A plugging slurry for oil and gas well drilling is given in Table 18–7 [1441]. It is formed by adding water to the Portland cement suspension containing the other three constituents. The bentonite clay powder is premixed with water

Table 18-7
Formulation for a Plugging Slurry for Oil and Gas Well Drilling

Compound	% by wt
Portland cement	100
Bentonite clay powder	5–12
Water glass (calculated on a dry wt basis)	0.4–0.8
Sodium chloride	9.4–16.3
Water	52–63

Table 18-8
Composition of the Silicone Additive

Components	% by wt
Sodium organosilicone	25–30
Silicoorganic component	15–18
Sodium hydroxide	10–14
Ethanol	15

glass to form a paste, and this paste is allowed to stand for at least 4 hours. The composition is then mixed to a paste with the NaCl solution. Premixing the clay powder with the water glass produces a protective layer on the clay particles to hinder the hydration of the clay. The formulation is used as a plugging slurry for drilling oil and gas wells. The moisture-combining capacity of the solidified rock is increased, while its permeability is simultaneously reduced.

Organosilicones

Silicones may improve the properties of a Portland cement formulation [1267] when added to the cement in amounts of 0.2% to 2%. An example of a formulation of the additive is given in Table 18-8.

The presence of the additive results in the formation of a homogeneous structure of the plugging rock, with an improved uniformity of the phase composition of the system and a more compact distribution of the dispersed particles. An increased strength of the cement rock is also obtained.

Formaldehyde Resin

Expanding additives improve the filtering properties of a cement [21,1720]. An expanding additive is based on calcium oxide (CaO) fired at high

Table 18–9
Plugging Solution with Portland Cement, Expanding Additive, and Formaldehyde Resin

Components	phr
Portland cement	100
Expanding additive, based on calcium oxide fired at high temperatures	2–10
Polymethylene urea, amino-formaldehyde resin	0.05–1.50
Water	45–55

Table 18–10
Portland Cement with Liquid Metal Alloy

Components	% by wt
Portland cement	63.4–65.0
Isobutyl alcohol	0.13–0.17
Liquid metal alloy of gallium, indium, and tin of melting point	0.13–0.17
Water	up to 100

temperatures. The expansion takes place on the uptake of water. An example for a formulation containing an expanding additive is shown in Table 18–9.

An amino-formaldehyde resin or acetone-formaldehyde resin has the capability to harden in alkaline media, in contact with a cement solution with a pH of 11 to 12. The presence of sintered CaO provides the required conditions for hardening of the methylol groups of the formaldehyde resin with Ca^{2+} ions and a further simultaneous reaction of the methylol groups that formed hydrate compounds, resulting in an improved dispersion and plastification of the solution.

Liquid Metal Alloy

A liquid metal alloy [36] containing gallium, indium, and tin has been proposed as an additive to Portland cement. A formulation is shown in Table 18–10. The liquid metal alloy has a melting point of 11° C. Its presence does not cause corrosion of stainless steel up to 250° C but causes corrosion of steel alloys at temperature above 35° C, and it dissolves aluminum at room temperature. The alloy is harmless to skin and mucous membranes.

Isobutanol is used as an alcohol component, to increase strength of produced cement stone. It should be noted, however, that gallium and indium are precious and better used in the nonferrous metal industries.

Coated Bentonite Particles

A composition for plugging wells that is biodegradable and environmentally safe is made of bentonite particles covered with a water-soluble coating of biodegradable natural resin [1541]. The bentonite may be in the form of chips or compressed pellets. When exposed to water, the soluble coating dissolves at a uniform rate, exposing the bentonite to water. As a result of the water contacts it expands to form a tough but flexible water-impermeable seal of a semisolid, gel-like mass. The well plugging composition is nontoxic, nonpolluting, and nonhazardous. It will not become sticky or commence expanding upon initial contact with water and thereby prevents binding, clumps, and faulty seals.

Bentonite Clay Powder

A two-component plugging material [1763] consists of an aqueous suspension of bentonite clay powder (20% to 25%), ground chalk (7% to 8%), sulfanol (0.10% to 0.15%), and carboxymethylcellulose (1.0% to 1.5%) as the first component. This solution is pumped into the formation. A gel is formed if diluted hydrochloric acid is pumped down and mixes with the first component. The hydrochloric acid is inhibited with a mixture of alkyl-polybenzyl pyridinium chloride and urotropin.

Each of the individual reagents has a low viscosity and good pumping properties. After mixing in the stratum, they produce a highly viscous nonfiltering plug as an end result of a chemical reaction.

Blast Furnace Slag

Blast furnace slag is used successfully as a mud-to-cement conversion technology in oil well cementing worldwide because of economic, technical, and environmental advantages [1429]. Slag-mix slurries were used as primary, temporary abandonment and sidetrack plug cements during prospect predrilling in the Gulf of Mexico. However, the penetration rates were slower than expected when these plugs were drilled out. Therefore a basic study concerning the drilling properties was initiated.

Slag-mix, solidified mud, and conventional class H Portland cement were evaluated under controlled laboratory conditions to better understand and quantify differences in drillability between these two types of cement under realistic down-hole conditions. The objectives of this study were to refine bit selection and drilling practices for more cost-effective slag-mix plug drilling.

Reinforcement by Fibers

Fibers can be added to a gelation solution [1212, 1213]. The fibers chosen must not interfere with the gelation process and must provide adequate

Table 18–11
Composition of Raw Fiber Material [1242, 1243]

Constituents	% by wt
SiO$_2$	38–42
Al$_2$O$_3$	6–15
CaO	15–38
MgO	6–15
Fe$_2$O$_3$	0.2–15
Total alkalis such as Na$_2$O	1–3
SO$_3$	.03–1
Loss on ignition	0.5
pH in water	<8.0
% Water soluble	<0.2
Water-soluble sulfate	<0.05

reinforcement. On the other hand, they should not adversely affect the ability of the solution to be pumped and injected. Glass fibers and cellulosic fibers are specifically disclosed as preferable reinforcing fibers.

Fiber-Reinforced Formulations for Cement

Acid-Soluble Mineral Fibers. Mineral fibers that are highly soluble in acid can be used to control the permeability [1242, 1243] of formations. The fibers are 5 to 15 μ in diameter and are formed into pellets of $\frac{1}{32}$ to $\frac{1}{2}$ inch diameter. A fluid-blocking layer formed of semidispersed pellets can bridge the face pores of the formation. After well rework, the plugging layer is treated with an acid solution to quickly dissolve the mineral fibers to the desired extent to control porosity.

The preferred fiber material typically consists of about 35.7% CaO, 9.6% MgO, 9.3% Al$_2$O$_3$, and 42.3% SiO$_2$. The composition of an example fiber material is shown in Table 18–11. This product is highly soluble in hydrochloric acid (HCl), forming a soluble silicic acid from the SiO$_2$, which is suspended in the HCl until, after a very long time, a residual silica gel may form and drop out of the solution. In a blend of 15% HCl and 10% acetic acid, the silica gel does not form or drop out of the solution. Thus a quick method of acid wash scavenging removes as much of the fibrous network in the sealing mat as desired to establish a porous formation face of predetermined permeability.

Asbestos Reinforcing. Asbestos has been proposed as a reinforcing component [583]. However, asbestos is believed to be a carcinogen.

Glass Fibers and Cellulosic Fibers. Glass fibers and cellulosic fibers are specifically disclosed as preferable reinforcing fibers for a gelling solution of partially hydrolyzed polyacrylamide with a crosslinking agent [1211]. The fibers are added to a gelation solution, and the mixture is pumped to a subterranean injection site, where it gels in place. The fibers do not interfere with the gelation process and can provide adequate reinforcement without adversely affecting the ability of the solution to be pumped and injected.

Gel from Aluminum Hydroxychloride

Aluminum hydroxychloride is used as a plugging agent [1394]. A weak-base activator such as sodium cyanate with an activator aid can establish delayed gelation.

Wastes

Several industrial waste products may be used as ingredients for plugging solutions. Some examples are summarized in Table 18–12; others are detailed in other sections. The use of industrial wastes for plugging solutions is regarded

Table 18–12
Industrial Waste Products as Ingredients for Plugging Solutions

Purpose	Waste	References
Clinker-less binder	Slag from melting of oxidized nickel ores	[1474]
Hemi-hydrated gypsum	Waste obtained in production of nitrilo-trimethyl phosphonic acid[a]	[1594]
	Waste from formic acid production	[1885]
Polyacrylamide	Manganese nitrate or waste from galvanizing, electronic works[b]	[998]
Hydrolyzed polyacrylonitrile	Waste from lanolin production treated with triethanol amine and water[c]	[1427]
	Industrial waste from nitric industry	[293]
Mineral binder	Waste from production of epoxide resins	[1097]
Portland cement	Sodium sulfate waste from sebacic acid production	[1428]

a) Waste for setting time adjustment for gypsum.
b) Two-component plugging solution.
c) Waste-complexing reagent.

Table 18–13
Compressive Strength of the Produced Plugging Rock

Temperature [° C]	Compressive strength [MPa]	Water absorption over 24 hr [%]
80	1.40–1.46	2.3–2.5
110	1.44–1.45	2.3–2.6
140	0.54–0.59	2.4–2.7

as a method of waste disposal and is proposed in eastern European countries. Waste material from other processes has been established to be active, such as waste from galvanizing processes [998]. In this case, the components must be placed in two portions, however. Iron and chromic salts from lignosulfonate are used for metal ions [1000]. Lignosulfonates are waste products from the paper industry.

Waste Oil Sludge

A plugging material with approximately 85% by weight oil sludge, 10% formaldehyde, and 2% sulfuric acid has been proposed [985]. The oil sludge was left behind during primary preparation of processing oil and contains 8% to 16% of oil. It is a pastelike mass and consists of finely crushed rock fractions and oil. The material is produced by direct mixing of the three components. For application, the mixture is pumped into the well followed by pressing-in air, which is preheated to 80° to 140° C.

The compressive strength of the produced plugging rock is shown in Table 18–13. The material is cheaper because industrial waste is used and it produces a plugging rock of increased strength.

Aluminum Trichloride

Aluminum trichloride, a cheap, abundant waste product of the chemical industry, forms a gel under certain conditions with carbonates and on mixing with alkalis. Laboratory and field tests showed that aluminum trichloride can be used as a gel-forming agent for reducing the permeability of water-conducting channels [674].

Aluminum Phosphate Ester Salts for Gelling Organic Liquids

Organic liquid gels are used for temporary plugging during fracturing operations. This type of gelling agent permits on-the-fly gelling of hydrocarbons, especially those used in hydraulic fracturing of subterranean formations to enhance oil and gas production.

A gel of an organic liquid, such as diesel or crude oil, can be formed using an aluminum phosphate diester in which all of the reagents are substantially free of water and pH-affecting substances [740, 779]. The diester may be prepared by the reaction of a triester with phosphorous pentoxide to produce a polyphosphate, which is then reacted with an alcohol to produce a phosphate diester. The latter diester is then added to the organic liquid along with a non-aqueous source of aluminum, such as aluminum isopropoxide in diesel oil, to produce the metal phosphate diester. The conditions in the two preceding reaction steps are controlled to provide a gel with good viscosity versus temperature and time characteristics. The gel is useful in fracturing subterranean formations by entraining a solid particulate proppant therein and pumping the resultant mixture into the subterranean rock formation at sufficient pressure to fracture the formation.

A similar process involves reacting triethyl phosphate and phosphorous pentoxide to form a polyphosphate in an organic solvent [871]. An excess of 1.3 moles of triethyl phosphate with respect to phosphorous pentoxide is preferred. In the second stage, a mixture of higher aliphatic alcohols from hexanol to decanol is added in an amount of 3 moles per 1 mole phosphorous pentoxide. Aluminum sulfate is used as a crosslinker. Hexanol results in high-temperature viscosity of the gel, while maintaining a pumpable viscosity at ambient temperatures [870].

Chapter 19
Oil Spill–Treating Agents

The most spectacular incidents concerning oil spills occur in coastal regions. Therefore most of this chapter is devoted to this topic. However, a section is devoted to subsurface and soil remediation jobs.

Chemical dispersants can be used to reduce the interfacial tension of floating oil slicks so that the oils disperse more rapidly into the water column and thus pose less of a threat to shorelines, birds, and marine mammals. The action of oil spill–treating agents goes beyond simply dispersing the spilled oil. In particular, oil spill–treating agents can be divided into four classes: solidifiers, demulsifying agents, surface-washing agents, and dispersants. The majority of oil spill–treating agents, however, are described as dispersants.

The terminology in the literature is not unique. Oil spill–treating agents are referred to as:

- Oil spill–treating agents
- Spill–treating agents
- Chemical shoreline cleaning agents (SCAs)
- Shoreline cleaning agents
- Chemical beach cleaners
- Oil spill dispersants
- Oil spill cleanup agents

History

Oil spill–treating agents have been a subject of controversy since their introduction during the Torrey Canyon oil spill off the coast of the United Kingdom in 1967. The dispersant policies of several European nations and Canada have been reviewed and compared with those of the United States [433].

Table 19–1
List of Major Oil Spills [263]

Type	Name of Vessel/ Platform/Region	Date	Location	Barrels
Various	Kuwait	01-19-91	Persian Gulf, Iran	9,000,000
Platform	IXTOC I	06-03-79	Bahia de Campeche, Mexico	3,522,400
Platform	Nowruz Oil Field	02-10-83	Persian Gulf, Iran	1,904,700
Tank vessel	Amoco Cadiz	03-16-78	Brittany, France	1,619,000
Tank vessel	Sea Star	12-19-72	Gulf of Oman	937,000
Tank vessel	Torrey Canyon	03-18-67	Land's End, England	860,000
Tank vessel	Urquiola	05-12-76	La Coruña, Spain	733,000
Tank vessel	Independenta	11-15-79	Istanbul, Turkey	687,700
Tank vessel	Jakob Maersk	01-29-75	Leixoes, Portugal	637,500
Tank vessel	Khark 55	12-19-89	400 miles north of Las Palmas, Canary Islands	452,400
Tank vessel	Metula	08-09-74	Strait of Magellan	398,000
Tank vessel	Assimi	01-07-83	Oman	379,000
Tank vessel	World Glory	06-13-68	65 miles east northeast of Durban, South Africa	334,000
Tank vessel	St. Peter	02-05-76	Cabo Manglares, Colombia	279,000
Tank vessel	Corinthos	01-31-75	Delaware River, Marcus Hook, Pennsylvania	266,000
Tank vessel	Burmah Agate	11-01-79	Galveston Bay, Texas	254,761
Tank vessel	Athenian Venture	04-22-88	Canada, southeast of Cape Race, Newfoundland	252,400
Tank vessel	Exxon Valdez	03-24-89	Bligh Reef, Prince William Sound, Alaska	240,500
Facility	Texaco Storage Tank	04-27-86	Bahia Las Minas, Panama	240,000

List of Major Oil Spills

The importance of spill-treating agents is reflected by a list of major oil spills in the past, which is given in Table 19–1.

General Requirements

There are some requirements that oil spill–treating agents should fulfill. Chemical dispersants are often used to disperse spilled oils, which threaten to pollute shoreline areas. Oil spill dispersants increase the surface area of the oil, which accelerates the process of biodegradation. However, the toxic properties of dispersants limit their use.

In particular, oil spill–treating agents should have a long shelf life and should be:

- Ecologically friendly
- Nontoxic
- Nonpolluting
- Biodegradable
- Highly active
- Noncorrosive
- Applicable from boats, aircraft, and helicopters

Special formulations that are suitable for various environments, that is, marine, shoreline, freshwater and saltwater, tropic, and arctic environments, have been developed.

Storage

Oil spill dispersant chemicals have often been stored for long periods awaiting their use in an emergency case. It is not uncommon to find stocks being stored for more than 5 years. Dispersants severely lose their efficiency or deteriorate in other ways during storage, so they no longer meet the performance specifications necessary. Therefore accelerated storage and corrosion tests have been performed [23]. The tests compiled background information on eight typical high-performance oil spill dispersants.

Mechanisms

Influence of the Dispersant Performance on the Crude Oil Type

Crude oils contain various amounts of indigenous surface-active agents that stabilize water-in-oil emulsions. Therefore crude oils may stabilize such emulsions. It has been shown that the effectiveness of a dispersant is dependent on both the dispersant type and the specific crude oil [309]. However, there is no apparent correlation between the degree of emulsion-forming tendency of the crude oil, which is a function of the indigenous surfactant content, and the effectiveness of the dispersant. In general, indigenous surfactants in crude oil reduce the effectiveness of the dispersant, but to an unpredictable level.

Surface Chemical Aspects of Oil Spill–Dispersant Behavior

Dispersants are widely used in many parts of the world to deal with oil spills on the ocean. The objective of adding the dispersant is to emulsify the oil slick into the water column. This prevents wind forces from moving the slick to shore and may increase the bioavailability of the oil because of the large increase in surface area caused by emulsification. Dispersants are surface-active agents whose behavior can be understood through the application of surface chemical principles [385].

Modern oil spill–dispersant formulations are concentrated blends of surface-active agents (surfactants) in a solvent carrier system. Surfactants are effective for lowering the interfacial tension of the oil slick and promoting and stabilizing oil-in-water dispersions. The solvent system has two key functions: (1) to reduce the viscosity of the surfactant blend to allow efficient dispersant application and (2) to promote mixing and diffusion of the surfactant blend into the oil film [601].

Photocatalytic Oxidation of Organic Compounds on Water

A method for treating an oil film floating on water is composed of two functions:

1. Dispersing a number of water-floatable particles on an oil film, the particles of a material that, under illumination and in the presence of air, accelerates the oxidation of organic compounds in the oil film
2. Allowing the particles to be exposed to solar illumination and ambient air [811]

The particles consist of a bead with an exterior surface that is at least partially coated with a material capable of accelerating the oxidation of organic compounds floating on water, under illumination, and in the presence of air. The coated bead is water-floatable and has a diameter of less than 2 mm. The bead consists of a plastic material coated with an intermediate layer of a material that will not accelerate the oxidation of the plastic material by air or by itself, oxidized under illumination and in the presence of air by the outer coating material.

Application

Oil spill–treating agents may be applied from boats, hydrofoils, aircraft, or helicopters in the case of large-scale pollution. For minor incidents such as car accidents, the application is done by hand.

Boat

A dispersant fan sprayer has been built and tested statically on land and demonstrated offshore on a supply vessel while spraying water. Coverage rates of 4 miles^2/day are possible by boat, using high-speed fans that create a focused airstream with maximal velocities of 90 miles/hr [39]. The dispersant is injected into and propelled by the airstream. With the airstream acting as a carrier for the dispersant, the spraying of smaller volumes of concentrate dispersant or dilute dispersant over a wide swath width is made possible. The water surface is gently agitated by the airstream and liquid impact.

Corexit 9527 is a water-and ethylene glycol monobutyl ether–dissolved dispersant. The nature of the surface-active agent has not been disclosed. Laboratory tests were conducted using 0.5-mm thick, fresh Alberta Sweet-Mixed Blend crude oil treated with Corexit 9527 dispersant applied from an overhead spray boom [165]. The effects on dispersion efficiency of mixing jet pressure, mixing jet flow rate, jet standoff distance, and vessel speed were evaluated. The system operates with a nozzle pressure of 7000 kPa, a flow rate of 55 liter/min per nozzle, and nozzles positioned approximately 0.6 m from the water surface. In laboratory tests, such a system was capable of dispersing 80% to 100% of the surface slick.

Herding Effect. In a series of trials, three dispersants were sprayed from a boat. It was concluded that a high level of energy at the sea surface mitigates the discrepancies in the efficiencies of the dispersants as measured in laboratory tests. Better results were obtained in relatively thick oil slicks. The low efficiency measured when treating downwind was attributed to the already-observed herding effect. These complementary results reinforce the actions that have been developed to optimize the application of dispersants by ship. Equipment for neat dispersant spraying is described. An operational treatment procedure has described how to map, mark out, prospect, and treat oil slicks according to the slick shape, estimated oil thickness, and wind direction [1209].

Hydrofoils

Ships are considered most suitable to apply a dispersant with spray booms, because of their large carrying capacity and their ability to navigate and operate even under bad weather conditions and at night. Experiments have shown that cleanup at a speed greater than 10 knots is unadvisable because the bow wave breaks up the oil film on the water. A high-speed craft such as a hydrofoil, when flying foilborne, solves this problem, however [1779]. In fact, upon reaching its take-off speed, it raises itself above the water. The hydrofoil has a special stability because it is kept above the water by the foil lift. Because it rises above the water, the hydrofoil avoids creating disturbing wave motion and thus long spray booms can be used.

Aircraft

A portable spray unit has been developed for the application of dispersants by large airplanes, such as the Hercules C–130. This spray unit can be placed rapidly in the cargo aircraft without any mechanical alterations. Tests spraying a dispersant concentrate have been performed [1107].

Campaigns of dispersant offshore trials were conducted from 1979 to 1985 off the French Mediterranean and Brittany coasts. Approximately 30 slicks were treated with several dispersants applied from ships, helicopters, and an aircraft by different spraying systems [199]. The experiments identified different effects of dispersants such as short-term dispersion of oil; delayed dissemination; and limiting parameters such as minimal energy of sea surface, ratio of dispersant to oil needed, and negative herding effect. Various techniques were tested to optimize the application of dispersants in different situations. Techniques included use of a variable flow rate system to spray neat concentrates from ships and varying methods of operating ships and aircraft to reach a selective distribution of dispersant and get good coverage of slicks.

A field test was conducted by spraying a commercial oil spill–dispersant (Corexit 9527) from aircraft [696]. Test objectives were to determine the efficiency of delivering the dispersant to a selected target using a large aircraft and to compare various measurement systems for droplet size and spray pattern distribution. The results indicated that aerial flights up to 46 m can produce droplet sizes and swath widths that would be operationally effective for an oil spill.

Corexit 9527, dyed with Rhodamine WT, was applied by aircrafts at a target dose rate of 5 gal/acre over a collection grid of metal trays, Kromekote cards, oil-sensitive cards, and a continuous trough [577]. Analysis of the collected dispersant was done colorimetrically, fluorometrically, and by image analysis. Correlations through the different methodologies demonstrated that high-speed, moderate-altitude application of oil dispersant could be successful in delivering the dispersant to the surface at an effective concentration and appropriate drop size. Environmental studies of the test area showed no residual dispersant in the soil following the cessation of spraying treatment.

Environmental Aspects

The rise in environmental concern, coupled with the enormity of some oil spills in the recent past, has led to the development of new generations of oil spill dispersants.

Biodegradation

Biodegradable oil spill dispersants with high efficiency and low toxicity have been prepared and tested. They consist of nonionic and low-toxicity surfactants with different molecular weights [2]. The relationship between interfacial tension and the efficiency and chemical structure of the prepared oil spill dispersants was also studied.

A test to determine the biodegradation rate of the dispersant and the biodegradation rate of the dispersant–oil mixture has been proposed [1302]. The test method is intended to supplement the toxicity tests and the effectiveness tests, which evaluate the performance of oil spill dispersants.

Standardized Measurement of Ecologic Effects

The number and variety of both toxicologic and analytic methodologies generating data on this topic are numerous, making it virtually impossible to compare datasets to arrive at a coherent conclusion. In 1994, the Chemical Response to Oil Spills Ecological Effects Research Forum (CROSERF) was formed. It is a working group composed of representatives from industry, academia, and government. The goals of CROSERF are dedicated to standardizing and improving the quality and usefulness of laboratory and mesocosmos research into the ecologic effects of oil spill–treating agents [109, 1631].

Toxicity

Seagrasses, Mangroves, and Corals. Jamaica's shoreline is at the intersection of five major petroleum tanker shipping routes and is a cargo transshipment point for the Caribbean. The island of Jamaica experiences six small- to medium-sized oil spills per year. Major ports of petroleum entry are close to mangroves, seagrass, and coral resources. One of the most critical habitats throughout the Atlantic subtropics and tropics is seagrass. Seagrasses, mangrove, and coral habitats function as fisheries' nursery habitats, food, and erosion control. If the seagrasses are removed, hundreds of fisheries' organisms disappear. Seven dispersants were tested for 100 hours on three seagrasses with respect to their toxicity [1745]. The results showed that the mortality differed among seagrasses and among dispersants. Based on these results, oil spill–cleanup plans were recommended that indicate exact dispersants and concentrations to be used in areas containing seagrasses. In general, the coral toxicity parallels the seagrass response to the dispersants [1746]. Recommendations for use of nontoxic dispersants, with primarily coral reef and fish sensitivity as paramount concerns, are Cold Clean, Corexit 9550, and Finasol OSR7 [1744].

Response on **Daphnia magna.** The use of dispersants for petroleum is often recommended in accidental aquatic pollution situations in which an oil layer is capable of reaching the banks of a river or water pond. The petroleum is then emulsified in the water, which makes it bioavailable for degrading organisms. However, this bioavailability may be responsible for an increase of the oil toxicity for the living organisms in the water. In addition, the dispersant itself is potentially toxic and its release in the environment must be controlled.

In the case of river streams, the effect of dispersing the oil creates a strong peak of pollution running along the river flow. The living organisms are submitted to short-term but intense pollution, leading to acute effects.

The time dependence of the acute toxicity of oil and dispersants on a sensitive freshwater organism, namely, *Daphnia magna,* was investigated [1805]. Two different oils were used: a crude oil from the southwest of France and a gas oil free from volatile substances after being equilibrated with atmosphere. Two commercial dispersants were used: British Petroleum Enersperse 1037 and Dasic Freshwater.

The response of marine macrophytes on oil dispersion is dependent on the type of both oil and oil dispersant [292]. Germination inhibition of the marine macrophyte *Phyllospora comosa* was used to assess and compare the effects of oil dispersants and dispersed diesel fuel and crude oil combinations. The inhibition of germination by the water-soluble fraction of diesel fuel increased after adding all dispersants investigated. This contrasted with crude oil, in which the addition of some dispersants resulted in an enhanced germination rate.

Implementation Application Programs

Guidelines

Ideally, the consideration of dispersant use should take place before an emergency to reach a timely decision [432]. Several states and regional response teams have active programs that address the planning and technical and environmental considerations affecting dispersant use. In several states where the use of dispersants is an emerging issue, there appears to be a willingness to consider their use on a case-by-case basis and a genuine interest in learning more about their effectiveness and toxicity.

A decision concerning the use of a certain dispersant involves several components, including considerations of operational feasibility and regulatory policy and environmental concerns. Eleven examples of major published procedures for making oil spill–response decisions, including decisions for or against the use of chemical dispersants, have been summarized and compared in a study [635].

Several guidelines have been given for the use of oil spill dispersants, among them, ASTM guidelines for use.* The guidelines include a variety of environments such as freshwater ponds, lakes, and streams, as well as land. The laboratory tests to measure dispersant effectiveness that are specified in federal regulations are not easy to perform, nor inexpensive, and generate a large quantity of oily waste water.

---

* References [399, 611, 634, 636, 1162, 1210, 1841].

Computerized Model

A computerized model has been developed for planning and implementing an effective dispersant application program [38]. The model makes it possible to conduct rapid assessment of specific oil–dispersant relationships, oil slick configurations, equipment types, and staging locations, as well as a broad range of dosages achievable within realistic operating constraints. Such constraints are provided for vessel, helicopter, and fixed-wing application systems. For a given spill scenario, the user can determine the amount of dispersant needed, the number of sorties required, the area and potential volume of oil treated per sortie, and the time required to treat a specified percentage of the slick.

Tests

Wave Basin

Many sea trials of dispersant chemicals to demonstrate the effectiveness of specific products or to elucidate the processes of oil dispersion into the water column have been described. Most tests have proved inconclusive, leading many to believe that dispersant chemicals are only marginally effective. Tests in a wave basin have been conducted to measure dispersant effectiveness under closely controlled conditions [261]. These tests show that dispersed oil plumes may be irregular and concentrated over small volumes, so extensive plume sampling was required to obtain accurate dispersant effectiveness measurements. In large-scale sea trials, dispersants have been shown effective, but only when sufficient sampling of the water column was done to detect small concentrated dispersed oil plumes and when it was known that the dispersant was applied primarily to the thick floating oil.

Broken Ice

Experiments have been conducted in a wave basin to determine the effectiveness of dispersants in breaking up oil spilled onto a mixture of broken ice and water. Forty-liter aliquots of a light crude oil were spilled into containment booms that had been frozen into the ice of a saltwater-filled wave tank [260]. The spills were treated with either Corexit 9527 or Corexit 9500, and then low-amplitude waves were generated for 2 hours. In a short time, the spills were dispersed by 90% or better. The oil-in-water dispersion was monitored by fluorometry, by video and still photography, and by measuring the oil remaining on the water and ice surface after the experiment. The size distribution of the ice floes had little effect on the amount of dispersion. The dispersion of oil spilled into a single straight lead in the ice sheet was also studied. It was found that oil spilled into a lead filled with slush ice and treated with dispersant rapidly dispersed into the water column.

Finite Difference Models

Finite difference models to simulate the diffusion and advection of oil in water have been developed and tested in wave basins [1748].

Small Scale Testing

There are various testing procedures, such as the Warren Spring Rotating Flask test (WSL test, Labofina test), Institute Francais du Petrole flow test (IFP test), Mackay–Nadeau–Steelman test (MNS test), EXDET, and other procedures.

Water Extraction Process. The traditional method of measuring dispersant effectiveness under laboratory conditions is to take a small aliquot of the dispersion test water; extract the oil, usually with methylene chloride; and then measure the color at a specific wavelength. This value is compared with a standard curve from which the effectiveness can be calculated. An error was found in the traditional approach of preparing standard curves because adding water to the extraction process produced some coloration in the methylene chloride [590]. Light oils have a low absorbance at the typical wavelengths chosen and tests were found to be erroneous by as much as 300% when traditional methods of analysis were used. More typical medium oils showed errors of only a few percent, but again heavy oils showed significant error because of the different wavelengths at which they absorb. Several methods of compensating for this effect were tried and found to be inadequate. Gas chromatography is suggested as the means to analyze for dispersant effectiveness in the laboratory.

Mackay–Nadeau–Steelman Test, Labofina Test, Oscillating Hoop Test, and Swirling Flask Test. Laboratory tests of the effectiveness of oil spill dispersants are used around the world to select dispersants for application to specific oils. The two most widely used tests are the Mackay test, otherwise known as the *Mackay–Nadeau–Steelman test,* and the Labofina test, otherwise known as the *Warren Springs* or *rotating flask test*. The Mackay test uses a high-velocity air stream to energize 6 liters of water, whereas the Labofina test uses rotation of a separatory funnel with 250 ml of water. Both tests apply a large amount of energy to the oil–water system. Two lesser-known devices are the oscillating hoop and the swirling flask [584]. The oscillating hoop apparatus uses a hoop moved up and down at the water surface. The concentric waves serve to energize the oil in the hoop and to contain it. Thirty-five liters of water are used in this test. The swirling flask test makes use of a 125-ml Erlenmeyer flask.

The flask is rotated using a standard chemical/biologic shaker to swirl the contents. Results show that all high-energy tests (the Mackay, the Labofina, and the oscillating hoop) produce unique dispersant effectiveness results that correlate poorly with the physical properties of the oil.

EXDET Test. A dispersant effectiveness test, named *EXDET*, was developed to address concerns associated with available laboratory dispersant effectiveness test procedures [160]. The EXDET procedure uses standard laboratory equipment (such as a Burrell Wrist–Action shaker) and small volumes of water, oil, and chemical dispersant. Other features include the capabilities to mass balance the dispersed and nondispersed oil and to generate replicate data for statistical analysis.

Portable Equipment. Chemical shoreline cleaning agents enhance removal of stranded oil from shoreline surfaces, but site-specific variables, physical and chemical properties of oil, and variations in substrate types influence the performance of shoreline cleaning agents. It is difficult to predict the performance for site-specific variables. Therefore on-site testing of shoreline cleaning agents with oil and local substrates is needed.

A portable field kit used to estimate quantitative and qualitative information for cleaning performance and dispersion of oil with shoreline cleaning agents has been described in literature [390]. The methodology was tested with three substrate types for field use of shoreline cleaning agents (gravel, rip-rap, and eelgrass), two oils (Bunker C and Bonny Light), and two shoreline cleaning agents (Corexit 9580 and PES-51). The results for cleaning performance and oil dispersion exhibit sufficient reproducibility to allow statistically significant differences to be identified in tests with and without shoreline cleaning agents or between shoreline cleaning agents.

Comparison of Effectiveness Tests. Three laboratory methods were compared: the revised standard dispersant effectiveness test used and required for regulation in the United States, the swirling flask test (developed by Environment Canada), and the IFP-dilution test (used in France and other European countries) [1693]. Six test oils and three dispersants were evaluated. It was concluded that the three tests gave similar precision results, but that the swirling flask test was fastest, cheapest, simplest, and required the least operator skill.

Correlations Among the Different Test Methods. Comparative studies revealed that the test results from different apparatus are not highly correlated, and often the effectiveness rank is not correlated [594]. The effect of the settling time and oil/water ratio are important in determining the final effectiveness value. Energy is important only to the extent that, when high energy is applied to an oil-dispersant system, dispersion is increased by an amount related to the oil's natural dispersibility.

A study on the efficiency of dispersants with various testing methods showed that some tests for the same oil dispersant system produced no correlation [1232]. In another study [1492], seven laboratory methods for testing dispersant effectiveness using commercial oil spill products and No. 2 and No. 6 fuel oils were evaluated. The tests included the EPA, Mackay, Russian, French, Warren Spring, and two interfacial tension test methods (one based on the du Nouy ring principle and the other on drop weight). These tests were reviewed in terms of type, scale, method of applying mixing energy, and the time required to conduct a product evaluation. The experimental results, compared in terms of test data precision and how effective the six nonionic dispersants were, demonstrate that the relative effectiveness found for the dispersants varies appreciably as a function of the testing method.

Effectiveness tests of dispersants have been performed according to two different methods: the WSL test and the IFP test [706]. The WSL test is a rotating flask test and the IFP test is a lower-energy test, with dilution by clean seawater. The results show mainly that there is no evident correlation between the methods, which may be because of their completely different designs. Another result is that the IFP test is much more selective than the WSL test. It can be concluded that the nature of the oil is as important as the design of the method. As a global conclusion, dispersants should be tested in different conditions because their effectiveness varies significantly in function of the test design and the test oil.

Effectiveness of Chemical Dispersants Under Real Conditions. It is believed that the effectiveness of dispersants is influenced by a number of factors, including the chemical natures of the dispersant and the nature of the oil, their relative amounts, and the microscopic mixing processes occurring as the dispersant lands on the oil and penetrates it while subject to turbulence originating in the air and water [1143]. In addition, the oil to be treated can also partly evaporate, form mousses, and spread into thick and sheen patches.

There is no doubt that effective dispersion takes place in laboratory conditions and under certain application conditions at sea. However, it is apparent that at sea, the effectiveness is often reduced by one or more factors:

- Underdosing and overdosing of the slick because of its variable thickness
- Underestimation of the effect of weathering
- The character of the energy available at the sea surface

Special Aspects

Arctic Conditions. The effectiveness of relevant dispersants for use under arctic conditions has been tested with a dilution test [235, 236]. Arctic conditions mean a low temperature of 0° C and water salinities of 0.5% to

3.3%. The results showed that many dispersants that previously showed excellent effectiveness at high salinity (3.3%) may have low effectiveness at low-salinity conditions (0.5%). The study emphasizes the need for development of dispersants with a high effectiveness both at low temperature and over a wide range of salinities.

Effectiveness in Salt Solutions. Effectiveness in calcium and magnesium salt solutions is different from that in sodium salt solutions [162]. In general, the effectiveness is lower at zero salinity.

Effectiveness Testing. Initially, it was emphasized that oil spill–treating agents can be divided into four classes: solidifiers, demulsifying agents, surface-washing agents, and dispersants.

Solidifiers, or gelling agents, solidify oil, requiring a large amount of agent to solidify oil—ranging from 16% to more than 200% by weight. Emulsion breakers prevent or reverse the formation of water-in-oil emulsions.

The effectiveness of a dispersant can be defined as the percentage of oil in the water column. Emulsion breakers have been tested for their performance [595, 596]. Among the tested products, only one highly effective formulation has been determined. However, the conclusion is not too discouraging. Many products will work but require large amounts of spill-treating agent. Surfactant-containing materials are of two types: surface-washing agents and dispersants. Testing has shown that an agent that is a good dispersant is, conversely, a poor surface-washing agent, and vice versa. Tests of surface-washing agents show that only a few agents have effectiveness of 25% to 40%, in which this effectiveness is the percentage of heavy oil removed from a test surface. The aquatic toxicity of these agents is an important factor and has been measured for many products [591].

Results using the swirling flask test for dispersant effectiveness have been reported. Heavy oils show effectiveness values of approximately 1%, medium crude oils of approximately 10%, light crude oils of approximately 30%, and very light oils of approximately 90%.

The effectiveness of a number of crude oil dispersants, measured using a variety of evaluation procedures, indicates that temperature effects result from changing viscosity, dispersants are most effective at a salinity of approximately 40 ppt (parts per thousand), and concentration of dispersant is critical to effectiveness. The mixing time has little effect on performance, and a calibration procedure for laboratory dispersant effectiveness must include contact with water in a manner analogous to the extraction procedure; otherwise, effectiveness may be inflated [587]. Compensation for the coloration produced by the dispersant alone is important only for some dispersants.

Natural Dispersion. In a study on the relationship of dispersant effectiveness with the factors of dispersant amount and mixing energy, the energy was varied by changing the rotational speed of a specially designed apparatus [592]. The effectiveness goes up linearly with energy, expressed as flask rotational speed. The natural dispersion shows a behavior similar to the chemical dispersion, except the thresholds occur at a higher energy and the effectiveness rises more slowly with increasing energy. The effect of the amount of dispersant is the same with both low and high energies. The effectiveness increases exponentially with increasing dispersant amount. Although a trade-off exists between dispersant amount and energy required to achieve high effectiveness values, energy is considered the more important factor.

Each oil-dispersant combination shows a unique threshold or onset of dispersion [589]. A statistic analysis showed that the principal factors involved are the oil composition, dispersant formulation, sea surface turbulence, and dispersant quantity [588]. The composition of the oil is very important. The effectiveness of the dispersant formulation correlates strongly with the amount of the saturate components in the oil. The other components of the oil (i.e., asphaltenes, resins, or polar substances and aromatic fractions) show a negative correlation with the dispersant effectiveness. The viscosity of the oil is determined by the composition of the oil. Therefore viscosity and composition are responsible for the effectiveness of a dispersant. The dispersant composition is significant and interacts with the oil composition. Sea turbulence strongly affects dispersant effectiveness. The effectiveness rises with increasing turbulence to a maximal value. The effectiveness for commercial dispersants is a Gaussian distribution around a certain salinity value.

The effect of water temperature variation is logarithmically correlated with dispersant effectiveness [585]. Dispersant/oil ratios greater than approximately 1:40 or 1:60 result in a low dispersant effectiveness. Dispersion experiments were conducted to investigate the effects of oil composition. The effectiveness is positively and strongly correlated with the saturate concentration in the oil and is negatively correlated with the contents of aromatic, asphaltene, and polar compounds in the oil. The effectiveness is weakly correlated with the viscosity of the oil. The dispersant effectiveness is limited primarily by the oil composition.

Studies have been conducted concerning the variances among several standard regulatory tests. Three main causes of differences have been identified: oil/water ratio, settling time, and energy [593]. The energy can be partially compensated for tests with high energy by correcting for the natural dispersion. With this correction and with high oil/water ratios and a settling time of at least 10 minutes, five test methods yield similar results for a variety of oils and dispersants. The repeatability of energy levels used in the instrumentation is largely responsible for the variation in dispersant-effectiveness values.

Analysis of Corexit 9527. Corexit 9527 in natural waters can be analyzed. The method is based on the formation of a *bis*(ethylenediamine) copper(II) complex, extraction of the complex into methylisobutylketone, and atomic absorption spectroscopy [1564]. The method is suitable for a concentration range of 2 to 100 mg/liter, with a precision as low as 5% relative to standard deviation for samples in the middle- to high range. Only a small sample volume (10 ml) is required. The sensitivity may be substantially increased for trace analysis by increasing the sample volume.

Subsurface, Soil, and Groundwater

Subsurface contamination by organic chemicals is a widespread and serious problem, restricted not only to oil spills, but also pertinent in former and still-operating industrial sites. Remarkably, chemical-enhanced oil recovery technology can be used to remove oily contaminants from soil; see Chapter 16, p. 232 for further explanation.

In Situ Chemical Oxidation

Chlorinated solvents, polyaromatic hydrocarbons, and other organics can be resistant to in situ biodegradation or may take exceedingly long periods of time to degrade in many subsurface settings.

Field experiences have demonstrated that the successful application of in situ chemical oxidation requires the consideration of several factors through an integrated evaluation and design practice. Matching the oxidant and in situ delivery system to the contaminants of concern and the site conditions is the key to successful implementation of such techniques [1778].

Groundwater

Groundwater contaminant plumes from accidental gasoline releases often contain methyl-tert-butyl ether. Experiments with certain soil microorganisms showed that a culture able to degrade methyl-tert-butyl ether did not degrade benzene and toluene. Further interactions were observed [468].

Chemicals in Detail

Oxyethylated Alkylphenol, Fatty Acid Amide, and Alkylphosphate Mixtures

A solution of a surfactant mixture in liquid paraffin, containing oil-soluble oxyethylated alkylphenol, having the formula with an C_8 to C_{12} alkyl rest, the alkylphosphate of a higher fatty acid alcohol $(RO)_2PO - OH$ where R is C_{10} to C_{20}, and a fatty acid amide of diethanolamine, are suitable for removing oils and petroleum products from water surfaces [350]. The composition has low toxicity, is not inactivated by freezing, and has high biologic activity, stimulating the growth of microflora and giving 80% to 83% dispersion in 5 minutes.

Sorbitan Oleates for Oil Slicks

Dispersant compositions for the treatment of oil spills at the surface of the water consist of a mixture of water, a hydrocarbon solvent, and a mixture of surfactants consisting of 55% to 65% by weight of emulsifiers and 35% to 45% by weight of dioctyl sodium sulfosuccinate. The emulsifying agents consist of a mixture of various sorbitan oleates [351–354].

Fatty Alcohols

Petroleum spillages can be removed from water surfaces more efficiently with the following detergent mixture [1692], which contains mainly oxyethylate fatty C_{10} to C_{20} alcohols and additional oxyethylated fatty C_{11} to C_{17} acids with an oxyethylene chain length of one to two units. It is used in the form of an aqueous 20% to 25% emulsion, which is sprayed onto a contaminated surface.

Proteins

A proteinaceous particulate material has been described that is effective as an oil spill–dispersant composition [1450]. The material is a grain product (such as oats) from which lipids are removed through organic solvent extraction. When such compositions are applied to an oil spill, they will adsorb oil, emulsify it, and finally, disperse it. Moreover, the compositions are substantially nontoxic.

Polymer

Functionalized copolymers from dienes and p-alkylstyrenes can serve as dispersants and viscosity index improvers. The functionalities are introduced via the aromatic units [233, 234]. The polymers are selectively hydrogenated to produce polymers that have highly controlled amounts of unsaturation, permitting a highly selective functionalization. The dispersant substances may also include a carrier fluid to provide concentrates of the dispersant.

Cyclic Monoterpenes

The recovery of sludging oil crudes from hydrocarbon-bearing formations during acid stimulation treatments can be enhanced using an antisludging agent that is basically a dispersant. Such an antisludging agent consists of an admixture of dicyclopentadiene and a mixture of naturally occurring cyclic monoterpenes isolated from *Pinus* species [621, 622]. The agent is added to the acid used for well stimulation treatment. Another dispersing agent active under these conditions is ethoxylated alkylphenol dissolved in a mixture of ethylene glycol, methanol, and water [620, 623, 624].

Special Chemicals for Oiled Shorelines

The use of chemical dispersant for oiled shorelines is one of the most controversial, complex, and time-critical issues facing officials responsible for making decisions about the response methods used on coastal oil spills [1810].

In general, the cleanup of oiled shorelines has been by mechanical, labor-intensive means. The use of surfactants to deterge and lift the oil from the surface results in more complete and rapid cleaning. Not only is the cleaning process more efficient, but it can also be less environmentally damaging because potentially less human intrusion and stress on the biologic community occurs and because the chemicals can make the washing more effective at a lower temperature.

Chemical beach cleaners can facilitate the cleanup of oiled shorelines by improving the efficiency of washing with water. A dispersant has been improved by reducing the adhesion of the oil coating, which makes it easier to remove from shoreline surfaces, thereby reducing washing time and lowering the temperature of the wash water needed to clean a given area [599]. These experiences resulted in the development of Corexit 9580 [310, 600]. Corexit 9580 consists of two surfactants and a solvent. It exhibits low fish toxicity, low dispersiveness, and effective rock cleaning capability. Experiments on mangroves to explore the potential use of Corexit 9580 to save and restore oiled vegetation has been considered.

Such a dispersant formulation for dispersing oil contains a mixture of a sorbitan monoester of an aliphatic monocarboxylic acid, a polyoxyethylene adduct of a sorbitant monoester of an aliphatic monocarboxylic acid, a water-dispersible salt of a dialkyl sulfosuccinate, a polyoxyethylene adduct of a sorbitan triester or a sorbital hexaester of an aliphatic monocarboxylic acid, and a propylene glycol ether as solvent [311, 312].

Coagulants

Glycerides (e.g., linseed oil); fatty acids; alkenes; and a polymer, such as polyisobutylmethacrylate, are treated in a thermal process to prepare an oil coagulant. The oil coagulant composition floats on the water surface and coagulates oil independent of both agitation and temperature and can be used in both saltwater and freshwater. After the coagulant has coagulated the spilled oil, at least 99.9% of the floating coagulated oil can be readily removed from the water by mechanical methods.

Chapter 20

Dispersants

The major fields of application for dispersants include cement slurries, drilling fluids, oil spill–treating agents, and transport applications.

Cement

Dispersants are used in well cement slurries. For this application, the dispersant should be water soluble. The dispersants prevent high initial cement slurry viscosities and friction losses when the slurries are pumped.

Polymelamine Sulfonate and Hydroxyethylcellulose

For oil field cement slurries containing microsilica, sodium polymelamine sulfonate and hydroxyethylcellulose [142–144] may be used as dispersing agents. The cement slurries may contain up to 30% by weight of microsilica (i.e., olloidal silica), silica flour, diatomaceous earth, or fly ash with particle dimensions between 0.05 and 5 µ. The slurries may further contain conventional additives, such as antifoaming agents and set retarding agents, etc.

Polyethyleneimine Phosphonate Derivatives

In oil and gas well cementing operations, polyethyleneimine phosphonate–derivative dispersants enhance the flow behavior of the cement slurry [422]. The slurry can be pumped in turbulent flow, thereby forming a bond between the well casing and the rock formation.

Acetone Formaldehyde Cyanide Resins

An aqueous solution of acetone and sodium cyanide is condensed by adding formaldehyde at 60° C. A resin with nitrile groups is obtained. A similar product with sodium sulfite can be obtained. These products are dispersants for cements [558, 559]. The dispersant properties of the composition can be enhanced by further reacting the composition with a hydroxide.

Napthalenosulfonic Acid Formaldehyde Condensates

A dispersing agent for a cement slurry is the magnesium salt from the condensation of napthalenosulfonic acid and formaldehyde [815, 816]. The additive eliminates free water, even at low temperatures and with those cements most susceptible to this phenomenon.

Sulfoalkylated Naphthols

Sulfoalkylated naphthol compounds are effective as dispersants in aqueous cement slurries. The compounds can also be applied in an admixture with water-soluble inorganic compounds of chromium to provide additives of increased overall effectiveness. Particularly suitable are sodium chromate or ammonium dichromate. α-Naphthol is reacted in an alkaline aqueous medium with formaldehyde to create condensation products. The aldehyde can be reacted with bisulfite to produce sulfoalkylated products [1404, 1410].

Azolignosulfonate

A cement dispersant has been described that is an azolignosulfonate formed from the coupling of a diazonium salt, made from sulfanilic acid or p-aminobenzoic acid, and a lignosulfonate [479]. The dispersant can reduce the aqueous cement slurry viscosity or thin the cement composition to make the cement slurry pumpable without signficantly retarding the set time of the cement slurry. The azo structure is formed by coupling lignosulfonate with diazonium salt. It masks the retardation effect of the phenolic group in the lignosulfonate molecule.

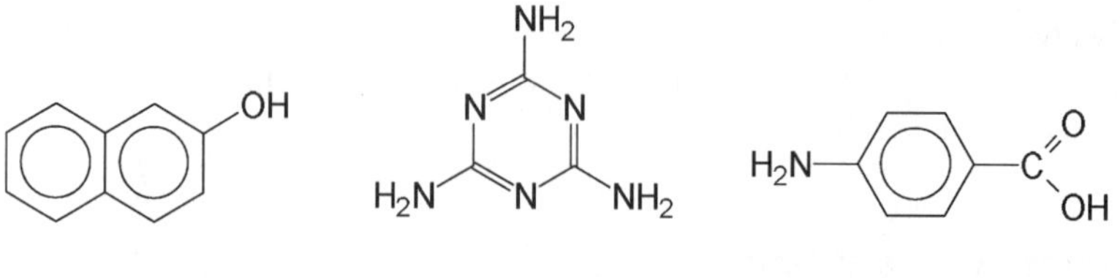

Figure 20–1. 2-Naphthol, melamine, and p-aminobenzoic acid.

Polymers from Allyloxybenzenesulfonate

Water-soluble polymers of allyloxybenzenesulfonate monomers can be used as dispersants in drilling fluids and in treating boiler waters in steamflooding and as plasticizers in cement slurries [1088, 1089]. The preferable molecular weight range is 1000 to 500,000 Dalton.

Sulfonated Isobutylene Maleic Anhydride Copolymer

A dispersant that can be used in drilling fluids, spacer fluids, cement slurries, completion fluids, and mixtures of drilling fluids and cement slurries controls the rheologic properties of and enhances the filtrate control in these fluids. The dispersant consists of polymers derived from monomeric residues, including low–molecular-weight olefins that may be sulfonated or phosphonated, unsaturated dicarboxylic acids, ethylenically unsaturated anhydrides, unsaturated aliphatic monocarboxylic acids, vinyl alcohols and diols, and sulfonated or phosphonated styrene. The sulfonic acid, phosphonic acid, and carboxylic acid groups on the polymers may be present in neutralized form as alkali metal or ammonium salts [192, 193].

Aqueous Drilling Muds

Low–Molecular-Weight Dispersants

Aqueous Drilling Muds Liquified By Means of Zirconium and Aluminum Complexes

Complexes of tetravalent zirconium and ligands selected from organic acids such as citric, tartaric, malic, and lactic acid and a complex of aluminum and citric acid are suitable as dispersants [288–291]. This type of dispersant is especially useful in dispersing bentonite suspensions. The muds can be used at pH values ranging from slightly acidic to strongly basic.

Synthetic Polymers

Polymers Containing Maleic Anhydride

A mixture of sulfonated styrene–maleic anhydride copolymer and polymers prepared from acrylic acid or acrylamide and their derivatives [759] are dispersants for drilling fluids. The rheologic characteristics of aqueous well drilling fluids are enhanced by incorporating into the fluids small amounts of sulfonated styrene–itaconic acid copolymers [761] and an acrylic acid or acrylamide polymer [755].

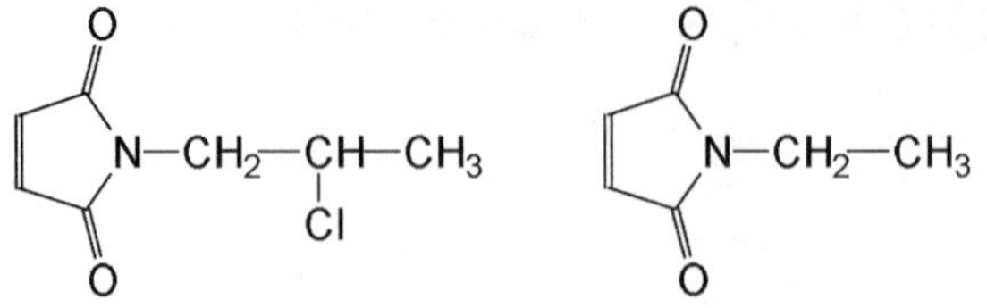

N-(2-Chloropropyl) maleimide N-Ethyl maleimide

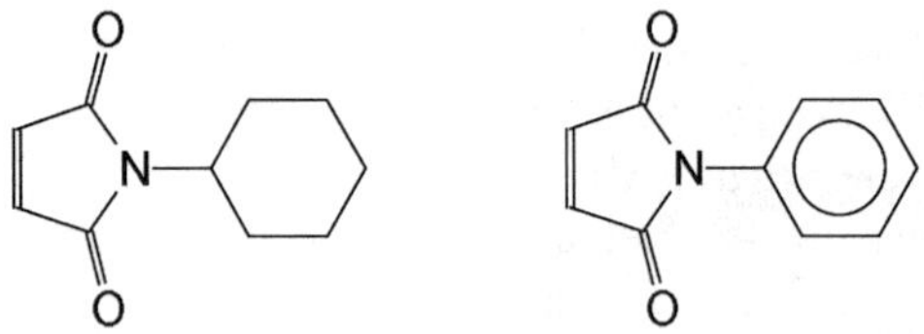

N-Cyclohexyl maleimide N-Phenyl maleimide

Figure 20–2. N-(2-Chloropropyl) maleimide, N-ethyl maleimide, N-cyclohexyl maleimide, and N-phenyl maleimide.

Sulfonated styrene–maleimide copolymers are similarly active [1073]. Examples of maleimide monomers are maleimide, N-phenyl maleimide, N-ethyl maleimide, N-(2-chloropropyl) maleimide, and N-cyclohexyl maleimide. N-aryl and substituted aryl maleimide monomers are preferred. The polymers are obtained by free radical polymerization in solution, in bulk, or by suspension.

In copolymers containing the styrene sulfonate moiety and maleic anhydride units, the maleic anhydride units can be functionalized with alkyl amine [1411–1416]. The water-soluble polymers impart enhanced deflocculation characteristics to the mud. Typically, the deflocculants are relatively low–molecular-weight polymers composed of styrene sodium sulfonate monomer; maleic anhydride, as the anhydride and/or the diacid; and a zwitterionic functionalized maleic anhydride. Typically the molar ratio of styrene sulfonate units to total maleic anhydride units ranges from 3:1 to 1:1. The level of alkyl amine functionalization of the maleic anhydride units is 75 to 100 mole-percent. The molar concentrations of sulfonate and zwitterionic units are not necessarily equivalent, because the deflocculation properties of these water-soluble polymers can be controlled via changes in their ratio.

Alternating Copolymers of Sodium Methallylsulfonate–Maleic Anhydride

Alternating 1:1 copolymers of sodium methallylsulfonate and maleic anhydride are useful as water-soluble dispersants [738]. The copolymers are produced by free radical polymerization in acetic acid solution. Because of their high solubility in water and the high proportion of sulfonate salt functional groups, these alternating polymers are useful as dispersing agents in water-based drilling fluids.

Copolymers of Acrylic Acid and Vinyl Sulfonic Acid

Low–molecular-weight copolymers of acrylic acid and salts of vinyl sulfonic acid have been described as dispersants and high-temperature deflocculants for the stabilization of the rheologic properties of aqueous, clay-based drilling fluids subjected to high levels of calcium ion contamination [1448, 1449]. Divalent ions, such as calcium ions or magnesium ions, can cause uncontrolled thickening of the mud and large increases in filtration of fluids from the mud into permeable formations. A flocculation of the mud can occur in high-temperature applications. This flocculation increases the thickening effects of certain chemical contaminants and deactivates or destroys many mud thinners, which are used to stabilize the muds with respect to these effects.

Polyacrylic Acid and Copolymers

Polyacrylic acid, or a water-soluble salt, having a molecular weight of 1500 to 5000, measured on the respective sodium salt and a polydispersity of 1.05 to 1.45, has been described as a dispersant for a drilling or packer fluid [576].

Copolymers or terpolymers of acrylic acid, which contain from 5 to 50 mole-percent of sulfoethyl acrylamide, acrylamide and sulfoethyl acrylamide, ethyl acrylate and sulfoethyl acrylamide, acrylamide and sulfophenyl acrylamide, and acrylamide and sulfomethyl acrylamide, are claimed to be calcium-tolerant deflocculants for drilling fluids [704]. In general, 0.1 to 2 lb of polymer per barrel of drilling fluid is sufficient to prevent flocculation of the additives in the drilling fluid.

Copolymer of Acrylic or Methacrylic Acid Neutralized with Alkanolamine or Alkylamine

A salt of a polymer or copolymer of acrylic or methacrylic acid, in which the acid is neutralized with alkanolamines, alkylamines, or lithium salts [677], is suitable as a dispersing agent.

Polymers with Amine Sulfide Terminal Moieties

Amine sulfide terminal moieties can be imparted into vinyl polymers by using aminethiols as chain transfer agents in aqueous radical polymerization [1182]. The polymers are useful as mineral dispersants. Other uses are as water-treatment additives for boiler waters, cooling towers, reverse osmosis applications, and geothermal processes and oil wells and as detergent additives acting as builders, antifilming agents, dispersants, sequestering agents, and encrustation inhibitors.

Polycarboxylated Polyalkoxylates

Polycarboxylated polyalkoxylates and their sulfate derivatives may be prepared by reacting an ethoxylated or propoxylated alcohol with a water-soluble, alkali or earth alkali metal salt of an unsaturated carboxylic acid [339]. The reaction occurs in aqueous solution in the presence of a free radical initiator and gives products of enhanced yield and reduced impurity levels, compared with the essentially anhydrous reactions with free carboxylic acids, which have been used otherwise. The method provides products that give solutions that are clear on neutralization, remain clear and homogeneous on dilution, and are useful as cleaning agents in drilling and other oil field operations.

Natural Modified Polymers

Phosphated, Oxidized Starch

Phosphated, oxidized starch with a molecular weight of 1500 to 40,000 Dalton, with a carboxyl degree of substitution of 0.30 to 0.96, is useful as a dispersant for drilling fluids [926].

Modified Polysaccharides

Physical mixtures consist of reversibly crosslinked and uncrosslinked hydrocolloid compositions and hydrocolloids. These show improved dispersion properties [1708].

Sulfonated Asphalt

Sulfonated asphalt can be produced as follows [1530]:

1. Heating an asphaltic material
2. Mixing the asphalt with a solvent, such as hexane
3. Sulfonating the asphalt with a liquid sulfonating agent, such as liquid sulfur trioxide
4. Neutralizing the sulfonic acids with a basic neutralizing agent, such as sodium hydroxide
5. Separating solvent from the sulfonated asphalt
6. Recovering the evaporated solvent for reuse
7. Drying the separated, sulfonated asphalt by passing it through a drum dryer

This is a batch-type process in which the rates of flow of the solvent, the asphaltic material, the sulfonating agent, and the neutralizing agent and the periods of time before withdrawal of the sulfonic acids and the sulfonated asphalt are coordinated according to a predetermined time cycle. The dried sulfonated asphalt can then be used in the preparation of drilling fluids, such as aqueous, oil-based, and emulsion types. Such drilling fluids have excellent rheologic properties, such as viscosity and gel strength, and exhibit a low rate of filtration or fluid loss.

Coal-Derived Humic Acids

Coal with a mean particle size of less than 3 mm is slurried with water and then oxidized with oxygen or mixtures of oxygen and air at temperatures ranging from 100° to 300° C, at partial oxygen pressures ranging from 0.1 to 10 MPa and reaction periods ranging from 5 to 600 minutes [425]. In the absence of catalysts, such as alkaline bases, the main products of oxidation are humic acids.

These humic acids are not dissolved because the pH of this slurry is in the range of 4 to 9. Small amounts of fulvic acids are formed, and these are soluble in the water of the slurry. The coal-derived humic acids find applications as drilling fluid dispersants and viscosity control agents, whereas the coal-derived fulvic acids may be used to produce plasticizers and petrochemicals.

Miscellaneous

A nonpolluting dispersing agent for drilling fluids [217–219] has been described. The agent is based on polymers or copolymers of unsaturated acids, such as acrylic acid or methacrylic acid, with suitable counter ions.

Dispersant for Sulfur. The deposition of elemental sulfur in conduits through which a sulfur-containing gas is flowing can be reduced by providing a sulfur dispersant. The dispersant is an adduct of a primary alcohol and epichlorohydrin, mixed with an aliphatic amine component [554].

Dispersants for Asphalts. Asphalt and asphaltene components can produce difficulties in various processes in recovering crude petroleum oils and preparing them for transportation through pipelines or in refining separation.

Certain alkyl-substituted phenol-formaldehyde resins can act as dispersants for asphalts and asphaltenes in crude oils [1681]. The dispersants help keep asphalt and asphaltenes in dispersion and inhibit fouling, precipitation, and buildup in the equipment.

Ethoxylated Sorbitol Oleate. Ethoxylated sorbitol oleate and mixtures are suitable for emulsifying or dispersing spilled petroleum products in either terrestrial or marine environments [1496].

Chapter 21

Defoamers

Defoaming is necessary in several industrial branches and is often a key factor for efficient operation. A review on defoamers is given by Owen [1381] in Kirk-Othmer.

Uses in Petroleum Technology
Gas–Oil Separation

Defoamers are used in oil extraction, such as in drilling muds and cementation, and also directly with crude oil itself. In its natural state, a crude oil contains dissolved gases at the pressure of the reservoir. When the pressure is reduced, the gases are liberated and troublesome foam can develop. There are three ways to prevent foaming in gas–oil separation:

1. Based on prior knowledge of crude oil foaming properties, a separator large enough to cope with foam formation may be installed.
2. The amount of foam can be reduced by injecting a defoamer.
3. The gas–oil separator can be equipped with a mechanical device to destroy or prevent a presumptive foam.

Understanding the factors that inhibit the foaming power is of great importance, because it yields a basic knowledge—how the materials produced will perform with respect to foaming. It also serves to predict how individual crude oil compositions would work with different defoamers [301].

Cleaning of Sour Gas

Desulfurization of natural gas can be achieved by bubbling it through an alkaline solution. Defoamers are added to avoid foaming.

Distillation and Petroleum Production

Air entrainment and foaming in hydrocarbon liquids can cause operational problems with high-speed machinery in physical–chemical processes such as petroleum production, distillation, cracking, coking, and asphalt processing.

Classification of Defoamers

Defoamer formulations currently contain numerous ingredients to meet the diverse requirements for which they are formulated. Various classification approaches are possible, including classification by application, physical form of the defoamer, and the chemical type of the defoamer. In general, defoamers contain a variety of active ingredients, both in solid and in liquid states, and a number of ancillary agents such as emulsifiers, spreading agents, thickeners, preservatives, carrier oils, compatibilizers, solvents, and water.

Active Ingredients

Active ingredients are the components of the formulation that control the actual foaming. These may be liquids or solids.

Liquid Components

Because lowering the surface tension is the most important physical property of a defoamer, it is reasonable to classify the defoamer by the hydrophobic operation of the molecule. In contrast, the classification of organic molecules by functional groups are often polar and hydrophilic (i.e., alcohol, acid, and salt are common in basic organic chemistry). Four classes of defoamers are known as liquid phase components:

- Hydrocarbons
- Polyethers
- Silicones
- Fluorocarbons

Synergistic Antifoam Action by Solid Particles

Often, dispersed solids are active in defoaming in suitable formulations. Some liquid defoamers are believed to be active only in the presence of a solid. It is believed that a surface-active agent present in the system will carry the solid particles in the region of the interface and the solid will cause a destabilization of the foam.

For example, a synergistic defoaming occurs when hydrophobic solid particles are used in conjunction with a liquid that is insoluble in the foamy solution [652]. Mechanisms for film rupture by either the solid or the liquid alone have been elucidated, along with explanations for the poor effectiveness, which are observed with many foam systems for these single-component defoamers.

Silicone Antifoaming Agents

Polydimethylsiloxane is active in nonaqueaous systems, but it shows little foam-inhibiting effect in aqueous systems. However, when it is compounded with a hydrophobic-modified silica, a highly active defoamer emerges.

Several factors contribute to the dual nature of silicone defoamers. For example, soluble silicones can concentrate at the air–oil interface to stabilize bubbles, while dispersed drops of silicone can accelerate the coalescence process by rapidly spreading at the gas–liquid interface of a bubble, causing film thinning by surface transport [1163].

Silicones exhibit an apparently low solubility in different oils. In fact, there is actually a slow rate of dissolution that depends on the viscosity of the oil and the concentration of the dispersed drops. The mechanisms of the critical bubble size and the reason a significantly faster coalescence occurs at a lower concentration of silicone can be explained in terms of the higher interfacial mobility, as can be measured by the bubble rise velocities.

Ancillary Agents

In addition to the defoamer itself, certain ancillary chemicals are incorporated into the formulation, for example, to effect emulsification or to enhance the dispersion.

Surface-Active Components

Emulsifiers are essential in oil–water emulsion systems. For example, oil-in-water emulsifiers are used to promote the dispersion in aqueous foaming systems.

Carriers

The formulation of a defoamer should be suitable for a prolonged storage time before use. A carrier system makes the defoamer easy to handle and dispersible for delivering the active defoamer components to the foaming system and also to stabilize the defoaming formulation.

Often, carriers are low-viscosity organic solvents. Aliphatic hydrocarbons are most commonly used as carriers. The carrier itself may also exhibit

Theory of Defoaming

Stability of Foams

Foams are thermodynamically unstable but are prevented from collapsing by the following properties:

- Surface elasticity
- Viscous drainage
- Reduced gas diffusion between bubbles
- Thin-film stabilization effects from the interaction of opposite surfaces

The stability of a foam can be explained by the Gibbs elasticity *(E)*. The Gibbs elasticity results from reducing the surface concentration of the active molecules in equilibrium when the film is extended. This causes an increase in the equilibrium surface tension σ, which acts as a restoring force.

$$E = 2A \frac{d\sigma}{dA} \tag{21-1}$$

A is the area of the surface. In a foam, where the surfaces are interconnected, the time-dependent Marangoni effect is important. A restoring force corresponding to the Gibbs elasticity will appear, because only a finite rate of absorption of the surface-active agent, which decreases the surface tension, can take place on the expansion and contraction of a foam. Thus the Marangoni effect is a kinetic effect.

The surface tension effects under nonequilibrium conditions are described in terms of dilatational moduli. The complex dilatational modulus ε of a single surface is defined in the same way as the Gibbs elasticity. The factor 2 is not used in a single surface.

$$\varepsilon = 2A \frac{d\sigma}{dA} \tag{21-2}$$

In a periodic dilatational experiment, the complex elasticity module is a function of the angular frequency:

$$\varepsilon(i\omega) = |\varepsilon| \cos\theta + i |\varepsilon| \sin\theta = \varepsilon_d(\omega) + \omega \eta_d(\omega) \tag{21-3}$$

ε_d is the dilatational elasticity, and η_d is the dilatational viscosity. It is characteristic for a stable foam to exhibit a high surface dilatational elasticity and a high dilatational viscosity. Therefore effective defoamers should reduce these properties of the foam.

Table 21-1
Dilatational Elasticities and Viscosities of Crude Oil at 1 mHz with Polydimethylsiloxanes (PDMS) [300]

Crude oil	Amount of PDMS (ppm)	ε_d (mNm^{-1})	η_d (mNsm^{-1})
North Sea	none added	1.34	153
North Sea	12,500	0.69	90
North Sea	60,000	0.51	33
Middle East	none added	1.63	105
Middle East	60,000	1.19	53

This has been verified for polydimethylsiloxanes added to crude oils. The effect of the dilatational elasticities and viscosities on crude oil by the addition of polydimethylsiloxanes is shown in Table 21–1. Under nonequilibrium conditions, both a high bulk viscosity and a surface viscosity can delay the film thinning and the stretching deformation, which precedes the destruction of a foam. There is another issue that concerns the formation of ordered structures. The development of ordered structures in the surface film may also stabilize the foams. Liquid crystalline phases in surfaces enhance the stability of the foam.

If the gas diffusion between bubbles is reduced, the collapse of the bubbles is delayed by retarding the bubble size changes and the resulting mechanical stresses. Therefore single films can persist longer than the corresponding foams. However, this effect is of minor importance in practical situations. Electric effects, such as double layers, form opposite surfaces of importance only for extremely thin films (less than 10 nm). In particular, they occur with ionic surfactants.

Action of Defoamers

At high bulk viscosity, lowering the surface tension is not relevant for the mechanism of stabilization of foams, but for all other mechanisms of foam stabilization a change of the surface properties is essential. A defoaming agent will change the surface properties of a foam upon activation. Most defoamers have a surface tension in the range of 20 to 30 mNm^{-1}. The surface tensions of some defoamers are shown in Table 21–2.

Two related antifoam mechanisms have been proposed for low surface tensions of certain defoamer formulations:

1. The defoamer is dispersed in fine droplets in the liquid. From the droplets, the molecules may enter the surface of the foam. The tensions created by this spreading result in the eventual rupture of the film.

Table 21–2
Surface Tensions of Some Defoamers

Material	Surface tension at 20°C (mNm^{-1})
Polyoxypropylene 3000 Dalton	31.2
Polydimethylsiloxane 3900 Dalton	20.2
Mineral oil	28.8
Corn oil	33.4
Peanut oil	35.5
Tributyl phosphate	25.1

2. Alternatively, it is suggested that the molecules will form a monolayer rather than spreading. The monolayer has less coherence than the original monolayer on the film and causes a destabilization of the film.

Spreading Coefficient. The spreading coefficient is defined as the difference of the surface tension of the foaming medium σ_f, the surface tension of the defoamer σ_d, and the interfacial tension of both materials σ_{df}.

$$S = \sigma_f - \sigma_d - \sigma_{df} \tag{21-4}$$

It can be readily seen that the spreading coefficient S becomes increasingly positive as the surface tension of the defoamer becomes smaller. This indicates the thermodynamic tendency of defoaming.

The above statements are adequate for liquid defoamers that are insoluble in the bulk. Experience has proven, however, that certain dispersed hydrophobic solids can greatly enhance the effectiveness of defoaming. A strong correlation between the effectiveness of a defoamer and the contact angle for silicone-treated silica in hydrocarbons has been established [300]. It is believed that the dewetting process of the hydrophobic silica causes the collapse of a foam by the direct mechanical shock occurring by this process.

Examples for Application

Aqueous Fluid Systems

Alcohols. Higher aliphatic alcohols with polyethyleneoxide and polypropyleneoxide are particularly effective at reducing the gas content of drilling solutions [1442].

An aliphatic alcohol with 8 to 32 carbon atoms can be used together with a solid carrier. The carrier is nonswelling in aqueous media and consists of small particles with an average size of less than 150 to 200 µ. The carrier adsorbs the alcohol [1439, 1440]. Some examples for solid carrier materials are sawdust, ground rice hulls, ground nut shells, and clays. Other carrier materials may include solids that are commonly added to drilling and other well fluids as fluid loss additives, bridging agents, and the like.

Fatty Acid Esters. Defoamers that are more environmentally acceptable than convential products are based on fatty acid esters of hydroxy alcohols, such as sorbitan monooleate [1908] or sorbitan monolaurate in combination with diethylene glycol monobutyl ether as a cosolvent [451]. These defoamer compositions are as effective as conventional materials, for example, those based on acetylenic alcohols are less toxic, especially to marine organisms, and are readily biodegradable. The defoamer compositions are used in water-based hydrocarbon well fluids during oil/gas well drilling, completion, and workover, especially in marine conditions.

Aerosil. Aerosil as a solid additive in combination with diesel is active as a defoamer [962]. The aerosil is modified with bifunctional silicoorganic compounds. The composition is added to drilling solution in the form of a 3% to 5% suspension, in an amount of 0.02% to 0.5% by weight of aerosil per weight of drilling solution.

Polyoxirane. Polyoxirane-containing formulations have a low cloud point, good ability to reduce surface and interfacial tensions, good wettability, and limited tendency to dispersion. The oxirane-methyloxirane copolymers [1176] are nontoxic and show high stability in both acidic and alkaline environments. Their poor stability against oxidation can be improved through chemical modification (blocking hydroxide groups) or by using an alkaline catalyst, which acts as an inhibitor of the self-oxidation process.

Polypropylene glycol, particulate hydrophobic silica, and a fatty acid methyl ester, or an olefin or linear paraffin as a liquid diluent, are proposed for well-stimulation jobs [357].

Plugging Agents

The foaming of plugging agents with a large content of lignosulfonate can be prevented by introducing a drilling mud antifoaming agent, consisting of polymethylsiloxane and tributyl phosphate [1905]. A synergistic effect is observed, in that polymethylsiloxane simultaneously displaces the adsorbed molecules of the foam stabilizer (lignosulfonate) from the foam film, and tributyl phosphate reduces the surface viscosity of the film.

Degassing Crude Oil

Fluorosilicones and Fluorocarbons. Early defoamers to remove gas from crude oil consisted of chlorofluorocarbons. The use of these compounds has essentially ceased. They were substituted gradually by pure fluorosilicones [302]. A formulation that is free of chlorofluorocarbons was described in the early 1990s.

A water-continuous emulsion, suitable for use as an antifoam additive, contains 85% to 98% by weight of a fluorosilicone oil and 2% to 15% by weight of an aqueous surfactant solution [1722]. The additive is suitable for use in separation of crude oil that contains associated gas. The additive may be used in both aqueous and nonaqueous systems and allows fluorosilicone oils to be used without the need for environmentally damaging chlorofluorocarbons.

Freshly extracted degassing crude oil can be defoamed with fluorinated norbornylsiloxanes [171]. The compounds are highly effective and show a broad area of application for defoaming degassing crude oils of different origins. The compounds can be used in concentrations as small as 20 ppm.

Polydienes. Polydienes that are modified with organosilicons have been described and find application as antifoaming and/or deaeration agents for oil field treating of crude oil [170].

High-Temperature Defoamers. Polyisobutylene compounds are particularly effective in high-temperature (300° to 1000° F) treatments of hydrocarbon fluids [786, 788], such as during the distillation of crude oil and coking of crude oil residues. Polyisobutylene compounds are less expensive than silicone-based compounds.

Natural Gas

A mixture of dialkylphthalate of higher isoalcohols, in excess of the respective isoalcohols, is used as an antifoaming composition in purification of natural gas where H_2S and CO_2 are removed by aqueous solution of amine [474].

Amyl alcohol and diethyldisulfide are used to improve the properties of a defoaming formulation for the removal of acidic components from natural gas [11]. The mixture contains 35% to 50% by weight tributyl phosphate and 20% to 25% by weight amyl alcohol. The rest, diethyldisulfide, is an industrial waste.

Esters, for example, dialkyl polypropyleneglycol adipate and dibutyl adipate, also find use as defoamers in the removal of H_2S and CO_2 from natural gas by bubbling it through an amine solution [659]. Use of the aforementioned components increases the efficiency of foam destruction.

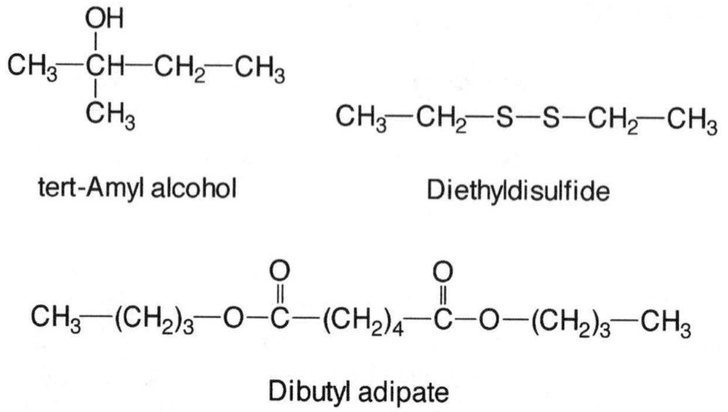

$$\underset{\text{tert-Amyl alcohol}}{CH_3-\underset{\underset{CH_3}{|}}{\overset{\overset{OH}{|}}{CH}}-CH_2-CH_3} \qquad \underset{\text{Diethyldisulfide}}{CH_3-CH_2-S-S-CH_2-CH_3}$$

$$\underset{\text{Dibutyl adipate}}{CH_3-(CH_2)_3-O-\overset{\overset{O}{\|}}{C}-(CH_2)_4-\overset{\overset{O}{\|}}{C}-O-(CH_2)_3-CH_3}$$

Figure 21–1. Amyl alcohol, diethyldisulfide, and dibutyl adipate.

Antimicrobial Antifoam Compositions

In addition to the typical constituents of a defoamer formulation (i.e., a primary antifoam agent with high surface area, such as silica, and a secondary antifoam agent for acting synergistically with the primary antifoam agent, such as polydimethylsiloxane), a water carrier with a quaternary ammonium salt silane compound [689] can be included, which acts as an antimicrobial agent. The silane is fixed to the surface of the silica. The composition makes the defoamer composition resistant to biologic degradation because of the presence of microorganisms in the system.

Chapter 22

Demulsifiers

Emulsions in Produced Crude Oil

In the production of crude oil, the greatest part of the crude oil occurs as a water-in-oil emulsion. The composition of the continuous phase depends on the water/oil ratio, the natural emulsifier systems contained in the oil, and the origin of the emulsion. The natural emulsifiers contained in crude oils have a complex chemical structure, so that, to overcome their effect, petroleum-emulsion demulsifiers must be selectively developed. As new oil fields are developed, and as the production conditions change at older fields, there is a constant need for demulsifiers that lead to a rapid separation into water and oil, as well as minimal-residual water and salt mixtures.

The emulsion must be separated by the addition of chemical demulsifiers before the crude oil can be accepted for transportation. The quality criteria for a delivered crude oil are the residual salt content and the water content. For the oil to have a pipeline quality, it is necessary to reduce the water content to less than approximately 1.0%.

The separated saltwater still contains certain amounts of residual oil, where now preferentially oil-in-water emulsions are formed. The separation of the residual oil is necessary in oil field water purification and treatment for ecologic and technical reasons, because the water is used for secondary production by waterflooding, and residual oil volumes in the water would increase the injection pressure.

The presence of water-in-oil emulsions often leads to corrosion and to the growth of microorganisms in the water-wetted parts of the pipelines and storage tanks.

At the refinery, before distillation, the salt content is often further reduced by a second emulsification with freshwater, followed by demulsification. Crude oils with high salt contents could lead to breakdowns and corrosion at the refinery. The object of using an emulsion breaker, or demulsifier, is to break the emulsion at the lowest possible concentration and, with little or no additional consumption of heat, to bring about a complete separation of the water and reduce the salt content to a minimum.

There are oil-soluble demulsifiers and water-soluble demulsifiers, the latter being widely used. Emulsions are variable in stability. This variability is largely dependent on oil type and degree of weathering. Emulsions that have a low stability will break easily with chemical emulsion breakers. Broken emulsions will form a foamlike material, called *rag,* which retains water that is not part of the stable emulsions. The most effective demulsifier must always be determined for the particular emulsion.

Demulsifiers are often added to the emulsion at the wellhead to take advantage of the temperature of the freshly raised emulsion to hasten the demulsification step.

Waterflooding

During improved oil-recovery processes, waterflooding of the oil is applied. The entrained water forms a water-in-oil emulsion with the oil. In addition, salts such as sodium chloride, calcium chloride, and magnesium chloride may be dissolved in the emulsified water.

Oil Spill Treatment

Demulsifiers (specifically, oil spill demulsifiers) can be applied to oil spills in low concentrations. They prevent mousse formation for significant periods of time and cause a large reduction in oil–water interfacial tension. The best of these was found to prevent emulsification at dosages as low as 1 part inhibitor to 20,000 parts of fresh oil at 20° C [273]. At dosages of 1:1000, at temperatures higher than 10° C, the chemical also results in significant and rapid dispersion of the oil. For very low temperatures or highly weathered oil, the performance of the chemical falls off sharply.

Desired Properties

Demulsifiers for crude oil emulsions should meet the following properties:

- Rapid breakdown into water and oil with minimal amounts of residual water
- Good shelflife
- Quick preparation

Mechanisms of Demulsification

Stabilization of Water–Oil Emulsions

The stabilization of water–oil emulsions happens as a result of the interfacial layers, which mainly consist of colloids present in the crude oil—asphaltenes and resins. By adding demulsifiers, the emulsion breaks up. With water-soluble demulsifiers, the emulsion stabilizers originally in the system will be displaced from the interface. In addition, a change in wetting by the formation of inactive complexes may occur. Conversely, using oil-soluble demulsifiers, the mechanism, in addition to the displacement of crude colloids, is based on neutralizing the stabilization effect by additional emulsion breakers and the breakup resulting from interface eruptions [1002].

Interfacial Tension Relaxation

The effectiveness of a crude oil demulsifier is correlated with the lowering of the shear viscosity and the dynamic tension gradient of the oil–water interface. The interfacial tension relaxation occurs faster with an effective demulsifier [1714]. Short relaxation times imply that interfacial tension gradients at slow film thinning are suppressed. Electron spin resonance experiments with labeled demulsifiers indicate that the demulsifiers form reverse micellelike clusters in the bulk oil [1275]. The slow unclustering of the demulsifier at the interface appears to be the rate-determining step in the tension relaxation process.

Performance Testing

The trial-and-error method of choosing an optimal demulsifier from a wide variety of demulsifiers to effectively treat a given oil field water-in-oil emulsion is time-consuming. However, there are methods to correlate and predict the performance of demulsifiers.

Spreading Pressure

The performance of demulsifiers can be predicted by the relationship between the film pressure of the demulsifier and the normalized area and the solvent properties of the demulsifier [1632]. The surfactant activity of the demulsifier is dependent on the bulk phase behavior of the chemical when dispersed in the crude oil emulsions. This behavior can be monitored by determining the demulsifier pressure-area isotherms for adsorption at the crude oil–water interface.

Characterization by Dielectric Constant

The dielectric constant can be used as a criterion for screening, ranking, and selecting demulsifiers for emulsion breaking. In a study, the dielectric constants of emulsions and demulsifiers were measured using a portable capacitance meter, and bottle tests were conducted according to the API specification [18]. The results showed that the dielectric constants can be used effectively to screen and rank demulsifiers, whereas a confirmatory bottle test should be conducted on the best demulsifiers to assist in the rapid selection of the most effective demulsifier.

Shaker Test Methods

A study by Environment Canada and the U.S. Minerals Management Service attempted to develop a standard test for emulsion breaking agents [586]. Nine types of shaker test methods were tried. Although the results are comparable with different tests, a stable water-in-oil emulsion must be used to yield reproducible results. Tests with unstable emulsions showed nonreproducible and inconsistent results.

Viscosity Measurements

Water content and viscosity measurements in certain systems show a correlation to emulsion stability [597]. The viscosity provides a more reliable measure of emulsion stability, but measurements of the water content are more convenient. Mixing time, agent amount, settling time, and mixing energy impact the effectiveness of an emulsifier.

Screening

Without knowing the structure in detail, nuclear magnetic resonance (NMR) information can be compiled by ^{13}C on the structure using the chemical shifts. The chemical shifts can be correlated with other data, such as bottle tests, and evaluated by statistical methods. In a series of experimental work [1140,1141] using principal component analysis, NMR data and bottle test data were used to cluster more than 100 demulsifiers into only a few distinctly different chemical groups as characterized by the NMR data. Similar chemical types had similar demulsification performance, which means that demulsifier evaluations can be made based on demulsifier chemistry. Only a few of the distinctly different emulsifiers need to be tested before the optimization procedeure can start. Because the chemical characterization by NMR imaging takes only a fraction of the time of a bottle test, it is possible to more rapidly focus on optimizing the dosage of the demulsifier.

Classification of Demulsifiers

The chemicals used as demulsifiers can be classified according to their chemical structure and their applications. With the latter respect, a main division for water-in-oil and oil-in-water applications exists. Furthermore, the demulsifiers can be classified according to the oil type used.

From the view of chemical classification, two major groups exist:

- Nonionic demulsifiers
- Ionic demulsifiers

Common Precursor Chemicals

Polyalkyleneoxides

Polyalkyleneoxides are substances of the following general structure:

$$HO - (CH_2 - CHR - O)_x - H$$

The most important additives are polyethyleneoxide, polypropyleneoxide, and polybutyleneoxide. They are also referred to as *polyalkylene glycols,* but this name is correct strictly for derivatives of 1,2-diols.

Polypropyleneoxide has a molar mass of 250 to 4000 Dalton. The lower molecular homologues are miscible with water, whereas the higher molecular polypropyleneoxides are sparingly soluble. They are formed by the polyaddition of, for example, propylene oxide to water or propanediol. The simplest examples are di- and tri-u tetrapropyleneglykol.

There are also block copolymers from ethylene oxide and propylene oxide.

$$HO - (R_1 - O)_x - (R_2 - O)_y - H$$

Polytetramethyleneglycol (polytetrahydrofuran) is formed by ring opening polyetherification of tetrahydrofuran. Branched polyalkyleneoxides are formed using polyfunctional alcohols such as trimethylolpropane and pentaerythrite. The products are liquids or waxes depending on the molar mass. Polyalkyleneoxides are often precursors for demulsifiers.

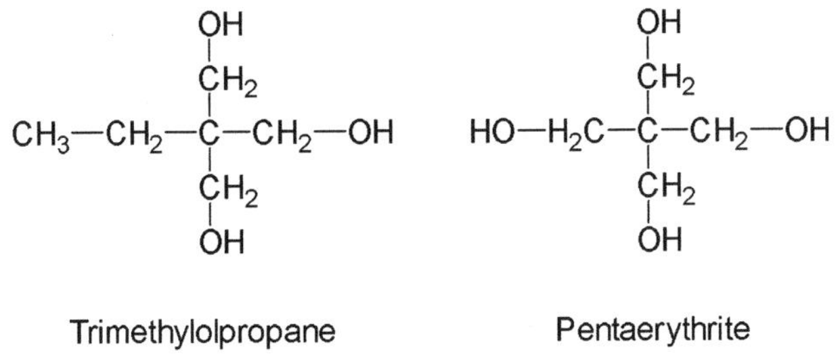

Figure 22–1. Trimethylolpropane and pentaerythrite.

Polyamines

Polyamines are usually open chain compounds with primary, secondary, or tertiary amino groups. Instead of polyamines, polyimines are used without a sharp difference. Actually, imines are compounds with the =N− group or cycles such as ethyleneimine. Examples of oligoamines and polyamines are ethylenediamine, propanediamine, and 1,4-butanediamine and the respective products of condensation such as diethyleneamine, dipropylenetriamine, and triethylenetetramine. The compounds are colorless to yellowish liquids or solids with alkaline reaction.

Polyamines can also be synthesized by cationic ring-opening polymerization of ethyleneimines (aziridines), trimethyleneimines (azetidines), and 2-oxazolines.

Polyalkyleneimines are polyamines whose structure is classified into linear and branched types as shown below.

Linear: $H_2N - (CH_2 - CH_2 - NH)_xH$
Branched: $H_2N - (CH_2 - CH_2 - N(CH_2 - CH_2 - NH_2))_xH_2$

Linear polyalkyleneimines have amino groups only in the main chain, and branched polyalkyleneimines have amino groups in both the main and side chains. In general, nitrogen atoms are at every third or fourth atom.

Linear polyethyleneimine is insoluble in benzene, diethyl ether, acetone, and water at room temperature and is soluble in hot water.

Ethoxylation

The reaction is also referred to as *ethoxylation, oxethylation,* or more generally, *oxalkylation.* Ethoxylation is an insertion of one or more $CH_2CH_2 - O-$ groups into a molecule with ethylene oxide. The reaction works with compounds with acid hydrogen atoms. Suitable compounds for ethoxylation are fatty alcohols, alkylphenols, fatty amines, fatty acid esters, mercaptans, and imidazolines. The reaction runs at 120° to 220° C under pressure (approximately 1 to 5 bar). The products are generally linear ethers and polyethers that have a hydroxyl functionality at one end of the chain. Depending on the amount of ethylene oxide used, a distribution of homologous ethoxylates are formed. Alkaline catalysts such as sodium methylate are used to obtain a broad Schulz-Flory distribution, whereas bivalent salts (calcium acetate, strontium phenolate) cause a narrow Poisson distribution. Acid catalysts (e.g., antimony pentachloride) also give a narrow distribution, but give 1,4-dioxan as an undesired byproduct.

Chemicals in Detail

Common demulsifiers are listed in Table 22–1.

Table 22-1
Demulsifiers

Demulsifier	Type	References
Blends containing (1) tannin or amino methylated tannin, (2) a cationic polymer, (3) polyfunctional amines	WiO	[1010]
Copolymer of diallyldimethyl ammonium chloride and quaternized amino alkylmethacrylates and (meth)acrylic esters (e.g., 2-ethylhexylacrylate)	OiW	[791, 1478, 1479]
Amphoteric acrylic acid copolymer	OiW	[224]
Branched polyoxyalkylene copolyesters	OiW	
Copolymer of esters of acrylic acid and the respective acids, methacrylic acid[a]	WiO	
Copolymer of polyglycol acrylate or methacrylate esters	OiW	[614]
Poly(1-acryloyl-4-methyl piperazine and copolymers of 1-acryloyl-4-methyl piperazine quaternary salts with acrylamide quaternary salts)	OiW	
Copolymers of acrylamidopropyltrimethyl ammonium chloride with acrylamide	OiW	
Vinyl phenol polymers[b]	OiW	
Ethoxylated or epoxidized polyalkylene glycol	WiO	
Polymers from dimethylaminoethyl methacrylate, dimethylaminopropyl methacrylamide	OiW	
Polymer of monoallylamine	OiW	
Copolymers of allyl-polyoxyalkylenes with acrylics		
Copolymer of diallyldimethyl ammonium chloride and vinyl trimethoxysilane	WiO	
Cationic amide-ester compositions	OiW	
Polyalkylenepolyamides-amines	OiW, WiO	
Fatty acid N,N-dialkylamides[c]	OiW	
Diamides from fatty amines[d]	WiO	
Polycondensates of oxalkylated fatty amine[d]	OiW	
Poly(diallyldimethyl ammonium chloride)	OiW	[787]
Alkoxylated fatty oil		
Oxalkylated polyalkylene polyamines	WiO	[546]
Crosslinked oxalkylated polyalkylene polyamines	OiW	
Phenol–formaldehyde resins, modified with benzylamine[e]		[3]
Alkoxylated alkylphenol–formaldehyde resins	WiO	
Phenol–formaldehyde polymer modified with ethylene carbonate	WiO	[1680]

Table 22–1 (continued)

Demulsifier	Type	References
Modified phenol–formaldehyde resins		
Polyalkylene polyamine salts	OiW	
Dithiocarbamate of bis-hexamethylenetriamine	OiW	[531]
Di- and tri-dithiocarbamic acid compounds		[1507, 1734–1743]
Polythioalkyloxides	WiO	[1572]
Polyether/polyurethanes	WiO	[1724]
Polyurea-modified polyetherurethanes	WiO	[1573]
Sulfonated polystyrenes	OiW	[1132]
Asphaltenes		
Acid-modified polyol	LS	[1072]

LS, Liquid-solid separation; OiW, oil-in-water use; WiO, water-in-oil use
a) Up to five comonomers.
b) Hydrolysis of polyacetoxystyrene.
c) Sludge or emulsions during the drilling or workover.
d) Act also as corrosion inhibitors.
e) Demulsifying at 65°–70° C.

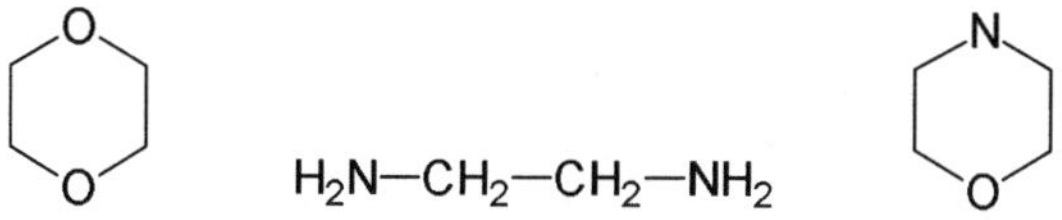

1,4-Dioxan   Ethylenediamine   Morpholine

Figure 22–2. 1,4-Dioxan, ethylenediamine, and morpholine.

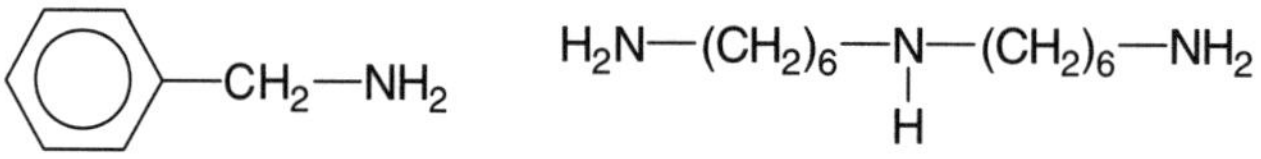

Benzylamine   bis-Hexamethylenetriamine

Figure 22–3. Benzylamine, bis-hexamethylenetriamine.

Polyoxyalkylenes

A process for separating crude oil emulsions of the water-in-oil type based on certain ethylene oxide–propylene oxide block polymers and certain polyglycidol ethers of phenol–formaldehyde–condensation products has been described [1026–1028].

Ethoxylated or Epoxidized Polyalkylene Glycol

Polypropylene glycol or a polybutylene glycol with a molecular weight in the range of 7000 to 20,000 Dalton is modified with ethylene oxide or a diglycidyl ether [1725]. Glycide derivates of polyethylene glycols can be prepared by the acid- or base-catalyzed reaction of polyethylene glycols with 0.5% to 10% by weight diepoxides [1186]. The diepoxides are aromatic or aliphatic precondensates that are used commonly as constituents for epoxide resins, such as the diglycidyl ether from bisphenol-A, and their oligomers. The modification leads to increased molecular weight.

Because of the high molecular weight of the modified polyalkylene glycol, the oil dehydration already occurs when used alone. The compound acts synergistically with other conventional demulsifiers.

Modified Polyalkyleneoxide Block Copolymers

Several patents are proposing polyalkyleneoxide block copolymers as demulsifiers. Variability exists in the mixtures.

For example, the block copolymers can be modified with a vinyl monomer [281]. In addition, diglycidyl ethers [282], that is, precondensates for epoxides, can be used as modifiers. Another possibility is modification with polyamines.

The preparation procedure can be quite complex. For example, a water-in-oil demulsifier is prepared by the following steps:

1. Reacting a high–molecular-weight polyalkylene glycol (PAG) with ethylene oxide (EO) to form a PAG/EO adduct
2. Esterifying the PAG/EO adduct with a diacid anhydride to form a diester
3. Reacting the diester with a vinyl monomer
4. Additionally esterifying the product of step (3) with a polyhydric material [1723]

Propoxylated-Ethoxylated Block Copolymers

A demulsifier composition that is a blend of (1) a propoxylated-ethoxylated block copolymer of a bis-hydroxyalkyl ether and (2) a propoxylated-ethoxylated block copolymer of 2-hydroxymethyl-1,3-propanediol has been described [1750, 1751].

The blend is partially crosslinked with a vinyl monomer when dissolved in an organic aprotic solvent and has a pH of 5.0 or lower. The first block copolymer is prepared by polycondensing a bis-hydroxyalkyl ether, such as dipropylene glycol, diethylene glycol, and the like, with propylene oxide. Next, the resulting propoxylated diol is reacted with ethylene oxide to produce the block copolymer. The second copolymer is prepared by polycondensing 2-amino-2-hydroxymethyl-1,3-propanediol, commonly known as *TRIS*, with

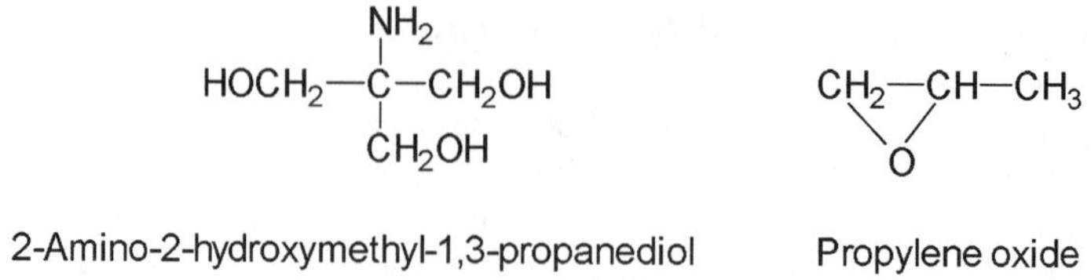

$$HOCH_2-\underset{\underset{CH_2OH}{|}}{\overset{\overset{NH_2}{|}}{C}}-CH_2OH \qquad CH_2-CH-CH_3$$
$$\text{2-Amino-2-hydroxymethyl-1,3-propanediol} \qquad \text{Propylene oxide}$$

Figure 22–4. 2-Amino-2-hydroxymethyl-1,3-propanediol and propylene oxide.

propylene oxide to provide a polymer having preferably 10 to 25 oxypropylene units. Next, the propoxylated diol is reacted with ethylene oxide to produce the block copolymer having preferably 5 to 10 ethylene oxide units. The two block copolymers are dissolved together in an organic aprotic solvent, such as toluene, xylene, trialkylbenzene, cyclohexane, heptane, and hexane.

Branched Polyoxyalkylene Copolyesters

Branched polyesters contain oxalkylated primary fatty amines or oxalkylated polyamines together with at least trivalent oxalkylated alkanol that is responsible for branching. The condensation is achieved with a dicarboxylic acid or a dicarboxylic acid anhydride [216]. In this way, branched polyoxyalkylene mixed polyesters are formed. Suitable solvents are water or organic solvents, such as methanol, isopropanol, butanol, or aromatic hydrocarbons (e.g., toluene, xylene).

The branched polyoxyalkylene mixed polyesters possess a high demulsifier effect. In the usual range of oil-processing temperatures, a complete water removal and a reduction of the salt content is achieved after a short time.

In heavy oil reservoirs with highly porous sands, cyclic steamflooding may cause the formation of stable emulsions. These emulsions can block the production paths in the wellbore. In steam cycle treatments, a blend of oxyalkylated alkanolamines and sulfonates showed a dramatic improvement over non–chemically enhanced steam cycles [331].

Branched Chain–Extended or Crosslinked Polyoxyalkylenes

Block polymers or copolymers of ethylene oxide and 1,2-propylene oxide can be chain extended or crosslinked, respectively, with diisocyanates, dicarboxylic acids, formaldehyde, and diglycidyl ethers [108].

Polyoxyalkylene-Polysiloxane Copolymer

A blend of a polyoxyalkylene-polysiloxane copolymer and an alkoxylated phenol–aldehyde resin is useful as a demulsifier [1457, 1458].

The polyoxyalkylene units in the copolymer have a molecular weight below 500, and the polysiloxane units have 3 to 50 silicon atoms. The resin has a phenol/aldehyde ratio of 2:1 to 1:5 and an average molecular weight of 500 to 20,000 Dalton. The composition shows synergistic demulsification activity when compared with the individual components. The siloxane units can be either in blocks [979, 980] of the polyoxyalkylene-polysiloxane copolymer or randomly distributed [728, 729].

Vinyl Polymers

Copolymer of Esters of Acrylics

A water-soluble demulsifier is an emulsion tetrapolymer of the following [176]:

- Methyl methacrylate
- Butyl acrylate
- Acrylic acid
- Methacrylic acid

Styrene may also be added to give a pentapolymer. The polymer is of random orientation and preferably has a molecular weight of approximately 10,000 Dalton. The polymer is a low-viscosity chalk-white fluid that is soluble in water at a pH of 6 to 7.

Copolymer of Polyglycol Acrylate or Methacrylate Esters

Emulsion breakers are made from acrylic acid or methacrylic acid copolymerized with hydrophilic monomers [148]. The acid groups of acrylic acid and methacrylic acid are oxalkylated by a mixture of polyglycols and polyglycol ethers to provide free hydroxy groups on the molecule. The copolymers are made by a conventional method, for example, by free radical copolymerization in solution, emulsion, or suspension. The oxalkylation is performed in the presence of an acid catalyst, the acid being neutralized by an amine when the reaction is complete.

Methacrylic Polymers

Hydrophobic polymers with some hydrophilic groups can be obtained with an emulsion polymerization technique. Suitable monomers are nitrogen-containing acrylics and methacrylics; allyl monomers such as dimethylaminoethyl methacrylate, dimethylaminopropyl methacrylamide, diethylaminoethyl methacrylate, dimethylaminoethyl acrylate, diethylaminoethyl acrylate; and nitrogen-containing allyl monomers (e.g., diallylamine and N,N-diallylcyclohexylamine) [225, 226].

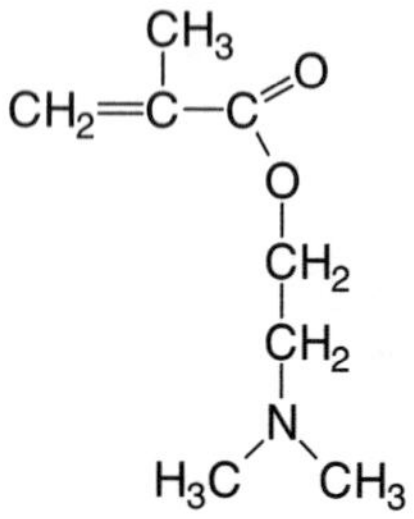

Dimethylaminoethyl methacrylate

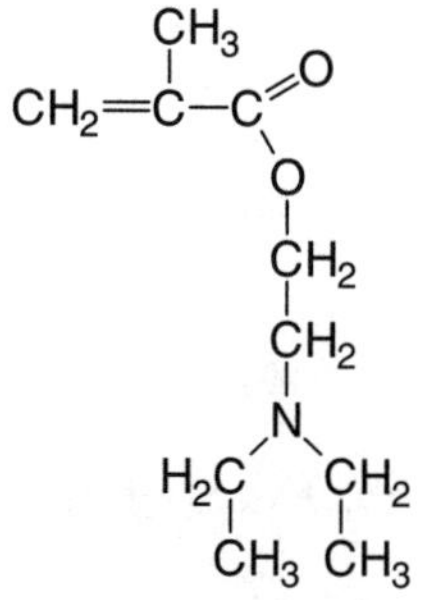

Diethylaminoethyl methacrylate

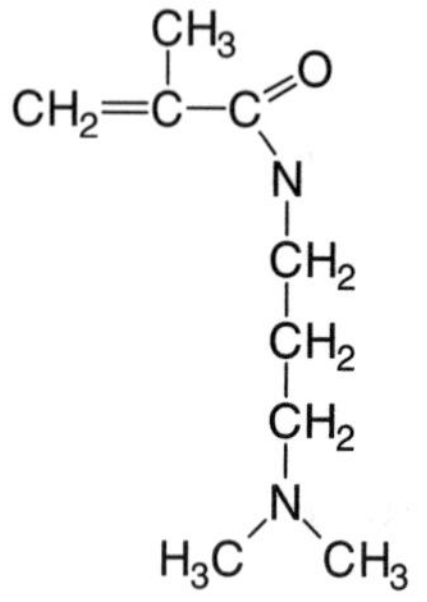

Dimethylaminopropyl methacrylamide

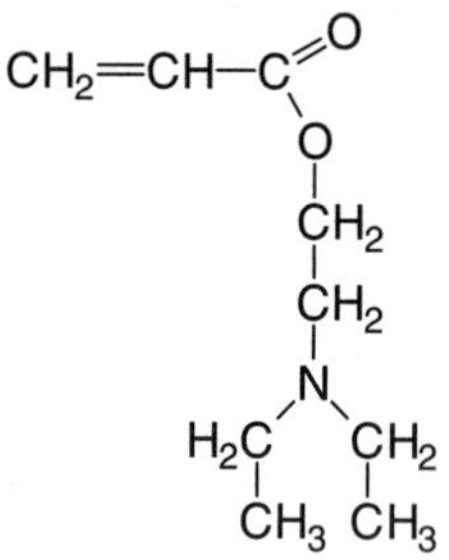

Diethylaminoethyl acrylate

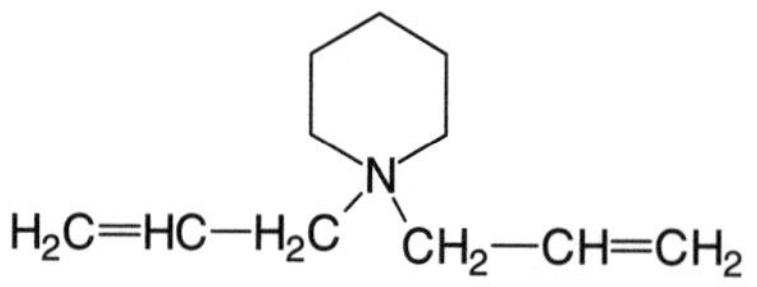

N,N-diallylcyclohexylamine

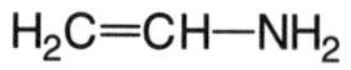

Diallylamine

Figure 22–5. Acrylic monomers for demulsifiers: dimethylaminoethyl methacrylate, diethylaminoethyl methacrylate, dimethylaminopropyl methacrylamide, diethylaminoethyl acrylate, N,N-diallylcyclohexylamine, diallylamine.

If the polymer is to remain in emulsion form, it is important that the pH of the water phase remain at or above 8. If the pH of the water phase falls below 8, the polymer will commence dissolving in water, and the polymer emulsion will break.

Optionally, the pH of the aqueous phase of the broken emulsion, after doing the job, can be adjusted to become alkaline. The salts of the polymers are converted into inactive species and the aqueous phase of the broken emulsion can be reinjected into a hydrocarbon-containing formation to recover additional hydrocarbons or bitumen [1187] as an improved oil-recovery process.

Alkylphenol–Polyethylene Oxide–Acrylate Polymer

Substrates coated with alkylphenol–polyethylene oxide–acrylate polymer [607] are useful for demulsifying naturally occurring crude oils. The monomers include oxethylated alkylphenols, such as oxethylated nonylphenol with 6 to 12 ethoxy units, which are esterified with acrylate.

Poly(1-Acryloyl-4-Methyl Piperazine) Quaternary Salts

Quaternary ammonium salts of 1-acryloyl-4-methyl piperazine can be prepared by methylation with methyl chloride and dimethyl sulfate. These monomers can be polymerized by means of radical polymerization, either alone or with a comonomer [617]. A useful comonomer with appropriate monomer reactivity ratios is acrylamide.

Certain basic polyamides can be further prepared by reacting piperazine derivatives with amines [814].

Copolymers of Acrylamidopropyltrimethyl Ammonium Chloride–Acrylamide

A terpolymer can be obtained from a water-soluble nonionic monomer, such as acrylamide; a cationic monomer, such as 3-acrylamidopropyltrimethyl ammonium chloride; and a hydrophobic monomer, such as an alkyl-acrylamide

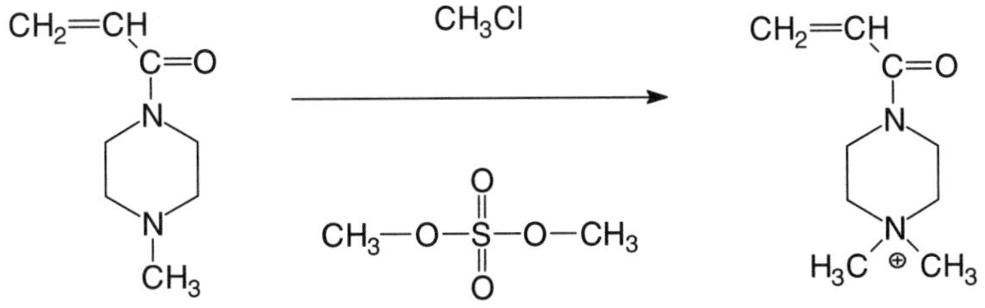

Figure 22–6. Quaternary ammonium salts of 1-acryloyl-4-methyl piperazine.

or alkyl-acrylate [895] or methacrylamide or methacrylate, respectively. The terpolymer is water dispersible.

Similarly, a cationic copolymer that exhibits efficacy in breaking oil-in-water and water-in-oil emulsions under a wide variety of conditions has been described. The cationic copolymer is a copolymer of acryloxyethyltrimethyl ammonium chloride and acrylamide [789, 790]. The preferred copolymer contains 40 to 80 mole-% acryloxyethyltrimethyl ammonium chloride. The copolymer is effective in a matrix that includes high percentages of oil at high temperatures.

Special polymerization techniques are described [298] in which the polymerization is performed in an aqueous solution together with a polyvalent anionic salt in the presence of a water-soluble cationic polymer acting as a dispersant polymer. Furthermore, a seed polymer that is water soluble and a cationic polymer that is insoluble in the aqueous solution of the polyvalent anionic salt are present.

Polymer of Monoallylamine

A polymer of monoallylamine is water soluble [1508, 1509]. It is used for breaking oil-in-water emulsions.

Poly(Diallyldimethyl Ammonium Chloride)

A combination of aluminum chlorohydrate and a polyamine, such as poly(diallyldimethyl ammonium chloride), in aqueous solution is effective at elevated temperatures for an oil-in-water emulsion [787].

Copolymer of Diallyldimethyl Ammonium Chloride and Vinyl Trimethoxysilane

Copolymers of a cationic monomer and a vinyl alkoxysilane may be prepared by conventional vinyl polymerization techniques. These techniques include solution polymerization in water and emulsion polymerization with either free radical initiators or redox initiators.

The cationic monomer can be a diallyldimethyl ammonium halide, a dimethylaminoethyl acrylate quaternary salt, or a dimethylaminoethyl methacrylate quaternary salt [1635]. The copolymers may be in solid, dispersion, latex, or solution form.

In particular, copolymers of diallyldimethyl ammonium chloride and vinyl trimethoxysilane will have a molecular weight in the range from 100,000 to 1,000,000 Dalton.

Copolymers of Allyl-Polyoxyalkylenes with Acrylics

The following are monomers for copolymers of allyl-polyoxyalkylenes with acrylics [614, 615]:

- Polyoxyalkylene ethers of allyl alcohol or methallyl alcohol
- Acrylic or methacrylic alkyl esters with up to 20 carbon atoms in the alkyl group
- Acrylic or methacrylic acid
- Acrylamide or methacrylamide

The copolymers may be mixed with other demulsifiers, in particular with alkoxylated novolaks and copolymers that are obtainable by copolymerization of one or more polyoxyalkylene ethers of allyl or methallyl alcohol with vinyl esters of alkyl monocarboxylic acids.

Vinyl Phenol Polymers

Polymers of vinyl phenol are obtained by hydrolyzing polyacetoxystyrene [286, 287]. The respective phenol salts can be used. The demulsifier is applicable to oil-in-water emulsions and does not require the use of zinc or other heavy metals. Therefore it does not cause the environmental problems inherent in such metals.

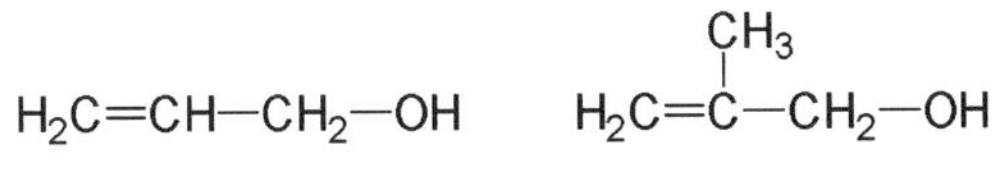

Allyl alcohol Methallyl alcohol

Figure 22–7. Allyl alcohol and methallyl alcohol.

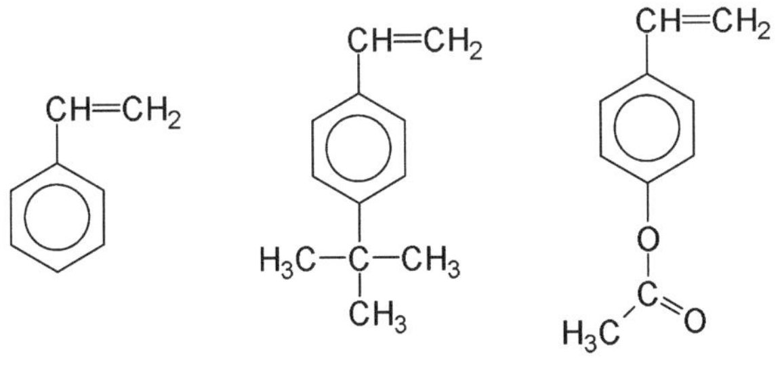

Styrene tert-Butylstyrene Acetoxystyrene

Figure 22–8. Styrene, tert-butylstyrene, and acetoxystyrene.

Polyamines

Polyalkylene Polyamine Salts

Polyalkylene polyamine salts are prepared by contacting polyamines with organic or inorganic acids. The polyamines have a molecular weight of at least 1000 Dalton and ranging up to the limits of water solubility [1185]. In a process of demulsification of the aqueous phase of the broken bitumen emulsions, the pH is adjusted to deactivate the demulsifier so that the water may be used in subsequent in situ hot water or steam floods of the tar sand formation.

Carboxylic-Sulfonic Acid Salts

To mitigate the effects of corrosion resulting from the presence of salts, it is advantageous to reduce the salt concentration to the range of 3 to 5 ppm. Typically, brine droplets in crude oil are stabilized by a mixture of surface-active components such as waxes, asphaltenes, resins, and naphthenic acids that are electrostatically bound to the droplets' surface. Such components provide an interfacial film over the brine droplet, resulting in a diminished droplet coalescence. Adding water to the crude oil can decrease the concentration of the surface-active components on the surface of each droplet, because the number of droplets is increased without increasing component concentration.

The amount of added water required for desalting may be minimized by adding a chemical emulsion breaker to the crude that is capable of displacing the surface-active components from the brine droplets. Quaternized carboxylic-sulfonic acid salts, shown in Figure 22–9, are useful for desalting [1791]. Preferably, the chemical emulsion breaker is used in combination with a delivery solvent, such as diethylene glycol monobutyl ether.

Oxalkylated Polyalkylene Polyamines

Alkoxylated polyethyleneimines are obtained by reacting polyethyleneimine with a molecular weight of 2500 to 35,000 with an excess of propylene oxide and ethylene oxide with respect to the ethyleneimine unit in the

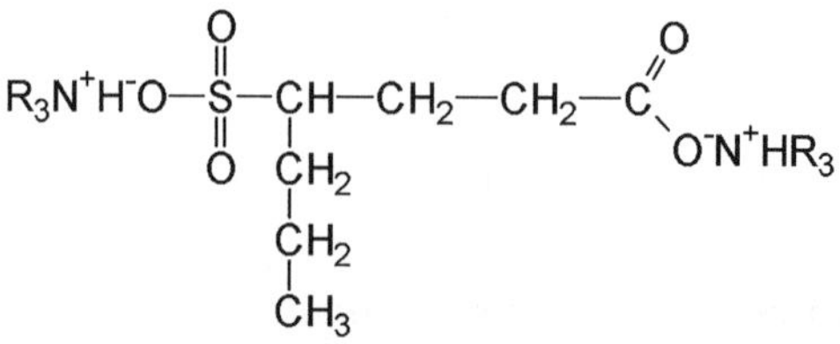

Figure 22–9. Quaternized carboxylic-sulfonic acid salts.

polyethyleneimine [546, 547]. The compounds can be used for the demulsification of petroleum emulsions in a temperature range of 10° to 130° C.

Alkylamine ethoxylates may be used as quaternary salts [785]. The amount necessary to break the emulsion is generally 1 to 100 ppm.

Crosslinked Oxalkylated Polyalkylene Polyamines

In the same way, a crosslinked oxalkylated polyalkylene polyamine can be obtained by preparing a completely oxalkylated polyalkylene polyamine with a degree of polymerization of 10 to 300, crosslinked with a polyalcohol. The demulsifying agent is made from a mixture of the crosslinked oxalkylated polyalkylene polyamine with 25% to 75% by weight of an oxethylated or oxypropylated isoalkylphenol–formaldehyde resin [154].

Crosslinking can be achieved also, if the polyamine is modified with a vinyl monomer [283–285]. Such mixtures are substantially free of copolymers derived from a polyoxyalkylene glycol and a diglycidyl ether.

Glycidyl Ethers

Glycidyl ether additives are obtained by esterification of alkoxylated primary fatty amines and additives of polyether-block polymers and glycidyl ethers with dicarboxylic acids [1025]. They are used as demulsifiers to break up oil emulsions and, as quaternized products, they are suitable as corrosion inhibitors.

Polyamides

Cationic Amide Ester Compositions

Cationic condensation products, namely, the reaction products of a dicarboxylic acid or an ester or acid halide thereof and an aminoalkylamine, that are quaternized are recommended for breaking crude oil emulsions from fireflooding [365].

Polyalkylene Polyamides-Amines

In general, polyalkylene polyamides-amines are obtained by the condensation of polyalkylenepolyamines with dicarboxylic acids. The materials are alkoxylated with an excess of ethylene oxide or propylene oxide or 1,2-butylene oxide [149].

Fatty Acid N,N-Dialkylamides

Compositions of a N,N-dialkylamide of a fatty acid in a hydrocarbon solvent and a mutual oil–water solvent are useful for the prevention of sludge formation or emulsion formation during the drilling or workover of producing oil wells [1526, 1528, 1529].

Diamides from Fatty Amines

Ordinary dicarboxylic acids or dimeric fatty acids are condensed with fatty amines to give emulsion breakers [822, 823, 1029, 1030]. Oxalkylated fatty amines and fatty amine derivatives have properties other than emulsion braking; in particular, they can act as corrosion inhibitors and pour-point depressants.

Polycondensates of Oxalkylated Fatty Amine

Quaternary oxalkylated polycondensates can be prepared by esterification of an oxalkylated primary fatty amine with a dicarbonic acid. An organometallic titanium compound is used as a catalyst for condensation [842]. The reaction product is then oxalkylated in the presence of a carbon acid [841]. These polycondensates can be used as demulsifiers for crude oil emulsions and as corrosion inhibitors in installations for the production of natural gas and crude oil; they can and also be used in processing.

Phenolics

Phenol–Formaldehyde Resins

A study on a commonly used demulsifier, namely, a phenol–formaldehyde resin, elucidated how various parameters such as interfacial tension, interfacial shear viscosity, dynamic interfacial-tension gradient, dilatational elasticity, and demulsifier clustering affect the demulsification effectiveness [1275].

Alkoxylated Alkylphenol–Formaldehyde Resins

Products from oxalkylated alkylphenol–formaldehyde resins, alcohols, bisphenols, or amines have been described as demulsifiers [548].

Modified Phenol–Formaldehyde Resins

A branched, high–molecular-weight condensation product of cardanol, an alkylphenol, and an aldehyde can be further ethoxylated and may be sulfonated by the addition of sodium bisulfite in the presence of a free radical

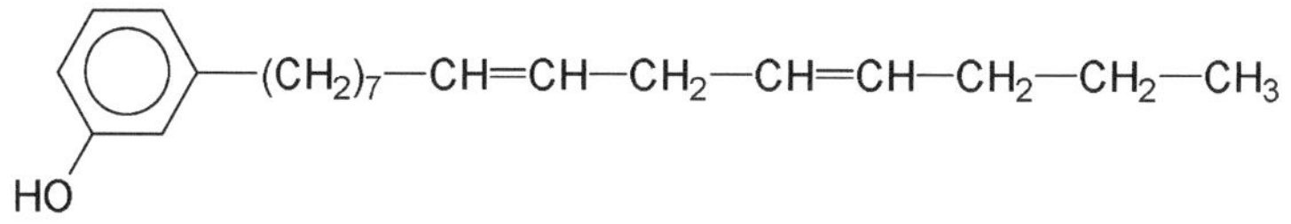

$$\underset{HO}{\bigcirc}-(CH_2)_7-CH=CH-CH_2-CH=CH-CH_2-CH_2-CH_3$$

Cardanol

Figure 22–10. Cardanol.

initiator. Cardanol is a naturally occurring phenol manufactured from cashew nut shell liquid. Unsaturated acids may be added to the phenolic hydroxides of these resins, and the resulting adduct may be copolymerized with acrylic acid [484]. The compositions show good emulsion breaking performance, especially when used in blends with other compositions.

Alkoxylated Fatty Oils

An alkoxylated fatty oil is used in a nonionic composition [1821]. The alkoxylated fatty oil has low solubility in the main emulsion phase. The process is used for breaking emulsions used in wellbore drilling fluids and in oil recovery.

Biodemulsifiers

Microbial deemulsifying agents (biodemulsifiers) represent microbial cell surfaces. Three bacteria, namely *Nocardia amaraebacteria, Corynebacterium petrophilum,* and *Rhodococcus aurantiacus,* were tested for their ability to break simple and complex water-in-oil and oil-in-water emulsions [994]. It was found that cells harvested at the early stage of growth were more active in deemulsification of water-in-oil emulsions, whereas the cells harvested at a stationary phase of growth were more active for oil-in-water emulsions. The deemulsifying capability was associated with the cell surface itself and could not be destroyed by heating and drastic chemical treatments, except for alkaline methanolysis, which destroyed the activity. Microbial aerobic and anaerobic sludges have also been shown to deemulsify water-in-oil emulsions, which opens interesting opportunities in conjunction with biologic wastewater treatment processes.

Cactus Extract. A biodemulsifier has been developed that is based on a cactus extract, an activator for the cactus extract, and a carrier liquid. The cactus extract is made from the leaves and stems of the prickly pear or *Opuntia* family of cactus [1021]. The leaves and stems of the cactus are brought to a rolling boil in water at the ratio of about 1 pound of prickly pear cactus parts to $1\frac{3}{4}$ gallons of water. The fibrous solids are then separated from the extract by screening, filtering, centrifuging, pressing, or other suitable techniques. The resulting liquid extract, which has been separated from the fibrous solids, is then preferably treated with citric acid to inhibit fermentation. The resulting cactus extract is most effective as a water–oil separator and is accompanied by a detergent activator.

Alkylpolyglycosides. Oxalkylated alkylpolyglycosides have a low toxicity and are biodegradable [172]. The amount of demulsifier to be used to break crude oil emulsions is related to the mass of the emulsion. The maximal mass is 1 to 5000 ppm, preferably 1 to 1000 ppm; the temperature is preferably 40° to 80° C.

References

[1] M. T. Abasov, T. V. Khismetov, A. S. Strekov, E. N. Mamalov, and A. M. Bernshtein. New EOR methods based on hydrogen peroxide and urea solutions application. In *Proceedings Volume,* volume 1, pages 272–276. 7th Eapg Imprj Oil Recovery Europe Symp (Moscow, Russia, 10/27–10/29), 1993.

[2] T. Abdel-Moghny and H. K. Gharieb. Biodegradable oil spill dispersants with high efficiency and low toxicity. In *Proceedings Volume,* pages 184–195. 1st Bahrain Soc Eng et al Environ Issues in the Petrol & Petrochem Ind Spec Conf (Manama, Bahrain, 12/4–12/6), 1995.

[3] Y. G. Abdullaev, R. G. Aliev, S. M. Akhmedov, S. M. Zejnally, N. A. Mekhtieva, K. A. Salaeva, and L. I. Sulejmanova. Breaking down of oil emulsion—by treating with potassium salt of phenol-formaldehyde resin, modified with benzylamine. Patent: SU 1705332-A, 1992.

[4] R. G. Abdulmazitov, R. K. Muslimov, A. F. Zakirov, and F. M. Khajretdinov. Oil recovery from carbonate oil-bearing seam—involves cyclic injection of acid solutions of increasing concentrations after reduction of seam pressure. Patent: RU 2073791-C, 1997.

[5] I. S. Abou-Sayed and R. D. Hazlett. Removing fracture fluid via chemical blowing agents. Patent: US 4832123, 1989.

[6] S. A. Abramov, A. P. Krezub, N. A. Mariampolskij, E. S. Bezrukova, and M. A. Egorov. Plugging solution for cementing of oil and gas wells—contains plugging Portland cement and granulated fly ash, of specified moisture content, as active additive. Patent: SU 1802089-A, 1993.

[7] D. B. Acker and F. Malekahmadi. Delayed release breakers in gelled hydrocarbons. Patent: US 6187720, 2001.

[8] C. W. Aften and R. K. Gabel. Clay stabilizing composition for oil and gas well treatment. Patent: US 5152906, 1992.

[9] C. W. Aften and R. K. Gabel. Clay stabilizing method for oil and gas well treatment. Patent: US 5099923, 1992.

[10] C. W. Aften and R. K. Gabel. Clay stabilizer. Patent: US 5342530, 1994.
[11] G. A. Agaev and T. A. Kuliev. Antifoam composition for amine desulphurisation of natural gas—contains tributylphosphate, and additional amyl alcohol and diethyl-disulphide to increase efficiency. Patent: SU 1736550-A, 1992.
[12] T. M. Aggour and M. J. Economides. Impact of fluid selection on high-permeability fracturing. In *Proceedings Volume,* volume 2, pages 281–287. SPE Europe Petrol Conf (Milan, Italy, 10/22–10/24), 1996.
[13] K. Aguilar, R. Colina, M. Borrell, A. Aponte, and Y. Rojas. Evaluation criteria to formulate foam as underbalanced drilling fluid. In *Proceedings Volume on CD ROM.* Iadc et al Underbalanced Drilling Conf (Houston, TX, 8/28–8/29), 2000.
[14] J. A. Ahlgren. Enzymatic hydrolysis of xanthan gum at elevated temperatures and salt concentrations. In *Proceedings Volume,* volume ·, pages ···–···, 6th Inst Gas Technol Gas, Oil, & Environ Biotechnol Int Symp (Colorado Springs, CO, 11/29–12/1), 1993.
[15] Y. S. Ahn and V. Jovancicevic. Mercaptoalcohol corrosion inhibitors. Patent: WO 0112878, 2001.
[16] B. R. Ainley and S. B. McConnell. Delayed borate crosslinked fracturing fluid. Patent: EP 528461, 1993.
[17] B. R. Ainley, K. H. Nimerick, and R. J. Card. High-temperature, borate-crosslinked fracturing fluids: A comparison of delay methodology. In *Proceedings Volume,* volume ·, pages 517–520. SPE Prod Oper Symp (Oklahoma City, 3/21–3/23), 1993.
[18] J. A. Ajienka, N. O. Ogbe, and B. C. Ezeaniekwe. Measurement of dielectric constant of oilfield emulsions and its application to emulsion resolution. *J Petrol Sci Eng,* 9(4):331–339, July 1993.
[19] V. E. Akhrimenko, E. A. Aleksandrova, Z. N. Tkachenko, E. A. Dmitriev, F. N. Derevenets, and S. A. Abramov. Effect of aluminum compounds on setting time of cement slurry and strength of set cement. *Stroit Neft Gaz Skvazhin Sushe More,* 10-11:30–32, October-November 1997.
[20] V. E. Akhrimenko, A. K. Kuksov, and E. M. Levin. Oil, gas well cementing additive—has added melamine formaldehyde resin for better bonding of cement to casing. Patent: SU 1709072-A, 1992.
[21] V. E. Akhrimenko, V. B. Levitin, L. V. Palij, Y. Ya. Taradymenko, and V. P. Timovskij. Plugging solution for cementing oil and gas wells—contains Portland cement, expanding additive based on sintered calcium oxide, amino-formaldehyde or acetone-formaldehyde resin and water. Patent: SU 1776765-A, 1992.
[22] A. H. Al-Khafaji. Implementations of enhanced oil recovery techniques in the Arab world are questioned. In *Proceedings Volume,*

pages 124–133. 6th Int Energy Found et al Mediter Petrol Conf (Tripoli, Libya, 11/23–11/25), 1999.
[23] D. J. Albone, M. G. Kibblewhite, L. E. Sansom, and P. R. Morris. The storage stability of oil spill dispersants. *Quart J Tech Pap,* pages 1–53, January-March 1990.
[24] P. Albonico and T. P. Lockhart. Aqueous gellable composition containing an anti-syneresis agent. Patent: EP 544377, 1993.
[25] F. I. Aleev. Oil displacement from heterogeneous seam by pumping sulphuric acid and water to seam—until sulpho-acids appear in liquid removed from wells. Patent: RU 2055166-C, 1996.
[26] F. I. Aleev, V. V. Andreev, S. V. Ivanov, P. P. Kivilev, N. A. Talalaev, V. A. Khodyrev, and I. F. Chernoshtanov. Oil deposit water flooding with improved efficiency—by pumping of sulphuric acid and water into seam prior to periodic pumping of water into seam through injection well and liquid removal from producing well. Patent: RU 2055165-C, 1996.
[27] J. A. Alford, P. G. Boyd, and E. R. Fischer. Polybasic acid esters as oil field corrosion inhibitors. Patent: US 5174913, 1992.
[28] J. A. Alford, P. G. Boyd, and E. R. Fischer. Acid-anhydride esters as oil field corrosion inhibitors (ester acide anhydride comme inhibiteur de corrosion dans le domaine des huiles). Patent: FR 2692283, 1993.
[29] J. A. Alford, P. G. Boyd, and E. R. Fischer. Polybasic acid esters as oil field corrosion inhibitors. Patent: CA 2075660, 1993.
[30] J. A. Alford, P. G. Boyd, and E. R. Fischer. Polybasic acid esters as oil field corrosion inhibitors (esters d'acides polybasiques utilises comme inhibiteurs de corrosion dans les champs petroliferes). Patent: FR 2681597, 1993.
[31] J. A. Alford, P. G. Boyd, and E. R. Fischer. Acid-anhydride esters as oil field corrosion inhibitors. Patent: GB 2268487, 1994.
[32] S. E. Alford. North Sea field application of an environmentally responsible water-base shale stabilizing system. In *Proceedings Volume,* pages 341–355. SPE/IADC Drilling Conf (Amsterdam, the Netherlands, 3/11–3/14), 1991.
[33] S. A. Ali, L. W. Sanclemente, B. C. Sketchler, and J. M. Lafontaine-McLarty. Acid breakers enhance open-hole horizontal completions. *Petrol Eng Int,* 65(11):20–23, November 1993.
[34] P. A. Alikin, N. N. Kasatkina, N. M. Makeev, and V. Yu. Vantsev. Plugging solution for deep wells—contains slag-sand cement, iron chloride, polyacrylamide, ethyl silicate and water, and has increased isolating efficiency. Patent: SU 1776761-A, 1992.
[35] B. A. M. O. Alink and B. T. Outlaw. Thiazolidines and use thereof for corrosion inhibition. Patent: WO 0140205A, 2001.

[36] R. A. Allakhverdiev, B. Khydyrkuliev, and N. V. Reznikov. Plugging solution for repairing oil and gas wells–contains plugging Portland cement, isobutanol, water and liquid metal alloy of gallium, indium and tin, to increase strength of cement stone. Patent: SU 1802082-A, 1993.

[37] M. L. Allan and L. E. Kukacka. Calcium phosphate cements for lost circulation control in geothermal. *Geothermics,* 24(2):269–282, April 1995.

[38] A. A. Allen and D. H. Dale. Dispersant mission planner: A computerized model for the application of chemical dispersants on oil spills. In *Proceedings Volume,* volume 1, pages 393–414. 18th Environ Can Arctic & Mar Oilspill Program Tech Seminar (Edmonton, Canada, 6/14–6/16), 1995.

[39] T. E. Allen. New concepts in spraying dispersants from boats. In *Proceedings Volume,* pages 3–6. 9th Bien API et al Oil Spill (Prev, Behav, Contr, Cleanup) Conf (Los Angeles, CA, 2/25–2/28), 1985.

[40] R. Kh. Almaev, A. G. Gabdrakhmanov, O. S. Kashapov, L. V. Bazekina, and S. E. Kostilevskij. Extraction of viscous oil from oil strata— involves pumping-in aqueous alkali-polymeric solution containing fraction of liquid hydrocarbon(s) from oil processing and using polyacrylamide. Patent: RU 2068084-C, 1996.

[41] N. E. Almond. Pipeline flow improvers. In *Proceedings Volume,* pages 307–311. API Pipeline Conf (Dallas, TX, 4/17–4/18), 1989.

[42] M. Alonso-Debolt and M. Jarrett. Synergistic effects of sulfosuccinate/ polymer system for clay stabilization. In *Proceedings Volume,* volume PD-65, pages 311–315. ASME Energy-Sources Technol Conf Drilling Technol Symp (Houston, TX, 1/29–2/1), 1995.

[43] M. Alonso-Debolt and M. A. Jarrett. Drilling fluid additive for water-sensitive shales and clays, and method of drilling using the same. Patent: EP 668339, 1995.

[44] M. A. Alonso-Debolt, R. G. Bland, B. J. Chai, P. B. Eichelberger, and E. A. Elphingstone. Glycol and glycol ether lubricants and spotting fluids. Patent: WO 9528455, 1995.

[45] M. A. Alonso-Debolt, R. G. Bland, B. J. Chai, P. B. Eichelberger, and E. A. Elphingstone. Glycol and glycol ether lubricants and spotting fluids. Patent: US 5945386, 1999.

[46] M. A. Alonso-Debolt and M. A. Jarrett. New polymer/surfactant systems for stabilizing troublesome gumbo shale. In *Proceedings Volume,* pages 699–708. SPE Int Petrol Conf of Mex (Veracruz, Mexico, 10/10– 10/13), 1994.

[47] L. L. Altpeter, Jr. Research recommended to develop odorant-fade model. *Pipe Line Gas Ind,* 80(2):39–40, February 1997.

[48] S. Ameri, K. Aminian, J. A. Wasson, and D. L. Durham. Improved CO_2 enhanced oil recovery—mobility control by in-situ chemical precipitation: Final report. US DOE Rep DOE/MC/22044-15, West Virginia Univ, June 1991.

[49] A. Anderko. Simulation of $FeCO_3$/FeS scale formation using thermodynamic and electrochemical models. Nace Int Corrosion Conf (Corrosion 2000) (Orlando, FL, 3/26–3/31), 2000.

[50] C. P. Anderson, S. A. Blenkinsopp, F. M. Cusack, and J. W. Costerton. Drilling mud fluid loss—an alternative to expensive bulk polymers. In *Proceedings Volume,* pages 481–489. 4th Inst Gas Technol Gas, Oil, & Environ Biotechnol Int Symp (Colorado Springs, CO, 12/9–12/11), 1991.

[51] B. A. Andreson, R. G. Abdrakhmanov, G. P. Bochkarev, V. N. Umutbaev, V. V. Fryazinov, V. N. Kudinov, and F. M. Valiakhmetov. Lubricating additive for water-based drilling solutions—contains products of condensation of monoethanolamine and tall oils, kerosene, monoethanolamine and flotation reagent. Patent: SU 1749226-A, 1992.

[52] W. H. Andrews. Process for making a new type of propping agent derived from bauxite for use in hydraulic fracturing, and material for use as a propping agent (procede pour la production d'un nouveau type d'agent de soutenement derive de la bauxite pour l'utilisation dans la fracturation hydraulique, et matiere propre a etre utilises comme agent de soutenement). Patent: FR 2582346, 1986.

[53] W. H. Andrews. Sintered bauxite pellets and their application as proppants in hydraulic fracturing. Patent: AU 579242, 1988.

[54] M. Angel and A. Negele. Utilization of vinyl formamide homopolymers or copolymers for inhibiting gas hydrates (Verwendung von Homo-oder Copolymerisaten des Vinylformamids zur Inhibierung von Gashydraten). Patent: WO 9964717, 1999.

[55] M. Angel, K. Neubecker, and A. Sanner. Grafted polymers as gas hydrate inhibitors (Pfropfpolymerisate als Gashydratinhibitoren). Patent: WO 0109271, 2001.

[56] O. K. Angelopulo, A. Kh. Ali, K. A. Dzhabarov, A. A. Rusaev, E. A. Konovalov, and I. V. Bojko. Plugging solution contains plugging Portland cement, waste from semiconductors production containing dispersed silica, chloride(s), carbonate(s), phosphate(s) and water. Patent: SU 1700202-A, 1991.

[57] Anonymous. Microbiologically influenced corrosion and biofouling in oilfield equipment. Nace TPC Publication TPC 3, 1990.

[58] Anonymous. Drilling fluids product directory. *Offshore Incorporating Oilman (Int Edition),* 51(9):43–44,46, September 1991.

[59] Anonymous. Drilling fluids product directory. *Offshore Incorporating Oilman (Int Edition),* 51(10):62, 64–65, 67–68, 70, 72–73, October 1991.
[60] Anonymous. World oil's 1991 guide to drilling, completion and workover fluids. *World Oil,* 212(6):75–112, June 1991.
[61] Anonymous. 1992–93 environmental drilling and completion fluids directory. *Offshore Incorporating Oilman (Int Ed),* 52(9):41–42, 45–46, 48, 50–56, September 1992.
[62] Anonymous. Cementing products and additives. *World Oil,* 216(3 [suppl]):C–3– C–18, March 1995.
[63] Anonymous. Well cements. *API Spec,* 10a 1/1/95, 1995.
[64] Anonymous. Chemical analysis of barite. *API Rp,* 13k, February 1996.
[65] Anonymous. World oil's 1996 drilling, completion and workover fluids. *World Oil,* 217(6):85–126, June 1996.
[66] Anonymous. Cementing products and additives. *World Oil,* 218(3 [suppl]):C3, C5–C6, C8, C10–C12, C14, C16–C18, March 1997.
[67] Anonymous. Shrinkage and expansion in oilwell cements: First edition. *API Tr,* 10(2), July 1997.
[68] Anonymous. Standard procedure for field testing water-based drilling fluids: Second edition. *API Rp,* pages 13b–1, September 1997.
[69] Anonymous. Cementing products and additives. *World Oil,* 219(3): 87–99, March 1998.
[70] Anonymous. Cementing products and additives. *World Oil,* 220(3): 87–102, March 1999.
[71] Anonymous. Fracturing products and additives. *World Oil,* 220(8): 135, 137, 139–145, August 1999.
[72] T. Arai and M. Ohkita. Application of polypropylene coating system to pipeline for high temperature service. In *Proceedings Volume,* pages 189–201. 8th Bhra Internal & External Protect of Pipes Int Conf (Florence, Italy, 10/24–10/26), 1989.
[73] M. J. Arco, J. G. Blanco, R. L. Marquez, S. M. Garavito, J. G. Tovar, A. F. Farias, and J. A. Capo. Field application of glass bubbles as a density-reducing agent. In *Proceedings Volume,* pages 115–126. Annu SPE Tech Conf (Dallas, TX, 10/1–10/4), 2000.
[74] J. F. Argillier, A. Audibert, P. Marchand, A. Demoulin, and M. Janssen. Lubricating composition including an ester-use of the composition and well fluid including the composition. Patent: US 5618780, 1997.
[75] J. F. Argillier, A. Demoulin, A. Audibert-Hayet, and M. Janssen. Borehole fluid containing a lubricating composition—method for verifying the lubrification of a borehole fluid—application with respect to fluids with a high pH (fluide de puits comportant une composition lubrifiante—procede pour controler la lubrification d'un fluide de puits—application aux fluides a haut pH). Patent: WO 9966006, 1999.

References

[76] J. F. Argillier and P. Roche. Drilling method using a reversible foaming composition (procede de forage utilisant une composition moussante reversible). Patent: EP 1013739, 2000.

[77] D. R. Armbruster. Precured coated particulate material. Patent: US 4694905, 1987.

[78] H. Ashjian, L. C. Peel, T. J. Sheerin, and R. S. Williamson. Non toxic, biodegradable well fluids. Patent: WO 9509215, 1995.

[79] S. S. Ashrawi. Hot water, surfactant, and polymer flooding process for heavy oil. Patent: US 5083612, 1992.

[80] ASTM C150-00: Standard specification for Portland cement. ASTM Book of Standards Volume: 04.01, 2001.

[81] ASTM C184-94E1: Standard test method for fineness of hydraulic cement by the 150-µm (No. 100) and 75-µm (No. 200) sieves. ASTM Book of Standards Volume: 04.01, 2001.

[82] ASTM C666-97: Standard test method for resistance of concrete to rapid freezing and thawing. ASTM Book of Standards Volume: 04.02, 2001.

[83] ASTM C786-96: Standard test method for fineness of hydraulic cement and raw materials by the 300-µm (No. 50), 150-µm (No. 100), and 75-µm (No. 200) sieves by wet methods. ASTM Book of Standards Volume: 04.01, 2001.

[84] ASTM E11-01: Standard specification for wire cloth and sieves for testing purposes. ASTM Book of Standards Volume: 14.02, 2001.

[85] S. E. Atwood. Identification of sulfonation by-products by ion chromatography. Patent: US 5133868, 1992.

[86] A. T. Au. Acyl derivatives of *tris*-hydroxy-ethyl-perhydro-1,3,5-triazine. Patent: US 4605737, 1986.

[87] A. T. Au and H. F. Hussey. Method of inhibiting corrosion using perhydro-*S*-triazine derivatives. Patent: US 4830827, 1989.

[88] M. L. Aubanel and J. C. Bailly. Amorphous high molecular weight copolymers of ethylene and α-olefins. Patent: EP 243127, 1987.

[89] R. Audebert, J. Janca, P. Maroy, and H. Hendriks. Chemically cross-linked polyvinyl alcohol (PVA), process for synthesizing same and its applications as a fluid loss control agent in oil fluids. Patent: GB 2278359, 1994.

[90] R. Audebert, J. Janca, P. Maroy, and H. Hendriks. Chemically cross-linked polyvinyl alcohol (PVA), process for synthesizing same and its applications as a fluid loss control agent in oil fluids. Patent: CA 2118070, 1996.

[91] R. Audebert, P. Maroy, J. Janca, and H. Hendriks. Chemically cross-linked polyvinyl alcohol (PVA), and its applications as a fluid loss control agent in oil fluids. Patent: EP 705850, 1998.

[92] A. Audibert and J. F. Argillier. Thermal stability of sulfonated polymers. In *Proceedings Volume,* pages 81–91. SPE Oilfield Chem Int Symp (San Antonio, TX, 2/14–2/17), 1995.

[93] A. Audibert and J. F. Argillier. Process and water-based fluid utilizing hydrophobically modified guar gums as filtrate (loss) reducer (procede et fluide a base d'eau utilisant des guars modifiees hydrophobiquement comme reducteur de filtrat). Patent: EP 722036, 1996.

[94] A. Audibert and J. F. Argillier. Process and water-based fluid utilizing hydrophobically modified guars as filtrate reducers. Patent: US 5720347, 1998.

[95] A. Audibert, J. F. Argillier, L. Bailey, and P. I. Reid. Procedure and water-based fluid utilizing hydrophobically modified cellulose derivatives as filtrate reducer. Patent: EP 670359, 1995.

[96] A. Audibert, J. F. Argillier, L. Bailey, and P. I. Reid. Process and water-base fluid utilizing hydrophobically modified cellulose derivatives as filtrate reducers. Patent: US 5669456, 1997.

[97] A. Audibert, J. F. Argillier, L. Bailey, and P. I. Reid. Process and water-base fluid utilizing hydrophobically modified cellulose derivatives as filtrate reducers. Patent: US 6040276, 2000.

[98] A. Audibert, J. F. Argillier, H. K. J. Ladva, P. W. Way, and A. O. Hove. Role of polymers on formation damage. In *Proceedings Volume,* pages 505–516. SPE Europe Formation Damage Conf (The Hague, Netherlands, 5/31–6/1), 1999.

[99] A. Audibert, J. F. Argillier, U. Pfeiffer, and G. Molteni. Well cementing method using HMHPG (hydrophobically modified hydroxy propyl guar) filtrate reducer. Patent: US 6257336, 2001.

[100] A. Audibert, J. Lecourtier, L. Bailey, P. L. Hall, and M. Keall. The role of clay/polymer interactions in clay stabilization during drilling. In *Proceedings Volume,* pages 203–209. 6th Inst Francais Du Petrole Explor & Prod Res Conf (Saint-Raphael, France, 9/4–9/6), 1992.

[101] A. Audibert, J. Lecourtier, L. Bailey, and G. Maitland. Method for inhibiting reactive argillaceous formations and use thereof in a drilling fluid. Patent: WO 9315164, 1993.

[102] A. Audibert, J. Lecourtier, L. Bailey, and G. Maitland. Process for inhibiting reactive argillaceous formations and application to a drilling fluid (procede d'inhibition de formations argileuses reactives et application a un fluide de forage). Patent: FR 2686892, 1993.

[103] A. Audibert, J. Lecourtier, L. Bailey, and G. Maitland. Method for inhibiting reactive argillaceous formations and use thereof in a drilling fluid. Patent: US 5677266, 1997.

[104] A. Audibert, J. Lecourtier, L. C. Bailey, and G. Maitland. Use of polymers having hydrophilic and hydrophobic segments for inhibiting the swelling of reactive argillaceous formations (l'utilisation d'un

polymere en solution aqueuse pour l'inhibition de gonflement des formations argileuses reactives). Patent: EP 578806, 1994.
[105] A. Audibert, L. Rousseau, and J. Kieffer. Novel high-pressure/high-temperature fluid loss reducer for water-based formulation. In *Proceedings Volume,* pages 235–242. SPE Oilfield Chem Int Symp (Houston, TX, 2/16–2/19), 1999.
[106] A. Audibert-Hayet, J. F. Argillier, and L. Rousseau. Filtrate reducing additive and well fluid (additif reducteur de filtrat et fluide de puits). Patent: WO 9859014, 1998.
[107] A. Audibert-Hayet, C. Noik, and A. Rivereau. Copolymer additives for cement slurries intended for well bores. Patent: GB 2359075, 2001.
[108] T. Augustin and R. Kehlenbach. Breaking of water-in-oil emulsions. Patent: CA 2057425, 1992.
[109] D. V. Aurand, R. Jamail, M. Sowby, R. R. Lessard, A. Steen, G. Henderson, and L. Pearson. Goals, objectives, and the sponsor's perspective on the accomplishments of the chemical response to oil spills: Ecological effects research forum (CROSERF). In *Proceedings Volume.* API et al Int Oil Spill Conf (Tampa, FL, 3/26–3/29), 2001.
[110] T. Austad, P. A. Bjorkum, T. A. Rolfsvag, and K. B. Oysaed. Adsorption: Pt 3: Nonequilibrium adsorption of surfactants onto reservoir cores from the North Sea: The effects of oil and clay minerals. *J Petrol Sci Eng,* 6(2):137–148, 1991.
[111] T. Austad, S. Ekrann, I. Fjelde, and K. Taugbol. Chemical flooding of oil reservoirs: Pt 9: Dynamic adsorption of surfactant onto sandstone cores from injection water with and without polymer present. *Colloids Surfaces, Sect A,* 127(1–3):69–82, 1997.
[112] T. Austad, B. Matre, J. Milter, A. Saevareid, and L. Oyno. Chemical flooding of oil reservoirs: Pt 8: Spontaneous oil expulsion from oil- and water-wet low permeable chalk material by imbibition of aqueous surfactant solutions. *Colloids Surfaces, Sect A,* 137(1–3):117–129, 1998.
[113] T. Austad, O. Rorvik, T. A. Rolfsvag, and K. B. Oysaed. Adsorption: Pt 4: An evaluation of polyethylene glycol as a sacrificial adsorbate towards ethoxylated sulfonates in chemical flooding. *J Petrol Sci Eng,* 4(6):265–276, January 1992.
[114] T. Austad, K. Veggeland, I. Fjelde, and K. Taugbol. Physicochemical principles of low tension polymer flood. In *Proceedings Volume,* volume 1, pages 208–219. 7th Eapg Impr Oil Recovery Europe Symp (Moscow, Russia, 10/27–10/29), 1993.
[115] P. W. Austin. Heterocyclic thione compounds and their use as biocides. Patent: EP 249328, 1987.
[116] P. W. Austin. Unsaturated, halogenated thiocyanates, the preparation thereof and use as a biocide. Patent: EP 316058, 1989.

[117] P. W. Austin and F. F. Morpeth. Composition and use. Patent: EP 500352, 1992.
[118] V. E. Avakov. Preventing absorption of drilling solution—by introduction of water-expandable material based on bentonite clay and polyacrylamide into circulating drilling solution. Patent: SU 1745123-A, 1992.
[119] C. Aviles-Alcantara, C. C. Guzman, and M. A. Rodriguez. Characterization and synthesis of synthetic drilling fluid shale stabilizer. In *Proceedings Volume*. SPE Int Petrol Conf in Mex (Villahermosa, Mexico, 2/1–2/3), 2000.
[120] D. Avlonitis, A. Danesh, A. C. Todd, and T. Baxter. The formation of hydrates in oil-water systems. In *Proceedings Volume,* pages 15–34. 4th Bhran Multi-Phase Flow Int Conf (Nice, France, 6/19–6/21), 1989.
[121] E. Babaian-Kibala. Naphthenic acid corrosion inhibitor. Patent: US 5252254, 1993.
[122] R. Bacskai and A. H. Schroeder. Alkylaniline/formaldehyde co-oligomers as corrosion inhibitors. Patent: US 4778654, 1988.
[123] R. Bacskai and A. H. Schroeder. Alkylaniline/formaldehyde oligomers as corrosion inhibitors. Patent: US 4780278, 1988.
[124] S. Bagci, M. V. Kok, and U. Turksoy. Determination of formation damage in limestone reservoirs and its effect on production. *J Petrol Sci Eng,* 28(1–2):1–12, October 2000.
[125] L. Bailey. Latex additive for water-based drilling fluids. Patent: GB 2351986, 2001.
[126] L. Bailey. Latex additive for water-based drilling fluids. Patent: WO 0104232A, 2001.
[127] L. Bailey, P. I. Reid, and J. D. Sherwood. Mechanisms and solutions for chemical inhibition of shale swelling and failure. In *Proceedings Volume,* pages 13–27. Recent Advances in Oilfield Chemistry, 5th Royal Soc Chem Int Symp (Ambleside, England, 4/13–4/15), 1994.
[128] S. Bailey, R. Bryant, and T. Zhu. A microbial trigger for gelled polymers. In *Proceedings Volume,* pages 611–619. 5th US DOE et al Microbial Enhanced Oil Recovery & Relat Biotechnol for Solving Environ Probl Int Conf (Dallas, TX, 9/11–9/14), 1995.
[129] C. Baillie and E. Wichert. Chart gives hydrate formation temperature for natural gas. *Oil Gas J,* 85(14):37–39, 1987.
[130] Y. A. Balakirov, A. I. Chernorubashkin, G. A. Makeev, I. P. Korolev, and V. N. Glushchenko. Composition for treatment of stratum head zone—contains oxyethylated isononyl phenol, aqueous solution of hydrofluoric acid and aqueous solution of hydrochloric acid. Patent: SU 1770555-A, 1992.
[131] S. D. Ball. *Comparison of transient interfacial tension behaviours of oil/alkaline systems as measured by the drop volume and spinning drop tensiometers.* PhD thesis, Ottawa Univ, 1995.

[132] S. D. Ball, V. Hornof, and G. H. Neale. Transient interfacial tension behavior between acidic oils and alkaline solutions. *Chem Eng Commun,* 147:145–156, May 1996.
[133] T. Ballard, S. Beare, and T. Lawless. Mechanisms of shale inhibition with water based muds. In *Proceedings Volume.* Ibc Tech Serv Ltd Prev Oil Discharge from Drilling Oper: The Options Conf (Aberdeen, Scotland, 6/23–6/24), 1993.
[134] T. J. Ballard, S. P. Beare, and T. A. Lawless. Shale inhibition with water-based muds: The influence of polymers on water transport through shales. In *Proceedings Volume,* pages 38–55. Recent Advances in Oilfield Chemistry, 5th Royal Soc Chem Int Symp (Ambleside, England, 4/13–4/15), 1994.
[135] J. Balzer, M. Feustel, M. Krull, and W. Reimann. Graft polymers, their preparation and use as pour point depressants and flow improvers for crude oils, residual oils and middle distillates. Patent: US 5439981, 1995.
[136] N. Banthia and S. Mindess. Water permeability of cement paste. *Cement Concrete Res,* 19(5):727–736, September 1989.
[137] V. P. Barannik, E. K. Kubyshkina, and N. M. Lezina. Corrosion and thermophysical properties of lithium chloride-based coolant. *Zashch Korrozii Okhrana Okruzhayushchej Sredy,* 8–9:12–14, August-September 1995.
[138] D. F. Bardoliwalla. Aqueous drilling fluids containing fluid loss additives. Patent: US 4622370, 1986.
[139] D. F. Bardoliwalla. Fluid loss control additives from AMPS (2-acrylamido-2-methylpropane sulfonic acid) polymers. Patent: US 4622373, 1986.
[140] J. F. Baret. Why cement fluid loss additives are necessary. In *Proceedings Volume,* pages 853–860. SPE Petrol Eng Int Mtg (Tianjin, China, 11/1–11/4), 1988.
[141] J. F. Baret, B. Dargaud, J. Villar, and M. Michaux. Cementing compositions and application of such compositions to cementing oil (or similar) wells. Patent: CA 2207885, 1997.
[142] J. F. Baret and P. Drecq. Additives for oilfield cements and corresponding cement slurries. Patent: EP 314242, 1989.
[143] J. F. Baret and P. Drecq. Dispersant for oil-well cement slurries and corresponding slurries (dispersant pour laitiers de ciment petroliers et laitiers correspondants). Patent: FR 2622572, 1989.
[144] J. F. Baret and P. Drecq. Dispersant for oilfield cement slurries and corresponding slurries (dispersant pour laitiers de ciment petroliers et laitiers correspondants). Patent: FR 2635469, 1990.
[145] O. Barkat. *Rheology of flowing, reacting systems: The crosslinking reaction of hydroxypropyl guar with titanium chelates.* PhD thesis, Tulsa Univ, 1987.

[146] V. Barlet-Gouedard and P. Maroy. Cementing compositions and application thereof to cementing oil or analogous wells. Patent: WO 9901397, 1999.

[147] K. A. Barsukov, V. Yu. Ismikhanov, A. A. Akhmetov, G. S. Pop, G. A. Lanchakov, and V. M. Sidorenko. Composition for hydro-bursting of oil and gas strata—consists of hydrocarbon phase, sludge from production of sulphonate additives to lubricating oils, surfactant-emulsifier and mineralised water. Patent: SU 1794082-A, 1993.

[148] K. Barthold, R. Baur, S. Crema, K. Oppenlaender, and J. Lasowski. Method of demulsifying crude oil and water mixtures with copolymers of acrylates or methacrylates and hydrophilic comonomers. Patent: US 5472617, 1995.

[149] K. Barthold, R. Baur, R. Fikentscher, J. Lasowski, and K. Oppenlaender. Alkoxylated polyamines containing amide groups and their use in breaking oil-in-water and water-in-oil emulsions (Alkoxylierte Amidgruppenhaltige Polyamine und deren Verwendung zum Brechen von Öl-in-Wasser- und Wasser-in-Öl-Emulsionen). Patent: EP 264755, 1991.

[150] R. T. Barthorpe. The impairment of scale inhibitor function by commonly used organic anions. In *Proceedings Volume,* pages 69–76. SPE Oilfield Chem Int Symp (New Orleans, LA, 3/2–3/5), 1993.

[151] F. P. Bashev and K. I. Kadantseva. Binder for cements used to plug oil and gas drilling holes—contains phosphoric slag, sodium hydroxide and trisodium phosphate. Patent: SU 704030-A, 1992.

[152] J. G. Batelaan and P. M. van der Horts. Method of making amide modified carboxyl-containing polysaccharide and fatty amide-modified polysaccharide so obtainable. Patent: WO 9424169, 1994.

[153] P. M. Bauer, D. J. Hanlon, and W. R. Menking. Process for producing bentonite clays exhibiting enhanced solution viscosity properties. Patent: US 5248641, 1993.

[154] R. Baur, K. Oppenlaender, and K. Barthold. Crosslinked oxyalkylated polyalkylenepolyamines and their use as crude oil demulsifiers (Vernetzte Oxalkylierte Polyalkylenpolyamine und ihre Verwendung als Erdölemulsionsspalter). Patent: EP 147743, 1989.

[155] M. Baviere and T. Rouaud. Solubilization of hydrocarbons in micellar solutions: Influence of structure and molecular weight (solubilisation des hydrocarbures dans les solutions micellaires: influence de la structure et de la masse moleculaire). *Rev Inst Franc Petrol,* 45(5):605–620, September-October 1990.

[156] J. H. Bayless. Hydrogen peroxide: A new thermal stimulation technique. *World Oil,* 219(5):75–78, May 1998.

[157] J. H. Bayless. Hydrogen peroxide applications for the oil industry. *World Oil,* 221(5):50–53, May 2000.

[158] B. B. Beall, H. D. Brannon, R. M. Tjon-Joe-Pin, and K. O'Driscoll. Evaluation of a new technique for removing horizontal wellbore damage attributable to drill-in filter cake. In *Proceedings Volume,* pages 53–65. 4th Annu India Oil & Nat Gas Corp India Oil & Gas Rev Symp (Mumbai, India, 8/18–8/19), 1997.

[159] J. R. Becker. *Corrosion and scale handbook.* Pennwell Publishing Co, Tulsa, 1998.

[160] K. W. Becker, M. A. Walsh, R. J. Fiocco, and M. T. Curran. A new laboratory method for evaluating oil spill dispersants. In *Proceedings Volume,* pages 507–510. 13th Bien API et al Oil Spill (Prev, Preparedness, Response & Coop) Int Conf (Tampa, FL, 3/29–4/1), 1993.

[161] S. L. A. Bel, V. V. Demin, N. G. Kashkarov, E. A. Konovalov, V. M. Sidorov, V. P. Bezsolitsen, M. V. Gorjacheva, S. G. Gorlov, A. M. Ivchenko, T. L. S. Mal, and Yu. N. Mojsa. Lubricating composition—for treatment of clay drilling solutions, contains additive in form of sulphonated fish fat. Patent: RU 2106381-C, 1998.

[162] J. L. Belk, D. J. Elliott, and L. M. Flaherty. The comparative effectiveness of dispersants in fresh and low salinity waters. In *Proceedings Volume,* pages 333–336. API et al Oil Spill (Prev, Behav, Contr, Cleanup) 20th Anniv Conf (San Antonio, TX, 2/13–2/16), 1989.

[163] S. Bell. Mud-to-cement technology converts industry practices. *Petrol Eng Int,* 65(9):51–52, 54–55, September 1993.

[164] H. E. Bellis and E. F. McBride. Composition and method for temporarily reducing the permeability of subterranean formations. Patent: EP 228196, 1987.

[165] R. Belore. Use of high-pressure water mixing for ship-based oil spill dispersing. In *Proceedings Volume,* pages 297–302. 10th Bien API et al Oil Spill (Prev, Behav, Contr, Cleanup) Conf (Baltimore, MD, 4/6–4/9), 1987.

[166] V. P. Belov, V. V. Zhivaeva, and K. K. Egorov. Use of caprolactam production wastes as a plasticizer of plugging slurries. *Izv Vyssh Ucheb Zavedenii, Neft Gaz,* (4):15–18, April 1989.

[167] S. S. Belyaev, I. A. Charakchian, and V. G. Kuznetsova. Strict anaerobic bacteria and their possible contribution to the enhancement of oil recovery. In E. C. Donaldson, editor, *Microbial enhancement of oil recovery: Recent advances: Proceedings of the 1990 International Conference on Microbial Enhancement of Oil Recovery,* volume 31 of *Developments in Petroleum Science,* pages 163–172. Elsevier Science Ltd, 1991.

[168] O. G. Benge and W. W. Webster. Evaluation of blast furnace slag slurries for oilfield application. In *Proceedings Volume,* pages 169–180. IADC/SPE Drilling Conf (Dallas, TX, 2/15–2/18), 1994.

[169] F. Bentiss, M. Lagrenee, and M. Traisnel. 2,5-*bis*(*N*-pyridyl)-1,3,4-oxadiazoles as corrosion inhibitors for mild steel in acidic media. *Corrosion*, 56(7):733–742, July 2000.

[170] R. Berger, H. F. Fink, W. Heilen, H. Holthoff, and C. Weitemeyer. Use of organosilicon modified polydienes as antifoaming and antibubble agents in organic systems (Verwendung von Organosiliciummodifizierten Polydienen als Entschäumungs und Entlüeftungsmittel für Organische Systeme). Patent: EP 293684, 1988.

[171] R. Berger, H. F. Fink, G. Koerner, J. Langner, and C. Weitemeyer. Use of fluorinated norbornylsiloxanes for defoaming freshly extracted degassing crude oil. Patent: US 4626378, 1986.

[172] R. Berkhof, H. Kwekkeboom, D. Balzer, and N. Ripke. Demulsifiers for breaking petroleum emulsions (Demulgatoren zur Spaltung von Erdölemulsionen). Patent: EP 468095, 1992.

[173] C. J. Bernu. High temperature stable modified starch polymers and well drilling fluids employing same. Patent: EP 852235, 1998.

[174] H. W. Bewersdorff and D. Ohlendorf. The behaviour of drag-reducing cationic surfactant solutions. *Colloid Polymer Sci*, 266(10):941–953, October 1988.

[175] P. Bharat. Well treating fluids and additives therefor. Patent: EP 372469, 1990.

[176] B. R. Bhattacharyya. Water soluble polymer as water-in-oil demulsifier. Patent: US 5100582, 1992.

[177] V. K. Bhupathiraju, P. K. Sharma, M. J. McInerney, R. M. Knapp, K. Fowler, and W. Jenkins. Isolation and characterization of novel halophilic anaerobic bacteria from oil field brines. In E. C. Donaldson, editor, *Microbial enhancement of oil recovery: Recent advances: Proceedings of the 1990 International Conference on Microbial Enhancement of Oil Recovery*, volume 31 of *Developments in Petroleum Science*, pages 131–143. Elsevier Science Ltd, 1991.

[178] D. Bhuyan. *Development of an alkaline/surfactant/polymer compositional reservoir simulator*. PhD thesis, Texas Univ, Austin, 1989.

[179] D. Bhuyan, G. A. Pope, and L. W. Lake. Simulation of high-pH coreflood experiments using a compositional chemical flood simulator. In *Proceedings Volume*, pages 307–316. SPE Oilfield Chem Int Symp (Anaheim, CA, 2/20–2/22), 1991.

[180] V. D. Bielewicz and L. Kraj. Laboratory data on the effectivity of chemical breakers in mud and filtercake (Untersuchungen zur Effektivität von Degradationsmitteln in Spülungen). *Erdöl Erdgas Kohle*, 114(2):76–79, February 1998.

[181] R. L. Bienvenu, Jr. Lightweight proppants and their use in hydraulic fracturing. Patent: US 5531274, 1996.

[182] R. L. Bienvenu, Jr. Lightweight proppants and their use in hydraulic fracturing. Patent: WO 9604464, 1996.
[183] S. Bin Ibrahim, R. Huizenga, D. Oakley, and K. Sithamparam. Multifunctional additive to cement slurries. Patent: WO 0166487, 2001.
[184] S. Bin Ibrahim, K. Sithamparam, R. Huizenga, and D. Oakley. Multifunctional additive to cement slurries. Patent: EP 1132354, 2001.
[185] J. Bissell, D. Acker, and L. R. Quaife. Pipeline leak-location technique utilizing a novel test fluid and trained dogs. In *Proceedings Volume*, number 9. 5th Pipe Line Ind & Pipes Pipelines Int Pipeline Pigging & Integrity Monit Int Conf (Houston, TX, 2/1–2/4), 1993.
[186] T. Bjornstad, O. B. Haugen, and I. A. Hundere. Dynamic behavior of radio-labelled water tracer candidates for chalk reservoirs. *J Petrol Sci Eng,* 10(3):223–238, February 1994.
[187] A. V. Blagov, Z. F. Prazdnikova, and V. I. Praporshchikov. Use of ecological factors for controlling biogenic sulfate reduction. *Neft Khoz,* (5):48–50, May 1990.
[188] V. A. Blazhevich, D. A. Khisaeva, and V. G. Umetbaev. Plugging solution—contains Portland cement, liquid waste from soda production and water and has reduced hardening time and increased strength. Patent: SU 1776762-A, 1992.
[189] V. A. Blazhevich, D. A. Khisaeva, V. G. Umetbaev, and I. V. Legostaeva. Polymer plugging solution for oil and gas wells—contains urea-formaldehyde resin, and aluminium chloride containing waste of isopropylbenzene production as acid hardener. Patent: SU 1763638-A, 1992.
[190] J. B. Bloys, R. B. Carpenter, and W. N. Wilson. Accelerating set of retarded cement. Patent: US 5005646, 1991.
[191] J. B. Bloys, R. B. Carpenter, and W. N. Wilson. Accelerating set of retarded cement. Patent: CA 2046714, 1992.
[192] J. B. Bloys, W. N. Wilson, E. Malachosky, R. D. Bradshaw, and R. A. Grey. Dispersant compositions comprising sulfonated isobutylene maleic anhydride copolymer for subterranean well drilling and completion. Patent: US 5360787, 1994.
[193] J. B. Bloys, W. N. Wilson, E. Malachosky, R. B. Carpenter, and R. D. Bradshaw. Dispersant compositions for subterranean well drilling and completion. Patent: EP 525037, 1993.
[194] J. B. Bloys and B. S. Wilton. Control of lost circulation in wells. Patent: US 5065820, 1991.
[195] G. C. Blytas and H. Frank. Copolymerization of polyethercyclicpolyols with epoxy resins. Patent: US 5401860, 1995.
[196] G. C. Blytas, H. Frank, A. H. Zuzich, and E. L. Holloway. Method of preparing polyethercyclicpolyols. Patent: EP 505000, 1992.

[197] G. C. Blytas, A. H. Zuzich, E. L. Holloway, and H. Frank. Method of preparing polyethercyclicpolyols. Patent: EP 505002, 1992.
[198] L. S. Boak, G. M. Graham, and K. S. Sorbie. The influence of divalent cations on the performance of $BaSO_4$ scale inhibitor species. In *Proceedings Volume*, pages 643–648. SPE Oilfield Chem Int Symp (Houston, TX, 2/16–2/19), 1999.
[199] C. Bocard, G. Castaing, J. Ducreux, C. Gatellier, J. Croquette, and F. Merlin. Protecmar: The French experience from a seven-year dispersant offshore trials program. In *Proceedings Volume*, pages 225–229. 10th Bien API et al Oil Spill (Prev, Behav, Contr, Cleanup) Conf (Baltimore, MD, 4/6–4/9), 1987.
[200] W. M. Bohon and G. R. Ruschau. Method for inhibiting external corrosion on an insulated pipeline. Patent: US 6273144, 2001.
[201] J. Boivin. Oil industry biocides. *Mater Performance,* 34(2):65–68, February 1994.
[202] J. W. Boivin, J. W. Costerton, E. J. Laishley, and R. Bryant. A new rapid test for microbial corrosion detection and biocide evaluation. In *Proceedings Volume*, pages 537–547. 2nd Inst Gas Technol Gas, Oil, Coal, & Environ Biotechnol Int Symp (New Orleans, LA, 12/11–12/13), 1989.
[203] J. W. Boivin, R. Shapka, A. E. Khoury, S. Blenkinsopp, and J. W. Costerton. An old and a new method of control for biofilm bacteria. In *Proceedings Volume*. Annu NACE Corrosion Conf (Corrosion 92) (Nashville, TN, 4/27–5/1), 1992.
[204] J. L. Boles and J. B. Boles. Cementing compositions and methods using recycled expanded polystyrene. Patent: US 5736594, 1998.
[205] J. L. Boles, A. S. Metcalf, and J. C. Dawson. Coated breaker for cross-linked acid. Patent: US 5497830, 1996.
[206] V. G. Boncan. Low temperature curing cement for wellbores. Patent: GB 2353523, 2001.
[207] V. G. Boncan and R. Gandy. Well cementing method using an AM/AMPS fluid loss additive blend. Patent: US 4632186, 1986.
[208] V. G. Boncan, D. T. Müller, M. J. Rogers, and W. S. Bray. Method and compositions for use in cementing. Patent: US 6032738, 2000.
[209] J. E. Bonekamp, G. D. Rose, D. L. Schmidt, A. S. Teot, and E. K. Watkins. Viscoelastic surfactant based foam fluids. Patent: US 5258137, 1993.
[210] J. K. Borchardt, D. B. Bright, M. K. Dickson, and S. L. Wellington. Surfactants for carbon dioxide foam flooding: Effects of surfactant chemical structure on one-atmosphere foaming properties. In *Proceedings Volume*, number 373, pages 163–180. 61st Annu Acs Colloid & Surface Sci Symp (Ann Arbor, MI, 6/21–6/24), 1987.
[211] J. K. Borchardt and A. R. Strycker. Olefin sulfonates for high temperature steam mobility control: Structure-property correlations.

In *Proceedings Volume,* pages 91–102. SPE Oilfield Chem Int Symp (Houston, TX, 2/18–2/21), 1997.

[212] S. W. Borenstein and G. J. Licina. An overview of monitoring techniques for the study of microbiologically influenced corrosion. In *Proceedings Volume*. 49th Annu Nace Int Corrosion Conf (Corrosion 94) (Baltimore, MD, 2/27–3/4), 1994.

[213] S. W. Borenstein and P. B. Lindsay. MIC (microbiologically influenced corrosion) failure analysis. *Mater Performance,* 33(4):43–45, April 1994.

[214] V. P. Bortsov, A. A. Baluev, and S. N. Bastrikov. Plugging solution for oil and gas wells—contains Portland cement, aluminum powder, fish scale, anionic and nonionic surfactant, plasticiser and water. Patent: RU 2078906-C, 1997.

[215] V. P. Bortsov, A. A. Baluev, and S. N. Bastrikov. Plugging solution for oil and gas wells—contains Portland cement, expanding additive, water and additionally aluminum powder, surfactants and plasticiser. Patent: RU 2082872-C, 1997.

[216] W. Böse, M. Hofinger, M. Hille, R. Böhm, and F. Staiss. Branched polyoxyalkylene polyesters, process for their preparation and application (Verzweigte Polyoxalkylenmischpolyester, Verfahren zu ihrer Herstellung und ihre Verwendung). Patent: EP 267517, 1988.

[217] P. Bouchut, Y. Kensicher, and J. Rousset. Non-polluting dispersing agent for drilling fluids based on freshwater or salt water. Patent: US 5099928, 1992.

[218] P. Bouchut, J. Rousset, and Y. Kensicher. A non-polluting dispersing agent for drilling fluids based on freshwater or salt water. Patent: AU 590248, 1989.

[219] P. Bouchut, J. Rousset, and Y. Kensicker. Non-polluting fluidizing agents for drilling fluids having soft or salt water base (agent fluidifiant non polluant pour fluides de forage a base d'eau douce ou saline). Patent: CA 1267777, 1990.

[220] D. L. Bour and J. D. Childs. Foamed well cementing compositions and methods. Patent: US 5133409, 1992.

[221] A. T. Bourgoyne. *Applied drilling engineering,* volume 2 of *SPE Textbook Series.* SPE Publications, 1986.

[222] M. N. Bouts, R. A. Trompert, and A. J. Samuel. Time delayed and low-impairment fluid-loss control using a succinoglycan biopolymer with an internal acid breaker. *SPE J,* 2(4):417–426, December 1997.

[223] P. G. Boyd, E. R. Fischer, and J. A. Alford. Polybasic acid esters as oil field corrosion inhibitors. Patent: GB 2259702, 1993.

[224] M. L. Braden. Preparation of amphoteric acrylic acid copolymers suitable as oil-in-water emulsion breakers. Patent: US 5552498, 1996.

[225] M. L. Braden and S. J. Allenson. Method for separating liquid from water using amine containing polymers. Patent: US 5021167, 1991.

[226] M. L. Braden and S. J. Allenson. Reverse emulsion breaking method using amine containing polymers. Patent: US 5032285, 1991.
[227] T. G. Braga, R. L. Martin, J. A. McMahon, B. A. Oude Alink, and B. T. Outlaw. Combinations of imidazolines and wetting agents as environmentally acceptable corrosion inhibitors. Patent: WO 0049204, 2000.
[228] J. R. Bragg. Oil recovery method using an emulsion. Patent: WO 9853181, 1998.
[229] J. R. Bragg. Oil recovery method using an emulsion. Patent: US 5855243, 1999.
[230] J. R. Bragg. Oil recovery method using an emulsion. Patent: US 6068054, 2000.
[231] B. G. Brake and J. Chatterji. Fluid loss reducing additive for cement compositions. Patent: EP 595660, 1994.
[232] H. Branch, III. Shale-stabilizing drilling fluids and method for producing same. Patent: US 4719021, 1988.
[233] E. B. Brandes and F. C. Loveless. Dispersants and dispersant viscosity index improvers from selectively hydrogenated polymers. Patent: WO 9640846, 1996.
[234] E. B. Brandes and F. C. Loveless. Dispersants and dispersant viscosity index improvers from selectively hydrogenated polymers. Patent: WO 9640845, 1996.
[235] P. J. Brandvik, O. O. Knudsen, M. O. Moldestad, and P. S. Daling. Laboratory testing of dispersants under arctic conditions. In *Proceedings Volume,* pages 191–206. 2nd ASTM Use of Chem in Oil Spill Response Symp (Victoria, Canada, 10/10–10/11), 1994.
[236] P. J. Brandvik, M. O. Moldestad, and P. S. Daling. Laboratory testing of dispersants under arctic conditions. In *Proceedings Volume,* pages 123–134. 15th Environ Can Arctic & Mar Oil Spill Program Tech Semin (Edmonton, Canada, 6/10–6/12), 1992.
[237] D. Brankling. Drilling fluid. Patent: WO 9402565, 1994.
[238] H. D. Brannon. Fracturing fluid slurry concentrate and method of use. Patent: EP 280341, 1988.
[239] H. D. Brannon, R. M. Hodge, and K. W. England. High temperature guar-based fracturing fluid. Patent: US 4801389, 1989.
[240] H. D. Brannon and Joe. Pin. R. M. Tjon. Characterization of breaker efficiency based upon size distribution of polymeric fragments. In *Proceedings Volume,* pages 415–429. Annu SPE Tech Conf (Dallas, TX, 10/22–10/25), 1995.
[241] H. D. Brannon and R. M. Tjon-Joe-Pin. Biotechnological breakthrough improves performance of moderate to high-temperature fracturing applications. In *Proceedings Volume,* volume 1, pages 515–530. 69th Annu SPE Tech Conf (New Orleans, LA, 9/25–9/28), 1994.

[242] W. S. Bray and W. R. Wood. Well cementing method using a dispersant and fluid loss intensifier. Patent: US 5105885, 1992.
[243] M. Brezinski, T. R. Gardner, W. M. Harms, J. L. Lane, Jr., and K. L. King. Controlling iron in aqueous well fracturing fluids. Patent: EP 599474, 1994.
[244] M. M. Brezinski. Metal corrosion inhibitor for use in aqueous acid solutions. Patent: US 5792420, 1998.
[245] M. M. Brezinski. Well acidizing compositions. Patent: EP 1132570-A, 2001.
[246] M. M. Brezinski and B. Desai. Method and composition for acidizing subterranean formations utilizing corrosion inhibitor intensifiers. Patent: EP 869258, 1998.
[247] M. M. Brezinski. Methods and acidizing compositions for reducing metal surface corrosion and sulfide precipitation. Patent: US 6315045, 2001.
[248] G. L. Briggs. Corrosion inhibitor for well acidizing treatments. Patent: US 4698168, 1987.
[249] G. L. Briggs. Corrosion inhibitor for well acidizing treatments. Patent: CA 1274379, 1990.
[250] M. Brod, P. Venables, and G. S. Lota. Crude and heavy fuel flow improvers. Patent: AU 603180, 1990.
[251] M. J. Bromley, S. H. Gaffney, and G. E. Jackson. Oilfield emulsion control, techniques and chemicals used to separate oil and water. In *Proceedings Volume,* volume 3. Inst Corrosion UK (Corrosion 91) Conf (Manchester, England, 10/22–10/24), 1991.
[252] L. E. Brothers. Method of reducing fluid loss in cement compositions containing substantial salt concentrations. Patent: US 4640942, 1987.
[253] L. E. Brothers. Method of reducing fluid loss in cement compositions which may contain substantial salt concentrations. Patent: US 4700780, 1987.
[254] L. E. Brothers. Method of reducing fluid loss in cement compositions. Patent: US 4806164, 1989.
[255] L. E. Brothers. Composition and method for inhibiting thermal thinning of cement. Patent: US 5135577, 1992.
[256] L. E. Brothers. Low temperature set retarded well cement compositions and methods. Patent: US 5472051, 1995.
[257] L. E. Brothers, D. C. Brennels, and J. D. Childs. Light weight high temperature well cement compositions and methods. Patent: US 5900053, 1999.
[258] L. E. Brothers and F. X. Deblanc. New cement formulation helps solve deep cementing problems. *J Petrol Technol,* 41(6):611–614, June 1989.

[259] L. E. Brothers, D. D. Onan, and R. L. Morgan. Well cement compositions containing rubber particles and methods of cementing subterranean zones. Patent: US 5779787, 1998.
[260] H. M. Brown and R. H. Goodman. The use of dispersants in broken ice. In *Proceedings Volume,* volume 1, pages 453–460. 19th Environ Can Arctic & Mar Oilspill Program Tech Seminar (Calgary, Canada, 6/12–6/14), 1996.
[261] H. M. Brown, R. H. Goodman, and G. P. Canevari. Where has all the oil gone? Dispersed oil detection in a wave basin and at sea. In *Proceedings Volume,* pages 307–312. 10th Bien API et al Oil Spill (Prev, Behav, Contr, Cleanup) Conf (Baltimore, MD, 4/6–4/9), 1987.
[262] J. M. Brown, J. R. Ohlsen, and R. D. McBride. Corrosion inhibitor composition and method of use. Patent: US 5512212, 1996.
[263] R. G. B. Brown, G. Curl, Jr., H. Curl, S. Christopherson, D. Dale, C. Hall, L. Harris, J. Kaperick, D. Kennedy, E. Levine, D. Mattson, B. McFarland, J. McGee, C. L. Merriam, J. Morris, J. Murphy, R. Pavia, E. Shaw, J. Snider, M. Straub, and I. C. White. *Oil spill case histories 1967–1991. Summaries of significant U.S. and international oil spills.* National Oceanic and Atmospheric Administration HMRAD 92-11, NOAA Hazardous Material Response and Assessment Division, Seattle, Washington, 1992.
[264] R. Bruckdorfer. Carbon dioxide corrosion resistance in cements. In *Proceedings Volume,* volume 2, pages 517–525. 36th Annu Cim Petrol Soc & Can Soc Petrol Geol Tech Mtg (Edmonton, Canada, 6/2–6/5), 1985.
[265] J. R. Bruton and H. C. McLaurine. Modified poly-amino acid hydration suppressant proves successful in controlling reactive shales. In *Proceedings Volume,* pages 127–135. 68th Annu SPE Tech Conf (Houston, TX, 10/3–10/6), 1993.
[266] E. Bryan, M. A. Veale, R. E. Talbot, K. G. Cooper, and N. S. Matthews. Biocidal compositions and treatments. Patent: EP 385801, 1990.
[267] R. S. Bryant and T. E. Burchfield. Review of microbial technology for improving oil recovery. In *Proceedings Volume.* Nat Inst Petrol Energy Res Microbial Enhanced Oil Recovery Short Course (Bartlesville, OK, 5/23), 1989.
[268] R. S. Bryant, T. E. Burchfield, R. E. Porter, D. M. Dennis, and D. O. Hitzman. Microbial enhanced waterflooding: A pilot study. In E. C. Donaldson, editor, *Microbial enhancement of oil recovery: Recent advances: Proceedings of the 1990 International Conference on Microbial Enhancement of Oil Recovery,* volume 31 of *Developments in Petroleum Science,* pages 399–419. Elsevier Science Ltd, 1991.

[269] B. Bubela. In situ biological production of surfactants for enhanced oil recovery. Australia Dep Resources Energy End of Grant Rep 151, March 1983.
[270] B. Bubela. Geobiology and microbiologically enhanced oil recovery. In E. C. Donaldson, G. V. Chilingarian, and T. F. Yen, editors, *Microbial enhanced oil recovery,* volume 22 of *Developments in Petroleum Science,* pages 75–97. Elsevier Science Ltd, 1989.
[271] R. G. Buchheit, T. E. Hinkebein, P. F. Hlava, and D. G. Melton. The effects of latex additions on centrifugally cast concrete for internal pipeline protection. In *Proceedings Volume,* volume 4, pages 2854–2864. 12th SPE/NACE Int Corrosion Congr (Houston, TX, 9/19–9/24), 1993.
[272] E. Buck, M. C. Allen, B. Sudbury, and B. Skjellerudsveen. Corrosion inhibitor detection by thin layer chromatography: Development of the technique. In *Proceedings Volume.* Annu NACE Corrosion Conf (Corrosion 93) (New Orleans, LA, 3/7–3/12), 1993.
[273] I. A. Buist and S. L. Ross. Emulsion inhibitors: A new concept in oil spill treatment. In *Proceedings Volume,* pages 217–222. 10th Bien API et al Oil Spill (Prev, Behav, Contr, Cleanup) Conf (Baltimore, MD, 4/6–4/9), 1987.
[274] N. I. Bulanov, V. A. Monastyrev, V. V. Balakin, A. N. Pavlenko, and V. E. Voropanov. Well bottom zone treatment with improved efficiency—includes injection of reagent breaking up clay crust for removal by flush solution stream. Patent: SU 1761944-A, 1992.
[275] J. L. Burba, III, E. F. Hoy, and A. E. Read, Jr. Adducts of clay and activated mixed metal oxides. Patent: WO 9218238, 1992.
[276] J. L. Burba, III and G. W. Strother. Mixed metal hydroxides for thickening water or hydrophylic fluids. Patent: US 4990268, 1991.
[277] C. Burcham, R. E. Fast, A. S. Murer, and P. S. Northrop. Oil recovery by enhanced imbibition in low permeability reservoirs. Patent: US 5411086, 1995.
[278] C. L. Burdick and J. N. Pullig. Sodium formate fluidized polymer suspensions process. Patent: US 5228908, 1993.
[279] E. D. Burger and G. R. Chesnut. Screening corrosion inhibitors used in acids for downhole scale removal. *Mater Performance,* 31(7):40–44, July 1992.
[280] E. D. Burger, A. B. Crews, and H. W. Ikerd, II. Inhibition of sulfate-reducing bacteria by anthraquinone in a laboratory biofilm column under dynamic conditions. NACE Int Corrosion Conf (Corrosion 2001) (Houston, TX, 3/11–3/16), 2001.
[281] R. S. Buriks and J. G. Dolan. Demulsifier composition and method of use thereof. Patent: US 4626379, 1986.

[282] R. S. Buriks and J. G. Dolan. Demulsifier composition and method of use thereof. Patent: EP 268713, 1988.
[283] R. S. Buriks and J. G. Dolan. Demulsifier compositions and methods of preparation and use thereof. Patent: EP 331323, 1989.
[284] R. S. Buriks and J. G. Dolan. Demulsifier compositions and methods of preparation and use thereof. Patent: US 4877842, 1989.
[285] R. S. Buriks and J. G. Dolan. Demulsifier compositions and methods of preparation and use thereof. Patent: CA 1318053, 1993.
[286] R. S. Buriks, A. R. Fauke, and D. J. Poelker. Vinyl phenol polymers for demulsification of oil-in-water emulsions. Patent: CA 2031122, 1991.
[287] R. S. Buriks, A. R. Fauke, and D. J. Poelker. Vinyl phenol polymers for demulsification of oil-in-water emulsions. Patent: US 5098605, 1992.
[288] G. Burrafato and S. Carminati. Aqueous drilling muds fluidified by means of zirconium and aluminum complexes. Patent: EP 623663, 1994.
[289] G. Burrafato and S. Carminati. Aqueous drilling muds fluidified by means of zirconium and aluminum complexes. Patent: CA 2104134, 1994.
[290] G. Burrafato and S. Carminati. Aqueous drilling muds fluidified by means of zirconium and aluminum complexes. Patent: US 5532211, 1996.
[291] G. Burrafato, A. Gaurneri, T. P. Lockhart, and L. Nicora. Zirconium additive improves field performance and cost of biopolymer muds. In *Proceedings Volume,* pages 707–710. SPE Oilfield Chem Int Symp (Houston, TX, 2/18–2/21), 1997.
[292] T. R. Burridge and M. A. Shir. The comparative effects of oil dispersants and oil/dispersant conjugates on germination of the marine macroalga phyllospora comosa (fucales: *Phaeophyta*). *Mar Pollut Bull,* 31(4–12):446–452, April-December 1995.
[293] M. A. Burshtejn and S. V. Logvinenko. Composition for temporary plugging of stratum during drilling process—comprises chemically precipitated chalk, or chalk-containing industrial waste, e.g., from nitric industry, as plugging component (eng). Patent: RU 2041340-C, 1995.
[294] B. D. Burts, Jr. Lost circulation material with rice fraction. Patent: US 5118664, 1992.
[295] B. D. Burts, Jr. Lost circulation material with rice fraction. Patent: US 5599776, 1997.
[296] B. D. Burts, Jr. Well fluid additive, well fluid made therefrom, method of treating a well fluid, method of circulating a well fluid. Patent: US 6323158, 2001.
[297] N. E. Byrne and J. D. Johnson. Water soluble corrosion inhibitors. Patent: US 5322640, 1994.

[298] N. E. Byrne, R. A. Marble, and M. Ramesh. New dispersion polymers for oil field water clarification. Patent: EP 595156, 1994.
[299] C. Cai, S. Yang, and W. Ai. Application of accelerant $CaCl_2$ in watercut wells in Liaohe oilfield. *Petrol Drilling Tech,* 22(1):9–11,68, March 1994.
[300] I. C. Callaghan, C. M. Gould, R. J. Hamilton, and E. L. Neustadter. The relationship between the dilatational rheology and crude oil foam stability: I. Preliminary studies. *Colloids and Surfaces,* 8(1):17–28, November 1983.
[301] I. C. Callaghan, S. A. Hickman, F. T. Lawrence, and P. M. Melton. Antifoams in gas-oil separation. In *Proceedings Volume,* number 59, pages 48–57. Royal Soc Chem Ind Div Ind Appl of Surfactants Symp (Salford, England, 4/15–4/17), 1986.
[302] I. C. Callaghan and A. S. Taylor. Fluorosilicone anti-foam additive composition for use in crude oil separation. Patent: GB 2234978, 1991.
[303] G. Calloni, N. Moroni, and F. Miano. Carbon black: A low cost colloidal additive for controlling gas migration in cement slurries. In *Proceedings Volume,* pages 563–574. Offshore Mediter Conf (Ravenna, Italy, 3/15–3/17), 1995.
[304] G. Calloni, N. Moroni, and F. Miano. Carbon black: A low cost colloidal additive for controlling gas migration in cement slurries. In *Proceedings Volume,* pages 145–153. SPE Oilfield Chem Int Symp (San Antonio, TX, 2/14–2/17), 1995.
[305] Y. Camberlin, J. Grenier, S. Poncet, A. Bonnet, J. P. Pascault, and H. Sautereau. Utilization of polymer compositions for the coating of surfaces, and surface coating containing such compositions (utilisation de compositions de polymeres pour le revetement de surfaces et revetement de surfaces comprenant ces compositions). Patent: EP 931819, 1999.
[306] Y. Camberlin, J. Grenier, J. Vallet, A. Bonnet, J. P. Pascault, and M. Taha. Polymer compositions, their preparation and their use (compositions de polymeres, leurs preparations et leurs utilisations). Patent: FR 2773809, 1999.
[307] Y. Camberlin, J. Grenier, J. Vallet, A. Bonnet, J. P. Pascault, and M. Taha. Polymer compositions, their preparation and uses (compositions de polymeres, leurs preparations et leurs utilisations). Patent: EP 931803, 1999.
[308] R. E. Campos and J. A. Hernandez. In-situ reduction of oil viscosity during steam injection process in EOR. Patent: US 5209295, 1993.
[309] G. P. Canevari. Basic study reveals how different crude oils influence dispersant performance. In *Proceedings Volume,* pages 293–296. 10th Bien API et al Oil Spill (Prev, Behav, Contr, Cleanup) Conf (Baltimore, MD, 4/6–4/9), 1987.

[310] G. P. Canevari, M. Fingas, R. J. Fiocco, and R. R. Lessard. Corexit 9580 shoreline cleaner: Development, application, and status. In *Proceedings Volume,* pages 227–239. 2nd ASTM Use of Chem in Oil Spill Response Symp (Victoria, Canada, 10/10–10/11), 1994.

[311] G. P. Canevari, R. J. Fiocco, K. W. Becker, and R. R. Lessard. Chemical dispersant for oil spills. Patent: WO 9413397, 1994.

[312] G. P. Canevari, R. J. Fiocco, K. W. Becker, and R. R. Lessard. Chemical dispersant for oil spills. Patent: US 5618468, 1997.

[313] L. A. Cantu and P. A. Boyd. Laboratory and field evaluation of a combined fluid-loss control additive and gel breaker for fracturing fluids. In *Proceedings Volume,* pages 7–16. SPE Oilfield Chem Int Symp (Houston, TX, 2/8–2/10), 1989.

[314] L. A. Cantu, E. F. McBride, and M. Osborne. Formation fracturing process. Patent: EP 401431, 1990.

[315] L. A. Cantu, E. F. McBride, and M. Osborne. Well treatment process. Patent: EP 404489, 1990.

[316] L. A. Cantu, E. F. McBride, and M. W. Osborne. Formation fracturing process. Patent: US 4848467, 1989.

[317] L. A. Cantu, E. F. McBride, and M. W. Osborne. Formation fracturing process. Patent: CA 1319819, 1993.

[318] R. J. Card, P. R. Howard, J. P. Feraud, and V. G. Constien. Control of particulate flowback in subterranean wells. Patent: US 6172011, 2001.

[319] R. Carpenter and D. Johnson. Method and cement-drilling fluid cement composition for cementing a wellbore. Patent: WO 9748655, 1997.

[320] R. B. Carpenter. Cement slurries for wells. Patent: US 5372641, 1994.

[321] R. B. Carpenter, J. B. Bloys, and D. L. Johnson. Cement composition containing synthetic hectorite clay. Patent: WO 9902464, 1999.

[322] R. B. Carpenter and W. N. Wilson. Method of accelerating set of cement by washover fluid containing alkanolamine. Patent: US 4976316, 1990.

[323] R. B. Carpenter and W. N. Wilson. Accelerating the set of cement compositions in wells. Patent: EP 448218, 1991.

[324] U. Cartalos, J. Lecourtier, and A. Rivereau. Use of scleroglucan as high temperature additive for well cements (utilisation du scleroglucane comme additif a haute temperature pour les laitiers de ciment). Patent: EP 526310, 1993.

[325] U. Cartalos, J. Lecourtier, and A. Rivereau. Use of scleroglucan as high temperature additive for well cements (utilisation du scleroglucane comme additif a haute temperature pour les laitiers de ciment). Patent: FR 2679899, 1993.

[326] U. Cartalos, J. Lecourtier, and A. Rivereau. Use of scleroglucan as high temperature additive for cement slurries. Patent: US 5301753, 1994.

[327] C. G. Carter, R. P. Kreh, and L. D. G. Fan. Composition and method for inhibiting scale and corrosion using naphthylamine polycarboxylic acids. Patent: EP 538969, 1998.

[328] J. M. Casabonne, M. Jowe, and E. Nelson. High-temperature petroleum-cement retarders, cement slurries, and associated cementation procedures (retardateurs haute temperature pour ciments petroliers, laitiers de ciments et procedes de cimentation correspondants). Patent: FR 2702472, 1994.

[329] B. M. Casad, C. R. Clark, L. A. Cantu, D. P. Cords, and E. F. McBride. Process for the preparation of fluid loss additive and gel breaker. Patent: US 4986355, 1991.

[330] H. A. Cash, A. S. Krupa, I. Vance, and B. V. Johnson. Laboratory testing of biocides against sessile oilfield bacteria. In *Proceedings Volume*. Inst Gas Technol Gas, Oil & Environ Biotechnol Symp (Chicago, IL, 9/21–9/23), 1992.

[331] L. U. Castro. Demulsification treatment and removal of in-situ emulsion in heavy-oil reservoirs. In *Proceedings Volume*. SPE West Reg Mtg (Bakersfield, CA, 3/26–3/30), 2001.

[332] A. Cavallaro, E. Curci, G. Galliano, M. Vicente, D. Crosta, and H. Leanza. Design of an acid stimulation system with chlorine dioxide for the treatment of water-injection wells. 7th SPE Latin Amer & Caribbean Petrol Eng Conf (Buenos Aires, Argentina, 3/25–3/28), 2001.

[333] A. N. Cavallaro, R. Baigorria, and E. Curci. Design of an acid stimulation system with chlorine dioxide for the treatment of water-injection wells. In *Proceedings Volume*. 51st Annu Cim Petrol Soc Tech Mtg (Calgary, Canada, 6/4–6/8), 2000.

[334] B. Caveny, D. Ashford, R. Hammack, and J. G. Garcia. Tires fuel oil field cement manufacturing. *Oil Gas J*, 96(35):64–67, 1998.

[335] W. J. Caveny, J. D. Weaver, and P. D. Nguyen. Control of particulate flowback in subterranean wells. Patent: US 5582249, 1996.

[336] K. E. Cawiezel and J. L. Elbel. A new system for controlling the crosslinking rate of borate fracturing fluids. In *Proceedings Volume*, pages 547–552. 60th Annu SPE Calif Reg Mtg (Ventura, CA, 4/4–4/6), 1990.

[337] K. E. Cawiezel, R. Navarrete, and V. Constien. Fluid loss control. Patent: GB 2291906, 1996.

[338] K. E. Cawiezel, R. C. Navarrete, and V. G. Constien. Fluid loss control. Patent: US 5948733, 1999.

[339] R. E. Chadwick and B. M. Phillips. Preparation of ether carboxylates. Patent: EP 633279, 1995.
[340] S. Chakrabarti and C. Z. Marczewski. Determining the concentration of a cross-linking agent containing zirconium. Patent: GB 2228996, 1990.
[341] S. Chakrabarti, J. P. Martins, and D. Mealor. Method for controlling the viscosity of a fluid. Patent: GB 2199408, 1988.
[342] K. S. Chan and T. J. Griffin, Jr. Low temperature, low rheology synthetic cement. Patent: US 5547027, 1996.
[343] S. Chandra, H. Justnes, and Y. Ohama. *Concrete polymer composites: The polymeric materials encyclopedia.* CRC Press, Inc, Boca Raton, FL, 1996.
[344] F. F. Chang, M. Bowman, M. Parlar, S. A. Ali, and J. Cromb. Development of a new crosslinked-HEC (hydroxyethylcellulose) fluid loss control pill for highly-overbalanced, high-permeability and/or high-temperature formations. In *Proceedings Volume,* pages 215–227. SPE Formation Damage Contr Int Symp (Lafayette, LA, 2/18–2/19), 1998.
[345] F. F. Chang and F. Civan. Practical model for chemically induced formation damage. *J Petrol Sci Eng,* 17(1–2):123–137, February 1997.
[346] F. F. Chang and M. Parlar. Method and composition for controlling fluid loss in high permeability hydrocarbon bearing formations. Patent: US 5981447, 1999.
[347] H. F. D. Chang and J. S. Meng. *Physicochem Hydrodyn,* 9:33, 1987.
[348] M. M. Chang and H. W. Gao. User's guide and documentation manual for "PC-Gel" simulator: Topical report. US DOE Fossil Energy Rep NIPER-705, NIPER, October 1993.
[349] R. F. Chanshev, S. V. Kovtunenko, F. D. Tsikunkov, R. A. Ismakov, G. V. Konesev, and R. A. Mulyukov. Lubricant for cutter bit bearings—contains ethylene-propylene synthetic rubber, zinc dioctyl-phenyl dithio-phosphate, polytetrafluoroethylene and mineral oil. Patent: SU 1778162-A, 1992.
[350] P. E. Chaplanov, A. D. Chernikov, O. G. Mironov, G. N. Semanov, A. G. Svinukhov, A. G. Yaremenko, F. V. Linchevskij, N. A. Melnik, I. A. Murashev, and I. S. Akhmetzhanov. Removing oils and petroleum products from water surfaces—using mixture of oxyethylated alkyl-phenol, alkyl-phosphate and fatty acid diethanolamine. Patent: SU 1325816-A, 1992.
[351] A. Charlier. Dispersant compositions for treating oil slicks on the surface of water (compositions dispersantes pour le traitement de nappes d'huile a la surface de l'eau). Patent: EP 254704, 1988.
[352] A. Charlier. Dispersant compositions for treating oil slicks on cold water. Patent: EP 398860, 1990.

[353] A. G. R. Charlier. Dispersant compositions for treating oil slicks. Patent: US 4830759, 1989.

[354] A. G. R. Charlier. Dispersant compositions for treating oil slicks. Patent: US 5051192, 1991.

[355] K. L. Chase, R. S. Bryant, T. E. Burchfield, K. M. Bertus, and A. K. Stepp. Investigations of microbial mechanisms for oil mobilization in porous media. In E. C. Donaldson, editor, *Microbial enhancement of oil recovery: Recent advances: Proceedings of the 1990 International Conference on Microbial Enhancement of Oil Recovery*, volume 31 of *Developments in Petroleum Science,* pages 79–94. Elsevier Science Ltd, 1991.

[356] J. Chatterji, D. C. Brennels, D. W. Gray, S. E. Lebo, and S. L. Dickman. Cement compositions and biodegradable dispersants therefor. Patent: US 6019835, 2000.

[357] J. Chatterji, R. S. Cromwell, and B. J. King. Defoaming compositions for well treatment fluids. Patent: EP 1018354, 2000.

[358] J. Chatterji, J. E. Griffith, P. L. Totten, and B. J. King. Lightweight well cement compositions and methods. Patent: US 5588489, 1996.

[359] J. Chatterji, B. J. King, P. L. Totten, and D. D. Onan. Resilient well cement compositions and methods. Patent: US 5688844, 1997.

[360] J. Chatterji, S. E. Lebo, C. D. Brenneis, S. L. Dickman, and D. W. Gray. Cement compositions comprising modified lignosulphonate. Patent: EP 985645, 2000.

[361] J. Chatterji, R. L. Morgan, and G. W. Davis. Set retarded cementing compositions and methods. Patent: US 5672203, 1997.

[362] J. Chatterji, P. L. Totten, B. J. King, and D. D. Onan. Well cement compositions. Patent: EP 816300, 1998.

[363] J. Chatterji, F. Zamora, B. J. King, and R. J. McKinley. Well cementing method. Patent: EP 1065186A, 2001.

[364] M. Chen, Z. Chen, and R. Huang. Hydration stress on wellbore stability. In *Proceedings Volume,* pages 885–888. 35th US Rock Mech Symp (Reno, NV, 6/5–6/7), 1995.

[365] R. G. Chen and A. J. Son. Cationic amide/ester compositions as demulsifiers. Patent: US 5117058, 1992.

[366] T. Chen, W. Pu, K. Ping, and Z. Ye. The rheological behavior of partially hydrolyzed polyacrylamide (HPAM) solutions in reservoir. *J Southwest Petrol Inst,* 19(3):2A–3A, 28–34, August 1997.

[367] X. Chen, C. P. Tan, and C. M. Haberfield. Wellbore stability analysis guidelines for practical well design. In *Proceedings Volume,* pages 117–126. SPE Asia Pacific Oil & Gas Conf (Adelaide, Australia, 10/28–10/31), 1996.

[368] P. S. Cheung. Expanding additive for cement composition. Patent: GB 2320246, 1998.

[369] P. S. Cheung. Expanding additive for cement composition. Patent: CA 2224201, 1998.
[370] P. S. Cheung. Expanding additive for cement composition. Patent: US 5942031, 1999.
[371] P. S. R. Cheung. Fluid loss additives for cementing compositions. Patent: US 5217531, 1993.
[372] M. Cheyrezy, J. Dugat, S. Boivin, G. Orange, and L. Frouin. Concrete comprising organic fibres dispersed in a cement matrix, concrete cement matrix and premixes (beton comportant des fibres organiques disperses dans une matrice cimentaire, matrice cimentaire du beton et premelanges). Patent: WO 9958468, 1999.
[373] N. P. Chilcott, D. A. Phillips, M. G. Sanders, I. R. Collins, and A. Gyani. The development and application of an accurate assay technique for sulphonated polyacrylate co-polymer oilfield scale inhibitors. 2nd Annu SPE Oilfield Scale Int Symp (Aberdeen, Scotland, 1/26–1/27), 2000.
[374] J. D. Childs and J. F. Burkhalter. Fluid loss reduced cement compositions. Patent: GB 2247234, 1992.
[375] W. C. Chin. *Formation invasion: With applications to measurement-while-drilling, time lapse analysis, and formation damage.* Gulf Publishing Co, Houston, 1995.
[376] C. I. Chiwetelu, V. Hornof, G. H. Neale, and A. E. George. Use of mixed surfactants to improve the transient interfacial tension behaviour of heavy oil/alkaline systems. *Can J Chem Eng,* 72(3):534–540, June 1994.
[377] S. Chou and J. Bae. Method for silica gel emplacement for enhanced oil recovery. Patent: US 5351757, 1994.
[378] K. Christensen, N. Davis, II, and M. Nuzzolo. Water-wettable drilling mud additives containing uintaite. Patent: EP 460067, 1991.
[379] K. C. Christensen, N. Davis, II, and M. Nuzzolo. Water-wettable drilling mud additives containing uintaite. Patent: WO 9010043, 1990.
[380] K. C. Christensen, N. Davis, II, and M. Nuzzolo. Water-wettable drilling mud additives containing uintaite. Patent: US 5030365, 1991.
[381] K. C. Christensen, N. Davis, II, and M. Nuzzolo. Water-wettable drilling mud additives containing uintaite. Patent: AU 636334, 1993.
[382] K. C. Christensen, N. Davis, II, and M. Nuzzolo. Water-wettable drilling mud additives containing uintaite. Patent: US RE35163, 1996.
[383] R. L. Christiansen, V. Bansal, and E. D. Sloan, Jr. Avoiding hydrates in the petroleum industry: Kinetics of formation. In *Proceedings Volume,* pages 383–393. SPE & Tulsa Univ Centen Petrol Eng Symp (Tulsa, OK, 8/29–8/31), 1994.
[384] R. L. Christiansen and E. D. Sloan, Jr. Mechanics and kinetics of hydrate formation. In *Proceedings Volume,* pages 283–305. New York

Acad Sci et al Natur Gas Hydrates Int Conf (New Paltz, NY, 6/20–6/24), 1993.

[385] C. A. Christopher. Surface chemical aspects of oil spill dispersant behavior. In *Proceedings Volume,* pages 375–389. 4th Annu Pennwell Conf & Exhibit Co Petro (Safe 93) Conf (Houston, TX, 1/26–1/28), 1993.

[386] D. K. Clapper and S. K. Watson. Shale stabilising drilling fluid employing saccharide derivatives. Patent: EP 702073, 1996.

[387] D. E. Clark and W. M. Dye. Environmentally safe lubricated well fluid method of making a well fluid and method of drilling. Patent: US 5658860, 1997.

[388] J. B. Clark and D. E. Langley. Biofilm control. Patent: US 4929365, 1990.

[389] M. D. Clark, P. L. Walker, K. L. Schreiner, and P. D. Nguyen. Methods of preventing well fracture proppant flow-back. Patent: US 6116342, 2000.

[390] J. R. Clayton, Jr., B. C. Stransky, A. C. Adkins, D. C. Lees, M. J. Schwartz, B. J. Snyder, J. Michel, and T. J. Reilly. Methodology for estimating cleaning effectiveness and dispersion of oil with shoreline cleaning agents in the field. In *Proceedings Volume,* volume 1, pages 423–451. 19th Environ Can Arctic & Mar Oilspill Program Tech Semin (Calgary, Canada, 6/12–6/14), 1996.

[391] P. J. Clewlow, A. J. Haslegrave, N. Carruthers, W. M. Hedges, B. I. Bourland, D. S. Sullivan, and H. T. R. Montgomerie. Amine adducts as corrosion inhibitors. Patent: EP 520761, 1992.

[392] J. A. Coates, J. M. Farrar, and M. H. Graham. Fluid loss-reducing additives for oil-based well working fluids. Patent: US 4941983, 1990.

[393] A. Coats, A. Sas-Jaworsky, II, and Clergy. J. St. Magnesian cement, CT (coiled tubing) offer benefits. *Amer Oil Gas Reporter,* 39(3):92–93, March 1996.

[394] S. Cobianco, M. Bartosek, and A. Guarneri. Non-damaging drilling fluids. Patent: EP 1104798A, 2001.

[395] J. K. Coleman, M. J. Brown, V. Moses, and C. C. Burton. Enhanced oil recovery. Patent: WO 9215771, 1992.

[396] K. S. Colle. Method for inhibiting hydrate formation. Patent: US 6222083, 2001.

[397] I. R. Collins. Scale inhibition at high reservoir temperatures. In *Proceedings Volume.* IBC Tech Serv Ltd Advances in Solving Oilfield Scaling Int Conf (Aberdeen, Scotland, 11/20–11/21), 1995.

[398] M. W. Conway and R. A. Schraufnagel. The effect of fracturing fluid damage on production from hydraulically fractured wells. In *Proceedings Volume,* pages 229–236. Ala Univ et al Int Unconven Gas Symp (Intergas 95) (Tuscaloosa, AL, 5/14–5/20), 1995.

[399] R. F. Corbin and G. L. Ott. Federal region II regional contingency planning for a dispersant decision process. In *Proceedings Volume*, pages 417–420. 9th Bien API et al Oil Spill (Prev, Behav, Contr, Cleanup) Conf (Los Angeles, CA, 2/25–2/28), 1985.

[400] J. Cossar and J. Carlile. A new method for oilfield corrosion inhibitor measurement. In *Proceedings Volume*. Annu NACE Corrosion Conf (Corrosion 93) (New Orleans, LA, 3/7–3/12), 1993.

[401] C. A. Costello, E. Berluche, R. H. Oelfke, and L. D. Talley. Maleimide copolymers and method for inhibiting hydrate formation. Patent: US 5744665, 1998.

[402] C. A. Costello, T. W. Lai, and R. K. Pinschmidt, Jr. Aqueous, clay-based drilling mud. Patent: GB 2225364, 1990.

[403] T. L. Cottrell, W. D. Spronz, and W. C. Weeks, III. Hugoton infill program uses optimum stimulation technique. *Oil Gas J*, 86(28):88–90, 1988.

[404] J. Courtois-Sambourg, B. Courtois, A. Heyraud, P. Colin-Morel, and M. Rinaudo-Duhem. Polymer compounds of the glycuronic acid, method of preparation and utilization particularly as gelifying, thickening, hydrating, stabilizing, chelating or flocculating means. Patent: WO 9318174, 1993.

[405] P. V. Coveney, M. Watkinson, A. Whiting, and E. S. Boek. Stabilising clay formations. Patent: GB 2332221, 1999.

[406] P. V. Coveney, M. Watkinson, A. Whiting, and E. S. Boek. Stabilizing clay formations. Patent: WO 9931353, 1999.

[407] J. C. Cowan, V. M. Granquist, and R. F. House. Organophilic polyphenolic acid adducts. Patent: US 4737295, 1988.

[408] K. M. Cowan and L. Eoff. Surfactants: Additives to improve the performance properties of cement. In *Proceedings Volume*, pages 317–327. SPE Oilfield Chem Int Symp (New Orleans, LA, 3/2–3/5), 1993.

[409] K. M. Cowan and A. H. Hale. Restoring lost circulation. Patent: US 5325922, 1994.

[410] K. M. Cowan and A. H. Hale. High temperature well cementing with low grade blast furnace slag. Patent: US 5379840, 1995.

[411] K. M. Cowan, A. H. Hale, and J. J. W. Nahm. Dilution of drilling fluid in forming cement slurries. Patent: US 5314022, 1994.

[412] K. M. Cowan, J. J. W. Nahm, R. E. Wyant, and R. N. Romero. Alumina wellbore cement composition. Patent: US 5488991, 1996.

[413] K. M. Cowan and T. R. Smith. Application of drilling fluids to cement conversion with blast furnace slag in Canada. In *Proceedings Volume*, number 93-601. CADE/CAODC Spring Drilling Conf (Calgary, Canada, 4/14–4/16) Proc, 1993.

[414] M. Crabtree, D. Eslinger, P. Fletcher, M. Miller, A. Johnson, and G. King. Fighting scale—removal and prevention. *Oilfield Rev*, 11(3):30–45, 1999.

[415] D. Craig and S. A. Holditch. The degradation of hydroxypropyl guar fracturing fluids by enzyme, oxidative, and catalyzed oxidative breakers: Pt 2: Crosslinked hydroxypropyl guar gels: Topical report (January–April 1992). Gas Res Inst Rep GRI-93/04192, Gas Res Inst, December 1993.

[416] D. Craig and S. A. Holditch. The degradation of hydroxypropyl guar fracturing fluids by enzyme, oxidative, and catalyzed oxidative breakers: Pt 1: Linear hydroxypropyl guar solutions: Topical report (February–December 1991). Gas Res Inst Rep GRI-93/04191, Gas Res Inst, December 1993.

[417] D. Craig, S. A. Holditch, and B. Howard. The degradation of hydroxypropyl guar fracturing fluids by enzyme, oxidative, and catalyzed oxidative breakers. In *Proceedings Volume,* pages 1–19. 39th Annu Southwestern Petrol Short Course Assoc Inc et al Mtg (Lubbock, TX, 4/22–4/23), 1992.

[418] D. P. Craig. *The degradation of hydroxypropyl guar fracturing fluids by enzyme, oxidative, and catalyzed oxidative breakers.* PhD thesis, Texas A & M Univ, 1991.

[419] D. Crawford. High pressure high temperature (HPHT) fluid loss control aid for drilling fluids. Patent: WO 0026322, 2000.

[420] J. P. Crawshaw, P. W. Way, and M. Thiercelin. A method of stabilizing a wellbore wall. Patent: GB 2363810, 2002.

[421] S. C. Crema and C. H. Kucera. Cementing compositions containing a copolymer as a fluid loss control additive. Patent: EP 444489, 1991.

[422] S. C. Crema, C. H. Kucera, C. Gousetis, and K. Oppenlaender. Cementing compositions containing polyethyleneimine phosphonate derivatives as dispersants. Patent: EP 444542, 1991.

[423] S. C. Crema, C. H. Kucera, G. Konrad, and H. Hartmann. Fluid loss control additives for oil well cementing compositions. Patent: US 5025040, 1991.

[424] S. C. Crema, C. H. Kucera, G. Konrad, and H. Hartmann. Fluid loss control additives for oil well cementing compositions. Patent: US 5228915, 1993.

[425] I. J. Cronje. Process for the oxidation of fine coal. Patent: EP 298710, 1989.

[426] R. J. Crook and D. G. Calvert. Cement technology: Pt 1: Improvements in techniques and equipment address cement issues. *Oil Gas J,* 98(40):60–66, 2000.

[427] C. W. Crowe. Laboratory study provides guidelines for selecting clay stabilizers. In *Proceedings Volume,* volume 1. Cim Petrol Soc/SPE Int Tech Mtg (Calgary, Canada, 6/10–6/13), 1990.

[428] C. W. Crowe. Laboratory study provides guidelines for selecting clay stabilizers. SPE Unsolicited Pap SPE-21556, January 1991.

[429] C. W. Crowe. Laboratory study provides guidelines for selecting clay stabilizers. In *Proceedings Volume,* pages 499–504. SPE Oilfield Chem Int Symp (Anaheim, CA, 2/20–2/22), 1991.
[430] D. K. Crump and D. A. Wilson. Cement compositions containing set retarders. Patent: CA 1234582, 1988.
[431] J. A. Cruze and D. O. Hitzman. Microbial field sampling techniques for MEOR (microbial enhanced oil recovery) processes. US DOE Fossil-energy Rep NIPER-351 CONF-870858, September 1987.
[432] J. Cunningham, M. Rojo, K. J. Kooyoomjian, and J. M. Jordan. Decision-making on the use of dispersants: The role of the states. In *Proceedings Volume,* pages 353–356. API et al Oil Spill (Prev, Behav, Contr, Cleanup) 20th Anniv Conf (San Antonio, TX, 2/13–2/16), 1989.
[433] J. M. Cunningham, K. A. Sahatjian, C. Meyers, G. Yoshioka, and J. M. Jordan. Use of dispersants in the United States: Perception or reality? In *Proceedings Volume,* pages 389–393. 12th Bien API et al Oil Spill (Prev, Behav, Contr, Cleanup) Int Conf (San Diego, CA, 3/4–3/7), 1991.
[434] F. M. Cusack, J. W. Costerton, and J. Novosad. Ultramicrobacteria enhanced oil recovery. In *Proceedings Volume,* pages 491–504. 4th Inst Gas Technol Gas, Oil, & Environ Biotechnol Int Symp (Colorado Springs, CO, 12/9–12/11), 1991.
[435] A. Dadgar. Corrosion inhibitors for clear, calcium-free high density fluids. Patent: US 4784779, 1988.
[436] A. Dadgar. Corrosion inhibitors for clear, calcium-free high density fluids. Patent: EP 290486, 1988.
[437] A. Dadgar. Corrosion inhibitors for clear, calcium-free high density fluids. Patent: WO 8802433, 1988.
[438] M. Dahanayake, J. Li, R. L. Reierson, and D. J. Tracy. Amphoteric surfactants having multiple hydrophobic and hydrophilic groups. Patent: EP 697244, 1996.
[439] R. Daharu, S. Thomas, and Ali. S. M. Farouq. Micellar flooding for tertiary recovery: Recent advances and potential. In *Proceedings Volume,* pages 414–428. 10th SPE Trinidad & Tobago Sect Tech Conf (Port of Spain, Trinidad, 6/26–6/28), 1991.
[440] J. Dahl, K. Harris, and K. McKown. Uses of small particle size cement in water and hydrocarbon based slurries. In *Proceedings Volume,* pages 25–29. 9th Kansas Univ et al Tertiary Oil Recovery Conf (Wichita, KS, 3/6–3/7), 1991.
[441] H. Dakhlia. *A simulation study of polymer flooding and surfactant flooding using horizontal wells.* PhD thesis, Texas Univ, Austin, 1995.
[442] E. D. Dalrymple, J. A. Dahl, L. E. East, and K. W. McOwn. A selective water control process. In *Proceedings Volume,* pages 225–230. SPE Rocky Mountain Reg Mtg (Casper, WY, 5/18–5/21), 1992.

[443] S. V. Danjuschewskij and R. Ghofrani. Volume changes in cement slurries and set cements (Volumenänderungen erhärtender Zementschlämme). *Erdöl Erdgas Kohle,* 107(11):447–455, November 1991.
[444] J. W. Darden and E. E. McEntire. Dicyclopentadiene dicarboxylic acid salts as corrosion inhibitors. Patent: EP 200850, 1986.
[445] J. W. Darden and E. E. McEntire. Dicyclopentadiene dicarboxylic acid salts as corrosion inhibitors. Patent: CA 1264541, 1990.
[446] J. W. Darden, C. A. Triebel, W. A. Van Neste, and J. P. Maes. Monobasic-dibasic acid/salt antifreeze corrosion inhibitor. Patent: EP 229440, 1987.
[447] T. R. Dartez and R. K. Jones. Method for selectively treating wells with a low viscosity epoxy resin-forming composition. Patent: US 5314023, 1994.
[448] T. R. Dartez and R. K. Jones. Method for selectively treating wells with a low viscosity epoxy resin-forming composition. Patent: WO 9532354, 1995.
[449] B. L. Dasinger and H. A. I. McArthur. Aqueous gel compositions derived from succinoglycan. Patent: EP 251638, 1988.
[450] T. E. Daubert. Evaluation of four methods for predicting hydrate equilibria. GPA Res Rep RR-134, Pennsylvania State Univ, July 1992.
[451] E. Davidson. Defoamers. Patent: WO 9509900, 1995.
[452] E. Davidson. Method and composition for scavenging sulphide in drilling fluids. Patent: WO 0109039A, 2001.
[453] S. N. Davies, G. H. Meeten, and P. W. Way. Water based drilling fluid additive and methods of using fluids containing additives. Patent: US 5652200, 1997.
[454] B. W. Davis. In situ chemical stimulation of diatomite formations. Patent: CA 1308550, 1992.
[455] G. W. Davis, J. Chatterji, and R. L. Morgan. Set retarded cementing compositions and methods. Patent: CA 2212600, 1998.
[456] K. P. Davis, A. C. Smith, and M. J. Williams. Cement setting modifiers. Patent: GB 2327417, 1999.
[457] K. P. Davis, A. C. Smith, and M. J. Williams. Phosphonocarboxylic acids and their use as cement setting retarders. Patent: EP 899246, 1999.
[458] J. C. Dawson. Method and composition for delaying the gellation (gelation) of borated galactomannans. Patent: US 5082579, 1992.
[459] J. C. Dawson. Method and composition for delaying the gellation of borated gallactomannans. Patent: CA 2037974, 1992.
[460] J. C. Dawson. Method for delaying the gellation of borated galactomannans with a delay additive such as glyoxal. Patent: US 5160643, 1992.

[461] J. C. Dawson. Method and composition for delaying the gellation of borated gallactomannans. Patent: GB 2253868, 1995.
[462] J. C. Dawson and H. V. Le. Controlled degradation of polymer based aqueous gels. Patent: US 5447199, 1995.
[463] J. C. Dawson and H. V. Le. Gelation additive for hydraulic fracturing fluids. Patent: US 5798320, 1998.
[464] J. C. Dawson and H. V. Le. Gelation additive for hydraulic fracturing fluids. Patent: US 5773638, 1998.
[465] F. De Larrard. Ultrafine particles for the making of very high strength concretes. *Cement Concrete Res,* 19(2):161–172, March 1989.
[466] G. D. Dean, C. A. Nelson, S. Metcalf, R. Harris, and T. Barber. New acid system minimizes post acid stimulation decline rate in the Wilmington field, Los Angeles county, California. In *Proceedings Volume.* 68th Annu SPE West Reg Mtg (Bakersfield, CA, 5/10–5/13), 1998.
[467] F. E. Debons and L. E. Whittington. Improved oil recovery surfactants based on lignin. *J Petrol Sci Eng,* 7(1–2):131–138, April 1991.
[468] R. A. Deeb, H. Y. Hu, J. R. Hanson, K. M. Scow, and L. Alvarez-Cohen. Substrate interactions in BTEX [benzene, toluene, ethylbenzene, and xylene] and MTBE [methyl tert-butyl ether] mixtures by an MTBE-degrading isolate. *Environ Sci Technol,* 35(2):312–317, 2001.
[469] C. K. Deem, D. D. Schmidt, and R. A. Molner. Use of MMH (mixed metal hydroxide)/propylene glycol mud for minimization of formation damage in a horizontal well. In *Proceedings Volume,* number 91-29. 4th CADE/CAODC Spring Drilling Conf (Calgary, Canada, 4/10–4/12) Proc, 1991.
[470] H. J. Delhommer and C. O. Walker. Encapsulated oil absorbent polymers as lost circulation additives for oil based drilling fluids. Patent: US 4704213, 1987.
[471] H. J. Delhommer and C. O. Walker. Method for controlling lost circulation of drilling fluids with hydrocarbon absorbent polymers. Patent: US 4633950, 1987.
[472] R. M. Delton and M. Hooper. Rock bit grease composition. Patent: GB 2276884, 1994.
[473] V. N. Demikhov, M. M. Gilyazov, E. K. Trutneva, Y. A. Levin, A. P. Rakov, and I. M. Shermergorn. New 1-N-alkyl-1,2,4-triazole bromide bactericidal compounds—prepared by alkylation of 1,2,4- triazole with N-hexadecyl or N-octadecyl. Patent: SU 1776653-A, 1992.
[474] M. A. Denisov, V. A. Matishev, K. K. Safronov, L. A. Smirnova, V. L. Sverdlik, L. Sh. Andre, A. A. Bienko, L. P. Gilyazetdinov, S. A. Kononov, L. V. Kulygin, G. P. Malyatova, V. I. Nasteka, V. S. Shcherbakov, A. P. Sidorenko, N. S. Sirotkin, and S. A. Slyushchenko. Antifoaming composition—contains higher 4-20 C iso-alcohol

dialkyl phthalate, and 8-12 C iso-alcohol. Patent: SU 1775125-A, 1992.
[475] R. M. Denton and Z. Fang. Rock bit grease composition. Patent: US 5589443, 1996.
[476] R. Derr, E. A. Morris, III, and D. H. Pope. Fate and persistence of glutaraldehyde in a natural gas storage facility. In *Proceedings Volume*. 7th Inst Gas Technol Gas, Oil, & Environ Biotechnol Int Symp (Colorado Springs, CO, 12/12–12/14), 1994.
[477] R. M. Derr, E. A. Morris, and D. H. Pope. Applicability and efficacy of iodine as a mitigation strategy for advanced microbiologically influenced souring (MIS). In *Proceedings Volume*. 8th Inst Gas Technol Gas, Oil, & Environ Biotechnol Int Symp (Colorado Springs, CO, 12/11–12/13), 1995.
[478] S. R. Deshmukh, P. N. Chaturved, and R. P. Singh. *J Appl Polym Sci*, 30:4013, 1985.
[479] W. J. Detroit. Oil well drilling cement dispersant. Patent: US 4846888, 1989.
[480] W. J. Detroit. Nitric acid oxidized lignosulfonates. Patent: US 5446133, 1995.
[481] W. J. Detroit and M. E. Sanford. Solubilized lignosulfonate derivatives. Patent: US RE32895, 1989.
[482] B. Dewprashad. Method of producing coated proppants compatible with oxidizing gel breakers. Patent: US 5420174, 1995.
[483] O. C. Dias and M. C. Bromel. Microbially induced organic acid underdeposit attack in a gas pipeline. *Mater Performance*, 29(4):53–56, April 1990.
[484] H. Diaz-Arauzo. Phenolic resins and method for breaking crude oil emulsions. Patent: US 5460750, 1995.
[485] W. R. Dill. Diverting material and method of use for well treatment. Patent: CA 1217320, 1987.
[486] W. R. Dill, W. G. F. Ford, M. L. Walker, and R. D. Gdanski. Treatment of iron-containing subterranean formations. Patent: EP 258968, 1988.
[487] R. L. Dillenbeck and S. G. Nelson. New clay-controlling technology addresses problems encountered in cement design and performance. In *Proceedings Volume*, pages 641–647. SPE Rocky Mountain Reg Mtg (Casper, WY, 5/18–5/21), 1992.
[488] R. L. Dillenbeck, S. G. Nelson, B. E. Hall, and D. S. Porter. Clay control additive for cement compositions and method of cementing. Patent: US 5232497, 1993.
[489] G. F. Dilullo Arias, P. J. Rae, and D. T. Müller. Multi-functional additive for use in well cementing. Patent: WO 9916723, 1999.
[490] A. Dindi, R. L. Johnston, Y. N. Lee, and D. F. Massouda. Slurry drag reducer. Patent: US 5539044, 1996.

[491] D. Dino and J. Thompson. Organophilic clay additives and oil well drilling fluids with less temperature dependent rheological properties containing said additives. Patent: EP 1138740A, 2001.

[492] D. J. Dino. Modified polygalactomannans as oil field shale inhibitors. Patent: US 5646093, 1997.

[493] D. J. Dino and A. Homack. Use of high purity imidazoline based amphoacetate surfactant as foaming agent in oil wells. Patent: US 5614473, 1997.

[494] R. N. Diyashev, F. M. Sattarova, K. G. Mazitov, V. M. Khusainov, F. N. Mannanov, I. R. Diyashev, and V. A. Burtov. Extraction of oil from lens-shaped deposits—involves cyclic and portion-wise pumping-in of solutions of potassium carbonate and inhibited hydrochloric acid. Patent: RU 2065942-C, 1996.

[495] R. N. Diyashev, F. M. Sattarova, K. G. Mazitov, V. M. Khusainov, K. I. Sulejmanov, G. S. Karimov, and I. R. Diyashev. Recovering oil not exploited from reservoir—by injecting alternating portions of ammonium carbonate and hydrochloric acid and displacing formed carbon dioxide with water. Patent: RU 2065940-C, 1996.

[496] J. B. Dobbs and J. M. Brown. An environmentally friendly scale inhibitor. NACE Int Corrosion Conf (Corrosion 99) (San Antonio, TX, 4/25–4/30), 1999.

[497] B. E. Dobroskok, Z. G. Gulyaeva, N. N. Kubareva, R. Kh. Muslimov, S. A. Nizova, V. V. Terekhova, E. E. Yanchenko, A. B. Zezin, V. A. Kabanov, R. Kh. Musabirov, and S. P. Valueva. Plugging composition for hydro-insulation of oil stratum—contains polydimethyl-diallyl ammonium chloride, sodium salt of carboxy methyl cellulose, sodium chloride and water. Patent: SU 1758209-A, 1992.

[498] J. W. Dobson and K. B. Mondshine. Method of reducing fluid loss of well drilling and servicing fluids. Patent: EP 758011, 1997.

[499] J. W. Dobson and T. C. Mondshine. Well drilling and servicing fluids which deposit an easily removable filter cake. Patent: EP 672740, 1995.

[500] J. W. Dobson, Jr., J. C. Harrison, III, and P. D. Kayga. Methods of reducing fluid loss and polymer concentration of well drilling and servicing fluids. Patent: AU 697559, 1998.

[501] J. W. Dobson, Jr. and P. D. Kayga. Magnesium peroxide breaker system improves filter cake removal. *Petrol Eng Int,* 68(10):49–50, October 1995.

[502] D. H. Doherty, D. M. Ferber, J. D. Marrelli, R. W. Vanderslice, and R. A. Hassler. Genetic control of acetylation and pyruvylation of xanthan based polysaccharide polymers. Patent: WO 9219753, 1992.

[503] S. Doleschall, G. Milley, and T. Paal. Control of clays in fluid reservoirs. In *Proceedings Volume,* pages 803–812. 4th BASF AG et al

Enhanced Oil Recovery Europe Symp (Hamburg, Germany, 10/27–10/29), 1987.
[504] S. I. Dolganskaya and A. U. Sharipov. Removal from hole of junk and stuck drilling equipment—by dissolving latter with mixture of nitric and hydrochloric acids with added sodium nitrate and ethanolamine. Patent: SU 1782271-A, 1992.
[505] E. C. Donaldson and T. Obeida. Enhanced oil recovery at simulated reservoir conditions. In E. C. Donaldson, editor, *Microbial enhancement of oil recovery: Recent advances: Proceedings of the 1990 International Conference on Microbial Enhancement of Oil Recovery*, volume 31 of *Developments in Petroleum Science,* pages 227–245. Elsevier Science Ltd, 1991.
[506] A. Donche, A. Vaussard, and P. Isambourg. Application of scleroglucane slurries to the drilling of deflected wells. Patent: WO 9211340, 1992.
[507] A. Donche, A. Vaussard, and P. Isambourg. Use of scleroglucane muds for drilling deviated wells (application des boues au scleroglucane au forage des puits devies). Patent: FR 2670794, 1992.
[508] A. Donche, A. Vaussard, and P. Isambourg. Application of scleroglucane slurries to the drilling of deflected wells (application des boues au scleroglucane au forage des puits devies). Patent: EP 521120, 1993.
[509] A. Donche, A. Vaussard, and P. Isambourg. Application of scleroglucan muds to drilling deviated wells. Patent: US 5330015, 1994.
[510] J. G. Doolan and C. A. Cody. Pourable water dispersible thickening composition for aqueous systems and a method of thickening said aqueous systems. Patent: US 5425806, 1995.
[511] J. A. Dougherty, B. T. Outlaw, and B. A. Oude Alink. Corrosion inhibition by ethoxylated fatty amine salts of maleated unsaturated acids. Patent: US 5582792, 1996.
[512] N. J. E. Dowling, J. Guezennec, and D. C. White. Facilitation of corrosion of stainless steel exposed to aerobic seawater by microbial biofilms containing both facultative and absolute anaerobes. In *Proceedings Volume*. Inst Petrol Microbiol Comm Microbial Problems in the Offshore Oil Ind Int Conf (Aberdeen, Scotland, 4/15–4/17), 1986.
[513] A. B. Downey, G. L. Willingham, and V. S. Frazier. Compositions comprising 4,5-dichloro-2-N-octyl-3-isothiazolone and certain commercial biocides. Patent: EP 680695, 1995.
[514] I. Drela, P. Falewicz, and S. Kuczkowska. New rapid test for evaluation of scale inhibitors. *Water Research,* 32(10):3188–3191, October 1998.
[515] E. Dreveton, J. Lecourtier, D. Ballerini, and L. Choplin. Process utilizing gellan as filtrate reducer for water-based drilling fluids (procede utilisant le gellane comme reducteur de filtrat pour les fluides de forage a base d'eau). Patent: EP 662563, 1995.

[516] E. Dreveton, J. Lecourtier, D. Ballerini, and L. Choplin. Process using gellan as a filtrate reducer for water-based drilling fluids. Patent: US 5744428, 1998.

[517] V. Drillet and D. Defives. Clay dissolution kinetics in relation to alkaline flooding. In *Proceedings Volume,* pages 317–326. SPE Oilfield Chem Int Symp (Anaheim, CA, 2/20–2/22), 1991. SPE Number: 21030.

[518] O. Dugstad, T. Bjornstad, and I. Hundere. Measurements and application of partition coefficients of compounds suitable for tracing gas injected into oil reservoirs. *Rev Inst Franc Petrol,* 47(2):205–215, March-April 1992.

[519] J. J. Duhon, Sr. Olive pulp additive in drilling operations. Patent: US 5801127, 1998.

[520] V. Kh. Dulaev, N. M. Bondarets, N. A. Polukhina, V. I. Petresku, R. G. Galiev, P. P. Kapustin, and O. S. Matrosov. Composition for preparation of aerated plugging solution—contains Portland cement, oxyethylated monoalkylphenol(s) of propylene trimers, glycol mixture, air, water, etc. Patent: SU 1745893-A, 1992.

[521] G. Duncan. Enhanced recovery engineering: Pt 1. *World Oil,* 215(9):95,97–100, September 1994.

[522] G. Duncan and P. Bulkowski. Enhanced recovery engineering: Pt 7. *World Oil,* 216(9):77–84, September 1995.

[523] S. N. Duncum, A. R. Edwards, A. R. Lucy, and C. G. Osborne. Method for inhibiting solids formation and blends for use therein. Patent: WO 9424413, 1994.

[524] S. N. Duncum, A. R. Edwards, and C. G. Osborne. Method for inhibiting hydrate formation. Patent: EP 536950, 1993.

[525] S. N. Duncum, P. K. G. Hodgson, K. James, and C. G. Osborne. Inhibitors and their uses in oils. Patent: WO 9821446, 1998.

[526] S. N. Duncum, K. James, and C. G. Osborne. Wax deposit inhibitors. Patent: GB 2323095, 1998.

[527] S. N. Duncum, K. James, and C. G. Osborne. Wax deposit inhibitors. Patent: US 6140276, 2000.

[528] C. Durand, A. Onaisi, A. Audibert, T. Forsans, and C. Ruffet. Influence of clays on borehole stability: A literature survey: Pt 2: Mechanical description and modelling of clays and shales drilling practices versus laboratory simulations. *Rev Inst Franc Petrol,* 50(3):353–369, May-June 1995.

[529] C. Durand, A. Onaisi, A. Audibert, T. Forsans, and C. Ruffet. Influence of clays on borehole stability: A literature survey: Pt 1: Occurrence of drilling problems physico-chemical description of clays and of their interaction with fluids. *Rev Inst Franc Petrol,* 50(2):187–218, March-April 1995.

References 383

[530] J. P. Durand, A. S. Baley, P. Gateau, and A. Sugier. Process to reduce the tendency of hydrates to agglomerate in production effluents (methode pour reduire la tendance a l'agglomeration des hydrates dans des effluents de production). Patent: EP 582507, 1994.
[531] D. K. Durham, U. C. Conkle, and H. H. Downs. Additive for clarifying aqueous systems. Patent: GB 2219291, 1989.
[532] A. M. Durr, Jr., J. Huycke, H. L. Jackson, B. J. Hardy, and K. W. Smith. An ester base oil for lubricant compounds and process of making an ester base oil from an organic reaction by-product. Patent: EP 606553, 1994.
[533] W. Dye, D. E. Clark, and R. G. Bland. Well fluid additive. Patent: EP 652271, 1995.
[534] S. J. Dyer, G. M. Graham, and K. S. Sorbie. Factors affecting the thermal stability of conventional scale inhibitors for application in high pressure/high temperature reservoirs. In *Proceedings Volume,* pages 167–177. SPE Oilfield Chem Int Symp (Houston, TX, 2/16–2/19), 1999.
[535] I. S. Dzhafarov, S. V. Brezitsky, A. K. Shakhverdiev, G. M. Panakhov, and B. A. Suleimanov. New in-situ carbon dioxide generation enhanced oil recovery technology. In *Proceedings Volume,* number 106. 10th EAGE Impr Oil Recovery Europe Symp (Brighton, UK, 8/18–8/20), 1999.
[536] S. D. Dzhanakhmedova, E. I. Pryanikov, S. A. Sulejmanova, K. K. Mamedov, E. G. Dubrovina, N. M. Indyukov, and A. B. Sulejmanov. Composition for preventing asphaltene-resin-paraffin deposits—contains waste from production of synthetic glycerine, in mixture with polyacrylamide. Patent: SU 1761772-A, 1992.
[537] R. G. Eagar, J. Leder, J. P. Stanley, and A. B. Theis. The use of glutaraldehyde for microbiological control in waterflood systems. *Mater Performance,* 27(8):40–45, August 1988.
[538] G. B. Eaton and M. J. Monahan. Composition of and process for forming polyalphaolefin drag reducing agents. Patent: US 5869570, 1999.
[539] P. Eaton and G. Sutton. The effect of flow on inhibitor film life. In *Proceedings Volume*. 49th Annu NACE Int Corrosion Conf (Corrosion 94) (Baltimore, MD, 2/27–3/4), 1994.
[540] C. D. Ebinger and E. Hunt. Keys to good fracturing: Pt 6: New fluids help increase effectiveness of hydraulic fracturing. *Oil Gas J,* 87(23):52–55, 1989.
[541] J. B. Egraz, H. Grondin, and J. M. Suau. Acrylic copolymer partially or fully soluble in water, cured or not and its use (copolymere acrylique partiellement ou totalement hydrosoluble, reticule ou non et son utilisation). Patent: EP 577526, 1994.

[542] V. S. Ekshibarov and T. R. Khasanov. Oil well casing cement—containing 20 to 60 weight per cent of lightweight additive by-product from hydraulic washing of brown coal. Patent: SU 1731939-A, 1992.

[543] I. M. El-Gamal and E. A. M. Gad. Low temperature rheological behavior of umbarka waxy crude and influence of flow improver. *Rev Inst Franc Petrol*, 52(3):369–379, May-June 1997.

[544] J. L. Elbel, R. C. Navarrete, and B. D. Poe, Jr. Production effects of fluid loss in fracturing high-permeability formations. In *Proceedings Volume*, pages 201–211. SPE Europe Formation Damage Contr Conf (The Hague, Netherlands, 5/15–5/16), 1995.

[545] M. Elboujdaini and V. S. Sastri. Field studies of microbiological corrosion in water injection plant. In *Proceedings Volume*. 50th Annu NACE Int Corrosion Conf (Corrosion 95) (Orlando, FL, 3/26–3/31), 1995.

[546] G. Elfers, W. Sager, H. H. Vogel, and K. Oppenlaender. Petroleum emulsion breakers. Patent: CA 2082287, 1993.

[547] G. Elfers, W. Sager, H. H. Vogel, and K. Oppenlaender. Petroleum emulsion breakers. Patent: US 5445765, 1995.

[548] G. Elfers, W. Sager, H. H. Vogel, and K. Oppenlaender. Demulsifier based on an alkoxylate and process for the preparation of the alkoxylate (Erdölemulsionsspalter auf der Basis eines Alkoxilats und Verfahren zur Herstellung Dieses Alkoxylats). Patent: EP 549918, 1996.

[549] P. D. Ellis and B. W. Surles. Chemically inert resin coated proppant system for control of proppant flowback in hydraulically fractured wells. Patent: US 5604184, 1997.

[550] E. A. Elphingstone and F. B. Woodworth. Dry biocide. Patent: US 6001158, 1999.

[551] J. W. Ely. *Fracturing fluids and additives*, volume 12 of *Recent advances in hydraulic fracturing (SPE Henry L Doherty Monogr Ser)*. SPE, Richardson, TX, 1989.

[552] J. W. Ely. How intense quality control improves hydraulic fracturing. *World Oil*, 217(11):59–60, 62–65, 68, November 1996.

[553] J. W. Ely, B. C. Wolters, and S. A. Holditch. Improved job execution and stimulation success using intense quality control. In *Proceedings Volume*, pages 101–114. 37th Annu Southwestern Petrol Short Course Assoc et al Mtg (Lubbock, TX, 4/18–4/19), 1990.

[554] D. H. Emmons. Sulfur deposition reduction. Patent: US 5223160, 1993.

[555] D. P. Enright, W. M. Dye, F. M. Smith, and A. C. Perricone. Drilling fluid methods and composition. Patent: US 5007489, 1991.

[556] M. V. Enzien, D. H. Pope, M. M. Wu, and J. Frank. Nonbiocidal control of microbiologically influenced corrosion using organic film-forming inhibitors. In *Proceedings Volume*. 51st Annu NACE Int Corrosion Conf (Corrosion 96) (Denver, CO, 3/24–3/29), 1996.

[557] L. Eoff. Set retarding additives, cement compositions and methods. Patent: US 5264470, 1993.
[558] L. Eoff. Acetone/formaldehyde/cyanide resins. Patent: US 5290357, 1994.
[559] L. Eoff. Dispersant for cement in water. Patent: EP 604000, 1994.
[560] L. Eoff, J. Chatterji, A. Badalamenti, and D. McMechan. Water-dispersible resin system for wellbore stabilization. In *Proceedings Volume*. SPE Oilfield Chem Int Symp (Houston, TX, 2/13–2/16), 2001.
[561] B. L. Estes and C. J. Bernu. New and improved drilling fluids and additives therefor. Patent: WO 9951701, 1999.
[562] B. Evans and S. Ali. Selecting brines and clay stabilizers to prevent formation damage. *World Oil*, 218(5):65–68, May 1997.
[563] A. L. Fabrichnaya, Y. V. Shamraj, R. G. Shakirzyanov, Z. Kh. Sadriev, V. N. Koshelev, L. P. Vakhrushev, and A. E. Tavrin. Additive for drilling solutions with high foam extinguishing properties—containing specified surfactant in hydrocarbon solvent with methyl-diethyl-alkoxymethyl ammonium methyl sulphate. Patent: RU 2091420-C, 1997.
[564] K. Faessler. Testing of corrosion inhibitors under pressure conditions in the presence of H_2S and CO_2 (Prüfung von Korrosionsinhibitoren Unter Druckbedingungen in Gegenwart von H_2S und CO_2). In *Proceedings Volume*. BASF et al Chem Prod in Petrol Prod Mtg H_2S—A Hazardous Gas in Crude Oil Recovery Discuss (Clausthal-Zellerfeld, Germany, 9/12–9/13), 1990.
[565] K. Fairchild, R. Tipton, J. F. Motier, and N. S. Kommareddi. Low viscosity, high concentration drag reducing agent and method therefor. Patent: WO 9701582, 1997.
[566] A. M. Fakhriev, F. R. Ismagilov, and M. M. Latypova. Odorant for imparting smell to natural gas—contains a mixture of ethyl-, propyl-, butyl- and amyl-mercaptans, and additionally prescribed dialkyldisulphides. Patent: RU 2041243-C, 1995.
[567] A. M. Fakhriev, M. M. Latypova, F. R. Ismagilov, and P. G. Navalikhin. Odorisation of liquefied hydrocarbon gases—using waste from oxidising demercaptanisation of light hydrocarbon material. Patent: RU 2000313-C, 1993.
[568] A. M. Fakhriev, M. M. Latypova, V. I. Nasteka, A. I. Berdnikov, and V. Ya. Klimov. Odorising agent for compressed hydrocarbon gas—contains ethyl-mercaptan or mixed mercaptans, and additionally waste from process of oxidising de-mercaptanisation of light hydrocarbons. Patent: RU 2009178-C, 1994.
[569] J. Falbe and M. Regitz. *CD Römpp Chemie-Lexikon, Version 1*. Thieme Verlag, Stuttgart, Germany, 1995.

[570] R. J. Falkiner and M. A. Poirier. Method for reducing elemental sulfur pick-up by hydrocarbon fluids in a pipeline. Patent: CA 2158789, 1996.
[571] G. Fan, L. P. Koskan, and R. J. Ross. Polyaspartates: An emerging green technology in oil production. In *Book of Abstr (ACS),* volume 1. 215th ACS Nat Mtg (Dallas, TX, 3/29–4/2), 1998.
[572] S. Fan, S. Yuan, and Y. Ye. A laboratory study of adhesion agent of er type for casing cementing. *J Jianghan Petrol Inst,* 18(2):77–80, June 1996.
[573] Z. Fang, S. Peterson, and R. Denton. O-ring seal with lubricant additives for rock bit bearings. Patent: GB 2318139, 1998.
[574] E. E. Fant. Odorization—a regulatory perspective. In *Odorization,* volume 3, pages 109–118. Institute of Gas Technology, Chicago, IL, 1993.
[575] G. B. Farquhar. A review of trends in MIC (microbiologically influenced corrosion). *Mater Performance,* 32(1):53–55, January 1990.
[576] D. Farrar, M. Hawe, and B. Dymond. Use of water soluble polymers in aqueous drilling or packer fluids and as detergent builders. Patent: EP 182600, 1992.
[577] R. R. Fay, C. P. Giammona, K. Binkley, and F. R. Engelhardt. Measuring the aerial application of oil dispersant from very large aircraft at moderate altitude. In *Proceedings Volume,* volume 2, pages 1057–1063. 16th Environ Can Arctic & Mar Oil Spill Program Tech Semin (Calgary, Canada, 6/7–6/9), 1993.
[578] M. S. Felix. A surface active composition containing an acetal or ketal adduct. Patent: WO 9600253, 1996.
[579] J. Felixberger. Mixed metal hydroxides (MMH)—an inorganic thickener for water-based drilling muds (mixed metal hydroxide [MMH]—Ein anorganisches Verdickungsmittel für wasserbasierte Bohrspülungen). In *Proceedings Volume,* pages 339–351. DMGK Spring Conf (Celle, Germany, 4/25–4/26), 1996.
[580] J. T. Fenton and J. F. Miller. Biocide for petroleum operations. Patent: US 4911923, 1990.
[581] J. P. Feraud, H. Perthuis, and P. Dejeux. Compositions for iron control in acid treatments for oil wells. Patent: US 6306799, 2001.
[582] M. Feustel, M. Krull, and H. J. Oschmann. Additives for improving the cold flow properties and the storage stability of crude oil (Additive zur Verbesserung von Kaltfließeigenschaften und Lagerstabilität von Rohölen). Patent: WO 0196503, 2001.
[583] V. G. Fil, L. A. Ezlova, V. D. Kovalenko, D. A. Kostenko, B. I. Navrotskij, V. V. Koptenko, and S. P. Dombrovskaya. Plugging solution for oil and gas wells—comprises Portland cement, asbestos reinforcing component, mixture of sodium chloride and sulphate mineral component, fly ash and water. (rus). Patent: RU 2013525-C, 1994.

[584] M. Fingas. Water-in-oil emulsion formation: A review of physics and mathematical modelling. *Spill Sci Technol Bull,* 2(1):55–59, March 1995.

[585] M. Fingas, I. Bier, M. Bobra, and S. Callaghan. Studies on the physical and chemical behavior of oil and dispersant mixtures. In *Proceedings Volume,* pages 419–426. 12th Bien API et al Oil Spill (Prev, Behav, Contr, Cleanup) Int Conf (San Diego, CA, 3/4–3/7), 1991.

[586] M. Fingas, B. Fieldhouse, I. Bier, D. Conrod, and E. Tennyson. Development of a test for water-in-oil emulsion breakers. In *Proceedings Volume,* volume 2, pages 909–954. 16th Environ Can Arctic & Mar Oil Spill Program Tech Semin (Calgary, Canada, 6/7–6/9), 1993.

[587] M. F. Fingas, B. Kolokowski, and E. J. Tennyson. Study of oil spill dispersants effectiveness and physical studies. In *Proceedings Volume,* pages 265–287. 13th Environ Can Arctic & Mar Oil Spill Program Tech Semin (Edmonton, Canada, 6/6–6/8), 1990.

[588] M. F. Fingas, D. Kyle, and E. Tennyson. Dispersant effectiveness: Studies into the causes of effectiveness variations. In *Proceedings Volume,* pages 92–132. 2nd ASTM Use of Chem in Oil Spill Response Symp (Victoria, Canada, 10/10–10/11), 1994.

[589] M. F. Fingas, D. A. Kyle, J. B. Holmes, and E. J. Tennyson. The effectiveness of dispersants: Variation with energy. In *Proceedings Volume,* pages 567–574. 13th Bien API et al Oil Spill (Prev, Preparedness, Response & Coop) Int Conf (Tampa, FL, 3/29–4/1), 1993.

[590] M. F. Fingas, D. A. Kyle, P. Lambert, Z. Wang, and J. V. Mullin. Analytical procedures for measuring oil spill dispersant effectiveness in the laboratory. In *Proceedings Volume,* volume 1, pages 339–354. 18th Environ Can Arctic & Mar Oilspill Program Tech Semin (Edmonton, Canada, 6/14–6/16), 1995.

[591] M. F. Fingas, D. A. Kyle, N. Laroche, B. Fieldhouse, G. Sergy, and G. Stoodley. The effectiveness testing of oil spill-treating agents. In *Proceedings Volume,* pages 286–298. 2nd ASTM Use of Chem in Oil Spill Response Symp (Victoria, Canada, 10/10–10/11), 1994.

[592] M. F. Fingas, D. A. Kyle, and E. J. Tennyson. Physical and chemical studies on dispersants: The effect of dispersant amount and energy. In *Proceedings Volume,* volume 2, pages 861–876. 16th Environ Can Arctic & Mar Oil Spill Program Tech Semin (Calgary, Canada, 6/7–6/9), 1993.

[593] M. F. Fingas, D. A. Kyle, Z. Wang, D. Handfield, D. Ianuzzi, and F. Ackerman. Laboratory effectiveness testing of oil spill dispersants. In *Proceedings Volume,* pages 3–40. 2nd ASTM Use of Chem in Oil Spill Response Symp (Victoria, Canada, 10/10–10/11), 1994.

[594] M. F. Fingas, D. L. Munn, B. White, R. G. Stoodley, and I. D. Crerar. Laboratory testing of dispersant effectiveness: The importance of

oil-to-water ratio and settling time. In *Proceedings Volume,* pages 365–373. API et al Oil Spill (Prev, Behav, Contr, Cleanup) 20th Anniv Conf (San Antonio, TX, 2/13–2/16), 1989.

[595] M. F. Fingas, R. Stoodley, and N. Laroche. Effectiveness testing of spill-treating agents. *Oil Chem Pollut,* 7(4):337–348, 1990.

[596] M. F. Fingas, R. Stoodley, N. Stone, R. Hollins, and I. Bier. Testing the effectiveness of spill-treating agents: Laboratory test development and initial results. In *Proceedings Volume,* pages 411–414. 12th Bien API et al Oil Spill (Prev, Behav, Contr, Cleanup) Int Conf (San Diego, CA, 3/4–3/7), 1991.

[597] M. F. Fingas, E. Tennyson, B. Fieldhouse, I. Bier, and D. Conrod. Laboratory effectiveness testing of water-in-oil emulsion breakers. In *Proceedings Volume,* pages 41–54. 2nd ASTM Use of Chem in Oil Spill Response Symp (Victoria, Canada, 10/10–10/11), 1994.

[598] M. Fink and J. Fink. Usage of pyrolysis products from organic materials to improve recovery of crude oil. In *Extended Abstr Volume,* volume 2, page P553. 60th EAGE Conf (Leipzig, Germany, 6/8-6/12), 1998.

[599] R. J. Fiocco, G. P. Canevari, J. B. Wilkinson, J. Bock, M. Robbins, H. O. Jahns, and R. K. Markarian. Development of Corexit 9580—a chemical beach cleaner. In *Proceedings Volume,* pages 395–400. 12th Bien API et al Oil Spill (Prev, Behav, Contr, Cleanup) Int Conf (San Diego, CA, 3/4–3/7), 1991.

[600] R. J. Fiocco, R. R. Lessard, and G. P. Canevari. Improving oiled shoreline cleanup with Corexit 9580. In *Proceedings Volume,* pages 276–280. ASME/API Energy Week 96 Conf (Houston, TX, 1/29–2/2), 1996.

[601] R. J. Fiocco, R. R. Lessard, G. P. Canevari, K. W. Becker, and P. S. Daling. The impact of oil dispersant solvent on performance. In *Proceedings Volume,* pages 299–309. 2nd ASTM Use of Chem in Oil Spill Response Symp (Victoria, Canada, 10/10–10/11), 1994.

[602] E. R. Fischer and J. E. Parker, III. Tall oil fatty acid anhydrides as corrosion inhibitor intermediates. In *Proceedings Volume.* 50th Annu NACE Int Corrosion Conf (Corrosion 95) (Orlando, FL, 3/26–3/31), 1995.

[603] E. R. Fischer and J. E. Parker, III. Tall oil fatty acid anhydrides as corrosion inhibitor intermediates. *Corrosion,* 53(1):62–64, January 1997.

[604] G. C. Fischer. Corrosion inhibitor compositions containing inhibitor prepared from amino substituted pyrazines and epoxy compounds. Patent: US 4895702, 1990.

[605] L. E. Fisher. Corrosion inhibitors and neutralizers: Past, present and future. In *Proceedings Volume.* Annu NACE Corrosion Conf (Corrosion 93) (New Orleans, LA, 3/7–3/12), 1993.

References 389

[606] T. E. Fisk and C. J. Tucker. N-(Hydrophobe aromatic)pyridinium compounds. Patent: US 5000873, 1991.

[607] P. H. Fitzgerald, N. O. Wolf, C. R. Clark, and D. P. Cords. Emulsion breaking using alkylphenol-polyethylene oxide-acrylate polymer coated coalescer material. Patent: US 5156767, 1992.

[608] J. J. Fitzgibbon. Use of uncalcined/partially calcined ingredients in the manufacture of sintered pellets useful for gas and oil well proppants. Patent: US 4623630, 1986.

[609] J. J. Fitzgibbon. Sintered, spherical, composite pellets prepared from clay as a major ingredient useful for oil and gas well proppants. Patent: CA 1232751, 1988.

[610] J. J. Fitzgibbon. Sintered spherical pellets containing clay as a major component useful for gas and oil well proppants. Patent: US 4879181, 1989.

[611] L. M. Flaherty, W. B. Katz, and S. Kaufmann. Dispersant use guidelines for freshwater and other inland environments. In *Proceedings Volume*, pages 25–30. Amer Soc Testing Mater Oil Dispersants: New Ecol Approaches Symp (Williamsburg, VA, 10/12–10/14), 1987.

[612] J. K. Fleming and H. C. Fleming. Invert emulsion drilling mud. Patent: WO 9504788, 1995.

[613] F. Fleyfel, K. Y. Song, A. Kook, R. Martin, and R. Kobayashi. ^{13}C NMR of hydrate precursors in metastable regions. In *Proceedings Volume*, pages 212–224. New York Acad Sci et al Natur Gas Hydrates Int Conf (New Paltz, NY, 6/20–6/24), 1993.

[614] J. Fock, E. Esselborn, and W. Hoehner. Use of copolymers of polyoxyalkylene ethers of allyl or methallyl alcohols and acrylic or methacrylic esters as emulsion breakers for crude oil containing water (Verwendung von Copolymerisaten von Polyoxyalkylenethern des Allyl- und/oder Methallylalkohols und Acryl- oder Methacrylestern als Dismulgatoren für Wasser enthaltendes Erdöl). Patent: DE 3513550, 1986.

[615] J. Fock, E. Esselborn, and W. Hohner. Copolymers of polyoxyalkylene ethers and acrylic or methacrylic esters and their use as demulsifiers for petroleum containing water. Patent: GB 2174096, 1986.

[616] D. W. Fodge, D. M. Anderson, and T. M. Pettey. Hemicellulase active at extremes of pH and temperature and utilizing the enzyme in oil wells. Patent: US 5551515, 1996.

[617] D. W. Fong and A. M. Halverson. Removal of dispersed oil from water. Patent: US 4802992, 1989.

[618] D. W. Fong and B. S. Shambatta. Hydroxamic acid containing polymers used as corrosion inhibitors. Patent: CA 2074535, 1993.

[619] D. W. Fong, C. F. Marth, and R. V. Davis. Sulfobetaine-containing polymers and their utility as calcium carbonate scale inhibitors. Patent: US6225430, 2001.

[620] W. G. Ford. Reducing sludging during oil well acidizing. Patent: US 4981601, 1991.
[621] W. G. Ford and K. H. Hollenbeak. Composition and method for reducing sludging during the acidizing of formations containing sludging crude oils. Patent: US 4663059, 1987.
[622] W. G. Ford and K. H. Hollenbeak. Composition and method for reducing sludging during the acidizing of formations containing sludging crude oils. Patent: CA 1280585, 1991.
[623] W. G. F. Ford. Reducing sludging during oil well acidizing. Patent: US 4823874, 1989.
[624] W. G. F. Ford. Reducing sludging during oil well acidizing. Patent: CA 1318122, 1993.
[625] G. T. Forrest. Drilling, completion, and workover fluid comprising ground peanut hulls. Patent: EP 873379, 1998.
[626] G. T. Forrest. Drilling, completion, and workover fluid comprising ground peanut hulls. Patent: US 5806592, 1998.
[627] G. T. Forrest. Drilling, completion, and workover fluid comprising ground peanut hulls. Patent: WO 9821290, 1998.
[628] J. W. Forsberg and R. W. Jahnke. Methods of drilling well boreholes and compositions used therein. Patent: US 5260268, 1993.
[629] J. W. Forsberg and R. W. Jahnke. Methods of drilling well boreholes and compositions used therein. Patent: WO 9302151, 1993.
[630] C. L. Fortenberry, Jr., N. J. Grahmann, C. D. Miller, and A. J. Son. Analysis of residual corrosion inhibitors in oilfield brines. In *Proceedings Volume,* pages 965–979. 68th Annu SPE Tech Conf (Houston, TX, 10/3–10/6), 1993.
[631] H. W. Frago and R. D. Powell. Re-use of certain oil exploration and production waste as supplemental fuel in the manufacture of cement. In *Abstracts Volume,* page 151. Annu Aapg-Sepm-Emd-Dpa-Deg Conv (Denver, CO, 6/12–6/15), 1994.
[632] H. P. Francis, E. D. Deboer, and V. L. Wermers. High temperature drilling fluid component. Patent: US 4652384, 1987.
[633] M. J. Franklin, D. E. Nivens, A. A. Vass, M. W. Mittelman, R. F. Jack, M. J. E. Dowling, and D. C. White. Effect of chlorine and chlorine/bromine biocide treatments on the number and activity of biofilm bacteria and on carbon steel corrosion. *Corrosion,* 47(2):128–134, February 1991.
[634] J. P. Fraser. Advance planning for dispersant use/non-use. In *Proceedings Volume,* pages 429–432. 9th Bien API et al Oil Spill (Prev, Behav, Contr, Cleanup) Conf (Los Angeles, CA, 2/25–2/28), 1985.
[635] J. P. Fraser. Methods for making dispersant use decisions. In *Proceedings Volume,* pages 321–330. API et al Oil Spill (Prev, Behav, Contr, Cleanup) 20th Anniv Conf (San Antonio, TX, 2/13–2/16), 1989.

[636] J. P. Fraser, S. A. Horn, L. J. Kazmierczak, M. L. Kinworthy, A. H. Lasday, and J. Lindstedt-Siva. Guidelines for use of dispersants on spilled oil—a model plan. In *Proceedings Volume*, pages 331–332. API et al Oil Spill (Prev, Behav, Contr, Cleanup) 20th Anniv Conf (San Antonio, TX, 2/13–2/16), 1989.

[637] E. C. French, W. F. Fahey, and J. G. Harte. Method of oil well corrosion inhibition via emulsions and emulsions therefor. Patent: CA 2019516, 1991.

[638] E. C. French, W. F. Fahey, and J. G. Harte. Method of oil well corrrosion inhibition via emulsions and emulsions therefor. Patent: US 5027901, 1991.

[639] T. R. French. Design and optimization of phosphate-containing alkaline flooding formulations: Topical report. US DOE Fossil Energy Rep NIPER-446, NIPER, February 1990.

[640] T. R. French and C. B. Josephson. Surfactant-enhanced alkaline flooding with weak alkalis. US DOE Rep NIPER-507, NIPER, February 1991.

[641] T. R. French and C. B. Josephson. Alkaline flooding injection strategy. US DOE Fossil Energy Rep NIPER-563, NIPER, March 1992.

[642] T. R. French and C. B. Josephson. The effect of polymer-surfactant interaction on the rheological properties of surfactant-enhanced alkaline flooding formulations: Topical report. US DOE Fossil Energy Rep NIPER-635, NIPER, February 1993.

[643] W. W. Frenier. Well treatment fluids comprising chelating agents. Patent: WO 0183639A, 2001.

[644] W. W. Frenier, C. N. Fredd, and F. Chang. Hydroxyaminocarboxylic acids produce superior formulations for matrix stimulation of carbonates. In *Proceedings Volume*. SPE Europe Formation Damage Conf (The Hague, Netherlands, 5/21–5/22), 2001.

[645] W. W. Frenier and F. B. Growcock. Mixtures of α,β-unsaturated aldehydes and surface active agents used as corrosion inhibitors in aqueous fluids. Patent: US 4734259, 1988.

[646] W. W. Frenier and F. B. Growcock. Process and composition for inhibiting iron and steel corrosion. Patent: EP 289665, 1988.

[647] A. I. Frolov, R. S. Khisamov, I. I. Rjabov, and M. Z. Taziev. Recovery of oil from reservoir—by injection of water and surfactant solution also additionally of wide hydrocarbon(s) fraction and of surfactant solution. Patent: RU 2103492-C, 1998.

[648] M. A. Frolov, I. V. Molyavko, A. I. Spivak, D. L. Rakhmankulov, V. R. Rakhmatullin, and N. A. Romanov. Lubricant for friction pairs working under heavy loads—contains mineral oil and additive in form of 2,4,8,10-tetra-oxaspiro-(5,5)-undecane, to improve anti-wear and anti-scratch properties. Patent: SU 1817788-A, 1993.

[649] L. M. Frostman and J. L. Przybylinski. Successful applications of antiagglomerant hydrate inhibitors. In *Proceedings Volume*. SPE Oilfield Chem Int Symp (Houston, TX, 2/13–16/2001), 2001.

[650] S. E. Fry, J. D. Childs, L. E. Brothers, and D. W. Lindsey. Method of reducing fluid loss in cement compositions which may contain substantial salt concentrations. Patent: US 4676317, 1987.

[651] S. E. Fry, P. L. Totten, J. D. Childs, and D. W. Lindsey. Chloride-free set accelerated cement compositions and methods. Patent: US 5127955, 1992.

[652] G. C. Frye and J. C. Berg. Mechanisms for the synergistic antifoam action by hydrophobic solid particles in insoluble liquids. *J Colloid Interface Sci*, 130(1):54–59, June 1989.

[653] M. Fu and X. Hu. An investigation into shale stability by utilizing copolymer of acrylamide and acrylonitrile and its derivatives. *J Jianghan Petrol Inst*, 19(1):70–73, March 1997.

[654] G. F. Fuh, N. Morita, D. L. Whitfill, and D. A. Strah. Method for inhibiting the initiation and propagation of formation fractures while drilling. Patent: US 5180020, 1993.

[655] G. S. Fung, P. E. Depalm, and P. Sharma. Pour point depression unit using mild thermal cracker. Patent: US 6337011, 2002.

[656] G. P. Funkhouser, J. M. Cassidy, J. L. Lane, K. Frost, T. R. Gardner, and K. L. King. Metal corrosion inhibitors, inhibited acid compositions and methods. Patent: US 6192987, 2001.

[657] G. P. Funkhouser and K. A. Frost. Method of plugging subterranean zone. Patent: EP 889197, 1999.

[658] G. P. Funkhouser and K. A. Frost. Polymeric compositions and methods for use in well applications. Patent: US 5960877, 1999.

[659] L. I. Gabidulina, V. A. Matishev, K. K. Safronov, V. L. Sverdlik, T. P. Zarubina, L. Sh. Andre, A. A. Bienko, L. P. Gilyazetdinov, S. A. Kononov, L. V. Kulygin, T. A. Marakaev, L. A. Marakaeva, V. I. Nasteka, V. S. Shcherbakov, N. S. Sirotkin, and Z. I. Syunyaev. Antifoam composition for use in purification of natural gas – contains dialkyl polypropylene-glycol adipate, and dibutyl adipate. Patent: SU 1775126-A, 1992.

[660] J. F. Gadberry, M. D. Hoey, R. Franklin, Carmen. Vale. G. Del, and F. Mozayeni. Surfactants for hydraulic fracturing compositions. Patent: US 5979555, 1999.

[661] M. M. Gajdarov and M. A. Tankibaev. Non-clayey drilling solution – contains organic stabiliser, caustic soda, water and mineral additive in form of zinc oxide, to improve its thermal stability. Patent: RU 2051946-C, 1996.

[662] P. P. Galchenko, A. P. Titareva, L. G. Kalashnik, and V. D. Krupenko. Plugging solution—contains phenol-formaldehyde resin,

formaldehyde, water and additionally bituminous emulsion of specified composition. Patent: SU 1795082-A, 1993.
[663] B. Gall. Use of sacrificial agents to reduce carboxymethylated ethoxylated surfactant loss during chemical flooding: Topical report. US DOE *Fossil Energy Rep,* 1989.
[664] B. L. Gall, D. R. Maloney, C. J. Raible, and A. R. Sattler. Permeability damage to natural fractures caused by fracturing fluid polymers. In *Proceedings Volume,* pages 551–560. SPE Rocky Mountain Reg Mtg (Casper, WY, 5/11–5/13), 1988.
[665] B. L. Gall and C. J. Raible. The use of size exclusion chromatography to study the degradation of water-soluble polymers used in hydraulic fracturing fluids. In *Proceedings 192nd ACS Nat Mtg,* volume 55, pages 572–576. Amer Chem Soc Polymeric Mater Sci Eng Div Tech Program (Anaheim, CA, 9/7–9/12), 1986.
[666] G. Gallino, A. Guarneri, G. Poli, and L. Xiao. Scleroglucan biopolymer enhances WBM (water-base mud) performances. In *Proceedings Volume,* pages 105–119. Annu SPE Tech Conf (Denver, CO, 10/6–10/9), 1996.
[667] B. Gampert and P. Wagner. *The influence of polymer additives on velocity and temperature fluid,* page 71. Springer-Verlag, Berlin, Germany, 1985.
[668] K. K. Ganguli. Control of gas flow through cement column. Patent: US 5099922, 1992.
[669] K. K. Ganguli. High temperature fluid loss additive for cement slurry and method of cementing. Patent: US 5116421, 1992.
[670] K. K. Ganguli, R. Kattenburg-Schuler, S. T. Leatherdale, A. J. Porat, and G. Zaslavsky. Lightweight thermally stable cement compositions and method of use. Patent: WO 9749644, 1997.
[671] J. Gao, D. Guo, J. Li, and Z. Qiu. The synthesis and properties of high temperature filtrate reducer, APS. *Drilling Fluid Completion Fluid,* 10(1):21–23,74, January 1993.
[672] V. Garg, M. Ralhan, H. C. Tewari, A. Srivastava, S. K. Nanda, and H. S. Rawat. Impact assessment of solid lubricants used in drilling fluids on the producing zone. In *Proceedings Volume,* volume 3, pages 55–60. 1st India Oil & Natur Gas Corp Ltd et al Int Petrol Conf (Petrotech 95) (New Delhi, India, 1/9–1/12), 1995.
[673] Deleted in proofs.
[674] Sh. S. Garifullin, I. M. Gallyamov, I. G. Plotnikov, and A. V. Shuvalov. Aluminum chloride-based gel-forming technologies. *Neft Khoz,* (2):32–35, February 1996.
[675] E. M. Gartner and R. P. Kreh. Cement additives and hydraulic cement mixes containing them. Patent: CA 2071080, 1993.

[676] C. M. Garvey, A. Savoly, and A. L. Resnick. Fluid loss control additives and drilling fluids containing same. Patent: US 4741843, 1988.
[677] C. M. Garvey, A. Savoly, and T. M. Weatherford. Drilling fluid dispersant. Patent: US 4711731, 1987.
[678] S. A. Garyan, L. P. Kuznetsova, and Yu. N. Moisa. Experience in using environmentally safe lubricating additive fk-1 in drilling muds during oil and gas well drilling. *Stroit Neft Gaz Skvazhin Sushe More,* (10):11–14, October 1998.
[679] R. J. Gay, C. C. Gay, V. M. Matthews, F. E. M. Gay, and V. Chase. Dynamic polysulfide corrosion inhibitor method and system for oil field piping. Patent: US 5188179, 1993.
[680] D. Gazaniol, T. Forsans, M. J. F. Boisson, and J. M. Piau. Wellbore failure mechanisms in shales: Prediction and prevention. In *Proceedings Volume,* volume 1, pages 459–471. SPE Europe Petrol Conf (London, England, 10/25–10/27), 1994.
[681] D. Gazaniol, T. Forsans, M. J. F. Boisson, and J. M. Piau. Wellbore failure mechanisms in shales: Prediction and prevention. *J Petrol Technol,* 47(7):589–595, July 1995.
[682] R. D. Gdanski. Fluid properties and particle size requirements for effective acid fluid-loss control. In *Proceedings Volume,* pages 81–94. SPE Rocky Mountain Reg Mtg/Low Permeability Reservoirs Symp (Denver, CO, 4/26–4/28), 1993.
[683] J. C. Gee, C. J. Lawrie, and R. C. Williamson. Drilling fluids comprising mostly linear olefins. Patent: WO 9521226, 1995.
[684] J. C. Gee, R. C. Williamson, C. J. Lawrie, and S. J. Miller. Skeletally isomerized linear olefins. Patent: US 5741759, 1998.
[685] J. C. Gee, R. C. Williamson, and C. J. Lawrie. Drilling fluids comprising mostly linear olefins. Patent: US 6057272, 2000.
[686] J. C. Gee, R. C. Williamson, C. J. Lawrie, and S. J. Miller. Skeletally isomerized linear olefins. Patent: US 6054415, 2000.
[687] G. G. Geib. Hydrocarbon gelling compositions useful in fracturing formations. Patent: US 6342468, 2002.
[688] L. Gelner. Protection of storage tank bottoms using volatile corrosion inhibitors (VCI). In *Proceedings Volume,* volume 1, pages 102–109. 7th NACE Int et al Middle East Corrosion Conf (Manama, Bahrain, 2/26–2/28), 1996.
[689] T. M. Gentle and W. C. White. Antimicrobial antifoam compositions and methods. Patent: EP 351828, 1990.
[690] B. Genuyt, M. Janssen, R. Reguerre, J. Cassiers, and F. Breye. Biodegradable lubricating composition and uses thereof, in particular in a bore fluid (composition lubrifiante biodegradable et ses utilisations, notamment dans un fluide de forage). Patent: WO 0183640A, 2001.

[691] J. M. Gerez and A. R. Pick. Heavy oil transportation by pipeline. In *Proceedings Volume,* volume 2, pages 699–710. 1st ASME et al Int Pipeline Conf (Calgary, Canada, 6/9–6/13), 1996.
[692] D. E. Gessell and P. H. Washecheck. Composition and method for friction loss reduction. Patent: US 4952738, 1990.
[693] J. M. Getliff and S. G. James. The replacement of alkyl-phenol ethoxylates to improve the environmental acceptability of drilling fluid additives. In *Proceedings Volume,* volume 2, pages 713–719. 3rd SPE et al Health, Safety & Environ Int Conf (New Orleans, LA, 6/9–6/12), 1996.
[694] D. Getzlaf. Field studies illustrate merits of FCRS (flexible corrosion-resistant sealant). *Amer Oil Gas Reporter,* 41(13):95–96,105, December 1998.
[695] D. Gevertz, J. R. Paterek, M. E. Davey, and W. A. Wood. Isolation and characterization of anaerobic halophilic bacteria from oil reservoir brines. Number 31, pages 115–129. 1991.
[696] R. Geyer, R. Fay, G. Denoux, C. Giammona, K. Binkley, and R. Jamail. Aerial dispersant application: Assessment of sampling methods and operational altitudes. Mar Spill Response Corp Tech Rep Ser 93-0091, 1992.
[697] R. Ghofrani. Development of CaO- and MgO-swelling cements into usage maturity for cementation of natural gas underground storage wells and natural gas production wells. DMGK Res Rep 444-3, Clausthal Tech Univ, 1997.
[698] R. Ghofrani, C. Marx, and H. Wolschendorf. Swelling (expanding) cements based on CaO and MgO (Quellzemente auf der Basis von CaO und MgO). *Erdöl Erdgas Kohle,* 108(1):9–11, January 1992.
[699] R. Ghofrani and H. Plack. Conditions for successful use of CaO and MgO expanding cements in annulus cementations (Voraussetzungen für den Erfolgreichen Einsatz von CaO- und MgO-Quellzementen bei Ringraumzementationen). In *Proceedings Volume,* pages 87–100. DMGK Spring Conf (Celle, Germany, 5/13–5/14), 1993.
[700] R. Ghofrani and K. C. Werner. Effect of the calcination temperature and the duration of calcination on the optimization of expanding efficiency of the additives CaO and MgO in swelling (expanding) cements (Optimierung der Expansionswirkung der Zusätze CaO und MgO in Quellzement). *Erdöl Erdgas Kohle,* 109(1):7–9, January 1993.
[701] J. L. Gibb, J. A. Laird, and L. G. Berntson. Novolac coated ceramic particulate. Patent: EP 308257, 1989.
[702] J. L. Gibb, J. A. Laird, G. W. Lee, and W. C. Whitcomb. Particulate ceramic useful as a proppant. Patent: CA 1232921, 1988.

[703] J. L. Gibb, J. A. Laird, G. W. Lee, and W. C. Whitcomb. Particulate ceramic useful as a proppant. Patent: US 4944905, 1990.
[704] D. M. Giddings and D. W. Fong. Calcium tolerant deflocculant for drilling fluids. Patent: US 4770795, 1988.
[705] D. M. Giddings and C. D. Williamson. Terpolymer composition for aqueous drilling fluids. Patent: US 4678591, 1987.
[706] A. Gillot, A. Charlier, and R. Van Elmbt. Correlation results between IFP (Institute Francais du Petrole) and WSL (Warren Spring laboratory) laboratory tests of dispersants. *Oil Chem Pollut,* 3(6):445–453, 1986.
[707] J. F. Girod, J. P. Leclerc, H. Muhr, G. Paternotte, and J. P. Corriou. Removing a small quantity of THT (tetrahydrothiophene) from gas storage groundwater through air stripping and gas-phase carbon adsorption. *Environ Progr,* 15(4):277–282, 1996.
[708] K. F. Gironda, G. H. Redlich, and R. B. Petigara. Bromate stabilization of nitrate-free 3- isothiazolones at pH 4-5.1. Patent: US 5478797, 1995.
[709] D. A. Glowka, G. E. Loeppke, P. B. Rand, and E. K. Wright. *Laboratory and field evaluation of polyurethane foam for lost circulation control,* volume 13 of *The geysers—three decades of achievement: A window on the future,* pages 517–524. Geothermal Resources Council, Davis, CA, 1989.
[710] A. D. Godwin and G. M. K. Mathys. Ester-free ethers. Patent: WO 9304028, 1993.
[711] A. D. Godwin and T. Sollie. Load bearing fluid. Patent: EP 532128, 1993.
[712] S. V. Goncharov, V. V. Neradovskij, M. E. Zevakov, M. A. Babets, E. I. Bektimirov, V. I. Bezdenezhnykh, and V. Sh. Shmavonyants. Sealing lubricant for profiled joints of e.g., casing strings—contains silicoorganic liquid, diethylene glycol, graphite powder, mixture of derivatives of synthetic fatty acids and solution of polyacrylamide. Patent: SU 1796648-A, 1993.
[713] M. E. Gonzalez and M. D. Looney. The use of encapsulated acid in acid fracturing treatments. Patent: WO 0075486A, 2000.
[714] M. E. Gonzalez and M. D. Looney. Use of encapsulated acid in acid fracturing treatments. Patent: US 6207620, 2001.
[715] S. Goodwin. Prediction, modelling and management of hydrates using low dosage additives: Pt 1: Additive types and operational implications. In *Proceedings Volume.* IBC Tech Serv et al Advances in Multiphase Oper Offshore Conf (London, England, 11/29–11/30), 1995.
[716] S. G. Goodyear and P. I. R. Jones. Assessment of foam for deep flow diversion. In *Proceedings Volume,* volume 2, pages 174–182. 8th EAPG Impr Oil Recovery Europe Symp (Vienna, Austria, 5/15–5/17), 1995.

[717] S. Gopalkirshnan and M. Roznowski. Additive composition for oil well cementing formulations. Patent: US 5258072, 1993.
[718] S. Gopalkrishnan. Additive composition for oil well cementing formulations. Patent: US 5258428, 1993.
[719] S. Gopalkrishnan. Additive composition for oil well cementing formulations. Patent: US 5252128, 1993.
[720] S. Yu. Gordienko and L. S. Lartseva. Plugging-solution—contains Portland cement, soda-sulphate waste from alumina production, glycerine and water and hardens at sub-zero temperatures. Patent: SU 1760087-A, 1992.
[721] V. A. Gorodilov, V. N. Shevchenko, S. I. Tipikin, A. D. Makurov, G. A. Makeev, and V. F. Fomichev. Method for high temperature seam oil deposit development—by pumping aluminium chloride and tri-sodium phosphate as sediment forming material. Patent: RU 2094599-C, 1997.
[722] V. P. Gorodnov, A. Yu. Ryskin, G. P. Kharlanov, A. A. Belov, and A. V. Shein. Enhancing oil recovery from boreholes—by injecting into seam aqueous solution of polyacrylamide, chrome alum and bentonite clay for improved flow characteristics. Patent: SU 1731942-A, 1992.
[723] V. P. Gorodnov, A. Yu. Ryskin, and M. V. Pavlov. Increasing oil extraction from stratum—by pumping-in hydrochloric acid solution or its mixture with hydrofluoric acid or ammonium fluoride, followed by pumping aqueous solution or surfactant. Patent: SU 1795092-A, 1993.
[724] K. F. Gotlieb, I. P. Bleeker, H. A. Van Doren, and A. Heeres. 2-Nitroalkyl ethers of native or modified starch, method for the preparation thereof, and ethers derived therefrom. Patent: EP 710671, 1996.
[725] C. F. L. Goudy. How flow improvers can reduce liquid line operating costs. *Pipe Line Ind,* 74(6):49–51, June 1991.
[726] R. N. Grabois and Y. N. Lee. Use of a water soluble drag reducer in a water/oil/gas system. Patent: US 5027843, 1991.
[727] E. Grabowski and J. E. Gillott. Effect of replacement of silica flour with silica fume on engineering properties of the oilwell cements at normal and elevated temperatures and pressures. *Cement Concrete Res,* 19(3):333–344, May 1989.
[728] D. E. Graham, W. A. Lidy, P. C. McGrath, and D. G. Thompson. Demulsifying process. Patent: CA 1221602, 1987.
[729] D. E. Graham, W. A. Lidy, P. C. McGrath, and D. G. Thompson. Demulsifying process using polysiloxane polyalkene oxide block copolymer. Patent: AU 565533, 1987.
[730] G. M. Graham, S. J. Dyer, and P. Shone. Potential application of amine methylene phosphonate based inhibitor species in HP/HT (high pressure/high temperature) environments for improved carbonate scale

inhibitor performance. 2nd Annu SPE Oilfield Scale Int Symp (Aberdeen, Scotland, 1/26–1/27), 2000.
[731] G. M. Graham, S. J. Dyer, K. S. Sorbie, W. Sablerolle, and G. C. Graham. Practical solutions to scaling in HP/HT (high pressure/high temperature) and high salinity reservoirs. In *Proceedings Volume*. 4th IBC UK Conf Ltd Advances In Solving Oilfield Scaling Int Conf (Aberdeen, Scotland, 1/28–1/29), 1998.
[732] G. M. Graham, S. J. Dyer, K. S. Sorbie, W. R. Sablerolle, P. Shone, and D. Frigo. Scale inhibitor selection for continuous and downhole squeeze application in HP/HT (high pressure/high temperature) conditions. In *Proceedings Volume*, pages 645–659. Annu SPE Tech Conf (New Orleans, LA, 9/27–9/30), 1998.
[733] G. M. Graham, M. M. Jordan, K. S. Sorbie, J. Bunney, G. C. Graham, W. Sablerolle, and P. Hill. The implication of HP/HT (high pressure/high temperature) reservoir conditions on the selection and application of conventional scale inhibitors: Thermal stability studies. In *Proceedings Volume*, pages 627–640. SPE Oilfield Chem Int Symp (Houston, TX, 2/18–2/21), 1997.
[734] B. D. Green. Method for creating dense drilling fluid additive and composition therefor. Patent: WO 0168787A, 2001.
[735] D. W. Green and G. P. Willhite. *Enhanced oil recovery*, volume 6 of *SPE Textbook Series*. SPE Publications, 1998.
[736] A. A. Gregoli and A. M. Olah. Low-temperature pipeline emulsion transportation enhancement. Patent: US 5156652, 1992.
[737] G. Gregory, D. Shuell, and J. E. Thompson, Sr. Overview of contemporary LFC (liquid frac concentrate) fracture treatment systems and techniques. In *Proceedings Volume*, number 91-01. 4th CADE/CAODC Spring Drilling Conf (Calgary, Canada, 4/10–4/12) Proc, 1991.
[738] R. A. Grey. Process for preparing alternating copolymers of olefinically unsaturated sulfonate salts and unsaturated dicarboxylic acid anhydrides. Patent: US 5210163, 1993.
[739] J. E. Griffith, P. L. Totten, B. L. King, and J. Chatterji. Well cementing methods and compositions for use in cold environments. Patent: US 5571318, 1996.
[740] J. M. Gross. Gelling organic liquids. Patent: EP 225661, 1987.
[741] F. B. Growcock. Surfactants can affect corrosion inhibition of oilfield steel. In *Proceedings Volume*. SPE Oilfield Chem Int Symp (San Antonio, TX, 2/4–2/6), 1987.
[742] F. B. Growcock and V. R. Lopp. The inhibition of steel corrosion in hydrochloric acid with 3-phenyl-2-propyn-1-ol. *Corrosion Sci*, 28(4):397–410, 1988.
[743] E. A. Grula, H. H. Russell, D. Bryant, and M. Kenaga. Oil displacement by anaerobic and facultatively anaerobic bacteria. In E. C. Donaldson,

G. V. Chilingarian, and T. F. Yen, editors, *Microbial enhanced oil recovery*, volume 22 of *Developments in petroleum science*, pages 113–123. Elsevier Science Ltd, 1989.

[744] E. A. Grula, H. H. Russell, and M. M. Grula. Potential health hazard of bacteria to be used in microbial enhanced oil recovery. In E. C. Donaldson, G. V. Chilingarian, and T. F. Yen, editors, *Microbial enhanced oil recovery*, volume 22 of *Developments in petroleum science*, pages 209–213. Elsevier Science Ltd, 1989.

[745] P. Guilbot, A. Valtz, and D. Richon. Partition coefficients at infinite dilution for different sulfur compounds in various solvents. In *Proceedings Volume*, pages 33–39. 76th Annu GPA Conv (San Antonio, TX, 3/10–3/12), 1997.

[746] P. I. C. Guimaraes, A. P. Monteiro, and F. B. Mainier. New corrosion inhibitors in solid form to protect carbon steel pipes in acidizing operations. In *Proceedings Volume*, volume 2. 5th Brazil Petrol Congr (Conexpo Arpel 94) (Rio De Janeiro, Brazil, 10/16–10/20), 1994.

[747] J. Gulbis, M. T. King, G. W. Hawkins, and H. D. Brannon. Encapsulated breaker for aqueous polymeric fluids. In *Proceedings Volume*, pages 245–254. 9th SPE Formation Damage Contr Symp (Lafayette, LA, 2/22–2/23), 1990.

[748] J. Gulbis, M. T. King, G. W. Hawkins, and H. D. Brannon. Encapsulated breaker for aqueous polymeric fluids. *SPE Prod Eng,* 7(1):9–14, February 1992.

[749] J. Gulbis, T. D. A. Williamson, M. T. King, and V. G. Constien. Method of controlling release of encapsulated breakers. Patent: EP 404211, 1990.

[750] P. D. Gullett and P. F. Head. Materials incorporating cellulose fibres, methods for their production and products incorporating such materials. Patent: WO 9318111, 1993.

[751] R. Gundersen, B. Johansen, P. O. Gartland, L. Fiksdal, I. Vintermyr, R. Tunold, and G. Hagen. The effect of sodium hypochlorite on the electrochemical properties of stainless steels in seawater with and without bacterial films. *Corrosion,* 47(10):800–807, October 1991.

[752] D. R. Guo, J. P. Gao, K. H. Lu, M. B. Sun, and W. Wang. Study on the biodegradability of mud additives. *Drilling Fluid Completion Fluid,* 13(1):10–12, 1996.

[753] D. V. S. Gupta and B. B. Prasek. Method for fracturing subterranean formations using controlled release breakers and compositions useful therein. Patent: US 5437331, 1995.

[754] B. M. Gustov, A. M. Khatmullin, V. S. Asmolovskii, V. G. Zyurin, F. Kh. Saifutdinov, and L. E. Lenchenkova. Field tests of gel technologies at the Arlan field. *Neft Khoz,* (2):36–38, February 1996.

[755] A. H. Hale. Well drilling fluids and process for drilling wells. Patent: US 4728445, 1988.
[756] A. H. Hale. Water base drilling fluid. Patent: GB 2216573, 1989.
[757] A. H. Hale. Well drilling cuttings disposal. Patent: US 5341882, 1994.
[758] A. H. Hale, G. C. Blytas, and A. K. R. Dewan. Water base drilling fluid. Patent: GB 2216574, 1989.
[759] A. H. Hale and H. F. Lawson. Well drilling fluids and process for drilling wells. Patent: US 4740318, 1988.
[760] A. H. Hale and R. E. Loftin. Glycoside-in-oil drilling fluid system. Patent: US 5494120, 1996.
[761] A. H. Hale and G. T. Rivers. Well drilling fluids and process for drilling wells. Patent: US 4721576, 1988.
[762] A. H. Hale and E. Van Oort. Efficiency of ethoxylated/propoxylated polyols with other additives to remove water from shale. Patent: US 5602082, 1997.
[763] B. E. Hall and C. A. Szememyei. Fluid additive and method for treatment of subterranean formations. Patent: US 5089151, 1992.
[764] W. Halliday, D. K. Clapper, and M. Smalling. New gas hydrate inhibitors for deepwater drilling fluids. In *Proceedings Volume,* pages 201–209. IADC/SPE Drilling Conf (Dallas, TX, 3/3–3/6), 1998.
[765] W. S. Halliday and D. K. Clapper. Purified paraffins as lubricants, rate of penetration enhancers, and spotting fluid additives for water-based drilling fluids. Patent: US 5837655, 1998.
[766] W. S. Halliday, D. K. Clapper, M. R. Smalling, and R. G. Bland. Glycol derivatives and blends thereof as gas hydrate inhibitors in water base drilling, drill-in, and completion fluids. Patent: WO 9840446, 1998.
[767] W. S. Halliday and D. Schwertner. Olefin isomers as lubricants, rate of penetration enhancers, and spotting fluid additives for water-based drilling fluids. Patent: US 5605879, 1997.
[768] W. S. Halliday and V. M. Thielen. Drilling mud additive. Patent: US 4664818, 1987.
[769] M. S. Hameed, E. J. Taha, and M. M. F. Al-Jarrah. Observations on the effect of low molecular weight polyethylene-oxides on the flow of water in closed pipes and standard fittings. *J Petrol Res,* 8(1):47–60, June 1989.
[770] B. E. Hamilton. Method of plugging openings in well conduits. Patent: US 4869321, 1989.
[771] W. A. Hamilton. Mechanisms of microbial corrosion. In *Proceedings Volume,* pages 1–11. Inst Petrol Microbiol Comm Microbial Problems in the Offshore Oil Ind Int Conf (Aberdeen, Scotland, 4/15–4/17), 1986.

[772] D. K. Han, C. Z. Yang, Z. H. Lou, Z. Q. Zhang, and Y. I. Chang. Recent development of enhanced oil recovery in china. *J Petrol Sci Eng,* 22(1–3):181–188, January 1999.
[773] S. Handa, P. K. G. Hodgson, and W. J. Ferguson. Asphaltene precipitation inhibiting polymer for use in oils. Patent: GB 2337522, 1999.
[774] E. Haque. *Physicochemical interactions between montmorillonite and polymerizing systems: Effect on particle-reinforced composites.* PhD thesis, Rice Univ, 1986.
[775] W. M. Harms. Catalyst for breaker system for high viscosity fluids. Patent: US 5143157, 1992.
[776] W. M. Harms and L. R. Norman. Concentrated hydrophilic polymer suspensions. Patent: US 4772646, 1988.
[777] W. M. Harms and E. Scott. Method for stimulating methane production from coal seams. Patent: US 5249627, 1993.
[778] W. M. Harms, M. Watts, J. Venditto, and P. Chisholm. Diesel-based HPG (hydroxypropyl guar) concentrate is product of evolution. *Petrol Eng Int,* 60(4):51–54, April 1988.
[779] L. E. Harris, M. D. Holtmyer, and R. W. Pauls. Method for fracturing subterranean formations. Patent: US 4622155, 1986.
[780] P. C. Harris. Fracturing-fluid additives. *J Petrol Technol,* 40(10):1277–1279, October 1988.
[781] P. C. Harris and S. J. Heath. Delayed release borate crosslinking agent. Patent: EP 594364, 1994.
[782] P. C. Harris and S. J. Heath. High-quality foam fracturing fluids. In *Proceedings Volume,* pages 265–273. SPE Gas Technol Symp (Calgary, Canada, 4/28–5/1), 1996.
[783] P. C. Harris, L. R. Norman, and K. H. Hollenbeak. Borate crosslinked fracturing fluids. Patent: EP 594363, 1994.
[784] R. E. Harris and R. J. Hodgson. Delayed acid for gel breaking. Patent: WO 9533914, 1995.
[785] P. R. Hart. Method of breaking water-in-oil emulsions by using quaternary alkyl amine ethoxylates. Patent: US 5250174, 1993.
[786] P. R. Hart. Control of foam in hydrocarbon fluids. Patent: US 5472637, 1995.
[787] P. R. Hart. Method of breaking reverse emulsions in a crude oil desalting system. Patent: CA 2126889, 1995.
[788] P. R. Hart. Methods for inhibiting foam in crude oils. Patent: US 5800738, 1998.
[789] P. R. Hart, J. M. Brown, and E. J. Connors. Method of resolving oil and water emulsions. Patent: CA 2126782, 1995.
[790] P. R. Hart, J. M. Brown, and E. J. Connors. Method of resolving oil and brine emulsions. Patent: CA 2156444, 1996.

[791] P. R. Hart, F. Chen, W. P. Liao, and W. J. Burgess. Copolymer formulations for breaking oil-and-water emulsions. Patent: US 5921912, 1999.
[792] J. A. Haslegrave and L. A. McDougall. Corrosion inhibitors. Patent: EP 286336, 1988.
[793] J. A. Haslegrave and D. S. Sullivan. N, S containing corrosion inhibitors. Patent: EP 243016, 1987.
[794] J. A. Haslegrave and D. S. Sullivan. N, S containing corrosion inhibitors. Patent: US 4673436, 1987.
[795] J. A. Haslegrave and D. S. Sullivan. N, S containing corrosion inhibitors. Patent: CA 1271323, 1990.
[796] K. Hatchman. Concentrates for use in structured surfactant drilling fluids. Patent: GB 2329655, 1999.
[797] K. Hatchman. Drilling fluid concentrates. Patent: EP 903390, 1999.
[798] R. H. Hausler. On the use of linear polarization measurements for the evaluation of corrosion inhibitors in concentrated HCL at 200° F (93° C). *Corrosion,* 42(12):729–739, December 1986.
[799] G. W. Hawkins. Molecular weight reduction and physical consequences of chemical degradation of hydroxypropylguar in aqueous brine solutions. In *Proceedings 192nd ACS Nat Mtg*, volume 55, pages 588–593. Amer Chem Soc Polymeric Mater Sci Eng Div Tech Program (Anaheim, CA, 9/7–9/12), 1986.
[800] S. He, A. T. Kan, and M. B. Tomson. Inhibition of calcium carbonate precipitation in NaCl brines from 25 to 90° C. *Applied Geochemistry,* 14(1):17–25, January 1999.
[801] J. A. Headley, T. O. Walker, and R. W. Jenkins. Environmentally safe water-based drilling fluid to replace oil-based muds for shale stabilization. In *Proceedings Volume,* pages 605–612. SPE/IADC Drilling Conf (Amsterdam, the Netherlands, 2/28–3/2), 1995.
[802] J. F. Heathman and R. S. Cromwell. Well cementing. Patent: EP 592217, 1994.
[803] J. F. Heathman and R. J. Crook. Fine particle size cement compositions. Patent: EP 611081, 1994.
[804] J. F. Heathman and R. J. Crook. Fine particle size cement compositions and methods. Patent: US 5346012, 1994.
[805] B. M. Hegarty and R. Levy. Control of oilfield biofouling. Patent: CA 2160305, 1996.
[806] B. M. Hegarty and R. Levy. Control of oilfield biofouling. Patent: EP 706759, 1996.
[807] B. M. Hegarty and R. Levy. Procedure for combatting biological contamination in petroleum production (procede pour combattre l'encrassement biologique dans la production de petrole). Patent: FR 2725754, 1996.

[808] K. H. Heier, R. Morschhaeuser, A. Tardi, S. Weber, and G. Botthof. Copolymers and their use as drilling aids. Patent: US 6380137, 2002.

[809] G. Heinrich. Process for recovering barite from drilling muds. Patent: CA 1310144, 1992.

[810] L. R. Heinze, N. Shahreyar, and B. M. Baruah. A review of past 50 years of paraffin prevention and removal techniques as presented in the SWPSC (Southwestern Petroleum Short Course). In *Proceedings Volume*, pages 230–238. 48th Annu Southwestern Petrol Short Course Assoc Inc et al Mtg (Lubbock, TX, 4/25–4/26), 2001.

[811] A. Heller and J. R. Brock. Materials and methods for photocatalyzing oxidation of organic compounds on water. Patent: AU 657470, 1995.

[812] J. Hen. Sulfonate-containing polymer/polyanionic cellulose combination for high temperature/high pressure filtration control in water base drilling fluids. Patent: US 5008025, 1991.

[813] D. F. Henderson. Large volume odorization, installation, operation, and maintenance. In *Odorization*, volume 3, pages 239–249. Institute of Gas Technology, Chicago, IL, 1993.

[814] U. W. Hendricks, B. Lehmann, and U. Litzinger. Process for the separation of oil-in-water emulsions, and basic polyamides (Verfahren zur Spaltung von Öl-in-Wasser-Emulsionen und Basische Polyamide). Patent: EP 691150, 1996.

[815] H. Hendriks. Dispersing agent for cement-slurry compounds, compounds containing this agent and procedures for the cementing of oil wells or similar (dispersant pour compositions de laitiers de ciment, compositions le contenant et procedes correspondants de cimentation de puits petroliers ou analogues). Patent: FR 2620442, 1989.

[816] H. Hendriks. Dispersing agent for cement-slurry compounds, compounds containing this agent and procedures for the cementing of oil wells or similar. Patent: EP 307997, 1989.

[817] E. R. Henson and P. A. Doty. Corrosion inhibitors for aqueous brines. Patent: US 4980074, 1990.

[818] G. Hernandez, C. Lemaitre, G. B. G. Mecanique, J. Guezennec, and J. P. Audouard. Biocorrosion of 316 l stainless steel modified by poison alloying elements in sea water. In *Proceedings Volume*. Annu NACE Corrosion Conf (Corrosion 92) (Nashville, TN, 4/27–5/1), 1992.

[819] M. I. Hernandez, M. Mas, R. J. Gabay, and L. Quintero. Thermally stable drilling fluid. Patent: US 5883054, 1999.

[820] M. I. Hernandez, M. Mas, R. J. Gabay, and L. Quintero. Thermally stable drilling fluid which includes styrene-butadiene copolymers. Patent: GB 2329657, 1999.

[821] I. Higuerey, P. Pereira, and V. Leon. Comparative study of compositional changes between thermal cracking and aquaconversion process. *ACS Petrol Chem Div Preprints,* 46(1):64–65, February 2001.

[822] M. Hille, R. Kupfer, and R. Böhm. Polyfunctional demulsifiers for crude oils. Patent: CA 2104506, 1994.

[823] M. Hille, R. Kupfer, and R. Böhm. Polyfunctional demulsifiers for crude oils (Polyfunktionelle Demulgatoren für Rohöle). Patent: EP 584708, 1994.

[824] M. Hille, H. Wittkus, J. Tonhauser, F. Engelhardt, and U. Riegel. Water-soluble copolymers useful in drilling fluids. Patent: US 5510436, 1996.

[825] M. Hille, H. Wittkus, and F. Weinelt. Application of acetal-containing mixtures (Verwendung von Acetal enthaltenden Mischungen). Patent: EP 702074, 1996.

[826] M. Hille, H. Wittkus, and F. Weinelt. Use of acetal-containing mixtures. Patent: US 5830830, 1998.

[827] M. Hille, H. Wittkus, B. Windhausen, H. J. Scholz, and F. Weinelt. Application of acetals (Verwendung von Acetalen). Patent: EP 512501, 1992.

[828] M. Hille, H. Wittkus, B. Windhausen, H. J. Scholz, and F. Weinelt. Use of acetals. Patent: US 5759963, 1998.

[829] R. E. Himes. Method for clay stabilization with quaternary amines. Patent: US 5097904, 1992.

[830] R. E. Himes, M. A. Parker, and E. G. Schmelzl. Environmentally safe temporary clay stabilizer for use in well service fluids. In *Proceedings Volume,* volume 3. CIM Petrol Soc/SPE Int Tech Mtg (Calgary, Canada, 6/10–6/13), 1990.

[831] R. E. Himes and E. F. Vinson. Fluid additive and method for treatment of subterranean formations. Patent: US 4842073, 1989.

[832] R. E. Himes and E. F. Vinson. Stabilizing clay-containing formations. Patent: EP 308138, 1989.

[833] R. E. Himes and E. F. Vinson. Environmentally safe salt replacement for fracturing fluids. In *Proceedings Volume,* pages 237–248. SPE East Reg Conf (Lexington, KY, 10/23–10/25), 1991.

[834] R. E. Himes, E. F. Vinson, and D. E. Simon. Clay stabilization in low-permeability formations. In *Proceedings Volume,* pages 507–516. SPE Prod Oper Symp (Oklahoma City, OK, 3/12–3/14), 1989.

[835] D. O. Hitzman and D. M. Dennis. Sulfide removal and prevention in gas wells. In *Proceedings Volume,* pages 433–438. SPE Prod Oper Symp (Oklahoma City, OK, 3/9–3/11), 1997.

[836] A. W. Ho. Derivatives of polyalkylenepolyamines as corrosion inhibitors. Patent: WO 9307307, 1993.

[837] A. W. Ho. Derivatives of polyalkylenepolyamines as corrosion inhibitors. Patent: US 5275744, 1994.

[838] M. H. Hodder, D. A. Ballard, and G. Gammack. Controlling drilling fluid enzyme activity. *Petrol Eng Int,* 64(11):31,33,35, November 1992.
[839] R. M. Hodge. Particle transport fluids thickened with acetylate free xanthan heteropolysaccharide biopolymer plus guar gum. Patent: US 5591699, 1997.
[840] G. G. Hoffmann and I. Steinfatt. Thermochemical sulfate reduction at steam flooding processes—a chemical approach. In *ACS Petrol Chem Div Preprints,* volume 38, pages 181–184. 205th ACS Nat Mtg Enhanced Oil Recovery Symp (Denver, CO, 3/28–4/2), February 1993.
[841] M. Hofinger, W. Böse, M. Hille, and R. Böhm. Quaternary oxyalkylated polycondensates, process for their manufacture and their utilization (Quaternäre oxalkylierte Polykondensate, Verfahren zu deren Herstellung und deren Verwendung). Patent: EP 212265, 1987.
[842] M. Hofinger and H. Schellenberg. Process for the preparation of polycondensation products from alkoxylated fatty amines, diols and aliphatic dicarboxylic acids (Verfahren zur Herstellung von Polykondensationsverbindungen aus oxalkylierten Fettaminen, Diolen und Aliphatischen Dicarbonsäuren). Patent: EP 299348, 1989.
[843] R. Hohlfeld. Longer life for glyco-based stationary engine coolants. *Pipeline Gas J,* 223(7):55–57, July 1996.
[844] S. A. Holditch, H. Xiong, Z. Rahim, and J. Rueda. Using an expert system to select the optimal fracturing fluid and treatment volume. In *Proceedings Volume,* pages 515–527. SPE Gas Technol Symp (Calgary, Canada, 6/28–6/30), 1993.
[845] K. Holing, J. Alvestad, and J. A. Trangenstein. The use of second-order godunov-type methods for simulating EOR processes in realistic reservoir models. In *Proceedings Volume,* pages 101–111. 2nd Inst Franc Du Petrol Math of Oil Recovery Europe Conf (Arles, France, 9/11–9/14), 1990.
[846] M. D. Holtmyer and C. V. Hunt. Method and composition for viscosifying hydrocarbons. Patent: US 4780221, 1988.
[847] M. D. Holtmyer and C. V. Hunt. Crosslinkable cellulose derivatives. Patent: EP 479606, 1992.
[848] C. Holzner, R. Kleinstueck, and A. Spaniol. Phosphonate-containing mixtures (phosphonathältige Mischungen). Patent: WO 0032610, 2000.
[849] D. G. Horstmann and D. S. Jones. Synergistic biocides of certain nitroimidazoles and aldehydes. Patent: US 4920141, 1990.
[850] D. H. Hoskin, T. O. Mitchell, and P. Shu. Oil reservoir permeability profile control with crosslinked welan gum biopolymers. Patent: US 4981520, 1991.
[851] D. H. Hoskin and L. D. Rollmann. Polysilicate esters for oil reservoir permeability control. Patent: EP 283602, 1988.

[852] D. E. Hostetler, R. J. Kostelnik, and In. Z. J. Shanti. Polymerization process. Patent: US 4845178, 1989.
[853] R. F. House and J. C. Cowan. Chitosan-containing well drilling and servicing fluids. Patent: US 6258755, 2001.
[854] R. F. House and V. M. Granquist. Polyphenolic acid adducts. Patent: US 4597878, 1986.
[855] R. F. House, A. H. Wilkinson, and C. Cowan. Well working compositions, method of decreasing the seepage loss from such compositions, and additive therefor. Patent: US 5004553, 1991.
[856] J. W. Hoyt and R. H. J. Sellin. Drag reduction by centrally-injected polymer "threads". *Rheol Acta,* 27(5):518–522, September–October 1988.
[857] J. W. Hoyt and R. H. J. Sellin. Polymer "threads" and drag reduction. *Rheol Acta,* 30(4):307–315, July–August 1991.
[858] M. J. Hrachovy. Hydraulic fracturing technique employing in situ precipitation. Patent: WO 9406998, 1994.
[859] M. J. Hrachovy. Hydraulic fracturing technique employing in situ precipitation. Patent: US 5322121, 1994.
[860] J. C. Hsu. Synergistic microbicidal combinations containing 3-isothiazolone and commercial biocides. Patent: US 4906651, 1990.
[861] J. C. Hsu. Synergistic microbicidal combinations containing 3-isothiazoline and commercial biocides. Patent: US 5190944, 1993.
[862] J. C. Hsu. Synergistic microbicidal combinations containing 3-isothiazolone and commercial biocides. Patent: US 5278178, 1994.
[863] J. C. Hsu. Biocidal compositions. Patent: EP 685158, 1995.
[864] J. C. Hsu. Biocidal compositions. Patent: EP 685159, 1995.
[865] J. F. Hsu, A. K. Al-Zain, K. U. Raju, and A. P. Henderson. Encapsulated scale inhibitor treatments experience in the Ghawar field, Saudi Arabia. 2nd Annu SPE Oilfield Scale Int Symp (Aberdeen, Scotland, 1/26–1/27), 2000.
[866] J. J. C. Hsu. Recovering hydrocarbons with a mixture of carbon dioxide and trichloroethane. Patent: US 5117907, 1992.
[867] F. Huang and R. Lee. Degradation of ethyl mercaptan in the presence of zero-valence iron. In *Proceedings Volume.* 4th US DOE, Tulsa Univ, et al Petrol Environ Conf (San Antonio, TX, 9/9–9/12), 1997.
[868] N. Huang. Synthesis of fluid loss additive of sulfonate tannic-phenolic resin. *Oil Drilling Prod Technol,* 18(2):39–42,106–107, 1996.
[869] M. K. Hubbert and D. G. Willis. *Mechanics of hydraulic fracturing.* Transactions of AIME, 1957.
[870] D. A. Huddleston. Hydrocarbon geller and method for making the same. Patent: US 4877894, 1989.
[871] D. A. Huddleston. Liquid aluminum phosphate salt gelling agent. Patent: US 5110485, 1992.

[872] D. A. Huddleston, R. K. Gabel, and C. D. Williamson. Method for reducing fluid loss from oilfield cement slurries using vinyl grafted wattle tannin. Patent: US 5134215, 1992.
[873] D. A. Huddleston and C. D. Williamson. Vinyl grafted lignite fluid loss additives. Patent: US 4938803, 1990.
[874] D. A. Huddleston and C. D. Williamson. Vinyl grafted lignite fluid loss additives. Patent: US 5028271, 1991.
[875] T. E. Hudson and J. W. Martin. Pyrolytic carbon coating of media improves gravel packing and fracturing capabilities. Patent: EP 301626, 1989.
[876] T. E. Hudson and J. W. Martin. Pyrolytic carbon coating of media improves gravel packing and fracturing capabilities. Patent: US 4796701, 1989.
[877] U. Huetz and P. Englezos. Measurement of structure h hydrate phase equilibrium and the effect of electrolytes. *Fluid Phase Equilibria,* 117(1–2):178–185, 1995.
[878] S. Hundebol. Method and plant for manufacturing cement clinker. Patent: WO 9429231, 1994.
[879] C. V. Hunt, R. J. Powell, M. L. Carter, S. D. Pelley, and L. R. Norman. Encapsulated enzyme breaker and method for use in treating subterranean formations. Patent: US 5604186, 1997.
[880] Z. Huo, E. Freer, M. Lamar, D. M. Knauss, E. D. Sloan, Jr., and B. Sannigrahi. Hydrate plug prevention by anti-agglomeration. *Chem Eng Sci,* 56(17):4979–4991, September 2001.
[881] B. G. Hurd. Method for using foams to improve alkaline flooding oil recovery. Patent: US 4981176, 1991.
[882] R. D. Hutchins and H. T. Dovan. Method for reducing water production from wells. Patent: US 5161615, 1992.
[883] R. D. Hutchins and D. L. Saunders. Tracer chemicals for use in monitoring subterranean fluids. Patent: US 5246860, 1993.
[884] R. J. Hwang and J. Ortiz. Mitigation of asphaltics deposition during CO_2 flood by enhancing CO_2 solvency with chemical modifiers. *Organic Geochem,* 31(12):1451–1462, 1999.
[885] F. B. Ibragimov, A. I. Kolesov, E. A. Konovalov, N. T. Rud, B. M. Gavrilov, Ju. N. Mojsa, A. A. Rjabokon, and O. M. Shcherbaeva. Preparation of lignosulphonate reagent—for drilling solutions, involves additional introduction of water-soluble salt of iron, and antifoaming agent. Patent: RU 2106383-C, 1998.
[886] V. E. Ignateva, A. G. Telin, N. I. Khisamutdinov, S. V. Safronov, V. N. Artemev, and Y. A. Ermilov. Composition for oil extraction—contains hydrocarbon- or alcohol-containing solvent, water and vat residue from production of glycerine or ethylene glycol. Patent: RU 2065941-C, 1996.

[887] M. J. Incorvia. Polythioether corrosion inhibition system. Patent: US 4759908, 1988.
[888] M. J. Incorvia. Thiol ester corrosion inhibition system. Patent: US 4744948, 1988.
[889] M. R. Islam and A. Chakma. Mathematical modelling of enhanced oil recovery by alkali solutions in the presence of cosurfactant and polymer. *J Petrol Sci Eng,* 5(2):105–126, February 1991.
[890] F. R. Ismagilov, A. V. Kazantsev, R. R. Akhunov, V. A. Rygalov, P. G. Navalikhin, V. V. Andrianov, and A. M. Fakhriev. Odour-inducing agent for compressed hydrocarbon gases—contains alkyl-mercaptane(s) and additionally industrial waste from oxidising de-mercaptanisation of propane-pentane fraction. Patent: RU 2051168-C, 1995.
[891] C. D. Ivan, L. D. Blake, and J. L. Quintana. Aphron-base drilling fluid: Evolving technologies for lost circulation control. In *Proceedings Volume.* Annu SPE Tech Conf (New Orleans, LA, 9/30–10/3), 2001.
[892] M. V. Ivanov and S. S. Belyaev. Biotechnology and enhanced oil recovery. *Neft Khoz,* (10):28–32, October 1989.
[893] V. I. Ivashov, R. U. Shafiev, T. Azizkhanov, K. Kh. Azizov, and M. A. Usmanova. Transporting highly viscous, water-laden oil through pipe—comprises addition to oil of 1:1 mixture of washing liquid and prescribed antifreeze (rus). Patent: SU 1827499-A, 1993.
[894] J. Ivory, M. Derocco, and N. Paradis. Investigation of the mechanisms involved in the steam-air injection process. In *Preprints,* number 21. 3rd CIM Petrol Soc Tech Conf (Regina, Canada, 9/25–9/27), 1989.
[895] D. F. Jacques, J. Bock, and P. L. Valint. Oil-in-water emulsion breaking with hydrophobically functionalized cationic polymers. Patent: US 4741835, 1988.
[896] J. A. Jakubowski. Admixtures of 2-bromo-2-bromomethylglutaro-nitrile and 2,2-dibromo-3-nitrilopropionamide. Patent: US 4604405, 1986.
[897] A. K. M. Jamaluddin and T. W. Nazarko. Process for increasing near-wellbore permeability of porous formations. Patent: US 5361845, 1994.
[898] J. Jamth. A cement slurry and a method for the production of such a cement slurry, as well as a use for a light weight filling material as an additive in a cement slurry. Patent: WO 9819976, 1998.
[899] L. K. Jang, T. F. Yen, G. V. Chilingarian, and E. C. Donaldson. Bacterial migration through nutrient-enriched sandpack columns for in-situ recovery of oil. In E. C. Donaldson, G. V. Chilingarian, and T. F. Yen, editors, *Microbial enhanced oil recovery,* volume 22 of *Developments in petroleum science,* pages 151–164. Elsevier Science Ltd, 1989.

[900] M. Jarrett. Nonionic alkanolamides as shale stabilizing surfactants for aqueous well fluids. Patent: WO 9632455, 1996.
[901] M. Jarrett. Amphoteric acetates and glycinates as shale stabilizing surfactants for aqueous well fluids. Patent: US 5593952, 1997.
[902] M. Jarrett. Nonionic alkanolamides as shale stabilizing surfactants for aqueous well fluids. Patent: US 5607904, 1997.
[903] G. E. Jenneman and J. B. Clark. Injection of scale inhibitors for subterranean microbial processes. Patent: US 5337820, 1994.
[904] G. E. Jenneman and J. B. Clark. Utilization of phosphite salts as nutrients for subterranean microbial processes. Patent: US 5327967, 1994.
[905] A. R. Jennings, Jr. Method of enhancing stimulation load fluid recovery. Patent: US 5411093, 1995.
[906] N. J. Jenvey, A. F. Maclean, A. F. Miles, and H. T. R. Montgomerie. The application of oil soluble scale inhibitors into the Texaco Galley reservoir: A comparison with traditional squeeze techniques to avoid problems associated with wettability modification in low water-cut wells. 2nd Annu SPE Oilfield Scale Int Symp (Aberdeen, Scotland, 1/26–1/27), 2000.
[907] D. Jiao and M. M. Sharma. Mechanism of cake buildup in crossflow filtration of colloidal suspensions. *J Colloid Interface Sci,* 162(2):454–462, February 1994.
[908] Z. Jiashen and Z. Jingmao. Control of corrosion by inhibitors in drilling muds containing high concentration of H_2S. *Corrosion,* 49(2):170–174, February 1993.
[909] C. K. Johnson and K. T. Tse. Bisphenol-containing resin coating articles and methods of using same. Patent: EP 735234, 1996.
[910] C. K. Johnson, K. T. Tse, and C. J. Korpics. Improved phenolic resin coated proppants with reduced hydraulic fluid interaction. Patent: EP 542397, 1993.
[911] C. K. Johnson, K. T. Tse, and C. J. Korpics. Phenolic coated proppants with reduced hydraulic fluid interaction. Patent: CA 2067261, 1993.
[912] D. M. Johnson and J. S. Ippolito. Corrosion inhibitor and sealable thread protector end cap for tubular goods. Patent: US 5352383, 1994.
[913] D. M. Johnson and J. S. Ippolito. Corrosion inhibitor and sealable thread protector end cap for tubular goods. Patent: US 5452749, 1995.
[914] J. D. Johnson, S. L. Fu, M. J. Bluth, and R. A. Marble. Inhibiting corrosion. Patent: GB 2299331, 1996.
[915] M. Johnson. Fluid systems for controlling fluid losses during hydrocarbon recovery operations. Patent: EP 691454, 1996.
[916] M. H. Johnson. Completion fluid-loss control using particulates. In *Proceedings Volume,* pages 319–320. SPE Formation Damage Contr Int Symp (Lafayette, LA, 2/9–2/10), 1994.

[917] M. H. Johnson and K. D. Smejkal. Fluid system for controlling fluid losses during hydrocarbon recovery operations. Patent: US 5228524, 1993.

[918] R. L. Johnston and Y. N. Lee. Nonaqueous drag reducing suspensions. Patent: WO 9816586, 1998.

[919] C. K. Jones and D. B. Acker. Oil-based drilling muds with increased viscosity. Patent: EP 922743, 1999.

[920] C. K. Jones, D. A. Williams, and C. C. Blair. Gelling agents comprising aluminium phosphate compounds. Patent: GB 2326882, 1999.

[921] R. R. Jones and R. B. Carpenter. New latex, expanding thixotropic cement systems improve job performance and reduce costs. In *Proceedings Volume*, pages 125–134. SPE Oilfield Chem Int Symp (Anaheim, CA, 2/20–2/22), 1991.

[922] M. M. Jordan, K. Sjursaether, R. Bruce, and M. C. Edgerton. Inhibition of lead and zinc sulphide scale deposits formed during production from high temperature oil and condensate reservoirs. In *Proceedings Volume*. SPE Asia Pacific Oil & Gas Conf (Brisbane, Australia, 10/16–10/18), 2000.

[923] M. M. Jordan, K. S. Sorbie, P. Chen, P. Armitage, P. Hammond, and K. Taylor. The design of polymer and phosphonate scale inhibitor precipitation treatments and the importance of precipitate solubility in extending squeeze lifetime. In *Proceedings Volume*, pages 641–651. SPE Oilfield Chem Int Symp (Houston, TX, 2/18–2/21), 1997.

[924] B. Ju, Z. Luan, Z. Wu, and G. Lu. A study of removal of organic formation damage by experiments and modeling approaches. In *Proceedings Volume*. SPE Asia Pacific Oil & Gas Conf (Jakarta, Indonesia, 4/17–4/19), 2001.

[925] S. Jullian, M. Thomas, and A. Rojey. Process for complete treatment of natural gas at a storage site (procede de traitement global de gaz naturel sur un site de stockage). Patent: EP 781832, 1997.

[926] E. K. Just and R. G. Nickol. Phosphated, oxidized starch and use of same as dispersant in aqueous solutions and coating for lithography. Patent: EP 319989, 1989.

[927] H. Justnes and E. Dahl-Jorgensen. Alternative cementing materials for completion of deep, hot oil-wells. Patent: WO 9412445, 1994.

[928] Y. T. Kalashnikov. Lubricant-sealer for profiled joints of casing pipes—contains soap plastic lubricant, polyacrylamide or carboxymethyl cellulose and additionally gypsum or cement powder, to increase sealing rate. Patent: RU 2007438-C, 1994.

[929] B. Kalpakci, T. G. Arf, D. M. Grist, S. B. Hyde, O. Vikane, and S. Espedal. A preliminary evaluation of an LTPF (low tension polymer flood) process for Statfjord Field, Norway. In *Proceedings Volume*,

volume 1, pages 193–207. 7th EAPG Impr Oil Recovery Europe Symp (Moscow, Russia, 10/27–10/29), 1993.
[930] R. D. Kanakamedala and M. R. Islam. A new method of petroleum sludge disposal and utilization. In *Proceedings Volume,* volume 2, pages 675–682. 6th Unitar et al Heavy Crude & Tar Sands Int Conf (Houston, TX, 2/12–2/17), 1995.
[931] S. Kanda and Z. Kawamura. Stabilization of xanthan gum in aqueous solution. Patent: GB 2192402, 1988.
[932] S. Kanda and Z. Kawamura. Stabilization of xanthan gum in aqueous solution. Patent: US 4810786, 1989.
[933] S. Kanda, M. Yanagita, and Y. Sekimoto. A method of stabilizing a fracturing fluid and a stabilized fracturing fluid. Patent: GB 2172007, 1986.
[934] S. Kanda, M. Yanagita, and Y. Sekimoto. Stabilized fracturing fluid and method of stabilizing fracturing fluid. Patent: US 4721577, 1988.
[935] C. Kang, W. P. Jepson, and M. Gopal. The effect of drag reducing agents on corrosion in multiphase flow. In *Proceedings Volume.* NACE Int Corrosion Conf (Corrosion 98) (San Diego, CA, 3/22–3/27), 1998.
[936] S. F. O. Karaev, S. O. O. Gusejnov, S. V. K. Garaeva, and G. M. O. Talybov. Producing propargyl ether for use as metal corrosion inhibitors—by condensing propargyl alcohol with olefin in presence of phosphotungstic acid. Patent: RU 2056401-C, 1996.
[937] E. V. Karaseva, S. N. Dedyukhina, and A. A. Dedyukhin. Treatment of water-based drilling solution to prevent microbial attack—by addition of dimethyl-tetrahydro-thiadiazine-thione bactericide. Patent: RU 2036216-C, 1995.
[938] A. Karydas. Use of organic fluorochemical compounds with oleophobic and hydrophobic groups in asphaltenic crude oils as viscosity reducing agents. Patent: EP 256979, 1988.
[939] A. I. Kashirskij, F. G. Mulyukov, V. V. Popov, Y. N. Rukhlin, A. Ya. Svetov, A. V. Glebov, and T. V. Puryga. Bitumen composition for insulation of pipelines—contains oxidised bitumen and hexamethylenetetramine as additive, giving insulation layer with good mechanical strength and corrosion protection. Patent: RU 2021309-C, 1994.
[940] N. G. Kashkarov, E. A. Konovalov, V. I. Vjakhirev, A. N. Gnoevykh, A. A. Rjabokon, and N. N. Verkhovskaja. Lubricant reagent for drilling muds—contains spent sunflower oil, and light tall oil and spent coolant-lubricant as modifiers. Patent: RU 2105783-C, 1998.
[941] N. G. Kashkarov, N. N. Verkhovskaya, A. A. Ryabokon, A. N. Gnoevykh, E. A. Konovalov, and V. I. Vyakhirev. Lubricating reagent for drilling fluids—consists of spent sunflower oil modified with additive in form of aqueous solutions of sodium alkylsiliconate(s). Patent: RU 2076132-C, 1997.

[942] D. L. Katz. *Handbook of natural gas engineering*. McGraw-Hill Series in Chemical Engineering. McGraw-Hill, New York, 1959.
[943] R. W. Keatch. Removal of sulphate scale from surface. Patent: GB 2314865, 1998.
[944] S. Kedzierski. The solubility of odorants in natural gas (rozpuszczalnosc preparatow nawaniajacych w gazie ziemnym). *Nafta Gaz (Pol)*, 52(8):357–360, August 1996.
[945] J. D. Kehoe and M. K. Joyce. Water soluble liquid alginate dispersions. Patent: US 5246490, 1993.
[946] G. Keilhofer and J. Plank. Solids composition based on clay minerals and use thereof. Patent: US 6025303, 2000.
[947] M. A. Kelland, T. M. Svartaas, and L. Dybvik. Studies on new gas hydrate inhibitors. In *Proceedings Volume*, pages 531–539. SPE Offshore Europe Conf (Aberdeen, Scotland, 9/5–9/8), 1995.
[948] S. R. Keller. Flow method and apparatus for well cementing. Patent: GB 2172629.
[949] S. R. Keller. Oscillatory flow method for improved well cementing. Patent: CA 1225018, 1987.
[950] M. A. Kennard and G. McNulty. Depositing corrosion inhibitors effectively. *Pipeline Gas J*, 220(4):66–71, April 1993.
[951] M. A. Kennard and J. G. McNulty. Conventional pipeline-pigging technology: Pt 2: Corrosion-inhibitor deposition using pigs. *Pipes Pipelines Int*, 37(4):14–20, July–August 1992.
[952] W. C. Kennedy, Jr. Corrosion inhibitors for cleaning solutions. Patent: US 4637899, 1987.
[953] T. Kenworthy. Polyurea coatings for offshore oil rigs in the Gulf of Mexico. *Mater Performance*, 40(10):40–42, October 2001.
[954] H. U. Khan, J. Handoo, K. M. Agrawal, and G. C. Joshi. Determination of wax separation temperature of crude oils from their viscosity behaviour. *Erdöl Erdgas Kohle*, 107(1):21–22, January 1991.
[955] A. G. Khanlarova, M. R. Musaev, A. M. Samedov, L. I. Kandinskaya, A. G. Gasanov, L. I. Alieva, K. G. Mirzoeva, and S. T. Alieva. Inhibiting growth of sulphate-reducing bacteria—involves introducing diammonium salts of tetrahydrophthalic acid or methyl-tetrahydrophthalic acid into bacteria-containing circulating water. Patent: SU 1828917-A, 1993.
[956] A. Khaund. Sintered low density gas and oil well proppants from a low cost unblended clay material of selected composition. Patent: US 4668645, 1987.
[957] A. K. Khaund. Improved stress-corrosion resistant proppant for oil and gas wells. Patent: EP 207427, 1987.
[958] A. K. Khaund. Stress-corrosion resistant proppant for oil and gas wells. Patent: US 4639427, 1987.

[959] R. Kh. Khazipov, N. N. Silishchev, T. A. Sokolova, A. F. Mufteev, and D. L. Rakhmankulov. Bactericidal activity of glycol monoesters. *Izv Vyssh Ucheb Zavedenii, Neft Gaz,* (1):45–47, January 1990.

[960] D. A. Khisaeva, V. A. Blazhevich, and V. G. Umetbaev. Plugging solution for isolation of absorption zones in boreholes—includes bentonite, calcium chloride, buckwheat husks and polymer reagent produced by interaction of polymethyl-methacrylate wastes with monoethanolamine. Patent: SU 1739005-A, 1992.

[961] V. N. Khlebnikov, V. G. Umetbaev, A. B. Loginov, R. M. Nazmetdinov, and R. M. Kamaletdinova. Use of salt-containing secondary resources as cement set accelerators. *Neftepromysl Delo,* (12):21–24, December 1997.

[962] M. I. Khoma. Composition of foam-extinguishing agent for drilling solutions—contains waste from production of hydrophilic and hydrophobic aerosil, modified with bifunctional silico-organic compound and diesel oil. Patent: SU 1795977-A, 1993.

[963] J. Kieffer, M. Michaux, and P. Rae. Additive for controlling the filtrate of well cementing slurries and corresponding cementing process (additif pour le controle du filtrat des laitiers de cimentation de puits, et procede de cimentation correspondant). Patent: FR 2592056, 1987.

[964] J. J. Kilbane, II, P. Chowdiah, K. J. Kayser, B. Misra, K. A. Jackowski, V. J. Srivastava, G. N. Sethu, A. D. Nikolov, and D. T. Wasan. In-situ remediation of contaminated soils using foams as carriers for chemicals, nutrients, and other amendments. In *Proceedings Volume*. 9th Inst Gas Technol Gas, Oil, & Environ Biotechnol Int Symp (Colorado Springs, CO, 12/9–12/11), 1996.

[965] J. J. Kilbane, II, P. Chowdiah, K. J. Kayser, B. Misra, K. A. Jackowski, V. J. Srivastava, G. N. Sethu, A. D. Nikolov, and D. T. Wasan. Remediation of contaminated soils using foams. In *Proceedings Volume*. 10th Inst Gas Technol Gas, Oil & Environ Biotechnol & Site Remediation Technol Int Symp (Orlando, FL, 12/8–12/10), 1997.

[966] J. E. Killough. Hydrocarbon recovery process. Patent: US 4678033, 1987.

[967] D. Kinchen, M. A. Peavy, T. Brookey, and D. Rhodes. Case history: Drilling techniques used in successful redevelopment of low pressure H_2S gas carbonate formation. In *Proceedings Volume,* volume 1, pages 392–403. SPE/IADC Drilling Conf (Amsterdam, the Netherlands, 2/27–3/1), 2001.

[968] B. J. King and P. L. Totten. Well cementing method using acid removable low density well cement compositions. Patent: US 5213161, 1993.

[969] M. T. King, J. Gulbis, G. W. Hawkins, and H. D. Brannon. Encapsulated breaker for aqueous polymeric fluids. In *Proceedings Volume,*

volume 2. CIM Petrol Soc/SPE Int Tech Mtg (Calgary, Canada, 6/10–6/13), 1990.
[970] C. L. Kissel. Method for reducing the pour point of an oil and compositions for use therein. Patent: US 5593955, 1997.
[971] C. L. Kissel. Process and composition for inhibiting corrosion (Verfahren und Zusammensetzung zur Inhibierung von Korrosion). Patent: EP 906969, 1999.
[972] U. C. Klomp, V. R. Kruka, R. Reijnhart, and A. J. Weisenborn. Method for inhibiting the plugging of conduits by gas hydrates. Patent: US 5460728, 1995.
[973] P. Klug, M. Feustel, and V. Frenz. Additives to inhibit the formation of gas hydrate. Patent: WO 9822557, 1998.
[974] P. Klug and M. Kelland. Additives for inhibiting formation of gas hydrates. Patent: WO 9823843, 1998.
[975] E. B. Klusmann. Odor monitoring at Southern California Gas Company. In *Odorization,* volume 3, pages 407–423. Institute of Gas Technology, Chicago, IL, Southern California Gas Co, 1993.
[976] A. A. Klyusov, M. M. Shalyapin, Y. N. Kalugin, and L. M. Kargapoltseva. Plugging solution for low temperature oil wells—contains plugging cement, water and hardening accelerator in form of sludge waste from production of metallic magnesium from carnallite. Patent: SU 884366-A, 1996.
[977] M. Knickrehm, E. Caballero, P. Romualdo, and J. Sandidge. Use of chlorine dioxide in a secondary recovery process to inhibit bacterial fouling and corrosion. In *Proceedings Volume.* NACE Corrosion 87 (San Francisco, CA, 3/9–3/13), 1987.
[978] E. E. Kochnev, G. I. Merentsova, T. L. Andreeva, and V. A. Ershov. Inhibitor solution to avoid inorganic salts deposition in oil drilling operations—contains water, carboxymethylcellulose or polyacrylamide and polyaminealkyl phosphonic acid and has improved distribution uniformity. Patent: SU 1787996-A, 1993.
[979] G. Koerner and D. Schaefer. Utilization of polyoxyalkylene-polysiloxane block mix polymers as demulsifiers for crude oil containing water (Verwendung von Polyoxyalkylen-Polysiloxan-Blockmischpolymerisaten als Demulgatoren für Wasser Enthaltendes Erdöl). Patent: DE 3622571, 1988.
[980] G. Koerner and D. Schaefer. Polyoxyalkylene-polysiloxane block copolymers as demulsifiers for water-containing oil. Patent: US 5004559, 1991.
[981] C. A. Koh, J. L. Savidge, and C. C. Tang. Time-resolved in-situ experiments on the crystallization of natural gas hydrates. *J Phys Chem,* 100(16):6412–6414, 1996.
[982] R. S. Kohn. Thixotropic aqueous solutions containing a divinylsulfone-crosslinked polygalactomannan gum. Patent: US 4752339, 1988.

[983] S. J. Kok, J. Guns, L. C. Kraan, G. E. Schuringa, and R. P. W. Kesselmans. Drilling fluids. Patent: WO 9952990, 1999.
[984] S. J. Kok, L. C. Kraan, G. E. Schuringa, J. Guns, and R. P. W. Kesselmans. Drilling fluids. Patent: EP 949311, 1999.
[985] I. P. Kolesnikova, N. A. Sushkova, and B. N. Ershov. Plugging material giving plugging rock of increased strength—contains oil sludge from primary oil processing, formaldehyde and sulphuric acid. Patent: SU 1781415-A, 1992.
[986] J. J. Kolle. Coiled tubing drilling with supercritical carbon dioxide. Patent: US 6347675, 2002.
[987] T. J. Koltermann and T. F. Willey. Lubricating grease. Patent: US 5891830, 1999.
[988] T. J. Koltermann and T. F. Willey. Lubricating grease. Patent: US 6056072, 2000.
[989] N. S. Kommareddi and L. J. Rzeznik. Microencapsulated drag reducing agents. Patent: US 6126872, 2000.
[990] N. Kommarredi and L. J. Rzeznik. Microencapsulated drag reducing agents. Patent: WO 9937396, 1999.
[991] M. Kondo and T. Sawada. Readily dispersible bentonite. Patent: US 5491248, 1996.
[992] E. A. Konovalov, Y. A. Ivanov, T. N. Shumilina, V. F. Pichugin, and N. N. Komarova. Lubricating reagent for drilling solutions—contains agent based on spent sunflower oil, water, vat residue from production of oleic acid, and additionally water glass. Patent: SU 1808861-A, 1993.
[993] E. A. Konovalov, A. L. Rozov, A. P. Zakharov, Yu. A. Ivanov, V. F. Pichugin, and N. N. Komarova. Lubricating reagent for drilling solutions—contains spent sunflower oil as active component, water, boric acid as emulsifier, and additionally water glass. Patent: SU 1808862-A, 1993.
[994] N. Kosaric. Bio-de-emulsifiers. In *Proceedings Volume,* pages 549–587. 2nd Inst Gas Technol Gas, Oil, Coal, & Environ Biotechnol Int Symp (New Orleans, LA, 12/11–12/13), 1989.
[995] K. C. Koshel, J. S. Bhatia, S. Kumar, and A. K. Samant. Corrosion problem in kalol injection water pipeline system and its control by using corrosion inhibitors. *Ongc Bull,* 25(2):115–133, December 1988.
[996] V. N. Koshelev, A. P. Krezub, D. M. Ponomarev, Y. N. Mojsa, N. V. Frolova, A. I. Penkov, and S. V. Vasilchenko. Drilling solution with improved lubricating and rheology—contains additionally oxyalkylated alkylphenol with nitrogen-containing additive, in aromatic solvent. Patent: SU 1797617-A, 1993.
[997] M. Kostic. On turbulent drag and heat transfer reduction phenomena and laminar heat transfer enhancement in non-circular duct flow of

certain non-Newtonian fluids. *Int J Heat Mass Transfer,* 37:133–147, March 1994.

[998] S. V. Kosyak, V. S. Danyushevskij, M. E. Pshebishevskij, and A. A. Trapeznikov. Plugging formation fluid transmitting channel—by successive injection of aqueous solution of polyacrylamide and liquid glass, buffer liquid and aqueous solution of polyacrylamide and manganese nitrate. Patent: SU 1797645-A, 1993.

[999] V. S. Kotelnikov, S. N. Demochko, V. G. Fil, and I. S. Marchuk. Drilling mud composition—contains carboxymethyl cellulose, acrylic polymer, ferrochrome lignosulphonate, cement and water. Patent: SU 1829381-A, 1996.

[1000] V. S. Kotelnikov, S. N. Demochko, V. G. Fil, and I. I. Rybchich. Polymeric composition for isolation of absorbing strata—contains ferrochromo-lignosulphonate, water-soluble acrylic polymer and water. Patent: SU 1730435-A, 1992.

[1001] V. S. Kotelnikov, S. N. Demochko, M. P. Melnik, and V. P. Mikitchak. Improving properties of drilling solution—by addition of ferrochrome-lignosulphonate and aqueous solution of cement and carboxymethyl cellulose. Patent: SU 1730118-A, 1992.

[1002] M. Kotsaridou-Nagel and B. Kragert. Demulsifying water-in-oil emulsions through chemical addition (Spaltungsmechanismus von Wasser-in-Erdöl-Emulsionen bei Chemikalienzusatz). *Erdöl Erdgas Kohle,* 112(2):72–75, February 1996.

[1003] F. S. Kovarik and J. P. Heller. Improvement of CO_2 flood performance. US DOE Rep DOE/MC/21136-24, New Mex Inst Mining Techn, August 1990.

[1004] R. M. Kowalik, I. Duvdevani, K. Kitano, and D. N. Schulz. Drag reduction agents for hydrocarbon solutions. Patent: US 4625745, 1986.

[1005] R. M. Kowalik, I. Duvdevani, R. D. Lundberg, D. N. Schulz, D. G. Peiffer, and K. Kitano. Enhanced drag reduction via interpolymer associations. *J Non-Newtonian Fluid Mech,* 24(1):1–10, April 1987.

[1006] T. C. Kowalski and R. W. Pike. Microencapsulated oil field chemicals. Patent: US 5922652, 1999.

[1007] T. C. Kowalski and R. W. Pike. Microencapsulated oil field chemicals. Patent: US 6326335, 2001.

[1008] W. Krass, A. Kittel, and A. Uhde. *Pipelinetechnik: Mineralölfernleitungen.* Verlag TÜV, Köln, 1979.

[1009] R. P. Kreh. Method of inhibiting corrosion and scale formation in aqueous systems. Patent: US 5073339, 1991.

[1010] L. N. Kremer. Demulsifying composition. Patent: US 5176847, 1993.

[1011] B. G. Kriel, A. B. Crews, E. D. Burger, E. Vanderwende, and D. O. Hitzman. The efficacy of formaldehyde for the control of biogenic sulfide production in porous media. In *Proceedings Volume,* pages 441–448. SPE Oilfield Chem Int Symp (New Orleans, LA, 3/2–3/5), 1993.

[1012] K. Kristiansen. Composition for use in well drilling and maintenance. Patent: WO 9409253, 1994.

[1013] M. Krull, S. P. von Halasz, W. Reimann, J. Balzer, and H. Geiss. Copolymers of ethylenically unsaturated carboxylic acid esters with polyoxyalkylene ethers of lower, unsaturated alcohols as flow-improving agents for paraffin containing oils. Patent: US 5718821, 1998.

[1014] E. Kubena, Jr., L. E. Whitebay, and J. A. Wingrave. Method for stabilizing boreholes. Patent: US 5211250, 1993.

[1015] C. H. Kucera, S. C. Crema, M. D. Roznowski, G. Konrad, and H. Hartmann. Fluid loss control additives for oil well cementing compositions. Patent: EP 342500, 1989.

[1016] Z. N. Kudryashova, B. V. Mikhajlov, A. P. Tarnavskij, S. R. Khajrullin, V. M. Mustafaev, and P. F. Tsytsymushkin. Plugging solution for cementing oil and gas wells—contains plugging cement, ash, additional clay powder and crude light pyridine bases, and water. Patent: SU 1765366-A, 1992.

[1017] L. E. Kukacka and T. Sugama. Lightweight CO_2-resistant cements for geothermal well completions. Brookhaven Nat Lab Rep BNL-60326, Brookhaven National Lab, 1994.

[1018] L. E. Kukacka and T. Sugama. Lightweight CO_2-resistant cements for geothermal well completions. Brookhaven Nat Lab Rep BNL-61259, Brookhaven National Lab, 1995.

[1019] A. K. Kuksov, A. P. Krezub, N. A. Mariampolskij, L. I. Ryabova, G. N. Lyshko, D. A. Loskutov, and E. I. Zhmurkevich. Oil and gas borehole plugging solution reagent—contains brown coal treated with alkali, organic silicon compound and lignosulphonate. Patent: SU 1719618-A, 1992.

[1020] K. Kulpa, R. Adkins, and N. S. Walker. New testing vindicates use of barite. *Amer Oil Gas Reporter,* 35(4):52–54, April 1992.

[1021] F. R. Kunkel. Water-oil separation method. Patent: US 5000857, 1991.

[1022] R. A. Kunzi, E. F. Vinson, P. L. Totten, and B. G. Brake. Low temperature well cementing compositions and methods. Patent: CA 2088897, 1993.

[1023] R. A. Kunzi, E. F. Vinson, P. L. Totten, and B. G. Brake. Low temperature well cementing compositions and methods. Patent: US 5346550, 1994.

[1024] R. A. Kunzi, E. F. Vinson, P. L. Totten, and B. G. Brake. Low temperature well cementing compositions and methods. Patent: US 5447198, 1995.

[1025] R. Kupfer, W. Böse, M. Hille, R. Böhm, and F. Staiss. Esterified glycidyl ether additives and their application (Veresterte Glycidylether-Additionsprodukte und deren Verwendung). Patent: EP 333135, 1989.
[1026] R. Kupfer, W. Böse, M. Hille, R. Böhm, and F. Staiss. Process for the separation of crude oil emulsions of the water-in-oil type (Verfahren zum Trennen von Erdölemulsionen vom Typ Wasser-in-Öl). Patent: EP 333141, 1989.
[1027] R. Kupfer, W. Böse, M. Hille, R. Böhm, and F. Staiss. Process for the separation of crude oil emulsions of the water-in-oil type. Patent: US 5039450, 1991.
[1028] R. Kupfer, W. Böse, M. Hille, R. Böhm, and F. Staiss. Process for the separation of crude oil emulsions of the water-in-oil type. Patent: CA 1319876, 1993.
[1029] R. Kupfer, M. Hille, R. Böhm, and F. Staiss. Process for separating petroleum emulsions of the water-in-oil type (Verfahren zum Trennen von Erdölemulsionen Vom Wasser-in-Öl-Typ). Patent: EP 572881, 1993.
[1030] R. Kupfer, M. Hille, R. Böhm, and F. Staiss. Process for separation of petroleum emulsions of the water-in-oil type. Patent: US 5385674, 1995.
[1031] Y. M. Kurbanov, S. E. Nikashin, and N. Ya. Kalugina. Plugging solution for cementing casing strings of oil wells—contains plugging cement, slag from production of high-alloyed aluminium, polyvalent metal salt in form of e.g. copper sulphate and water. Patent: SU 1795083-A, 1993.
[1032] Y. M. Kurbanov, S. E. Nikashin, and N. Ya. Kalugina. Plugging solution for cementing oil and gas wells—contains plugging cement, aluminium slag from highly alloyed aluminium production, acrylic polymer or cellulose ether and water. Patent: SU 1789665-A, 1993.
[1033] B. M. Kurochkin, S. N. Khannanov, R. Z. Saitgareev, S. A. Kashapov, and I. A. Sagidullin. Ways of effective use of rubber crumbs during plugging works in cased wells. *Neftepromysl Delo*, (12):19–21, December 1996.
[1034] B. M. Kurochkin, L. V. Kolesov, and M. B. Biryukov. Use of ellipsoidal glass granules as an antifriction mud additive. *Neft Khoz*, (12):61–64, December 1990.
[1035] B. M. Kurochkin, L. V. Kolesov, V. I. Masich, N. V. Stepanov, V. F. Tselovalnikov, V. T. Alekperov, I. N. Kerimov, O. N. Ibragimov, and B. M. Bulanov. Solution for drilling gas and oil wells—contains ellipsoidal glass beads as additive reducing friction between walls of well and casing string. Patent: SU 1740396-A, 1992.
[1036] B. M. Kurochkin, N. T. Moskvicheva, V. N. Lobanova, I. S. Kateev, and R. A. Khabibullin. Plugging solution for isolating absorption zones in oil and gas wells—contains plugging Portland cement, heterogeneous

bi-quaternary ammonium salt, plastic waste as filler, additional calcined soda, and water. Patent: SU 1726733-A, 1992.

[1037] B. M. Kurochkin, E. A. Simonyan, A. A. Simonyan, E. F. Khirazov, P. A. Ozarchuk, V. O. Voloshinivskii, and A. Ya. Glushakov. New technology of drilling with the use of glass granules. *Neft Khoz,* (7):9–11, July 1992.

[1038] B. M. Kurochkin and V. F. Tselovalnikov. Use of ellipsoidal glass granules for drilling under complicated conditions. *Neft Khoz,* (10): 7–13, October 1994.

[1039] O. B. Kutergin, V. P. Melnikov, and A. N. Nesterov. Effect of surface-active agents on the mechanism and kinetics of hydrate formation of gases. *Dokl Akad Nauk Sssr,* 323(3):549–553, 1992.

[1040] H. C. Kuzee and H. W. C. Raaijmakers. Method for preventing deposits in oil extraction. Patent: WO 9964716, 1999.

[1041] V. L. Kuznetsov, G. A. Lyubitskaya, E. I. Kolesnik, E. N. Kazakova, B. M. Kurochkin, and V. N. Lobanova. Plugging solution for isolating absorption zones in oil and gas wells—contains prescribed synthetic latex, water soluble salt of methacrylate-methacrylic acid copolymer as additive, and water. Patent: RU 2024734-C, 1994.

[1042] V. L. Kuznetsov, G. A. Lyubitskaya, E. I. Krayushkina, E. I. Kolesnik, E. N. Kazakova, V. K. Vystorop, S. D. Nechaeva, A. D. Polikarpov, B. M. Kurochkin, O. V. Makhanko, and V. N. Lobanova. Plugging solution for oil and gas wells—contains polyisoprene or butadiene-styrene latex, bentonite or chalk as filler, polyoxypropylene as additive, and water. Patent: SU 1733624-A, 1992.

[1043] B. Kvamme. Mechanisms for initiation of hydrate from liquid water: Liquid phase clustering, surface adsorbtion, or what? In *Proceedings Volume,* pages 306–310. New York Acad Sci et al Natur Gas Hydrates Int Conf (New Paltz, NY, 6/20–6/24), 1993.

[1044] V. B. Kvashenkin. Plugging solution for cementing low pressure oil and gas wells—contains plugging Portland cement, waste of silicon production as the lightening additive and calcium chloride as mineral salt, and water. Patent: SU 1832149-A, 1993.

[1045] A. Lacret and A. Donche. Use of scleroglucan muds for the drilling of large diameter holes (application des boues au scleroglucane au forage des puits a gros diametre). Patent: FR 2662447, 1991.

[1046] A. Ladret and A. Donche. Use of scleroglucan muds for the drilling of large diameter holes (application des boues au scleroglucane au forage des puits a gros diametre). Patent: EP 459881, 1991.

[1047] A. Ladret and A. Donche. Application of muds containing scleroglucan to drilling large diameter wells. Patent: US 5525587, 1996.

[1048] H. K. J. Ladva, C. A. Sawdon, and P. R. Howard. Additive for wellbore fluids. Patent: EP 1041242A, 2000.

[1049] H. K. J. Ladva, C. A. Sawdon, and P. R. Howard. Wellbore service fluids. Patent: GB 2348447, 2000.
[1050] T. W. Lai and B. R. Vijayendran. Cement composition for oil well drilling holes containing high molecular weight poly(vinylamines). Patent: EP 331045, 1989.
[1051] J. A. Laird and W. R. Beck. Ceramic spheroids having low density and high crush resistance. Patent: EP 207668, 1989.
[1052] I. Lakatos, K. Bauer, and J. Lakatos-Szabó. Potential application of oxygen containing gases to enhance gravity drainage in heavy oil bearing reservoirs. *Erdöl Erdgas Kohle,* 113(6):260–263, June 1997.
[1053] I. Lakatos and J. Lakatos-Szabó. Effect of IOR/EOR (improved oil recovery/enhanced oil recovery) chemicals on interfacial rheological properties of crude oil/water systems. In *Proceedings Volume.* SPE Oilfield Chem Int Symp (Houston, TX, 2/13–2/16), 2001 SPE Number: 65391.
[1054] I. Lakatos and S. J. Lakatosne. Dynamical characteristics of the polymer flooding: Pt 2 (a polimeres elarasztas kinamikus jellemzoi: 2. resz). *Koolaj Foldgaz,* 24(6):170–178, June 1991.
[1055] J. Lakatos-Szabó and I. Lakatos. Effect of sodium hydroxide on interfacial rheological properties of oil-water systems. In *Colloids Surfaces, Sect A,* volume 149, pages 507–513. 9th Surface & Colloid Sci Int Conf (Sofia, Bulgaria, 7/6–7/12), 1997.
[1056] J. Lakatos-Szabó, I. Lakatos, and B. Kosztin. Role of interfacial rheological properties of oil/water systems in mechanism and design of EOR/IOR technologies. In *Proceedings Volume,* number 057. 9th EAGE Impr Oil Recovery Europe Symp (The Hague, Netherlands, 10/20–10/22) Proc, 1997.
[1057] T. M. Lamarre and C. H. Martin. Synergistic biocide of tributyl tetradecyl phosphonium chloride and 1.5-pentanedial. Patent: CA 1269300, 1990.
[1058] D. K. Landry and T. J. Koltermann. Bearing grease for rock bit bearings. Patent: CA 2018779, 1991.
[1059] D. K. Landry and T. J. Koltermann. Bearings grease for rock bit bearings. Patent: US 5015401, 1991.
[1060] P. Lange and J. Plank. Mixed metal hydroxide (MMH) viscosifier for drilling fluids: properties and mode of action (mixed metal hydroxide [MMH]—Eigenschaften und Wirkmechanismus als Verdickungsmittel in Bohrspülungen). *Erdöl Erdgas Kohle,* 115(7-8):349–353, July-August 1999.
[1061] W. Lange and B. Boehmer. Water-soluble polymers and their use as flushing liquid additives for drilling. Patent: US 4749498, 1988.
[1062] P. W. Langemeier, M. A. Phelps, and M. E. Morgan. Method for reducing the viscosity of aqueous fluids. Patent: EP 330489, 1989.

[1063] B. Langlois. Fluid comprising cellulose nanofibrils and its use for oil mining. Patent: WO 9802499, 1998.

[1064] B. Langlois. Fluids useful for oil mining comprising de-acetylated xanthan gum and at least one compound increasing the medium ionic strength (fluides utilisables dans l'exploitation du petrole comprenant de la gomme xanthane desacetylee et au moins un com un compose augmentant la force ionique du milieu). Patent: WO 9903948, 1999.

[1065] B. Langlois, G. Guerin, A. Senechal, R. Cantiani, I. Vincent, and J. Benchimol. Fluid comprising cellulose nanofibrils and its use for oil mining (fluide comprenant des nanofibrilles de cellulose et son application pour l'exploitation de gisements petroliers). Patent: EP 912653, 1999.

[1066] S. B. Laramay, R. J. Powell, and S. D. Pelley. Perphosphate viscosity breakers in well fracture fluids. Patent: US 5386874, 1995.

[1067] A. L. Larsen. Process for inhibiting corrosion in oil production fluids. Patent: EP 446616, 1991.

[1068] J. Larsen, P. F. Sanders, and R. E. Talbot. Experience with the use of tetrakishydroxymethylphosphonium sulfate (THPS) for the control of downhole hydrogen sulfide. NACE Int Corrosion Conf (Corrosion 2000) (Orlando, FL, 3/26–3/31), 2000.

[1069] H. C. Lau. Laboratory development and field testing of succinoglycan as a fluid-loss-control fluid. *SPE Drilling Completion,* 9(4):221–226, December 1994.

[1070] H. C. Lau. Laboratory development and field testing of succinoglycan as fluid-loss-control fluid. *SPE Peer Approved Pap,* 1994.

[1071] H. C. Lau, A. H. Hale, and L. A. Bernardi, Jr. Drilling fluid. Patent: US H1685, 1997.

[1072] R. S. Lauer, L. N. Kremer, J. L. Stark, and A. McCallum. Method for separating solids from hydrocarbon slurries. Patent: EP 1108775-A, 2001.

[1073] H. F. Lawson and A. H. Hale. Well drilling fluids and process for drilling wells. Patent: US 4812244, 1989.

[1074] I. Lazar. The microbiology of MEOR (microbial enhanced oil recovery): practical experience in Europe. In *Proceedings Volume,* volume 2, pages 329–338. Minerals, Metals & Mater Soc et al Biohydromet Technol Int Symp (Jackson Hole, WY, 8/22–8/25), 1993.

[1075] I. Lazar, A. Voicu, G. Archir, T. Toma, I. G. Lazar, L. Blanck, and V. Constantin. Investigations on a new Romanian biopolymer (pseudozan) for use in enhanced oil recovery (EOR). In *Proceedings Volume,* volume 2, pages 357–364. Minerals, Metals & Mater Soc et al Biohydromet Technol Int Symp (Jackson Hole, WY, 8/22–8/25), 1993.

[1076] H. V. Le, S. Kesavan, J. C. Dawson, D. J. Mack, and S. G. Nelson. Compositions and methods for hydraulic fracturing. Patent: CA 2239599, 1998.

[1077] S. Le Roy-Delage, B. Dargaud, and J. F. Baret. Cementing compositions and use of such compositions for cementing oil wells or the like. Patent: WO 9958467, 1999.

[1078] S. Le Roy-Delage, B. Dargaud, J. F. Baret, and M. Thiercelin. Cementing compositions and the use of such compositions for cementing oil wells or the like. Patent: WO 0020350, 2000.

[1079] M. C. P. Leblanc, J. A. Durrieu, J. P. Binon, J. J. Fery, and G. G. L. Provin. Process for treating an aqueous solution of acrylamide resin in order to enable it to gel slowly even at high temperature. Patent: GB 2197655, 1988.

[1080] M. C. P. Leblanc, J. A. Durrieu, J. P. Binon, G. G. L. Provin, and J. J. Fery. Process for treating an aqueous solution of acrylamide resin in order to enable it to gel slowly even at high temperature (procede de traitement d'une solution aqueuse de resine acrylamide pour en permettre une gelification lente meme a temperature elevee). Patent: EP 267835, 1988.

[1081] M. C. P. Leblanc, J. A. Durrieu, J. P. P. Binon, G. G. Provin, and J. J. Fery. Process for treating an aqueous solution of acrylamide resin in order to enable it to gel slowly even at high temperature. Patent: US 4975483, 1990.

[1082] N. Lecocumichel and C. Amalric. Concentrated aqueous compositions of alkylpolyglycosides, and applications thereof. Patent: WO 9504592, 1995.

[1083] J. Leder. Antimicrobial composition and method of use. Patent: EP 364789, 1990.

[1084] J. Leder. Antimicrobial composition and method of use in oil well flooding. Patent: US 5055493, 1991.

[1085] L. J. Lee, A. Patel, and E. Stamatakis. Glycol based drilling fluid. Patent: WO 9710313, 1997.

[1086] R. L. Lee and D. La Verne Katz. *Natural gas engineering: production and storage*. McGraw-Hill Chemical Engineering Series, McGraw-Hill Economics Dept, New York, 1991.

[1087] Y. N. Lee and F. R. Wiggins. Activation of water-in-oil emulsions of friction reducing polymers for use in saline fluids. Patent: US 5067508, 1991.

[1088] J. C. Leighton and M. J. Sanders. Water soluble polymers containing allyloxybenzenesulfonate monomers. Patent: EP 271784, 1988.

[1089] J. C. Leighton and M. J. Sanders. Water soluble polymers containing allyloxybenzenesulfonate monomers. Patent: US 4892898, 1990.

[1090] G. M. Lein, Jr. Preparation of isothiazolones. Patent: EP 318194, 1989.

[1091] Z. R. Lemanczyk. The use of polymers in well stimulation: performance, availability and economics. In *Proceedings Volume*. Plast Rubber Inst Use of Polymers in Drilling & Oilfield Fluids Conf (London, England, 12/9), 1991.
[1092] Z. R. Lemanczyk. The use of polymers in well stimulation: an overview of application, performance and economics. *Oil Gas Europe Mag*, 18(3):20–26, October 1992.
[1093] P. R. Lemieux and D. S. Rumpf. Low density proppant and methods for making and using same. Patent: EP 353740, 1990.
[1094] P. R. Lemieux and D. S. Rumpf. Lightweight oil and gas well proppant and methods for making and using same. Patent: AU 637576, 1993.
[1095] P. R. Lemieux and D. S. Rumpf. Lightweight proppant for oil and gas wells and methods for making and using same. Patent: AU 637575, 1993.
[1096] P. R. Lemieux and D. S. Rumpf. Lightweight proppants for oil and gas wells and methods for making and using same. Patent: CA 1330255, 1994.
[1097] Y. R. Leonov, M. E. Lamosov, S. A. Ryabokon, V. A. Mosin, B. G. Dzetl, F. G. Mamulov, O. G. Bobrov, and V. M. Savoskin. Plugging material for wells in the oil and gas industry—contains mineral binder and powder waste from production of epoxide resins as epoxide resin-based additive. Patent: RU 2036297-C, 1995.
[1098] Y. R. Leonov, M. E. Lamosov, S. A. Ryabokon, V. A. Mosin, F. G. Mamulov, A. R. Akhmadzhanov, and G. D. Varlamov. Plugging material for oil and gas wells—contains Portland cement and additive in form of product of reaction between furfurol and ammonia and has improved corrosion resistance. Patent: SU 1818463-A, 1993.
[1099] Y. R. Leonov, V. A. Mosin, K. M. Potapov, M. E. Lamosov, and V. N. Ryzhov. Plugging material—contains furfurol-acetone monomer, oligooxymethyl hydride silmethylene-siloxy-silane, and acidic or alkali hardener. Patent: SU 1821550-A, 1993.
[1100] Y. Li, I. C. Y. Yang, K. I. Lee, and T. F. Yen. Subsurface application of *Alcaligenes eutrophus* for plugging of porous media. In E. T. Premuzic and A. Woodhead, editors, *Microbial enhancement of oil recovery: recent advances: Proceedings of the 1992 International Conference on Microbial Enhanced Oil Recovery*, volume 39 of *Developments in Petroleum Science*, pages 65–77. Elsevier Science Ltd, 1993.
[1101] Y. G. Li, S. L. Li, and Z. L. Wang. New high temperature filtration reducer fla for drilling fluid. *Drilling Fluid Completion Fluid*, 13(3):33–35, 1996.
[1102] Y. H. Li, G. R. Chesnut, R. D. Richmond, G. L. Beer, and V. P. Caldarera. Laboratory tests and field implementation of gas drag reduction

chemicals. In *Proceedings Volume,* pages 457–469. SPE Oilfield Chem Int Symp (Houston, TX, 2/18–2/21), 1997.

[1103] Z. Li. Effect of fluid viscoelasticity on isolated eddy transmission. *J Univ Petrol, China,* 15(5):33–38, October 1991.

[1104] G. Liao, Y. Yang, and H. Wang. Effect on anti corrosion ability of oil well slurry using silica flour. *Oil Drilling Prod Technol,* 18(4):31–34,43,106, 1996.

[1105] K. F. Lin. Synthetic paraffinic hydrocarbon drilling fluid. Patent: US 5569642, 1996.

[1106] S. C. Lin, J. C. Goursaud, P. J. Kramer, G. Georgiou, and M. M. Sharma. Production of biosurfactant by *Bacillus licheniformis* strain JF-2. In E. C. Donaldson, editor, *Microbial enhancement of oil recovery: recent advances: Proceedings of the 1990 International Conference on Microbial Enhancement of Oil Recovery,* volume 31 of *Developments in Petroleum Science,* pages 219–226. Elsevier Science Ltd, 1991.

[1107] G. P. Lindblom. Measurement and prediction of depositional accuracy in dispersant spraying from large airplanes. In *Proceedings Volume,* pages 325–328. 10th Bien API et al Oil Spill (Prev, Behav, Contr, Cleanup) Conf (Baltimore, MD, 4/6–4/9), 1987.

[1108] K. O. Lindstrom and W. D. Riley. Soil-cement compositions and their use. Patent: EP 605075, 1994.

[1109] M. R. Lindstrom and R. P. Louthan. Inhibiting corrosion. Patent: US 4670163, 1987.

[1110] M. R. Lindstrom and H. W. Mark. Inhibiting corrosion: benzylsulfinylacetic acid or benzylsulfonylacetic acid. Patent: US 4637833, 1987.

[1111] F. J. Liotta, Jr. and S. A. Schwartz. Supported carbonic acid esters useful as set accelerators and thixotropic agents in cement. Patent: WO 0002828, 2000.

[1112] M. I. Lipkes, A. O. Mezhlumov, L. A. Shits, G. E. Avdeev, V. I. Fomenko, and A. M. Shvetsov. Carbonate weighting material for drilling-in producing formations and well overhaul. *Stroit Neft Gaz Skvazhin Sushe More,* (5–6):34–41, May-June 1996.

[1113] E. S. Littmann and T. L. McLean. Chemical control of biogenic H_2S in producing formations. In *Proceedings Volume,* pages 339–342. SPE Prod Oper Symp (Oklahoma City, OK, 3/8–3/10), 1987.

[1114] W. Littmann. *Polymer flooding,* volume 24 of *Developments in Petroleum Science.* Elsevier Science Publishing Co Inc, New York, 1988.

[1115] H. Liu and Y. Zhang. Rheological property of the xanthan biopolymer flooding systems. *J Univ Petrol, China,* 19(4):41–44, August 1995.

[1116] J. Liu, G. A. Pope, and K. Sepehrnoori. A high-resolution, fully implicit method for enhanced oil recovery simulation. In *Proceedings Volume,* pages 35–50. 13th SPE Reservoir Simulation Symp (San Antonio, TX, 2/12–2/15), 1995.

[1117] Y. Y. Liu, W. G. Hou, D. J. Sun, T. Wu, and C. G. Zhang. Study on function mechanism of filtration reducer: research and development of new filtration reducer. *Drilling Fluid Completion Fluid,* 13(4):12–14, 1996.

[1118] I. Livsey and R. Shaunak. Cement setting retarding agents. Patent: GB 2182031, 1987.

[1119] E. Ljosland, T. Bjornstad, O. Dugstad, and I. Hundere. Perfluorocarbon tracer studies at the Gullfaks field in the North Sea. *J Petrol Sci Eng,* 10(1):27–38, October 1993.

[1120] G. R. Lloyd and P. W. Neail. Biocides for minimum environmental impact. In *Proceedings Volume,* volume 2. 2nd Shell Co Australia et al Australian Int Oil, Gas & Petrochem Conf (Offshore Australia 93) (Melbourne, Australia, 11/23–11/26), 1993.

[1121] T. P. Lockhart and G. Burrafato. Gellable buffered aqueous composition and its use in enhanced petroleum recovery. Patent: EP 390281, 1990.

[1122] T. P. Lockhart and G. Burrafato. Method and composition for reducing the permeability of a high permeability zone in an oil reservoir. Patent: EP 390280, 1990.

[1123] T. P. Lockhart and G. Burrafato. Method and composition for reducing the permeability of a high-permeability zone in an oil reservoir. Patent: CA 2013468, 1990.

[1124] S. V. Logvinenko, D. E. Bateev, Y. V. Chernyak, V. V. Osipov, and V. P. Potapkin. Low density cement slurry preparation—comprises mixing cement with specified weight percentage of aqueous solution of polyacrylnitrile. Patent: RU 2072026-C, 1997.

[1125] D. L. Lord, P. S. Vinod, S. Shah, and M. L. Bishop. An investigation of fluid leakoff phenomena employing a high-pressure simulator. In *Proceedings Volume,* pages 465–474. Annu SPE Tech Conf (Dallas, TX, 10/22–10/25), 1995.

[1126] P. B. Lorenz. The effect of alkaline agents on retention of EOR chemicals. US DOE Rep NIPER-535, NIPER, 1991.

[1127] F. E. Lowther. Method for treating tubulars with a gelatin pig. Patent: US 5215781, 1993.

[1128] N. Lu, Z. Jia, H. Ma, W. Wang, and D. Wu. Study and applications of low density furnace slag cement slurry system. *Oil Drilling Prod Technol,* 19(2):37–40,53,107, 1997.

[1129] C. A. Lukach and J. Zapico. Thermally stable hydroxyethylcellulose suspension. Patent: EP 619340, 1994.

[1130] A. O. Lundan, P. H. Anas, and M. J. Lahteenmaki. Stable CMC (carboxymethyl cellulose) slurry. Patent: WO 9320139, 1993.

[1131] A. O. Lundan and M. J. Lahteenmaki. Stable CMC (carboxymethyl cellulose) slurry. Patent: US 5487777, 1996.
[1132] R. D. Lundberg, D. G. Peiffer, J. C. Newlove, and E. R. Werlein. Drilling fluids. Patent: EP 259111, 1988.
[1133] B. S. Lyadov. Gel-forming composition for isolating works in well—contains polyacrylamide, urotropin, water-soluble chromate(s) and water. Patent: SU 1730432-A, 1992.
[1134] B. S. Lyadov. Polymeric solution for isolation of absorption strata—contains urea-formaldehyde and/or phenol formaldehyde resin and lignosulphonate. Patent: SU 1730434-A, 1992.
[1135] W. C. Lyons. *Standard handbook of petroleum and natural gas engineering*, volume 1–2. Gulf Publishing Co, 1996.
[1136] I. A. Lyubinin, A. S. Gubarev, V. V. Butovets, and A. V. Torgashov. Modern lubricants for roller bit bearings. *Stroit Neft Gaz Skvazhin Sushe More*, (3):14–25, March 1995.
[1137] G. Ma. Laboratory study on polymer flooding in oil reservoir with high salinity. *Oil Gas Recovery Technol*, 3(2):I,1–4,33, 1996.
[1138] G. Ma, K. Zuo, and Z. Xu. Polymer flooding test in low permeability and high salinity reservoir of maling oilfield. *Oil Drilling Prod Technol*, 21(1):89–93,109–110, 1999.
[1139] H. Ma. Study on the stability of low density cement with hollow microsphere. *Drilling Fluid Completion Fluid*, 10(1):61–65,76, January 1993.
[1140] C. Macconnachie, R. J. Mikula, R. J. Scoular, and L. J. Kurucz. Optimizing demulsifier performance: a fundamental approach. In *Preprints*, volume 1. 45th Annu CIM Petrol Soc et al Tech Mtg (Calgary, Canada, 6/12–6/15), 1994.
[1141] C. A. Macconnachie, R. J. Mikula, L. Kurucz, and R. J. Scoular. Correlation of demulsifier performance and demulsifier chemistry. In *Proceedings Volume*, number 39. 5th CIM Petrol Soc et al Saskatchewan Petrol Conf (Regina, Canada, 10/18–10/20) Preprints, 1993.
[1142] M. P. Mack. Improved use of flow improvers. Patent: EP 196350, 1986.
[1143] D. Mackay and A. Chau. The effectiveness of chemical dispersants: a discussion of laboratory and field test results. *Oil Chem Pollut*, 3(6):405–415, 1986.
[1144] E. J. Mackay and K. S. Sorbie. Modelling scale inhibitor squeeze treatments in high crossflow horizontal wells. *J Can Petrol Technol*, 39(10):47–51, October 1998.
[1145] E. J. Mackay and K. S. Sorbie. An evaluation of simulation techniques for modelling squeeze treatments. In *Proceedings Volume*, pages 373–387. Annu SPE Tech Conf (Houston, TX, 10/3–10/6), 1999.
[1146] E. J. Mackay, K. S. Sorbie, M. M. Jordan, A. P. Matharu, and R. Tomlins. Modelling of scale inhibitor treatments in horizontal wells:

application to the Alba field. In *Proceedings Volume,* pages 337–348. SPE Formation Damage Contr Int Symp (Lafayette, LA, 2/18–2/19), 1998.

[1147] N. Macleod, E. Bryan, A. J. Buckley, R. E. Talbot, and M. A. Veale. Control of reservoir souring by a novel biocide. In *Proceedings Volume.* 50th Annu NACE Int Corrosion Conf (Corrosion 95) (Orlando, FL, 3/26–3/31), 1995.

[1148] F. Mainier, C. A. Saliba, and G. Gonzalez. Effectiveness of acid corrosion inhibitors in the presence of alcohols. SPE Unsolicited Paper, SPE-20404, July 1990.

[1149] F. B. Mainier, W. Lazaro, and Rosario. F. F. Do. Silicate-based corrosion-inhibitor in drilling fluids: an environmentally friendly option (inibidor de corrosao a base de silicato em fluidos de perfuracao: uma opcao nao agressiva ao meio ambiente). In *Proceedings Volume,* volume 1, pages 467–475. 8th Petrobras et al Latin Amer Drilling Congr (Rio De Janeiro, Brazil, 10/14–10/16), 1992.

[1150] S. Majumdar, S. H. Holey, and R. D. Singh. *Eur Polym J,* 16:1201, 1980.

[1151] T. Y. Makogon, A. P. Mehta, and E. D. Sloan, Jr. Structure H and structure I hydrate equilibrium data for 2,2-dimethylbutane with methane and xenon. *J Chem Eng Data,* 41(2):315–318, March-April 1996.

[1152] Y. F. Makogon. *Hydrates of natural gas.* Pennwell, Tulsa, OK, 1981.

[1153] Y. F. Makogon. Formation of hydrates in shut-down pipelines in offshore conditions. In *Proceedings Volume,* volume 4, pages 749–756. 28th Annu SPE et al Offshore Technol Conf (Houston, TX, 5/6–5/9), 1996.

[1154] A. Malandrino, M. Andrei, F. Gagliardi, and T. P. Lockhart. A thermodynamic model for PPCA (phosphino-polycarboxylic acid) precipitation. In *Proceedings Volume.* 4th IBC UK Conf Ltd Advances in Solving Oilfield Scaling Int Conf (Aberdeen, Scotland, 1/28–1/29), 1998.

[1155] G. A. Malchow, Jr. Friction modifier for oil-based (invert) well drilling fluids and methods of using the same. Patent: US 5593953, 1997.

[1156] S. Malik and R. A. Mashelkar. Hydrogen bonding mediated shear stable clusters as drag reducers. *Chem Eng Sci,* 50(1):105–116, January 1995.

[1157] S. Malik, S. N. Shintre, and R. A. Mashelkar. Process for the preparation of a new polymer useful for drag reduction in hydrocarbon fluids in exceptionally dilute polymer solutions. Patent: US 5080121, 1992.

[1158] S. Malik, S. N. Shintre, and R. A. Mashelkar. A polymer useful for drag reduction in hydrocarbon fluids and its preparations. Patent: EP 471116, 1996.

[1159] B. M. Malyarchuk, V. Yu. Pavlyuk, F. V. Nijger, B. G. Tarasov, and P. V. Tarabarinov. Composition for plugging flooded porous strata—contains clay, alumina cement, natural zeolite, calcined soda, polyacrylamide, alkyl resorcinol resin, pentaerythritol production waste and natural saline. Patent: SU 1776764-A, 1992.

[1160] R. A. Mamleev, E. M. Yulbarisov, R. N. Fakhretdinov, L. N. Zagidullina, A. N. Kulikov, and Z. R. Kutushev. Composition for pumping into oil stratum—contains specified biopolymer, polydimethyldiallyl ammonium chloride, water and formaldehyde. Patent: SU 1828161-A, 1997.

[1161] P. V. Manalastas, E. N. Drake, E. N. Kresge, W. A. Thaler, L. A. McDougall, J. C. Newlove, V. Swarup, and A. J. Geiger. Breaker chemical encapsulated with a crosslinked elastomer coating. Patent: US 5110486, 1992.

[1162] C. A. Manen, L. B. Fox, Jr., C. Getter, J. Whitney, L. Harris, P. S. O'Brien, B. Hahn, H. Metsker, and D. Kennedy. Oil dispersant guidelines: Alaska. In *Proceedings Volume,* pages 141–151. Amer Soc Testing Mater Oil Dispersants: New Ecol Approaches Symp (Williamsburg, VA, 10/12–10/14), 1987.

[1163] R. J. Mannheimer. Factors that influence the coalescence of bubbles in oils that contain silicone antifoamants. *Chem Eng Commun,* 113:183–196, March 1992.

[1164] F. Mansfeld and B. Little. The application of electrochemical techniques for the study of MIC (microbiologically influenced corrosion)—a critical review. In *Proceedings Volume.* NACE Int Corrosion Forum (Corrosion 90) (Las Vegas, NV, 4/23–4/27), 1990.

[1165] A. R. Mansour and T. Aldoss. Drag reduction in pipes carrying crude oil using an industrial cleaner. SPE Unsolicited Paper, SPE-17918, 1988.

[1166] W. Mardis, J. Sanchaz, and H. Basson. Organoclay compositions manufactured with organic acid ester-derived quaternary ammonium compounds, their preparation and non-aqueous fluid systems containing such compositions. Patent: EP 798267, 1997.

[1167] R. L. Martin. Multifunctional corrosion inhibitors. Patent: US 4722805, 1988.

[1168] R. L. Martin. The reaction product of nitrogen bases and phosphate esters as corrosion inhibitors. Patent: EP 567212, 1993.

[1169] F. D. Martischius, B. Raab, and D. Karau. Process for improving the drag reducing properties of high-molecular weight polymer solutions in crude oil or refined products (Verfahren zur Verbesserung der fließwiderstandsvermindernden Eigenschaften hochmolekularer Polymerlösungen in Rohöl oder Raffinerieprodukten). Patent: EP 397002, 1990.

[1170] S. V. Martyanova, A. A. Chezlov, A. G. Nigmatullina, L. A. Piskareva, and R. D. Shamsutdinov. Production of lignosulphonate reagent for drilling muds—by initial heating with sulphuric acid, condensation with formaldehyde, and neutralisation of mixture with sodium hydroxide. Patent: RU 2098447-C, 1997.

[1171] J. A. Mason. Use of chlorous acid in oil recovery. Patent: US 4892148, 1990.

[1172] J. A. Mason. Use of chlorous acid in oil recovery. Patent: CA 2007218, 1991.

[1173] J. A. Mason. Use of chlorous acid in oil recovery. Patent: GB 2239867, 1991.

[1174] R. M. Matherly, J. Jiao, J. S. Ryman, and D. J. Blumer. Determination of imidazoline and amido-amine type corrosion inhibitors in both crude oil and produced brine from oilfield production. In *Proceedings Volume*. 50th Annu NACE Int Corrosion Conf (Corrosion 95) (Orlando, FL, 3/26–3/31), 1995.

[1175] J. R. Mattox. Stabilized isothiazolone compositions. Patent: EP 315464, 1989.

[1176] H. Matyschok and R. Janik. Defoamers in drilling muds (srodki przeciwpianowe w pluczkach wiertniczych). *Nafta (Pol),* 46(7–9):120–123, July-September 1990.

[1177] G. F. Matz, A. L. Melby, R. J. Loeffler, N. F. Vozza, and S. R. T. Chen. Water soluble polymer composition and method of use. Patent: WO 0105365A, 2001.

[1178] R. E. Maughmer and D. Dalrymple. Treatment helps to decrease water. *Amer Oil Gas Reporter,* 35(6):114,116,118, June 1992.

[1179] R. Maurer and M. Landry. Delayed-gelling compositions and their use for plugging subterranean formations. Patent: GB 2226066, 1990.

[1180] S. Maxwell, K. M. McLean, and J. Kearns. Biocide application and monitoring in a waterflood system. In *Proceedings Volume,* pages 209–218. Inst Petrol Microbiol Comm Microbial Problems in Offshore Oil Ind Int Conf (Aberdeen, Scotland, 4/15–4/17), 1986.

[1181] M. A. McCabe, J. M. Wilson, J. D. Weaver, and J. J. Venditto. Biocidal well treatment method. Patent: US 5016714, 1991.

[1182] T. F. McCallum, III and B. Weinstein. Amine-thiol chain transfer agents. Patent: US 5298585, 1994.

[1183] S. M. McCarthy, D. J. Daulton, and S. J. Bosworth. Blast furnace slag use reduces well completion cost. *World Oil,* 216(4):87–88,90,92,94,96, April 1995.

[1184] C. L. McCormick and R. D. Hester. Polymers for mobility control in enhanced oil recovery: final report. US DOE Fossil Energy Rep DOE/BC/10844-20, Southern Mississippi Univ, 1990.

[1185] D. R. McCoy. Demulsification of bitumen emulsions using polyalkylene polyamine salts. Patent: CA 1220151, 1987.
[1186] D. R. McCoy, M. Cuscurida, and G. P. Speranza. Demulsification of bitumen emulsions using water soluble epoxy-containing polyethers. Patent: CA 1225004, 1987.
[1187] D. R. McCoy, E. E. McEntire, and R. M. Gipson. Demulsification of bitumen emulsions. Patent: CA 1225003, 1987.
[1188] T. M. McCullough. Emulsion minimizing corrosion inhibitor for naphtha/water systems. Patent: US 5062992, 1991.
[1189] W. J. McDonald, J. H. Cohen, and C. M. Hightower. New lightweight fluids for underbalanced drilling. DOE/FETC Rep 99-1103, Maurer Engineering Inc, 1999.
[1190] L. A. McDougall, F. Malekahmadi, and D. A. Williams. Method of fracturing formations. Patent: EP 540204, 1993.
[1191] E. E. McEntire and J. F. Knifton. Process for formation of dialkyl-aminomethylated internal olefin polymers. Patent: US 4657984, 1987.
[1192] R. E. McGlothlin and F. B. Woodworth. Well drilling process and clay stabilizing agent. Patent: US 5558171, 1996.
[1193] M. J. McInerney, A. D. Montgomery, and K. L. Sublette. Microbial control of the production of sulfide. In E. C. Donaldson, editor, *Microbial enhancement of oil recovery: recent advances: Proceedings of the 1990 International Conference on Microbial Enhancement of Oil Recovery,* volume 31 of *Developments in Petroleum Science,* pages 441–449. Elsevier Science Ltd, 1991.
[1194] E. H. McKerrell and A. Lynes. Development of an HPLC (high performance liquid chromatography) method for the determination of nitrogen containing corrosion inhibitors in a mixed hydrocarbon/glycol matrix. In *Proceedings Volume,* number 67, pages 212–222. 3rd Royal Soc Chem Ind Chem in the Oil Ind Int Symp (Manchester, England, 4/19–4/20), 1988.
[1195] J. M. McLennan, K. D. Brunt, and W. G. Guthrie. Solid antibacterial compositions. Patent: GB 2183477, 1987.
[1196] A. J. McMahon and D. Harrop. Green corrosion inhibitors: an oil company perspective. In *Proceedings Volume.* 50th Annu NACE Int Corrosion Conf (Corrosion 95) (Orlando, FL, 3/26–3/31), 1995.
[1197] A. J. McMahon, P. S. Smith, and Y. Lee. Drag reducing chemical enables increased sea water injection without increasing the oxygen corrosion rate. In *Proceedings Volume.* NACE Int Corrosion Conf (Corrosion 97) (New Orleans, LA, 3/9–3/14), 1997.
[1198] K. McNally, H. Nae, and J. Gambino. Oil well drilling fluids with improved anti-settling properties and methods of preparing them. Patent: EP 906946, 1999.

[1199] G. H. Medley, Jr., J. E. Haston, R. L. Montgomery, I. D. Martindale, and J. R. Duda. Field application of lightweight, hollow-glass-sphere drilling fluid. *J Petrol Technol,* 49(11):1209–1211, November 1997.

[1200] G. H. Medley, Jr., J. E. Haston, R. L. Montgomery, I. D. Martindale, and J. R. Duda. Field application of lightweight hollow glass sphere drilling fluid. In *Proceedings Volume,* pages 699–707. Annu SPE Tech Conf (San Antonio, TX, 10/5–10/8), 1997.

[1201] G. H. Medley, Jr., W. C. Maurer, and A. Y. Garkasi. Use of hollow glass spheres for underbalanced drilling fluids. In *Proceedings Volume,* pages 511–520. Annu SPE Tech Conf (Dallas, TX, 10/22–10/25), 1995.

[1202] A. P. Mehta and E. D. Sloan. Structure H hydrates: data and industrial potential. In *Proceedings Volume,* pages 67–75. 72nd Annu GPA Conv (San Antonio, TX, 3/15–3/17), 1993.

[1203] A. P. Mehta and E. D. Sloan. Structure H hydrates: implications for the petroleum industry. In *Proceedings Volume,* pages 607–613. Annu SPE Tech Conf (Denver, CO, 10/6–10/9), 1996.

[1204] D. H. Mei, J. Liao, J. T. Yang, and T. M. Guo. Experimental and modeling studies on the hydrate formation of a methane + nitrogen gas mixture in the presence of aqueous electrolyte solutions. *Ind Eng Chem Res,* 35(11):4342–4347, November 1996.

[1205] A. F. Meinhold. Framework for a comparative environmental assessment of drilling fluids. Brookhaven Nat Lab Rep BNL-66108, Brookhaven Nat Lab, November 1998.

[1206] A. L. Mendoza. Surface active composition for conditioning a gas containing entrained asphaltenes. Patent: GB 2279964, 1995.

[1207] G. Mensa-Wilmot, R. L. Garrett, and R. S. Stokes. PAO (polyalphaolefin) lubricant inhibits bit balling, speeds drilling. *Oil Gas J,* 95(16):68–70, 1997.

[1208] J. D. Mercer and L. L. Nesbit. Oil-base drilling fluid comprising branched chain paraffins such as the dimer of 1-decene. Patent: US 5096883, 1992.

[1209] F. Merlin. Optimization of dispersant application, especially by ship. In *Proceedings Volume,* pages 337–342. API et al Oil Spill (Prev, Behav, Contr, Cleanup) 20th Anniv Conf (San Antonio, TX, 2/13–2/16), 1989.

[1210] F. Merlin, C. Bocard, R. Cabridenc, J. Oudot, and E. Vindimian. Toward a French approval procedure for the use of dispersants in inland waters. In *Proceedings Volume,* pages 401–404. 12th Bien API et al Oil Spill (Prev, Behav, Contr, Cleanup) Int Conf (San Diego, CA, 3/4–3/7), 1991.

[1211] L. S. Merrill. Fiber reinforced gel for use in subterranean treatment process. Patent: WO 9319282, 1993.

[1212] L. S. Merrill. Fiber reinforced gel for use in subterranean treatment process. Patent: GB 2277112, 1994.
[1213] L. S. Merrill. Fiber reinforced gel for use in subterranean treatment processes. Patent: US 5377760, 1995.
[1214] J. U. Messenger. *Lost circulation*, pages 44–56. PennWell Publishing Co, Tulsa, OK, 1981.
[1215] S. F. Messner. Cleaning of pipelines with gel pigs (csotavvezetek tisztitasa geles csomalacokkal). *Koolaj Foldgaz,* 24(7):219–222, July 1991.
[1216] M. Metwally. Effect of gaseous additives on steam processes for Lindbergh Field, Alberta. *J Can Petrol Technol,* 29(6):26–30, November-December 1990.
[1217] G. Meyer, D. Kessel, and I. Rahimian. The effect of pour point depressants of the polyacrylate-type on crude oil (Wirkung von Stockpunkterniedrigern des Polyacrylat-Typs auf Rohöle). *Erdöl Kohle-Erdgas-Petrochem,* 48(3):135–137, March 1995.
[1218] G. R. Meyer. Corrosion inhibiting compositions. Patent: GB 2353793, 2001.
[1219] V. Meyer, A. Audibert-Hayet, J. P. Gateau, J. P. Durand, and J. F. Argillier. Water-soluble copolymers containing silicon. Patent: GB 2327946, 1999.
[1220] F. Miano, G. Calloni, N. Moroni, and A. Marcotullio. Cementitious composition for the cementation of oil wells. Patent: EP 636591, 1995.
[1221] M. Michaux. Controlling setting in a high-alumina cement. Patent: WO 9933763, 1999.
[1222] A. Michelson and H. Vattement. Bentonite-based drilling mud and drilling method making use thereof (boue de forage a base de bentonite et procede de forage la mettant en oeuvre). Patent: EP 936263, 1999.
[1223] S. A. Mikhailov, E. P. Khmeleva, E. V. Moiseeva, and T. M. Sleta. Determination of the optimal dose of salt deposition inhibitors. *Neft Khoz,* (7):43–45, July 1987.
[1224] N. B. Milestone, T. Sugama, L. E. Kukacka, and N. Carciello. Carbonation of geothermal grouts: Pt 1: CO_2 attack at 150° C. *Cement Concrete Res,* 16(6):941–950, November 1986.
[1225] D. Miller, A. Vollmer, M. Feustel, and P. Klug. Synergistic mixtures of phosphoric acid esters and carboxylic acid derivatives as asphaltene dispersants (Synergistische Mischungen von Phosphorsäureestern mit Carbonsäurederivaten als Asphalten-Dispergatoren). Patent: EP 967361, 1999.
[1226] J. F. Miller, C. O. Sheely, J. W. Wimberley, and R. A. Howard. Use of nonradioactive complex metal anion as tracer in subterranean reservoirs. Patent: US 5246861, 1993.
[1227] W. K. Miller, II, G. A. Roberts, and S. J. Carnell. Fracturing fluid loss and treatment design under high shear conditions in a partially depleted, moderate permeability gas reservoir. In *Proceedings Volume,*

pages 451–460. SPE Asia Pacific Oil & Gas Conf (Adelaide, Australia, 10/28–10/31), 1996.
[1228] L. V. Minevski and J. A. Gaboury. Thiacrown ether compound corrosion inhibitors for alkanolamine units. Patent: EP 962551, 1999.
[1229] T. O. Mitchell, R. J. Card, and A. Gomtsyan. Cleanup additive. Patent: US 6242390, 2001.
[1230] Y. Miyajima, Y. Kariyazono, S. Funatsu, and E. Endo. Durability of polyethylene coated steel pipe at elevated temperature. In *Proceedings Volume,* number 7, pages 183–190. 10th Bhr Group Ltd et al Pipe Protect Int Conf (Amsterdam, Netherlands, 11/10–11/12), 1993.
[1231] F. K. Mody and A. H. Hale. A borehole stability model to couple the mechanics and chemistry of drilling fluid shale interaction. In *Proceedings Volume,* pages 473–490. SPE/IADC Drilling Conf (Amsterdam, Netherlands, 2/23–2/25), 1993.
[1232] A. Moet, M. Y. Bakr, M. Abdelmonim, and O. Abdelwahab. Factors affecting measurements of the efficiency of spilled oil dispersion. *ACS Petrol Chem Div Preprints,* 40(4):564–566, August 1995.
[1233] S. Mohanty and S. Khataniar. Sodium orthosilicate: an effective additive for alkaline steamflood. *J Petrol Sci Eng,* 14(1–2):45–49, December 1995.
[1234] M. Mohitpour, H. Golshan, and A. Murray. *Pipeline design & construction: a practical approach.* American Society of Mechanical Engineers, New York, 2000.
[1235] D. Momeni, J. R. Chen, and T. F. Yen. MEOR (microbial enhanced oil recovery) studies in a radial flow system—the research outlook. US DOE Fossil Energy Rep NIPER-351 CONF-870858, September 1988.
[1236] D. Momeni, T. F. Yen, L. K. Jang, R. McDavid, J. F. Kuo, V. Huang, K. Lee, J. E. Findley, J. R. Chen, and P. W. Chang. *Microbial enhanced oil recovery: principle and practice.* CRC Press, Inc, Boca Raton, FL, 1990.
[1237] T. C. Mondshine. Crosslinked fracturing fluids. Patent: WO 8700236, 1987.
[1238] T. C. Mondshine. Process for decomposing polysaccharides in alkaline aqueous systems. Patent: EP 559418, 1993.
[1239] T. C. Mondshine and G. R. Benta. Process and composition to enhance removal of polymer-containing filter cakes from wellbores. Patent: US 5238065, 1993.
[1240] N. Monfreux, P. Perrin, F. Lafuma, and C. Sawdon. Invertible emulsions stabilised by amphiphilic polymers and application to bore fluids (emulsions inversables stabilisees par des polymeres amphiphiles et application a des fluides de forage). Patent: WO 0031154, 2000.
[1241] A. Monjoie and C. Schroeder. Flooding of oil in chalk (deplacement des hydrocarbures dans la craie). *Soc Geol Nord Ann,* 5(4):325–329, October 1997.

[1242] F. Montgomery, S. Montgomery, and P. Stephens. Method of controlling porosity of well fluid blocking layers and corresponding acid soluble mineral fiber well facing product. Patent: US 5222558, 1993.
[1243] F. Montgomery, S. Montgomery, and P. Stephens. Method of controlling porosity of well fluid blocking layers and corresponding acid soluble mineral fiber well facing product. Patent: US 5354456, 1994.
[1244] S. S. Moody and H. T. R. Montgomerie. Control of oilfield biofouling. Patent: EP 706974, 1996.
[1245] C. H. Moore. Computer simulation of formation damage resulting from thermal recovery. In *Proceedings Volume*. SPE Int Therm Oper & Heavy Oil Symp (Margarita Island, Venezuela, 3/12–3/14), 2001.
[1246] W. R. Moore, B. B. Beall, and S. A. Ali. Formation damage removal through the application of enzyme breaker technology. In *Proceedings Volume*, pages 135–141. SPE Formation Damage Contr Int Symp (Lafayette, LA, 2/14–2/15), 1996.
[1247] A. Moradi-Araghi. Gelling compositions useful for oil field applications. Patent: US 5432153, 1995.
[1248] A. Moradi-Araghi and G. A. Stahl. Gelation of acrylamide-containing polymers with furfuryl alcohol and water dispersible aldehydes. Patent: EP 447967, 1991.
[1249] A. Moradi-Araghi and G. A. Stahl. Gelation of acrylamide-containing polymers with hydroxyphenyl alkanols. Patent: EP 446865, 1991.
[1250] L. K. Moran and L. L. Moran. Composition and method to control cement slurry loss and viscosity. Patent: CA 2184548, 1997.
[1251] L. K. Moran and L. L. Moran. Composition and method to control cement slurry loss and viscosity. Patent: US 5850880, 1998.
[1252] L. K. Moran and L. L. Moran. Composition and method to control cement slurry loss and viscosity. Patent: US 5728210, 1998.
[1253] L. K. Moran and T. R. Murray. Well cement fluid loss additive and method. Patent: US 5009269, 1991.
[1254] G. Moritis. EOR survey and analysis. *Oil Gas J*, 94(16):39–42,44–61, 1996.
[1255] G. Moritis. EOR oil production up slightly. *Oil Gas J*, 96(16):49–77, 1998.
[1256] N. Moroni, G. Calloni, and A. Marcotullio. Gas impermeable carbon black cements: analysis of field performances. In *Proceedings Volume*, volume 2, pages 781–792. E&P Forum et al Offshore Mediter Conf (OMC 97) (Ravenna, Italy, 3/19–3/21), 1997.
[1257] F. F. Morpeth. Biocide composition and its use. Patent: EP 542721, 1993.
[1258] F. F. Morpeth and M. Greenhalgh. Composition and use. Patent: EP 390394, 1990.

[1259] E. A. Morris, III, D. M. Dziewulski, D. H. Pope, and S. T. Paakkonen. Field and laboratory studies into the detection and treatment of microbiologically influenced souring (MIS) in natural gas storage facilities. In *Proceedings Volume*. 49th Annu NACE Int Corrosion Conf (Corrosion 94) (Baltimore, MD, 2/27–3/4), 1994.
[1260] E. A. Morris, III and D. H. Pope. Field and laboratory investigations into the persistence of glutaraldehyde and acrolein in natural gas storage operations. In *Proceedings Volume*. 49th Annu NACE Int Corrosion Conf (Corrosion 94) (Baltimore, MD, 2/27–3/4), 1994.
[1261] I. Morris and G. Perry. High pressure storage and transport of natural gas containing added C_2 or C_3, or ammonia, hydrogen fluoride or carbon monoxide. Patent: US 6217626, 2001.
[1262] I. Morris and G. Perry. Pipeline transmission method. Patent: US 6201163, 2001.
[1263] B. J. Morris-Sherwood and E. C. Brink, Jr. Corrosion inhibiting composition. Patent: EP 221212, 1987.
[1264] B. J. Morris-Sherwood and E. C. Brink, Jr. Corrosion inhibiting composition and method. Patent: CA 1264539, 1990.
[1265] V. G. Mosienko, Y. I. Petrakov, V. F. Nagornova, and V. N. Nikiforova. Complex additive for plugging solutions—contains modifying reagent in form of waste from production of sebacic acid, from stage of neutralising of sodium salts of fatty acids. Patent: RU 2074310-C, 1997.
[1266] V. G. Mosienko, Y. I. Petrakov, A. M. Pedus, and V. N. Nikiforova. Complex additive to plugging solution—contains waste from production of sebacic acid, ammonium sulphate and carboxymethyl cellulose, reducing water separation, etc. Patent: RU 2078908-C, 1997.
[1267] N. T. Moskvicheva, V. G. Zhzhonov, and R. I. Kateeva. Plugging mixture improving cement rock quality—contains plugging Portland cement and aqueous alcoholic solution of sodium organo-siliconate as modifying additive. Patent: RU 2002040-C, 1993.
[1268] J. T. Moss, Jr. and J. T. Moss. Enhanced oil recovery using hydrogen peroxide injection. US DOE Fossil Energy Rep NIPER/BDM-0086, Tejas Petrol Engineers Inc, 1994.
[1269] R. J. Mouche and E. B. Smyk. Noncorrosive scale inhibitor additive in geothermal wells. Patent: US 5403493, 1995.
[1270] T. Moussa and C. Tiu. Factors affecting polymer degradation in turbulent pipe flow. *Chem Eng Sci,* 49(10):1681–1692, May 1994.
[1271] D. T. Mueller, V. G. Boncan, and J. P. Dickerson. Stress resistant cement compositions and methods for using same. Patent: US 6230804, 2001.
[1272] F. F. Muganlinskij, M. M. Lyushin, A. M. O. Samedov, and V. Kh. U. Akosta. Suppressing activity of sulphate-reducing bacteria on

petroleum extraction—by treating flooding water with di-(tri-N-butyl)-(1,4-benzodioxan-6,7-dimethyl) diammonium dichloride. Patent: RU 2033393-C, 1995.

[1273] D. J. Muir and M. J. Irwin. Encapsulated breakers, compositions and methods of use. Patent: WO 9961747, 1999.

[1274] H. Mukherjee and G. Cudney. Extension of acid fracture penetration by drastic fluid-loss control. *SPE Unsolicited Paper*, 1992.

[1275] S. Mukherjee and A. P. Kushnick. Effect of demulsifiers on interfacial properties governing crude oil demulsification. In *Proceedings Volume*. Annu AICHE Mtg (New York, 11/15–11/20), 1987.

[1276] G. A. Mullen and A. Gabrysch. Synergistic mineral blends for control of filtration and rheology in silicate drilling fluids. Patent: US 6248698, 2001.

[1277] D. T. Müller. Performance characteristics of vinylsulfonate-based cement fluid-loss additives. In *Proceedings Volume*, pages 609–617. SPE Rocky Mountain Reg Mtg (Casper, WY, 5/18–5/21), 1992.

[1278] D. T. Müller and R. L. Dillenbeck, III. The versatility of silica fume as an oilwell admixture. In *Proceedings Volume*, pages 529–536. SPE Prod Oper Symp (Oklahoma City, OK, 4/7–4/9), 1991.

[1279] H. Müller, W. Breuer, C. P. Herold, P. Kuhm, and S. von Tapavicza. Mineral additives for setting and/or controlling the rheological properties and gel structure of aqueous liquid phases and the use of such additives. Patent: US 5663122, 1997.

[1280] H. Müller, C. P. Herold, F. Bongardt, N. Herzog, and S. von Tapavicza. Lubricants for drilling fluids (Schmiermittel für Bohrspülungen). Patent: WO 0029502, 2000.

[1281] H. Müller, C. P. Herold, and S. von Tapavicza. Drilling fluids. Patent: ZA 9002669, 1990.

[1282] H. Müller, C. P. Herold, and S. von Tapavicza. Monocarboxylic acid-methyl esters in invert-emulsion muds (Monocarbonsäure-Methylester in Invert-Bohrspülschlämmen). Patent: EP 382071, 1990.

[1283] H. Müller, C. P. Herold, and S. von Tapavicza. Oleophilic alcohols as components of invert emulsion drilling fluids (Oleophile Alkohole als Bestandteil von Invert-Bohrspülungen). Patent: EP 391252, 1990.

[1284] H. Müller, C. P. Herold, and S. von Tapavicza. Oleophilic basic amine derivatives as additives in invert emulsion muds (Oleophile basische Aminverbindungen als Additiv in Invert-Bohrspülschlämmen). Patent: EP 382070, 1990.

[1285] H. Müller, C. P. Herold, and S. von Tapavicza. Use of hydrated castor oil as a viscosity promoter in oil-based drilling muds. Patent: WO 9116391, 1991.

[1286] H. Müller, C. P. Herold, and S. von Tapavicza. Use of selected fatty alcohols and their mixtures with carboxylic acid esters as lubricant

components in water-based drilling fluid systems for soil exploration (Verwendung ausgewählter Fettalkohole und ihrer Abmischungen mit Carbonsäureestern). Patent: EP 948576, 1999.
[1287] H. Müller, C. P. Herold, S. von Tapavicza, and J. F. Fues. Fluid borehole-conditioning agent based on polycarboxylic acid esters. Patent: WO 9119771, 1991.
[1288] H. Müller, C. P. Herold, S. von Tapavicza, D. J. Grimes, J. M. Braun, and S. P. T. Smith. Use of selected ester oils in drilling muds, especially for offshore oil or gas recovery (Verwendung ausgewählter Esteröle in Bohrspülungen insbesondere zur Off-Shore-Erschließung von Erdöl- bzw Erdgasvorkommen [I]). Patent: EP 374671, 1990.
[1289] H. Müller, C. P. Herold, S. von Tapavicza, D. J. Grimes, J. M. Braun, and S. P. T. Smith. Use of selected ester oils in drilling muds, especially for offshore oil or gas recovery (Verwendung ausgewählter Esteröle in Bohrspülungen insbesondere zur Off-Shore-Erschließung von Erdöl- bzw Erdgasvorkommen [II]). Patent: EP 374672, 1990.
[1290] H. Müller, C. P. Herold, S. von Tapavicza, M. Neuss, and F. Burbach. Use of selected ester oils of low carboxylic acids in drilling fluids. Patent: US 5318954, 1994.
[1291] H. Müller, C. P. Herold, S. von Tapavicza, M. Neuss, W. Zöllner, and F. Burbach. Esters of medium chain length carboxylic acids as oil-phase components in invert drilling muds (Ester von Carbonsäuren mittlerer Kettenlänge als Bestandteil der Ölphase in Invert-Bohrspülschlämmen). Patent: EP 386636, 1990.
[1292] H. Müller, C. P. Herold, S. von Tapavicza, M. Neuss, W. Zöllner, and F. Burbach. Invert drilling muds. Patent: WO 9010681, 1990.
[1293] H. Müller, C. P. Herold, A. Westfechtel, and S. von Tapavicza. Free-flowing drill hole treatment agents based on carbonic acid diesters. Patent: WO 9118958, 1991.
[1294] H. Müller, C. P. Herold, A. Westfechtel, and S. von Tapavicza. Fluid drill-hole treatment agents based on carbonic acid diesters. Patent: ZA 9104341, 1992.
[1295] H. Müller, C. P. Herold, A. Westfechtel, and S. von Tapavicza. Free-flowing drill hole treatment agents based on carbonic acid diesters (Fließfähige Bohrlochbehandlungsmittel auf Basis von Kohlensäurediestern). Patent: EP 532570, 1993.
[1296] H. Müller, C. P. Herold, A. Westfechtel, and S. von Tapavicza. Free-flowing drill hole treatment agents based on carbonic acid diesters. Patent: AU 643299, 1993.
[1297] H. Müller, C. P. Herold, A. Westfechtel, and S. von Tapavicza. Fluid-drill-hole treatment agents based on carbonic acid diesters. Patent: US 5461028, 1995.
[1298] H. Müller, S. Podubrin, C. P. Herold, and A. Heidbreder. Dispersions containing homopolymers or copolymers of hydroxy carboxylic acids

as a rheological additive (Dispersionen enthaltend Homo-oder Copolymere von Hydroxycarbonsäuren als rheologisches Additiv). Patent: WO 9952623, 1999.
[1299] H. Müller, G. Stoll, C. P. Herold, and S. von Tapavicza. Application of selected ethers of monofunctional alcohols in drilling fluids (Verwendung ausgewählter Ether monofunktioneller Alkohole in Bohrspülungen). Patent: EP 391251, 1990.
[1300] H. Müller, G. Stoll, C. P. Herold, and S. von Tapavicza. Drilling fluids. Patent: ZA 9002665, 1990.
[1301] H. Müller, C. P. Herold, and J. F. Fues. Use of surface-active α-sulfofatty acid di-salts in water and oil based drilling fluids and other drillhole treatment agents. Patent: US 5508258, 1996.
[1302] M. Mulyono, E. Jasjfi, and M. Maloringan. Biodegradation test for oil spill dispersant (OSD) and OSD-oil mixture. In *Proceedings Volume,* pages 355–363. 5th Asian Counc Petrol Conf (Ascope 93) (Bangkok, Thailand, 11/2–11/6), 1993.
[1303] N. A. Mumallah. Chromium (iii) propionate: a crosslinking agent for water-soluble polymers in real oilfield waters. In *Proceedings Volume.* SPE Oilfield Chem Int Symp (San Antonio, TX, 2/4–2/6), 1987.
[1304] N. A. Mumallah. Altering subterranean formation permeability. Patent: US 4917186, 1990.
[1305] N. A. Mumallah. Altering subterranean formation permeability. Patent: CA 2009117, 1990.
[1306] N. A. Mumallah. Methods and compositions for altering subterranean formation permeability. Patent: EP 383337, 1990.
[1307] N. A. K. Mumallah and T. K. Shioyama. Process for preparing a stabilized chromium (iii) propionate solution and formation treatment with a so prepared solution. Patent: EP 194596, 1986.
[1308] R. Munro, G. Hanni, and A. Young. The economics of a synthetic drilling fluid for exploration drilling in the UK sector of the North Sea. In *Proceedings Volume.* IBC Tech Serv Ltd Prev Oil Discharge Drilling Oper—The Options Conf (Aberdeen, Scotland, 6/23–6/24), 1993.
[1309] C. L. Muth and S. M. Kolby. Cost saving by use of flow improver. In *Proceedings Volume,* pages 353–357. 13th Int Pipeline Technol Conf (Houston, TX, 2/5–2/7), 1985.
[1310] H. Nae, W. Reichert, and A. C. Eng. Organoclay compositions containing two or more cations and one or more organic anions, their preparation and use in non-aqueous systems. Patent: EP 681990, 1999.
[1311] H. Nae, W. W. Reichert, and A. C. Eng. Organoclay compositions prepared with a mixture of two organic cations and their use in non-aqueous systems. Patent: EP 542266, 1993.
[1312] H. N. Nae, W. W. Reichert, and A. C. Eng. Organoclay compositions containing two or more cations and one or more organic anions, their

preparation and use in non-aqueous systems. Patent: US 5429999, 1995.

[1313] H. N. Nae, W. W. Reichert, and A. C. Eng. Organoclay compositions prepared with a mixture of two organic cations and their use in non-aqueous systems. Patent: CA 2082712, 1999.

[1314] J. J. W. Nahm, H. J. Vinegar, J. M. Karanikas, and R. E. Wyant. High temperature wellbore cement slurry. Patent: US 5226961, 1993.

[1315] M. I. Naiman and J. C. Chang. Methods and compositions for reduction of drag in hydrocarbon fluids. Patent: US 4983186, 1991.

[1316] S. K. Nanda, R. Kumar, K. L. Goyal, and K. L. Sindhwani. Characterization of polyacrylamine-Cr^{6+} gels used for reducing water/oil ratio. In *Proceedings Volume*. SPE Oilfield Chem Int Symp (San Antonio, TX, 2/4–2/6), 1987.

[1317] A. Naraghi. Corrosion inhibitor containing phosphate groups. Patent: US 5611991, 1997.

[1318] A. Naraghi and N. Grahmann. Corrosion inhibitor blends with phosphate esters. Patent: US 5611992, 1997.

[1319] A. R. Naraghi and R. S. Rozell. Method for reducing torque in downhole drilling. Patent: US 5535834, 1996.

[1320] I. S. Nashawi. Laboratory investigation of the effect of brine composition on polymer solutions: Pt 2: Xanthan gum (XG) case. SPE Unsolicited Paper, SPE-23534, United Arab Emirates Univ, August 1991.

[1321] T. N. Nasr and E. E. Isaacs. Process for enhancing hydrocarbon mobility using a steam additive. Patent: US 6230814, 2001.

[1322] T. N. Nasr and E. E. Isaacs. Process for enhancing hydrocarbon mobility using a steam additive. Patent: WO 0127439A, 2001.

[1323] H. A. Nasr-El-Din, J. D. Lynn, and K. C. Taylor. Lab testing and field application of a large-scale acetic acid-based treatment in a newly developed carbonate reservoir. In *Proceedings Volume*. SPE Oilfield Chem Int Symp (Houston, TX, 2/13–2/16), 2001.

[1324] R. C. Navarrete, J. E. Brown, and R. P. Marcinew. Application of new bridging technology and particulate chemistry for fluid-loss control during fracturing highly permeable formations. In *Proceedings Volume,* volume 2, pages 321–325. SPE Europe Petrol Conf (Milan, Italy, 10/22–10/24), 1996.

[1325] R. C. Navarrete and J. P. Mitchell. Fluid-loss control for high-permeability rocks in hydraulic fracturing under realistic shear conditions. In *Proceedings Volume,* pages 579–591. SPE Prod Oper Symp (Oklahoma City, OK, 4/2–4/4), 1995.

[1326] R. C. Navarrete, J. M. Seheult, and M. D. Coffey. New biopolymers for drilling, drill-in, completions, spacer, and coil-tubing fluids: Pt 2. In *Proceedings Volume*. SPE Oilfield Chem Int Symp (Houston, TX, 2/13–2/16), 2001.

[1327] R. C. Navarrete, J. M. Seheult, and R. E. Himes. Applications of xanthan gum in fluid-loss control and related formation damage. In *Proceedings Volume*. SPE Permian Basin Oil & Gas Recovery Conf (Midland, TX, 3/21–3/23), 2000.

[1328] R. C. Navarrete and S. N. Shah. New biopolymer for coiled tubing applications. In *Proceedings Volume*. SPE/Int Coiled Tubing Assoc Coiled Tubing Roundtable (Houston, TX, 3/7–3/8), 2001.

[1329] K. R. Neel. *Gilsonite,* volume 11, 3 edition, pages 802–806. J. Wiley & Sons, New York, 1980.

[1330] K. D. Neemla, R. C. Saxena, and A. Jayaraman. Corrosion inhibition of oil-well equipment during acidization. *Corrosion Prev Contr,* 39(3):69–73, June 1992.

[1331] V. R. Negomedzyanov, V. P. Bortsov, V. S. Denisov, V. V. Slepov, and S. S. Volkova. Plugging composition for use in oil and gas extraction industry—contains Portland cement and aluminium-containing additive in form of slag dust waste from aluminium production process. Patent: RU 2029067-C, 1995.

[1332] E. B. Nelson. Well treating process and composition. Patent: CA 1216742, 1987.

[1333] E. B. Nelson, editor. *Well cementing,* volume 28 of *Elsevier Developments in Petroleum Science.* Elsevier Science Publishing Co, New York, 1990.

[1334] A. R. Nerheim, E. K. Samuelson, and T. M. Svartaas. Investigation of hydrate kinetics in the nucleation and early growth phase by laser light scattering. In *Proceedings Volume,* volume 1, pages 620–627. 2nd Soc Offshore Polar Eng et al Offshore & Polar Eng Int Conf (San Francisco, CA, 6/14–6/19), 1992.

[1335] A. R. Nerheim, T. M. Svartaas, and E. J. Samuelsen. Laser light scattering studies of natural gas hydrates. In *Proceedings Volume,* volume 2, pages 303–309. 69th Annu SPE Tech Conf (New Orleans, LA, 9/25–9/28), 1994.

[1336] A. D. Nevers. Odor-fading prevention from organosulfur-odorized liquefied petroleum gas. Patent: CA 1274692, 1990.

[1337] J. A. Newton. Applications of spectroscopic quality control techniques for oilfield chemicals. NACE Corrosion 88 (St Louis, MO, 3/21–3/25), 1988.

[1338] P. D. Nguyen and K. L. Schreiner. Preventing well fracture proppant flow-back. Patent: US 5908073, 1999.

[1339] P. D. Nguyen and J. D. Weaver. Method of controlling particulate flowback in subterranean wells and introducing treatment chemicals. Patent: US 6209643, 2001.

[1340] P. D. Nguyen, J. D. Weaver, and J. L. Brumley. Stimulating fluid production from unconsolidated formations. Patent: US 6257335, 2001.

[1341] P. D. Nguyen, J. D. Weaver, R. C. Cole, and C. R. Schulze. Development and field application of a new fluid-loss control material. In *Proceedings Volume,* pages 933–941. Annu SPE Tech Conf (Denver, CO, 10/6–10/9), 1996.
[1342] P. D. Nguyen, J. D. Weaver, M. A. Parker, and D. G. King. Thermoplastic film prevents proppant flowback. *Oil Gas J,* 94(6):60–62, 1996.
[1343] P. D. Nguyen, J. D. Weaver, M. A. Parker, D. G. King, R. L. Gillstrom, and D. W. Van Batenburg. Proppant flowback control additives. In *Proceedings Volume,* pages 119–131. Annu SPE Tech Conf (Denver, CO, 10/6–10/9), 1996.
[1344] L. F. Nicora and W. M. McGregor. Biodegradable surfactants for cosmetics find application in drilling fluids. In *Proceedings Volume,* pages 723–730. IADC/SPE Drilling Conf (Dallas, TX, 3/3–3/6), 1998.
[1345] K. Nimerick. Fracturing fluid and method. Patent: GB 2291907, 1996.
[1346] K. H. Nimerick and C. L. Boney. Method of fracturing high temperature wells and fracturing fluid therefore. Patent: US 5103913, 1992.
[1347] K. H. Nimerick, C. W. Crown, S. B. McConnell, and B. Ainley. Method of using borate crosslinked fracturing fluid having increased temperature range. Patent: US 5259455, 1993.
[1348] K. H. Nimerick and J. J. Hinkel. Enhanced methane production from coal seams by dewatering. Patent: EP 444760, 1991.
[1349] K. H. Nimerick, S. B. McConnell, and M. L. Samuelson. Compatibility of resin-coated proppants with crosslinked fracturing fluids. In *Proceedings Volume,* pages 245–250. 65th Annu SPE Tech Conf (New Orleans, LA, 9/23–9/26), 1990.
[1350] J. H. Y. Niu, J. G. Edmondson, and S. E. Lehrer. Method of inhibiting corrosion of metal surfaces in contact with a corrosive hydrocarbon containing medium. Patent: EP 256802, 1988.
[1351] Y. Niu, J. Ouyang, Z. Zhu, G. Wang, G. Sun, and J. Shi. Research on hydrophobically associating water-soluble polymer used for EOR. In *Proceedings Volume.* SPE Oilfield Chem Int Symp (Houston, TX, 2/13–2/16), 2001.
[1352] C. Noik and A. Rivereau. Method and material for well cementing (methode et materiau pour la cimentation de puits). Patent: EP 881353, 1998.
[1353] L. Noran, S. Vitthal, and J. Terracina. New breaker technology for fracturing high-permeability formations. In *Proceedings Volume,* pages 187–199. SPE Europe Formation Damage Contr Conf (The Hague, Netherlands, 5/15–5/16), 1995.
[1354] L. R. Norman and S. B. Laramay. Encapsulated breakers and method for use in treating subterranean formations. Patent: US 5373901, 1994.

[1355] L. R. Norman, R. Turton, and A. L. Bhatia. Breaking fracturing fluid in subterranean formation. Patent: EP 1152121A, 2001.
[1356] P. S. Northrop. Method for disposing of waste gas in subterranean formations. Patent: US 5267614, 1993.
[1357] P. S. Northrop. Imbibition process using a horizontal well for oil production from low permeability reservoirs. Patent: US 5411094, 1995.
[1358] N. Obeyesekere, A. Naraghi, and J. S. McMurray. Synthesis and evaluation of biopolymers as low toxicity corrosion inhibitors for North Sea oil fields. NACE Int Corrosion Conf (Corrosion 2001) (Houston, TX, 3/11–3/16), 2001.
[1359] K. Oh-Kil and C. Ling-Siu. Drag reducing polymers. In *The polymeric materials encyclopedia*. CRC Press, Inc, Boca Raton, FL, 1996.
[1360] J. R. Ohlsen, J. M. Brown, G. F. Brock, and V. K. Mandlay. Corrosion inhibitor composition and method of use. Patent: US 5459125, 1995.
[1361] N. A. Okishev, A. G. Ivanov, and I. V. Karpenko. Plugging solution for oil and gas wells with increased sedimentation stability—containing Portland cement, nitrile trimethyl phosphonic acid, polyoxyethylene, water-soluble cationic polyelectrolyte and water. Patent: RU 2039207-C, 1995.
[1362] D. K. Olsen. Use of amine oxide surfactants for chemical flooding EOR: topical report. US DOE Fossil Energy Rep NIPER-417, November 1989.
[1363] D. D. Onan, D. T. Terry, and B. G. Brake. Downhole cement composition. Patent: EP 618344, 1994.
[1364] D. D. Onan, D. T. Terry, and W. D. Riley. Set-activated cementitious compositions and methods. Patent: US 5332041, 1994.
[1365] D. D. Onan, W. W. Webster, and J. E. Griffith. Well cementing method. Patent: EP 787698, 1997.
[1366] K. Oppenlaender, B. Wegner, and W. Slotman. Ammonium salt of an alkenylsuccinic half-amide and the use thereof as corrosion inhibitor in oil and/or gas production technology. Patent: US 5250225, 1993.
[1367] D. Oppong and C. G. Hollis. Synergistic antimicrobial compositions containing (thiocyanomethylthio)benzothiazole and an organic acid. Patent: WO 9508267, 1995.
[1368] Deleted in proofs.
[1369] D. Oppong and V. M. King. Synergistic antimicrobial compositions containing a halogenated acetophenone and an organic acid. Patent: WO 9520319, 1995.
[1370] Deleted in proofs.

[1371] D. Osborne. The development in the use of FRE (fiber reinforced epoxy) pipe systems for industrial and offshore applications. In *Proceedings Volume,* number 7, pages 27–45. 10th BHR Group Ltd et al Pipe Protect Int Conf (Amsterdam, Netherlands, 11/10–11/12), 1993.

[1372] W. T. Osterloh. Long chain alcohol additives for surfactant foaming agents. Patent: US 5333687, 1994.

[1373] W. T. Osterloh and M. J. Jante. Low-cost foam surfactant from wood pulping by-products. In *Proceedings Volume,* volume 2, pages 191–199. 8th EAPG Impr Oil Recovery Europe Symp (Vienna, Austria, 5/15–5/17), 1995.

[1374] W. T. Osterloh and M. J. Jante, Jr. Surfactant-polymer flooding with anionic PO/EO surfactant microemulsions containing polyethylene glycol additives. In *Proceedings Volume,* volume 1, pages 485–494. 8th SPE/DOE Enhanced Oil Recovery Symp (Tulsa, OK, 4/22–4/24), 1992.

[1375] G. M. Ostryanskaya, Y. D. Abramov, V. N. Makarov, S. N. Osipov, and A. V. Razhkevich. Gel-forming plugging composition—contains ligno-sulphonate, modified carboxymethyl cellulose, bichromate, calcium chloride and water. Patent: SU 1776766-A, 1992.

[1376] R. J. Oswald, R. Morschhaeuser, K. H. Heier, A. Tardi, J. Tonhauser, C. Kayser, and D. Patterson. Water-soluble copolymers and their use in the exploration for and recovery of oil and gas. Patent: EP 1059316, 2000.

[1377] B. A. Oude Alink. Water soluble 1,2-dithio-3-thiones. Patent: EP 415556, 1991.

[1378] B. A. Oude Alink. Water soluble 1,2-dithio-3-thiones. Patent: US 5252289, 1993.

[1379] P. Oudman. Odorization and odorant monitoring practices at Canadian Western Natural Gas Company Limited. In *Odorization,* volume 3, pages 389–405. Institute of Gas Technology, Chicago, IL, Southern California Gas Co, 1993.

[1380] V. P. Ovchinnikov, A. A. Shatov, N. Yu. Shulgina, and P. V. Ovchinnikov. Salt composition for regulating hardening processes of plugging agents. *Neftepromysl Delo,* (11–12):32–33, November-December 1995.

[1381] M. J. Owen. *Defoamers,* volume 7, 4 edition, pages 929–945. J. Wiley & Sons, New York; Chichester, Brisbane, 1996.

[1382] A. Padron. Stable emulsion of viscous crude hydrocarbon in aqueous buffer solution and method for forming and transporting same. Patent: CA 2113597, 1994.

[1383] A. Padron. Stable emulsion of viscous crude hydrocarbon in aqueous buffer solution and method for forming and transporting same. Patent: EP 672860, 1995.

[1384] A. Padron, L. Castro, and G. Zamora. Transportable and stable hydrocarbons in buffer solution dispersions. In *Proceedings Volume,* volume 2, pages 587–596. 6th Unitar et al Heavy Crude & Tar Sands Int Conf (Houston, TX, 2/12–2/17), 1995.

[1385] D. G. Pafitis, G. C. Maitland, and S. N. Davies. Thixotropic materials for oilwell applications. Patent: GB 2296713, 1996.

[1386] M. K. Pakulski and B. T. Hlidek. Slurried polymer foam system and method for the use thereof. Patent: WO 9214907, 1992.

[1387] T. Palermo, A. Sinquin, H. Dhulesia, and J. M. Fourest. Pilot loop tests of new additives preventing hydrate plugs formation. In *Proceedings Volume,* pages 133–147. 8th Bhr Group Ltd et al Multiphase 97 Int Conf (Cannes, France, 6/18–6/20), 1997.

[1388] L. V. Palij, V. E. Akhrimenko, A. K. Kuksov, V. M. Medentsev, and V. I. Panov. Plugging solution for oil and gas wells—contains Portland cement, fly ash, sodium or calcium sulphate and additionally hydrosil, to increase adhesion of cement rock to casing string. Patent: SU 1802087-A, 1993.

[1389] F. H. Palmer and A. Wright. Polyurethanes as insulating systems for high temperature marine pipelines. In *Proceedings Volume,* pages 169–187. Bhr Group Ltd et al Pipe Protect Conf (Cannes, France, 9/23–9/25), 1991.

[1390] S. Palumbo, D. Giacca, M. Ferrari, and P. Pirovano. The development of potassium cellulosic polymers and their contribution to the inhibition of hydratable clays. In *Proceedings Volume,* pages 173–182. SPE Oilfield Chem Int Symp (Houston, TX, 2/8–2/10), 1989.

[1391] A. K. Pandey and N. P. Joshi. Effectivity of additives in reducing down hole friction and preventing sticking. In *Proceedings Volume,* volume 3, pages 45–54. 1st India Oil & Natur Gas Corp Ltd et al Int Petrol Conf (Petrotech 95) (New Delhi, India, 1/9–1/12), 1995.

[1392] H. J. Panneman, R. C. Pot-Gerritsen, E. M. Kuiper-Van Loo, H. Pastoor, and R. Janssen-Van Rosmalen. UV (ultraviolet)-oxidation process for water treatment at gas plant sites. In *Proceedings Volume, Pt B,* pages 269 271–285. 20th Int Gas Union World Gas Conf (Copenhagen, Denmark, 6/10–6/13), 1997.

[1393] J. S. Parent and P. R. Bishnoi. Investigations into the nucleation behaviour of methane gas hydrates. *Chem Eng Commun,* 144:51–64, February 1996.

[1394] A. Parker. Process for delaying and controlling the formation of gels or precipitates derived from aluminum and corresponding compositions together with its applications particularly those concerning the operation of oil wells. Patent: EP 266808, 1988.

[1395] C. F. Parks, P. E. Clark, O. Barkat, and J. Halvaci. Characterizing polymer solutions by viscosity and functional testing. In *Proceedings*

192nd ACS Nat Mtg, volume 55, pages 880–888. Amer Chem Soc Polymeric Mater Sci Eng Div Tech Program (Anaheim, CA, 9/7–9/12), 1986.

[1396] A. V. Parusyuk, I. N. Galantsev, V. N. Sukhanov, T. A. Ismagilov, A. G. Telin, L. N. Barinova, M. Z. Igdavletova, and A. G. Skorokhod. Gelforming compositions for leveling of injectivity profile and selective water inflow shutoff. *Neft Khoz,* (2):64–68, February 1994.

[1397] A. D. Patel. Silicone based fluids for drilling applications. Patent: EP 764709, 1997.

[1398] A. D. Patel. Water-based drilling fluids with high temperature fluid loss control additive. Patent: US 5789349, 1998.

[1399] A. D. Patel and H. C. McLaurine. Drilling fluid additive. Patent: EP 427107, 1991.

[1400] A. D. Patel and H. C. McLaurine. Drilling fluid additive and method for inhibiting hydration. Patent: US 5149690, 1992.

[1401] A. D. Patel and H. C. McLaurine. Drilling fluid additive and method for inhibiting hydration. Patent: CA 2088344, 1993.

[1402] A. D. Patel, H. C. McLaurine, E. Stamatakis, and C. J. Thaemlitz. Drilling fluid additive and method for inhibiting hydration. Patent: EP 634468, 1995.

[1403] A. D. Patel, E. Stamatakis, and E. Davis. Shale hydration inhibition agent and method of use. Patent: US 6247543, 2001.

[1404] B. Patel and M. Stephens. Well cement slurries and dispersants therefor. Patent: US 5041630, 1991.

[1405] B. B. Patel. Fluid composition comprising a metal aluminate or a viscosity promoter and a magnesium compound and process using the composition. Patent: EP 617106, 1994.

[1406] B. B. Patel. Tin/cerium compounds for lignosulfonate processing. Patent: EP 600343, 1994.

[1407] B. B. Patel. Liquid additive comprising a sulfonated asphalt and processes therefor and therewith. Patent: US 5502030, 1996.

[1408] B. B. Patel. Drilling fluid additive and process therewith. Patent: WO 0020527, 2000.

[1409] B. B. Patel and G. T. Muller. Compositions comprising an acrylamide-containing polymer and process therewith. Patent: EP 728826, 1996.

[1410] B. B. Patel and M. Stephens. Well cement slurries and dispersants therefor. Patent: US 4923516, 1990.

[1411] D. G. Peiffer, J. Bock, and J. Elward-Berry. Zwitterionic functionalized polymers as deflocculants in water based drilling fluids. Patent: US 5026490, 1991.

[1412] D. G. Peiffer, J. Bock, and J. Elward-Berry. Thermally stable hydrophobically associating rheological control additives for water-based drilling fluids. Patent: US 5096603, 1992.

[1413] D. G. Peiffer, J. Bock, and J. Elward-Berry. Zwitterionic functionalized polymers. Patent: GB 2247240, 1992.

[1414] D. G. Peiffer, J. Bock, and J. Elward-Berry. Zwitterionic functionalized polymers as deflocculants in water based drilling fluids. Patent: CA 2046669, 1992.

[1415] D. G. Peiffer, J. Bock, and J. Elward-Berry. Thermally stable hydrophobically associating rheological control additives for water-based drilling fluids. Patent: CA 2055011, 1993.

[1416] D. G. Peiffer, J. Bock, and J. Elward-Berry. Zwitterionic functionalized polymers as deflocculants in water based drilling fluids. Patent: AU 638917, 1993.

[1417] D. G. Peiffer, R. M. Kowalik, and R. D. Lundberg. Drag reduction with novel hydrocarbon soluble polyampholytes. Patent: US 4640945, 1987.

[1418] D. G. Peiffer, R. D. Lundberg, L. Sedillo, and J. C. Newlove. Fluid loss control in oil field cements. Patent: US 4626285, 1986.

[1419] J. J. M. Pelissier and S. Biasini. Biodegradable drilling mud (boue de forage biodegradable). Patent: FR 2649988, 1991.

[1420] A. I. Penkov, L. P. Vakhrushev, and E. V. Belenko. Characteristics of the behavior and use of polyalkylene glycols for chemical treatment of drilling muds. *Stroit Neft Gaz Skvazhin Sushe More,* (1–2):21–24, January-February 1999.

[1421] G. S. Penny. Method of increasing hydrocarbon production from subterranean formations. Patent: US 4702849, 1987.

[1422] G. S. Penny. Method of increasing hydrocarbon productions from subterranean formations. Patent: EP 234910, 1987.

[1423] G. S. Penny and J. E. Briscoe. Method of increasing hydrocarbon production by remedial well treatment. Patent: CA 1216416, 1987.

[1424] G. S. Penny and M. W. Conway. Coordinated studies in support of hydraulic fracturing of coalbed methane: final report (July 1990–May 1995). Gas Res Inst Rep GRI-95/0283, Gas Res Inst, September 1995.

[1425] G. S. Penny, R. S. Stephens, and A. R. Winslow. Method of supporting fractures in geologic formations and hydraulic fluid composition for same. Patent: US 5009797, 1991.

[1426] G. S. Penny and M. W. Conway. *Fluid leakoff,* volume 12 of *Recent advances in hydraulic fracturing (SPE Henry L. Doherty Monogr Ser),* pages 147–176. SPE, Richardson, TX, 1989.

[1427] A. A. Perejma and L. V. Pertseva. Complex reagent for treating plugging solutions—comprises hydrolysed polyacrylonitrile, ferrochromolignosulphonate Cr-containing additive, waste from lanolin production treated with triethanolamine and water. Patent: RU 2013524-C, 1994.

[1428] A. A. Perejma, K. M. Tagirov, V. I. Ilyaev, and A. A. Kovalev. Plugging solution for conducting well repair works, etc.—contains Portland cement, polyacrylamide, specified stabilising additive, sodium sulphate waste from sebacic acid production and water. Patent: RU 2035585-C, 1995.

[1429] R. C. Pessier, K. Javanmardi, J. J. Nahm, J. M. Leimkuhler, J. W. Dudley, II, and F. K. Moody. Evaluating slag solidified mud drillability and bit performance. *World Oil,* 215(10):65,67–68,70,72,74, October 1994.

[1430] H. Petersen, J. Martens, W. Harms, and U. Kramer. Process for avoiding the formation of condensates, hydrates or ice in the decompression of natural gas stored in salt caverns (Verfahren zur Vermeidung der Bildung von Kondensaten, Hydraten oder Eis bei der Entspannung von in Salzkavernen gespeichertem Erdgas). Patent: DE 3927905, 1991.

[1431] P. R. Petersen, L. G. Coker, and D. S. Sullivan. Method of inhibiting corrosion using N-S containing compounds. Patent: GB 2221458, 1990.

[1432] N. J. Phillips, J. P. Renwick, J. W. Palmer, and A. J. Swift. The synergistic effect of sodium thiosulphate on corrosion inhibition. In *Proceedings Volume,* volume 1, pages 110–137. 7th NACE Int et al Middle East Corrosion Conf (Manama, Bahrain, 2/26–2/28), 1996.

[1433] J. S. Pic, J. M. Herri, and M. Cournil. Experimental influence of kinetic inhibitors on methane hydrate particle size distribution during batch crystallization in water. *Can J Chem Eng,* 79(3):374–383, June 2001.

[1434] W. Pittman, W. C. Maurer, W. G. Deskins, F. Sabins, and D. Müller. Carbon-fiber technology for improved downhole cement performance: final report. Gas Res Inst Rep GRI-96/0276, Gas Res Inst, August 1996.

[1435] M. J. Pitts, K. Wyatt, T. C. Sale, and K. R. Piontek. Utilization of chemical-enhanced oil recovery technology to remove hazardous oily waste from alluvium. In *Proceedings Volume,* pages 33–44. SPE Oilfield Chem Int Symp (New Orleans, LA, 3/2–3/5), 1993.

[1436] J. Plank. Field results with a novel fluid loss polymer for drilling muds. *Oil Gas Europe Mag,* 16(3):20–23, October 1990.

[1437] J. Plank. Drilling mud composition and process for reducing the filtrate of metal hydroxide mixtures containing drilling mud compositions. Patent: WO 9312194, 1993.

[1438] J. P. Plank and F. A. Gossen. Visualization of fluid-loss polymers in drilling mud filter cakes. In *Proceedings Volume,* pages 165–176. 64th Annu SPE Tech Conf (San Antonio, TX, 10/8–10/11), 1989.

[1439] M. L. Pless, J. D. Kercheville, and J. J. Augsburger. Defoamer composition for use in water based drilling fluids. Patent: EP 339762, 1989.

[1440] M. L. Pless, J. D. Kercheville, and J. J. Augsburger. Defoamer composition for use in water based drilling fluids. Patent: AU 608681, 1991.
[1441] V. M. Podgornov, I. A. Vedishchev, and M. Akhmad Mokhamad. Plugging slurry for oil and gas well drilling—contains cement, bentonite clay powder, water glass, sodium chloride and water. Patent: SU 1756537-A, 1992.
[1442] D. M. Ponomarev, L. P. Vakhrushev, B. M. Gavrilov, S. G. Gorlov, V. N. Koshelev, Y. N. Mojsa, V. I. Ryabchenko, S. G. Tishchenko, N. I. Khusht, and N. T. Rud. Foam-extinguishing agent for drilling solutions—contains individual higher alcohol(s) or their mixture and block copolymer of ethylene and propylene oxide. Patent: SU 1720681-A, 1992.
[1443] O. Ponsati, A. Trius, C. P. Herold, H. Müller, C. Nitsch, and S. von Tapavicza. Use of selected oleophilic compounds with quaternary nitrogen to improve the oil wettability of finely divided clay and their use as viscosity promoters. Patent: WO 9219693, 1992.
[1444] O. Ponsati, A. Trius, C. P. Herold, H. Müller, C. Nitsch, and S. von Tapavicza. Use of selected oleophilic compounds with quaternary nitrogen to improve the oil wettability of finely divided clay and their use as viscosity promoters (Verwendung ausgewählter oleophiler Verbindungen mit quartärem Stickstoff zur Verbesserung der Ölbenetzbarkeit feinteiliger Tone und deren Anwendung als Viskositätsbildner). Patent: EP 583285, 1994.
[1445] D. H. Pope. Concern over MIC (microbiologically-influenced corrosion) expanding among corrosion engineers. *Pipe Line Gas Ind*, 80(2):23-25, February 1997.
[1446] D. H. Pope, D. Dziewulski, and J. R. Frank. Microbiologically influenced corrosion in the gas industry. *Pipeline*, 62(5):8-9, October 1990.
[1447] D. H. Pope, D. M. Dziewulski, S. F. Lockwood, D. P. Werner, and J. R. Frank. Microbiological corrosion concerns for pipelines and tanks. In *Proceedings Volume*, pages 290-321. API Pipeline Conf (Houston, TX, 4/7-4/8), 1992.
[1448] R. C. Portnoy. Anionic copolymers for improved control of drilling fluid rheology. Patent: GB 2174402, 1986.
[1449] R. C. Portnoy. Anionic copolymers for improved control of drilling fluid rheology. Patent: US 4680128, 1987.
[1450] R. Potter. Proteinaceous oil spill dispersant. Patent: WO 9419942, 1994.
[1451] P. J. Powell, R. D. Gdanski, M. A. McCabe, and D. C. Buster. Controlled-release scale inhibitor for use in fracturing treatments.

In *Proceedings Volume,* pages 571–579. SPE Oilfield Chem Int Symp (San Antonio, TX, 2/14–2/17), 1995.
[1452] R. J. Powell, A. R. Fischer, R. D. Gdanski, M. A. McCabe, and S. D. Pelley. Encapsulated scale inhibitor for use in fracturing treatments. In *Proceedings Volume,* pages 557–563. Annu SPE Tech Conf (Dallas, TX, 10/22–10/25), 1995.
[1453] R. J. Powell, A. R. Fischer, R. D. Gdanski, M. A. McCabe, and S. D. Pelley. Encapsulated scale inhibitor for use in fracturing treatments. In *Proceedings Volume,* pages 107–113. SPE Permian Basin Oil & Gas Recovery Conf (Midland, TX, 3/27–3/29), 1996.
[1454] R. Prasad. Pros and Cons of ATP (adenosine triphosphate) measurement in oil field waters. In *Proceedings Volume.* NACE Corrosion 88 (St Louis, MO, 3/21–3/25), 1988.
[1455] B. B. Prasek. Interactions between fracturing fluid additives and currently used enzyme breakers. In *Proceedings Volume,* pages 265–279. 43rd Annu Southwestern Petrol Short Course Assoc Inc et al Mtg (Lubbock, TX, 4/17–4/18), 1996.
[1456] N. M. Prokhorov, L. N. Smirnova, and V. Z. Luban. Neutralisation of hydrogen sulphide in drilling solution—by introduction of additive consisting of iron sulphate and additionally sodium aluminate, to increase hydrogen sulphide absorption. Patent: SU 1798358-A, 1993.
[1457] A. F. Psaila. Demulsifiers. Patent: EP 222587, 1987.
[1458] A. F. Psaila. Demulsifiers. Patent: CA 1267584, 1990.
[1459] R. J. Purinton, Jr. and S. Mitchell. Practical applications for gelled fluid pigging. *Pipe Line Ind,* 66(3):55–56, March 1987.
[1460] A. G. Putz, B. M. Pedron, and B. Bazin. Commercial polymer injection in the Courtenay field, 1994 update. In *Proceedings Volume,* volume 1, pages 403–413. 69th Annu SPE Tech Conf (New Orleans, LA, 9/25–9/28), 1994.
[1461] D. E. Putzig. Zirconium chelates and their use for cross-linking. Patent: EP 278684, 1988.
[1462] D. E. Putzig and K. C. Smeltz. Organic titanium compositions useful as cross-linkers. Patent: EP 195531, 1986.
[1463] L. R. Quaife and K. J. Moynihan. A new pipeline leak-locating technique utilizing a novel odourized test-fluid (patent pending) and trained domestic dogs. In *Proceedings Volume,* pages 647–657. 1st US Environ Protect Agency et al Oil & Gas Explor & Prod Waste Manage Pract Int Symp (New Orleans, LA, 9/10–9/13), 1990.
[1464] L. R. Quaife, K. J. Moynihan, and D. A. Larson. A new pipeline leak-location technique utilizing a novel (patented) test-fluid and trained domestic dogs. In *Proceedings Volume,* pages 154–161. 71st Annu GPA Conv (Anaheim, CA, 3/16–3/18), 1992.

[1465] L. R. Quaife, J. Szarka, K. J. Moynihan, and M. E. Moir. Test-fluid composition and method for detecting leaks in pipelines and associated facilities. Patent: US 5049312, 1991.
[1466] L. R. Quaife, J. Szarka, K. J. Moynihan, and M. E. Moir. Test-fluid composition and method for detecting leaks in pipelines and associated facilities. Patent: CA 2052242, 1993.
[1467] C. Quet, P. Cheneviere, M. Bourrel, and G. Glotin. Core surface analysis for wettability assessment. In *Proceedings Volume,* pages 119–131. Advances in Core Evaluation II: Reservoir Appraisal (2nd Soc Core Anal et al Europe Core Analysis Symp (EuroCAS II) (London, England, 5/20–5/22), Gordon & Breach Science Publishers, 1991.
[1468] C. Quet, P. Cheneviere, G. Glotin, and M. Bourrel. Pore surface chemistry and wettability. In *Proceedings Volume,* pages 81–88. 6th Inst Francais Du Petrole Explor & Prod Res Conf (Saint-Raphael, France, 9/4–9/6), 1992.
[1469] P. Rae and E. Brown. Polymer additives improve cementing in salt formations. *Oil Gas J,* 86(49):31–32,35–36,38, 1988.
[1470] P. J. Rae, N. Johnston, and G. Dilullo. Storable liquid systems for use in cementing oil and gas wells. Patent: US 6173778, 2001.
[1471] R. Rae and G. Di Lullo. Chemically-enhanced drilling with coiled tubing in carbonate reservoirs. In *Proceedings Volume.* SPE/Int Coiled Tubing Assoc Coiled Tubing Roundtable (Houston, TX, 3/7–3/8), 2001.
[1472] C. Raible. Improvement in oil recovery using cosolvents with CO_2 gas floods. US DOE Fossil Energy Rep NIPER-559, NIPER, January 1992.
[1473] R. H. Raines. Use of low M.S. (molar substitution) hydroxyethyl cellulose for fluid loss control in oil well applications. Patent: US 4629573, 1986.
[1474] T. K. Rakhmatullin, F. A. Agzamov, V. V. Ivanov, N. Kh. Karimov, M. A. Tankibaev, and N. V. Trenkenshu. Clinker-less binder composition—contains mixture of slag from melting of oxidised nickel ores and tails from enrichment of phosphorite ores. Patent: SU 1777617-A, 1992.
[1475] W. G. Rakitsky and D. D. Richey. Rapidly hydrating welan gum. Patent: EP 505096, 1992.
[1476] T. A. Ramanarayanan and H. L. Vedage. Inorganic/organic inhibitor for corrosion of iron containing materials in sulfur environment. Patent: US 5279651, 1994.
[1477] M. Rame and M. Delshad. A compositional reservoir simulator on distributed memory parallel computers. In *Proceedings Volume,* pages 89–100. 13th SPE Reservoir Simulation Symp (San Antonio, TX, 2/12–2/15), 1995.

[1478] M. Ramesh and A. Sivakumar. Hydrophobic demulsifiers for oil-in-water systems. Patent: CA 2124301, 1994.
[1479] M. Ramesh and A. Sivakumar. Hydrophobically-modified demulsifiers for oil-in-water systems. Patent: US 5635112, 1997.
[1480] J. A. Ramsay, D. G. Cooper, and R. J. Neufeld. Effects of oil reservoir conditions on the production of water-insoluble levan by *Bacillus licheniformis*. *Geomicrobiol J*, 7(3):155–165, July 1989.
[1481] S. Rangus, D. B. Shaw, and P. Jenness. Cellulose ether thickening compositions. Patent: WO 9308230, 1993.
[1482] B. A. Rastegaev, B. A. Andreson, and Yu. L. Raizberg. Bactericidal protection of chemical agents from biodegradation while drilling deep wells. *Stroit Neft Gaz Skvazhin Sushe More*, (7–8):32–34, July-August 1999.
[1483] S. R. Rayudu and J. D. Pera. A method of inhibiting the growth of microorganisms in aqueous liquids. Patent: EP 313272, 1989.
[1484] A. Razzaq. *Characterization of Surfactants in the Presence of Oil for Steam Foam Applications*. PhD thesis, Stanford Univ, 1992.
[1485] A. Razzaq and L. M. Castanier. Characterization of surfactants in the presence of oil for steam foam application. US DOE Fossil Energy Rep DOE/BC/14600-37, Stanford Univ, 1992.
[1486] B. R. Reddy, R. J. Crook, J. Chatterji, B. J. King, D. W. Gray, R. M. Fitzgerald, R. J. Powell, and B. L. Todd. Controlling the release of chemical additives in well treating fluids. Patent: US 6209646, 2001.
[1487] M. G. Reed. Permeability of fines-containing earthen formations by removing liquid water. Patent: CA 2046792, 1993.
[1488] A. L. Reid and H. A. Grichuk. Polymer composition comprising phosphorous-containing gelling agent and process thereof. Patent: US 5034139, 1991.
[1489] J. M. Reizer, M. G. Rudel, C. D. Sitz, R. M. S. Wat, and H. Montgomerie. Scale inhibitors. Patent: US 6379612, 2002.
[1490] Z. C. Ren, A. P. Shi, S. L. Leng, W. S. Zhang, and L. P. Qin. Formulation of crosslinking polyacrylamide solution for EOR. *Oilfield Chem*, 15(2):146–149, June 1998.
[1491] W. A. Rendall, C. Ayasse, and J. Novosad. Surfactant-stabilized foams for enhanced oil recovery. Patent: US 5074358, 1991.
[1492] R. T. Rewick, K. A. Sabo, J. Gates, J. H. Smith, and L. T. McCarthy, Jr. Project summary: evaluation of oil spill dispersant testing requirements. US Environ Protect Agency Rep EPA/600/S2-87/070, SRI International, EPA, 1988.
[1493] M. J. Reynhout, C. E. Kind, and U. C. Klomp. A method for preventing or retarding the formation of hydrates. Patent: EP 526929, 1993.
[1494] W. C. Richardson and K. R. Kibodeaux. Chemically assisted thermal flood process. Patent: US 6305472, 2001.

[1495] J. Ridland and D. A. Brown. Organo-metallic compounds. Patent: CA 2002792, 1990.
[1496] P. Riley. Compositions and methods for dispersing and biodegrading spilled petroleum oils and fuels. Patent: US 5753127, 1998.
[1497] D. P. Rimmer, A. A. Gregoli, J. A. Hamshar, and E. Yildirim. Pipeline emulsion transportation for heavy oils. In L. L. Schramm, editor, *Emulsions: fundamentals and applications in the petroleum industry*, volume 231 of *ACS Advances in Chemistry Series*, pages 295–312. American Chemical Society, Washington DC, 1992.
[1498] D. L. Ripley, J. W. Goetzinger, and M. L. Whisman. Human response research evaluation of alternate odorants for lp-gas. *GPA Res*, 1990.
[1499] W. Ritter and C. P. Herold. New aqueous emulsion copolymerizates for improving flow properties of crude oils. Patent: WO 9002766, 1990.
[1500] W. Ritter, C. Meyer, W. Zöllner, C. P. Herold, and S. von Tapavicza. Use of selected acrylic and/or methacrylic acid ester copolymers as flow enhancers in paraffin-rich crude oil and crude oil fraction (II) (Verwendung ausgewählter Copolymeren der Acryl- und/oder Methacrylsäureester als Fließverbesserer in paraffinreichen Erdölen und Erdölfraktionen [II]). Patent: EP 332002, 1989.
[1501] W. Ritter, C. Meyer, W. Zöllner, C. P. Herold, and S. von Tapavicza. Copolymers of acrylic acid and/or methacrylic acid esters as flow improvers. Patent: AU 611265, 1991.
[1502] W. Ritter, C. Meyer, W. Zöllner, C. P. Herold, and S. von Tapavicza. Copolymers of (meth)acrylic acid esters as flow improvers in oils. Patent: US 5039432, 1991.
[1503] W. Ritter, O. Pietsch, W. Zöllner, C. P. Herold, and S. von Tapavicza. Use of selected acrylic and/or methacrylic acid ester copolymer versions as flow enhancers in paraffin-rich crude oil and crude oil fractions (I) (Verwendung ausgewählter Copolymertypen der Acryl- und/oder Methacrylsäureester als Fließverbesserer in paraffinreichen Erdölen und Erdölfraktionen [I]). Patent: EP 332000, 1989.
[1504] W. Ritter, O. Pietsch, W. Zöllner, C. P. Herold, and S. von Tapavicza. Copolymers of acrylic and/or methacrylic acid esters as flow improvers. Patent: AU 610700, 1991.
[1505] W. Ritter, O. Pietsch, W. Zöllner, C. P. Herold, and S. von Tapavicza. Copolymers of (meth) acrylic acid esters as flow improvers in petroleum oils. Patent: CA 1334013, 1995.
[1506] H. Rivas, S. Acevedo, and X. Gutierrez. Emulsion of viscous hydrocarbon in aqueous buffer solution and method for preparing same. Patent: GB 2274254, 1994.
[1507] G. T. Rivers. Water clarifier. Patent: US 5152927, 1992.

[1508] D. N. Roark. Demulsification of oil-in-water emulsions. Patent: US 4614593, 1986.
[1509] D. N. Roark. Demulsification of oil-in-water emulsions. Patent: CA 1264263, 1990.
[1510] D. N. Roark, A. Nugent, Jr., and B. K. Bandlish. Fluid loss control and compositions for use therein. Patent: EP 201355, 1986.
[1511] D. N. Roark, A. Nugent, Jr., and B. K. Bandlish. Fluid loss control in well cement slurries. Patent: US 4698380, 1987.
[1512] F. Robinson. Polymers useful as pH responsive thickeners and monomers therefor. Patent: WO 9610602, 1996.
[1513] F. Robinson. Polymers useful as pH responsive thickeners and monomers therefor. Patent: US 5874495, 1999.
[1514] C. A. Rocha, D. W. Green, G. P. Willhite, and M. J. Michnick. An experimental study of the interactions of aluminum citrate solutions and silica sand. In *Proceedings Volume,* pages 403–413. SPE Oilfield Chem Int Symp (Houston, TX, 2/8–2/10), 1989.
[1515] P. S. Rocha, M. A. Miller, and K. Sepehrnoori. A succession-of-states front-tracking model for the in-situ combustion recovery process. *In Situ,* 21(1):65–100, February 1997.
[1516] P. M. Rodger, T. R. Forester, and W. Smith. Simulations of the methane hydrate/methane gas interface near hydrate forming conditions. *Fluid Phase Equilibria,* 116(1–2):326–332, 1995.
[1517] K. A. Rodrigues. Cement set retarding additives, compositions and methods. Patent: US 5341881, 1994.
[1518] K. A. Rodrigues. Set retarded cement compositions. Patent: EP 607039, 1994.
[1519] K. A. Rodrigues. Cement set retarding additives, compositions and methods. Patent: US 5421879, 1995.
[1520] K. A. Rodrigues. Set retarded cement compositions, additives and methods. Patent: CA 2127346, 1995.
[1521] K. A. Rodrigues. Set retarding cement additive. Patent: EP 633390, 1995.
[1522] K. A. Rodrigues and L. S. Eoff. Functionalized polymers containing amine groupings and their use as retarders in cement slurries. Patent: US 5368642, 1994.
[1523] D. Rodriguez, L. Quintero, M. T. Terrer, G. E. Jimenez, F. Silva, and P. J. Salazar. Hydrocarbon dispersions in water. Patent: GB 2231284, 1990.
[1524] A. Rojey, M. Thomas, and S. Jullian. Process for treatment of natural gas at a storage site. Patent: US 5803953, 1998.
[1525] J. Romocki. Application of N,N-dialkylamides to reduce precipitation of asphalt from crude oil. Patent: WO 9418430, 1994.

[1526] J. Romocki. Application of N,N-dialkylamides to control the formation of emulsions or sludge during drilling or workover of producing oil wells. Patent: WO 9522585, 1995.
[1527] J. Romocki. Application of N,N-dialkylamides to reduce precipitation of asphalt from crude oil. Patent: US 5388644, 1995.
[1528] J. Romocki. Application of N,N-dialkylamides to control the formation of emulsions or sludge during drilling or workover of producing oil wells. Patent: EP 745111, 1996.
[1529] J. Romocki. Application of N,N-dialkylamides to control the formation of emulsions or sludge during drilling or workover of producing oil wells. Patent: US 5567675, 1996.
[1530] P. Rooney, J. A. Russell, and T. D. Brown. Production of sulfonated asphalt. Patent: US 4741868, 1988.
[1531] G. D. Rose, A. S. Tetot, and P. A. Doty. Process for reversible thickening of a liquid. Patent: GB 2191716, 1987.
[1532] R. A. Rose. Method of drilling with fluid including nut cork and drilling fluid additive. Patent: US 5484028, 1996.
[1533] A. Rossi, J. E. Chandler, and R. Barbour. Polymers and additive compositions. Patent: WO 9319106, 1993.
[1534] Plugging solution for low temperature wells—contains plugging cement, water and molten chloride melt obtained as waste from titanium-magnesium production as complex chloride additive. Patent: RU 1091616-C, 1995.
[1535] Working out of oil stratum—comprises pumping sulphuric acid into stratum through pressing-in wells, pressing-in with water and collecting oil through extraction wells. Patent: RU 1480411-C, 1994.
[1536] V. A. Runov, Y. N. Mojsa, T. V. Subbotina, K. S. Pak, A. P. Krezub, V. N. Pavlychev, N. N. Istomin, Z. A. Evdokimova, and V. I. Borzenko. Lubricating additive for clayey drilling solution—is obtained by esterification of tall oil or tall pitch with hydroxyl group containing agent, e.g., low mol. wt. glycol or ethyl cellulose. Patent: SU 1700044-A, 1991.
[1537] V. A. Runov, T. V. Subbotina, Y. N. Mojsa, A. P. Krezub, A. K. Samotoj, and A. N. Morgunov. Lubricant additive for clayey drilling muds—contains chalk, carbon black or graphite as mineral component, and glycol ester(s) of synthetic higher fatty acids as organic component. Patent: SU 1726491-A, 1992.
[1538] L. M. Ruzin, O. E. Pleshkova, and L. V. Konovalova. Generation of carbon dioxide during thermal steam treatment of carbonate reservoirs containing high-viscosity oil. *Neft Khoz*, (11):59–62, November 1990.
[1539] S. A. Ryabokon and V. V. Goldshtein. Prospects of developing polymer plugging agents. *Neft Khoz*, (5):7–13, May 1989.

[1540] L. I. Ryabova, N. B. Savenok, N. A. Mariampolskij, L. M. Surova, and D. F. Kovalev. Plugging solution—contains Portland cement, aluminium oxychloride, water and phosphonium complexone in form of e.g., oxy-ethylidene diphosphonic acid. Patent: SU 1802086-A, 1993.

[1541] R. G. Ryan. Environmentally safe well plugging composition. Patent: US 5476543, 1995.

[1542] V. M. Ryzhov, V. S. Mironyuk, T. Ya. Mazepa, and V. V. Muravev. Controlling water absorption of drilling solution—based on carboxymethyl cellulose, involves introduction of sodium and/or potassium chloride into aqueous solution of carboxymethyl cellulose. Patent: RU 2066684-C, 1996.

[1543] A. Saasen, H. Hoset, E. J. Rostad, A. Fjogstad, O. Aunan, E. Westgard, and P. I. Norkyn. Application of ilmenite as weight material in water based and oil based drilling fluids. In *Proceedings Volume*. Annu SPE Tech Conf (New Orleans, LA, 9/30–10/3), 2001.

[1544] A. Saasen, B. Salmelid, N. Blomberg, K. Hansen, S. P. Young, and H. Justnes. The use of blast furnace slag in North Sea cementing applications. In *Proceedings Volume*, volume 1, pages 143–153. SPE Europe Petrol Conf (London, England, 10/25–10/27), 1994.

[1545] G. Saether, K. Kubberud, S. Nuland, and M. Lingelem. Drag reduction in two-phase flow. In *Proceedings Volume*, pages 171–184. 4th Bhra Multi-Phase Flow Int Conf (Nice, France, 6/19–6/21), 1989.

[1546] J. P. Salanitro, M. P. Williams, and G. C. Langston. Growth and control of sulfidogenic bacteria in a laboratory model seawater flood thermal gradient. In *Proceedings Volume*, pages 457–467. SPE Oilfield Chem Int Symp (New Orleans, LA, 3/2–3/5), 1993.

[1547] A. K. Samant, K. C. Koshel, and S. S. Virmani. Azoles as corrosion inhibitors for mild steel in a hydrochloric acid medium. SPE Unsolicited Paper, SPE-19022, February 1989.

[1548] K. A. Sandbeck and D. O. Hitzman. Biocompetitive exclusion technology: a field system to control reservoir souring and increase production. In *US DOE Rep*, number CONF-9509173, pages 311–319. 5th US DOE et al Microbial Enhanced Oil Recovery & Relat Biotechnol for Solving Environ Probl Int Conf (Dallas, TX, 9/11–9/14), 1995.

[1549] S. Saneie. *Alkaline Assisted Thermal Oil Recovery: Kinetic and Displacement Studies*. PhD thesis, Southern California Univ, 1992.

[1550] S. Saneie and Y. C. Yortsos. Alkaline assisted thermal oil recovery: kinetic and displacement studies: topical report. US DOE Rep DOE/BC/14600-45, Southern California Univ, June 1993.

[1551] B. D. Sanford, C. R. Dacar, and S. M. Sears. Acid fracturing with new fluid-loss control mechanisms increases production, Little Knife field, North Dakota. In *Proceedings Volume*, pages 317–324. SPE Rocky Mountain Reg Mtg (Casper, WY, 5/18–5/21), 1992.

[1552] T. Sanner, A. P. Kightlinger, and J. R. Davis. Borate-starch compositions for use in oil field and other industrial applications. Patent: US 5559082, 1996.
[1553] M. Sano. Polypropylene glycol (PPG) used as drilling fluids additive. *Sekiyu Gakkaishi,* 40(6):534–538, November 1997.
[1554] M. Santhanam and K. Macnally. Oil and oil invert emulsion drilling fluids with improved anti-settling properties. Patent: EP 1111024-A, 2001.
[1555] C. A. Sanz and G. A. Pope. Alcohol-free chemical flooding: from surfactant screening to coreflood design. In *Proceedings Volume,* pages 117–128. SPE Oilfield Chem Int Symp (San Antonio, TX, 2/14–2/17), 1995.
[1556] G. Sartori, D. W. Savage, D. C. Dalrymple, B. H. Ballinger, S. C. Blum, and W. E. Wales. Esterification of acidic crudes. Patent: US 6251305, 2001.
[1557] S. Sasnanand. *Adsorption of Tetrahydrothiophene in Porous Media: An Experimental Approach.* PhD thesis, New Mex Inst Mining Techn, 1993.
[1558] D. V. Satyanarayana Gupta and A. Cooney. Encapsulations for treating subterranean formations and methods for the use thereof. Patent: WO 9210640, 1992.
[1559] Deleted in proofs.
[1560] N. B. Savenok, N. A. Mariampolskij, P. N. Mariampolskij, and L. A. Loshmankina. Plugging solution—contains Portland cement, calcium chloride, phosphonium complexone containing amine-group and water and has high adhesion to stratal rock. Patent: SU 1802088-A, 1993.
[1561] A. Savoly and D. P. Elko. Foaming agent composition and process. Patent: WO 9516515, 1995.
[1562] A. Savoly, J. L. Villa, C. M. Garvey, and A. L. Resnick. Fluid loss agents for oil well cementing composition. Patent: US 4674574, 1987.
[1563] C. Sawdon, M. Tehrani, and P. Craddock. Electrically conductive non-aqueous wellbore fluids. Patent: GB 2345706, 2000.
[1564] G. M. Scelfo and R. S. Tjeerdema. A simple method for determination of Corexit 9527 in natural waters. *Mar Environ Res,* 31(1):69–78, 1991.
[1565] R. F. Scheuerman and B. M. Bergersen. Injection water salinity, formation pretreatment, and well operations fluid selection guidelines. In *Proceedings Volume,* pages 33–49. SPE Oilfield Chem Int Symp (Houston, TX, 2/8–2/10), 1989.
[1566] J. A. Schield, M. I. Naiman, and G. A. Scherubel. Polyimide quaternary salts as clay stabilization agents. Patent: GB 2244270, 1991.

[1567] P. Schilling. Aminated sulfonated or sulfomethylated lignins as cement fluid loss control additives. Patent: US 4990191, 1991.
[1568] P. Schilling. Polyamine condensates of styrene-maleic anhydride copolymers as corrosion inhibitors. Patent: US 5391636, 1995.
[1569] P. Schilling and D. V. Braddon. Corrosion inhibitors. Patent: US 4614600, 1986.
[1570] P. Schilling and P. E. Brown. Cationic and anionic lignin amines corrosion inhibitors. Patent: US 4789523, 1988.
[1571] S. Schilling. Lignin-based cement fluid loss control additive. Patent: US 4926944, 1990.
[1572] T. Scholl, W. Oberkirch, and H. Perrey. Application of condensation products of thioalkyloxides or polythioalkyloxides with polyalkyloxides as breakers of water-in-oil emulsions (Verwendung von Kondensationsprodukten von Thioalkylenoxiden oder Polythioalkylenoxiden mit Polyalkylenoxiden als Spalter für Wasser-in-Öl-Emulsionen). Patent: EP 192999, 1989.
[1573] T. Scholl, H. Perrey, T. Augustin, and C. Wegner. Polyurea-modified polyetherurethanes and their use as emulsion breakers for water-in-oil emulsions. Patent: US 4870151, 1989.
[1574] L. L. Schramm, C. Ayasse, K. Mannhardt, and J. Novosad. Method for improving enhanced recovery of oil using surfactant-stabilized foams. Patent: CA 2006482, 1991.
[1575] L. L. Schramm, C. Ayasse, K. Mannhardt, and J. Novosad. Method for improving enhanced recovery of oil using surfactant-stabilized foams. Patent: US 5060727, 1991.
[1576] L. L. Schramm, C. Ayasse, K. Mannhardt, and J. Novosad. Recovery of oil using surfactant-stabilized foams. Patent: GB 2239278, 1991.
[1577] R. J. Schreiber, Jr. and C. Yonley. The use of spent catalyst as a raw material substitute in cement manufacturing. *ACS Petrol Chem Div Preprints,* 38(1):97–99, February 1993.
[1578] G. W. Schriver, A. O. Patil, D. J. Martella, and K. Lewtas. Substituted fullerenes as flow improvers. Patent: US 5454961, 1995.
[1579] D. N. Schulz, J. J. Maurer, J. Bock, and R. M. Kowalik. Process for the formation of novel acrylamide acrylate copolymers. Patent: CA 1213606, 1986.
[1580] D. N. Schulz, J. J. Maurer, J. Bock, and R. M. Kowalik. A process for the formation of novel acrylamide acrylate copolymers. Patent: EP 116779, 1987.
[1581] H. U. Schutt. Reducing stress corrosion cracking in treating gases with alkanol amines. Patent: US 4959177, 1990.
[1582] A. N. Scian, Lopez. J. M. Porto, and E. Pereira. Mechanochemical activation of high alumina cements—hydration behavior: Pt 1. *Cement Concrete Res,* 21(1):51–60, January 1991.

[1583] E. G. Scovell, N. Grainger, and T. Cox. Maintenance of oil production and refining equipment. Patent: WO 0174966A, 2001.
[1584] J. T. Sears, R. Müller, and M. A. Reinsel. Inhibition of sulfate-reducing bacteria via nitrite production. Patent: WO 9612867, 1996.
[1585] L. P. Sedillo, J. C. Newlove, and R. C. Portnoy. Fluid loss control in oil field cements. Patent: US 4659750, 1987.
[1586] I. Sekine, M. Yuasa, T. Shimode, and K. Takaoka. Inhibition of corrosion. Patent: GB 2234501, 1991.
[1587] A. M. Selikhanovich, G. A. Chuprina, and A. N. Olejnikov. Reagent for treating drilling muds—contains acrylic polymer, sodium hydroxide, additional diethylene glycol, and water. Patent: RU 2087515-C, 1997.
[1588] K. P. A. Senaratne and K. C. Lilje. Preparation of branched chain carboxylic esters. Patent: US 5322633, 1994.
[1589] S. N. Shah, P. C. Harris, and H. C. Tan. Rheological characterization of borate crosslinked fracturing fluids employing a simulated field procedure. In *Proceedings Volume*. SPE Prod Technol Symp (Hobbs, NM, 11/7–11/8), 1988.
[1590] S. S. Shah, W. F. Fahey, and B. A. Oude Alink. Corrosion inhibition in highly acidic environments by use of pyridine salts in combination with certain cationic surfactants. Patent: CA 2066797, 1992.
[1591] S. S. Shah, W. F. Fahey, and B. A. Oude Alink. Corrosion inhibition in highly acidic environments by use of pyridine salts in combination with certain cationic surfactants. Patent: US 5336441, 1994.
[1592] S. Sharif. Process for preparation and composition of stable aqueous solutions of boron zirconium chelates for high temperature frac fluids. Patent: US 5217632, 1993.
[1593] S. Sharif. Process for preparation of stable aqueous solutions of zirconium chelates. Patent: US 5466846, 1995.
[1594] A. U. Sharipov, S. I. Dolganskaya, K. S. Rayanov, M. V. Rudomino, E. G. Kopejko, N. I. Krutikova, E. K. Kolova, A. G. Shkuro, V. V. Lezhenin, P. Ya. Reshetnikov, M. A. Bogachev, N. K. Malinin, and A. I. Lipatov. Composition for isolation of absorbing strata in oil and gas extraction contains hemi-hydrated gypsum and product obtained by reacting ammonia with waste obtained in production of nitrilotrimethyl phosphonic acid. Patent: SU 1745892-A, 1992.
[1595] A. U. Sharipov, S. I. Dolganskaya, L. I. Ryabova, L. A. Chumachenko, and G. I. Petrukhin. Plugging solution for gas extraction wells—contains cement, polymethyl methacrylate modified with monoethanolamine, and water, and produces plugging stone of increased strength. Patent: SU 1818464-A, 1993.
[1596] A. U. Sharipov, S. I. Dolganskaya, T. V. Zajkovskaya, and L. I. Ryabova. Composition for isolation of oil stratum—contains plugging

cement, aluminium oxychloride and poly-methyl methacrylate modified with mono-ethanolamine. Patent: SU 1709073-A, 1992.
[1597] A. G. Sharov, R. A. Terteryan, L. I. Burova, and L. A. Shulgaitser. Effect of a copolymer inhibitor of paraffin deposits on oils of various fields. *Neft Khoz,* (9):55–58, September 1989.
[1598] A. A. Shatov, N. Kh. Karimov, M. R. Mavlyutov, F. A. Agzamov, A. T. Gareev, and N. N. Nazarenko. Plugging solution for cementing oil and gas wells—contains Portland cement, distiller liquid—waste from calcined soda production and additionally baryta. Patent: RU 2059793-C, 1996.
[1599] A. A. Shatov, N. Kh. Karimov, M. R. Mavlyutov, F. A. Agzamov, V. M. Titov, and V. V. Yakimtsev. Plugging solution for oil and gas wells—containing Portland cement, waste from calcined soda production and power station fly ash giving reduced cost and improved expansion properties. Patent: RU 2059792-C, 1996.
[1600] A. A. Shatov, N. Kh. Karimov, M. R. Mavlyutov, F. A. Agzamov, A. V. Voronin, and I. D. Maltseva. Plugging solution for cementing oil and gas wells—contains Portland cement, waste from production of calcined soda, water and additionally slag waste from metallurgical industry. Patent: RU 2059791-C, 1996.
[1601] A. A. Shatov, V. N. Sergeev, V. M. Titov, I. D. Maltseva, and V. P. Ovchinnikov. Plugging solution for oil and gas wells—includes cement, water, and additive in form of waste from calcined soda production containing calcium and sodium salts. Patent: RU 2030557-C, 1995.
[1602] A. A. Shatov and Z. Z. Sharafutdinov. Plugging solution—contains Portland cement and specified additive based on waste from production of calcined soda. Patent: RU 2072027-C, 1997.
[1603] V. P. Shchipanov, D. M. Batalov, N. E. Shcherbich, Z. P. Khomyakova, and T. A. Machulis. Plugging solution for oil and gas wells—contains Portland cement, additive in form of pentaerythritol, and water and has reduced hardening time. Patent: RU 2042785-C, 1995.
[1604] A. J. Sheehy. Microbial physiology and enhanced oil recovery. 3rd US DOE Microbial Enhancement of Oil Recovery Int Conf (Norman, OK, 5/27–6/1), 1990.
[1605] A. J. Sheehy. Microbial physiology and enhanced oil recovery. In E. C. Donaldson, editor, *Microbial enhancement of oil recovery: recent advances: Proceedings of the 1990 International Conference on Microbial Enhancement of Oil Recovery,* volume 31 of *Developments in Petroleum Science,* pages 37–44. Elsevier Science Ltd, 1991.
[1606] J. Shen and A. S. Al-Saeed. Study of oil field chemicals by combined field desorption/collision-activated dissociation mass spectrometry via linked scan. *Analy Chem,* 62(2):116–120, 1990.

[1607] W. Shen, H. Pan, T. Du, and D. Jia. Preparation and applications of oleophilic modified barite. *J Univ Petrol, China*, 22(1):66–69,114, February 1998.

[1608] W. Shen, H. F. Pan, and Y. Q. Qin. Advances in chemical surface modification of barite. *Oilfield Chem*, 16(1):86–90, 1999.

[1609] J. J. Sheu and R. G. Bland. Drilling fluid additive. Patent: GB 2245579, 1992.

[1610] J. J. Sheu and R. G. Bland. Drilling fluid with browning reaction anionic carbohydrate. Patent: US 5106517, 1992.

[1611] J. J. Sheu and R. G. Bland. Drilling fluid with stabilized browning reaction anionic carbohydrate. Patent: US 5110484, 1992.

[1612] T. Shimokawatoko, K. Sumida, and H. Ueda. Method and apparatus for measuring odorant concentration and oderant adding system. Patent: US 5844124, 1998.

[1613] T. Shimomura, Y. Irie, H. Takahashi, K. Kajikawa, J. Saga, T. Fujiwara, and T. Hatsuda. Process for production of acrylate and acrylate-containing polymer. Patent: EP 372706, 1990.

[1614] C. C. Shin. Corrosion inhibiting composition for zinc halide-based clear, high density fluids. Patent: WO 8802432, 1988.

[1615] T. Shinomura. Method of reducing friction losses in flowing liquids. Patent: US 4751937, 1988.

[1616] T. Shomokawatoko, K. Sumida, and H. Ueda. Method and apparatus for measuring odorant concentration and odorant adding system. Patent: EP 836091, 1998.

[1617] P. Shu. Programmed gelation of polymers using melamine resins. Patent: US 4964461, 1990.

[1618] P. Shu, R. C. Ng, and C. H. Phelps. In-situ cementation for profile control. Patent: US 5211231, 1993.

[1619] P. Shu, C. H. Phelps, and R. C. Ng. In-situ silica cementation for profile control during steam injection. Patent: US 5211232, 1993.

[1620] W. R. Shu and P. Shu. Method for steam flooding profile control. Patent: CA 2041584, 1991.

[1621] C. E. Shuchart, J. M. Terracina, B. F. Slabaugh, and M. A. McCabe. Method of treating subterranean formation. Patent: EP 916806, 1999.

[1622] M. W. Shuey and R. S. Custer. Quebracho-modified bitumen compositions, method of manufacture and use. Patent: US 5401308, 1995.

[1623] P. J. Shuler and W. H. Jenkins. Prevention of downhole scale deposition in the Ninian field. In *Proceedings Volume*, volume 2. SPE Offshore Europe Conf (Aberdeen, Scotland, 9/5–9/8), 1989.

[1624] R. D. Shupe and T. D. Baugh. Thermal stability and degradation mechanism of alkylbenzene sulfonates in alkaline media. *J Colloid Interface Sci*, 145(1):235–254, August 1991.

References 461

[1625] R. Siegmeier, M. Kirschey, and M. Voges. Acrolein based polymers as scale inhibitors. In *Proceedings Volume*, number PAP 70. NACE Int Corrosion Conf (Corrosion 98) (San Diego, CA, 3/22–3/27), 1998.

[1626] T. R. Sifferman, J. M. Swazey, C. B. Skaggs, N. Nguyen, and D. B. Solarek. Fluid loss control additives and subterranean treatment fluids containing the same. Patent: WO 9905235, 1999.

[1627] C. S. Sikes and A. Wierzbicki. Stereospecific and nonspecific inhibition of mineral scale and ice formation. In *Proceedings Volume*. 51st Annu NACE Int Corrosion Conf (Corrosion 96) (Denver, CO, 3/24–3/29), 1996.

[1628] D. Sikora. Hydrazine—a universal oxygen scavenger (hydrazyna—uniwersalny inhibitor korozji tlenowej w pluczkach wiertniczych). *Nafta Gaz (Pol)*, 50(4):161–168, April 1994.

[1629] D. C. Silverman, D. J. Kalota, and F. S. Stover. Effect of pH on corrosion inhibition of steel by polyaspartic acid. In *Proceedings Volume*. 50th Annu NACE Int Corrosion Conf (Corrosion 95) (Orlando, FL, 3/26–3/31), 1995.

[1630] G. D. Simpson, R. F. Miller, G. D. Laxton, and W. R. Clements. A focus on chlorine dioxide: the "ideal" biocide. In *Proceedings Volume*. Annu NACE Corrosion Conf (Corrosion 93) (New Orleans, LA, 3/7–3/12), 1993.

[1631] M. Singer, R. Tjeerdema, D. Aurand, J. Clark, G. Sergy, and M. Sowby. CROSERF (chemical response to oil spills ecological effects research forum): toward a standardization of oil spill cleanup agent ecological effects research. In *Proceedings Volume*, volume 2, pages 1263–1270. 18th Environ Can Arctic & Mar Oilspill Program Tech Semin (Edmonton, Canada, 6/14–6/16), 1995.

[1632] B. P. Singh. Performance of demulsifiers: prediction based on film pressure-area isotherms and solvent properties. *Energy Sources*, 16(3):377–385, July-September 1994.

[1633] M. A. Singleton, J. A. Collins, N. Poynton, and H. J. Formston. Developments in phosphonomethylated polyamine (PMPA) scale inhibitor chemistry for severe $BaSO_4$ scaling conditions. 2nd Annu SPE Oilfield Scale Int Symp (Aberdeen, Scotland, 1/26–1/27), 2000.

[1634] P. M. Sirenko, N. G. Bublikova, V. D. Kovalenko, and D. A. Kostenko. Light heat-resistant plugging material—contains lightening-stabilising additive in form of bentonite clay powder, and additionally water absorption reducing agent. Patent: RU 2043481-C, 1995.

[1635] A. Sivakumar and M. Ramesh. Demulsification of oily waste waters using silicon containing polymers. Patent: US 5560832, 1996.

[1636] P. Skovborg, H. J. Ng, P. Rasmussen, and U. Mohn. Measurement of induction times for the formation of methane and ethane gas hydrates. *Chem Eng Sci*, 48(3):445–453, February 1993.

[1637] E. D. Sloan and F. Fleyfel. Hydrate dissociation enthalpy and guest size. *Fluid Phase Equilibria,* 76:123–140, 1991.

[1638] E. D. Sloan, Jr. Method for controlling clathrate hydrates in fluid systems. Patent: WO 9412761, 1994.

[1639] E. D. Sloan, Jr. and R. L. Christiansen. Prediction and inhibition of hydrate and wax formation: final report (January 1991–December 1993). Gas Res Inst Rep GRI-94/0404, Gas Res Inst, March 1995.

[1640] B. F. Sloat. Nitrogen stimulation of a potassium hydroxide wellbore treatment. Patent: US 4844169, 1989.

[1641] B. F. Sloat. Nitrogen stimulation of a potassium hydroxide wellbore treatment. Patent: CA 1291419, 1991.

[1642] M. E. Slodki and M. C. Cadmus. High-temperature, salt-tolerant enzymic breaker of xanthan gum viscosity. In E. C. Donaldson, editor, *Microbial enhancement of oil recovery: recent advances: Proceedings of the 1990 International Conference on Microbial Enhancement of Oil Recovery,* volume 31 of *Developments in Petroleum Science,* pages 247–255. Elsevier Science Ltd, 1991.

[1643] C. K. Smith and T. G. Balson. Shale-stabilizing additives. Patent: GB 2340521, 2000.

[1644] D. H. Smith and J. R. Comberiati. Chemical alteration of the rock surfaces by asphaltenes or surfactants, and its effect on oil recovery by CO_2 flooding. In *Proceedings Volume.* Annu AICHE Mtg (Chicago, IL, 11/11–11/16), 1990.

[1645] D. K. Smith. *Cementing.* SPE Publications, 2nd Printing, 1976.

[1646] J. E. Smith. Performance of 18 polymers in aluminum citrate colloidal dispersion gels. In *Proceedings Volume,* pages 461–470. SPE Oilfield Chem Int Symp (San Antonio, TX, 2/14–2/17), 1995.

[1647] K. W. Smith, L. V. Haynes, and D. F. Massouda. Solvent free oil soluble drag reducing polymer suspension. Patent: US 5449732, 1995.

[1648] K. W. Smith and L. J. Persinski. Hydrocarbon gels useful in formation fracturing. Patent: US 5417287, 1995.

[1649] K. W. Smith and T. R. Thomas. Method of treating shale and clay in hydrocarbon formation drilling. Patent: EP 680504, 1995.

[1650] K. W. Smith and T. R. Thomas. Method of treating shale and clay in hydrocarbon formation drilling. Patent: WO 9514066, 1995.

[1651] K. W. Smith and T. R. Thomas. Method of treating shale and clay in hydrocarbon formation drilling. Patent: US 5607902, 1997.

[1652] R. J. Smith and D. R. Jeanson. Dehydration of drilling mud. Patent: US 6216361, 2001.

[1653] P. Smolarchuk and W. Dill. Iron control in fracturing and acidizing operations. In *Proceedings Volume,* volume 1, pages 391–397. 37th Annu CIM Petrol Soc Tech Mtg (Calgary, Canada, 6/8–6/11), 1986.

[1654] M. J. Snowden, B. Vincent, and J. C. Morgan. Conformance control in underground reservoirs. Patent: GB 2262117, 1993.
[1655] P. Somasundaran. Surfactant loss control in chemical flooding: spectroscopic and calorimetric study of adsorption and precipitation on reservoir minerals—annual report for the reporting period September 30, 1992 to September 30, 1993. US DOE Fossil Energy Rep DOE/BC/14884-5, Columbia Univ, 1994.
[1656] P. Somasundaran. Surfactant loss control in chemical flooding: spectroscopic and calorimetric study of adsorption and precipitation on reservoir minerals–annual report (September 30, 1993–September 30, 1994). US DOE Fossil Energy Rep DOE/BC/14884-12, Columbia Univ, 1995.
[1657] A. G. Sommese and R. Nagarajan. Settling stabilization of polymer containing particle dispersions in oil. Patent: US 5438088, 1995.
[1658] A. J. Son and J. Chakravarty. Analysis of residual corrosion inhibitors by fluorescence and ultraviolet spectrophotometry. In *Proceedings Volume*. 51st Annu NACE Int Corrosion Conf (Corrosion 96) (Denver, CO, 3/24–3/29), 1996.
[1659] J. Song. *A New Flux Correcting Method for Reducing Numerical Dispersion—Application to EOR (Enhanced Oil Recovery) By Chemical Processes (Reduction de la Dispersion Numerique par Correction des flux Massiques—Application au Probleme de la Recuperation d'Hydrocarbures par Procedes Chimiques)*. PhD thesis, Paris VI Univ, 1992.
[1660] T. M. Sopko and R. E. Lorentz. Method of using polymers of amidosulfonic acid containing monomers and salts as drilling additive. Patent: US 5039433, 1991.
[1661] K. S. Sorbie. *Polymer-improved oil recovery*. CRC Press, Inc, Boca Raton, FL, 1991.
[1662] M. Soreau and D. Siegel. Injection composition for filling or reinforcing grounds. Patent: GB 2170838, 1986.
[1663] M. Soreau and D. Siegel. Application of a gelling agent for an alkalisilicate solution for sealing and consolidating soils (Verwendung eines Geliermittels für zum Abdichten und Verfestigen von Böden bestimmte Alkalisilikatlösung). Patent: DE 3506095, 1990.
[1664] V. A. Sorokin. Microbiological technology of oil recovery. *Neft Khoz,* (6):49–50, June 1989.
[1665] E. Souto, B. Bazin, and M. Sardin. Ion exchange between hydrogen and homoionic brines related to permeability reduction. In *Proceedings Volume*, pages 491–500. SPE Oilfield Chem Int Symp (New Orleans, LA, 3/2–3/5), 1993.
[1666] S. J. Spencer. Governments, operators eyeing effects of synthetic-based drilling fluids. *Oil Gas J,* 98(39):88–89, 2000.

[1667] G. T. Sperl and P. L. Sperl. Enhanced oil recovery using denitrifying microorganisms. Patent: US 5044435, 1991.
[1668] J. W. Squyres and H. Lopez. Silica fume as a strength enhancer in low density slurries. In *Proceedings Volume,* pages 47–53. 37th Annu Southwestern Petrol Short Course Assoc et al Mtg (Lubbock, TX, 4/18–4/19), 1990.
[1669] A. L. Stacy and R. B. Weber. Method for reducing deleterious environmental impact of subterranean fracturng processes. Patent: US 5424285, 1995.
[1670] L. A. Starukhina, V. V. Deriabin, and V. J. Titov. New biopolymer for EOR Moscow. In *Proceedings Volume,* volume 1, pages 371–393. 6th Europe Impr Oil Recovery Symp (Stavanger, Norway, 5/21–5/23), 1991.
[1671] B. A. Stefl and K. L. George. Antifreezes and deicing fluids. In *Kirk-Othmer, Encyclopedia of Chemical Technology*, volume 3, pages 347–366. J. Wiley & Sons, New York, 1996.
[1672] G. L. Stegemeier and G. E. Perry. Method utilizing spot tracer injection and production induced transport for measurement of residual oil saturation. Patent: US 5168927, 1992.
[1673] G. S. Stepanova, Y. G. Mamedov, I. A. Babayeva, A. A. Mosina, T. L. Nenartovich, and A. A. Li. Thin under gas oil rims—new technology for efficient oil production using foam generating oil/water polymers. In *Proceedings Volume,* number 050. 9th EAGE Impr Oil Recovery Europe Symp (The Hague, Netherlands, 10/20–10/22) Proc, 1997.
[1674] G. S. Stepanova, M. D. Rozenberg, O. A. Boksha, G. F. Gubkina, and T. L. Nenartovich. Chemical treatment of oil strata—comprises pumping-in ammonium carbonate and nitrate solution followed by aqueous solution of ammonium nitrate-sulphuric acid mixture, and water. Patent: RU 2021496-C, 1994.
[1675] G. S. Stepanova, M. D. Rozenberg, O. A. Boksha, G. F. Gubkina, T. L. Nenartovich, and S. V. Safronov. Development of oil field—comprises pumping down well portion of aqueous solution of ammonium carbonate and displacing solution into seam with water, giving increased oil recovery. Patent: RU 2021495-C, 1994.
[1676] M. Stephens. Fluid loss additives for well cementing compositions. Patent: GB 2202526, 1988.
[1677] M. Stephens. Low density well cement compositions and method of use. Patent: US 6176314, 2001.
[1678] M. Stephens and B. Pereira. Low density well cement compositions. Patent: WO 0105727A, 2001.
[1679] M. Stephens and B. L. Swanson. Drilling mud comprising tetrapolymer consisting of N-vinyl-2-pyrrolidone, acrylamidopropanesulfonic acid, acrylamide, and acrylic acid. Patent: US 5135909, 1992.

[1680] W. K. Stephenson and J. D. Deshazo. Method of breaking crude oil emulsions using ethylene carbonate adducts of alkylphenol-formaldehyde resins. Patent: US 5205964, 1993.
[1681] W. K. Stephenson and M. Kaplan. Asphaltene dispersants-inhibitors. Patent: CA 2029465, 1991.
[1682] R. B. Stewart, W. C. M. Lohbeck, D. S. Gill, and M. N. Baaijens. An expandable slotted tubing, fibre-cement wellbore lining system. In *Proceedings Volume,* volume 1, pages 133–140. SPE Europe Petrol Conf (Milan, Italy, 10/22–10/24), 1996.
[1683] R. B. Stewart, W. C. M. Lohbeck, D. S. Gill, and M. N. Baaijens. An expandable slotted tubing, fibre-cement wellbore lining system. In *Proceedings Volume,* pages 127–134. Annu SPE Tech Conf (Denver, CO, 10/6–10/9), 1996.
[1684] R. B. Stewart, W. C. M. Lohbeck, D. S. Gill, and M. N. Baaijens. An expandable-slotted-tubing, fiber-cement wellbore-lining system. *SPE Drilling Completion,* 12(3):163–167, September 1997.
[1685] W. S. Stewart, G. G. Dixon, J. M. Elsen, and B. L. Swanson. Drilling fluid additives for use in hard brine environments. Patent: US 4743383, 1988.
[1686] C. Stowe, R. G. Bland, D. Clapper, T. Xiang, and S. Benaissa. Water-based drilling fluids using latex additives. Patent: GB 2363622, 2002.
[1687] A. Strycker. Selection and design of ethoxylated carboxylates for chemical flooding: topical report. Government Report NIPER-449, US Dept Energy, 1990.
[1688] Plugging solution contains Portland cement, sodium sulphate, potassium ferricyanide and nitrile-tri-methyl-phosphonic acid. Patent: SU 1700204-A, 1991.
[1689] S. Subramanian, M. Islam, and C. R. Burgazli. Quaternary ammonium salts as thickening agents for aqueous systems. Patent: WO 0118147-A, 2001.
[1690] S. Subramanian, Y. P. Zhu, C. R. Bunting, and R. E. Stewart. Gelling system for hydrocarbon fluids. Patent: WO 0109482A, 2001.
[1691] T. Sugama, L. E. Kukacka, N. Carciello, and B. Galen. Oxidation of carbon fiber surfaces for improvement in fiber-cement interfacial bond at a hydrothermal temperature of 300° C. *Cement Concrete Res,* 18(2):290–300, March 1988.
[1692] A. B. Sulejmanov, T. B. Geokchaev, and R. A. Dashdiev. Removal of petroleum spillages from water surface—using a detergent mixture containing oxyethylated fatty 10-20 C alcohols and additional oxyethylated fatty 11-17 C acids to improve properties. Patent: SU 1803418-A, 1993.
[1693] D. Sullivan, J. Farlow, and K. A. Sahatjian. Evaluation of three oil spill laboratory dispersant effectiveness tests. In *Proceedings Volume,* pages

515–520. 13th Bien API et al Oil Spill (Prev, Preparedness, Response & Coop) Int Conf (Tampa, FL, 3/29–4/1), 1993.
[1694] E. Sunde, J. Beeder, R. K. Nilsen, and T. Torsvik. Aerobic microbial enhanced oil recovery for offshore use. In *Proceedings Volume*, volume 2, pages 497–502. 8th SPE/DOE Enhanced Oil Recovery Symp (Tulsa, OK, 4/22–4/24), 1992.
[1695] E. Sunde and H. Olsen. Removal of H_2S in drilling mud. Patent: WO 0023538, 2000.
[1696] E. Sunde, T. Thorstenson, and T. Torsvik. Growth of bacteria on water injection additives. In *Proceedings Volume*, pages 727–733. 65th Annu SPE Tech Conf (New Orleans, LA, 9/23–9/26), 1990.
[1697] E. Sunde, T. Thorstenson, T. Torsvik, J. E. Vaag, and M. S. Espedal. Field-related mathematical model to predict and reduce reservoir souring. In *Proceedings Volume*, pages 449–456. SPE Oilfield Chem Int Symp (New Orleans, LA, 3/2–3/5), 1993.
[1698] M. N. Sunil Kumar. Review on polymeric and copolymeric pour point depressants for waxy crude oils and studies on bombay high crude oil. *Quart J Tech Pap (Inst Petrol)*, pages 47–71, October-December 1989.
[1699] I. Suratman. *A Study of the Laws of Variation (Kinetics) and the Stabilization of Swelling of Clay (Contribution a L'etude De La Cinetique et De La Stabilisation Du Gonflement Des Argiles)*. PhD thesis, Malaysia, 1985.
[1700] L. M. Surguchev, A. Koundin, and D. Yannimaras. Air injection—cost effective IOR method to improve oil recovery from depleted and waterflooded fields. In *Proceedings Volume*. SPE Asia Pacific Impr Oil Recovery Conf (APIORC 99) (Kuala Lumpur, Malaysia, 10/25–10/26), 1999, SPE Number: 57296.
[1701] V. Swarup, D. G. Peiffer, and M. L. Gorbaty. Encapsulated breaker chemical. Patent: US 5580844, 1996.
[1702] P. G. Sweeny. Hydantoin-enhanced halogen efficacy in pulp and paper applications. Patent: WO 9611882, 1996.
[1703] L. Sweet. Method of fracturing a subterranean formation with a lightweight propping agent. Patent: US 5188175, 1993.
[1704] R. D. Sydansk. Lost circulation treatment for oil field drilling operations. Patent: US 4957166, 1990.
[1705] R. D. Sydansk. Enhanced liquid hydrocarbon recovery process. Patent: US 5129457, 1992.
[1706] R. D. Sydansk. Hydrocarbon recovery process utilizing a gel prepared from a polymer and a preformed crosslinking agent. Patent: US 5415229, 1995.
[1707] A. R. Syrinek and L. B. Lyon. Low temperature breakers for gelled fracturing fluids. Patent: US 4795574, 1989.

[1708] K. Szablikowski, W. Lange, R. Kiesewetter, and E. Reinhardt. Easily dispersible blends of reversibly crosslinked and uncrosslinked hydrocolloids, with aldehydes as crosslinker (Abmischungen von mit Aldehyden Reversibel Vernetzten Hydrokolloidmischungen und Unvernetzten Hydrokolloiden mit Verbesserter Dispergierbarkeit). Patent: EP 686666, 1995.

[1709] T. Hervot, A. Leverot, B. Lesgent, J. N. Philippot, M. Martin, F. Schaeffer, and G. Wolsfelt. *Cementing technology and procedures.* Editions Techniq, Paris, France, Total, Gaz De France, Inst Francais Du Petrole, 1993.

[1710] S. Talabani and G. Hareland. Expansion-contraction cycles for cement optimized as a function of additives. In *Proceedings Volume,* pages 277–284. SPE Prod Oper Symp (Oklahoma City, OK, 4/2–4/4), 1995.

[1711] S. Talabani and G. Hareland. New cement additives that eliminate cement body permeability. In *Proceedings Volume,* pages 169–176. SPE Asia Pacific Oil & Gas Conf (Kuala Lumpur, Malaysia, 3/20–3/22), 1995.

[1712] S. Talabani, G. Hareland, and M. R. Islam. New additives for minimizing cement body permeability. *Energy Sources,* 21(1–2):163–176, January-March 1999.

[1713] L. D. Talley and G. F. Mitchell. Application of proprietary kinetic hydrate inhibitors in gas flowlines. In *Proceedings Volume,* volume 3, pages 681–689. 31st Annu SPE et al Offshore Technol Conf (Houston, TX, 5/3–5/6), 1999.

[1714] D. Tambe, J. Paulis, and M. M. Sharma. Factors controlling the stability of colloid-stabilized emulsions: Pt 4: Evaluating the effectiveness of demulsifiers. *J Colloid Interface Sci,* 171(2):463–469, May 1995.

[1715] C. P. Tan, B. G. Richards, S. S. Rahman, and R. Andika. Effects of swelling and hydrational stress in shales on wellbore stability. In *Proceedings Volume,* pages 345–349. SPE Asia Pacific Oil & Gas Conf (Kuala Lumpur, Malaysia, 4/14–4/16), 1997.

[1716] D. Tan. Test and application of drilling fluid filtrate reducer polysulfonated humic acid resin. *Oil Drilling Prod Technol,* 12(1):27–32,97–98, 1990.

[1717] J. S. Tang and B. C. Harker. Use of tracers to monitor in situ miscibility of solvent in oil reservoirs during EOR. Patent: US 5111882, 1992.

[1718] J. S. Tang and B. C. Harker. Use of tracers to monitor in situ miscibility of solvent in oil reservoirs during EOR. Patent: CA 1310140, 1992.

[1719] Y. Tang, Z. Han, H. Wang, and H. Chen. Sp-2 acid corrosion inhibitor. *J Univ Petrol, China,* 19(1):98–101, February 1995.

[1720] Y. Ya. Taradymenko, V. P. Timovskij, V. A. Kushu, V. E. Akhrimenko, and Y. G. Karpenko. Plugging solution—contains Portland cement, calcium oxide-based expanding additive, water and additionally polymethylene-urea, to improve efficiency. Patent: SU 1799999-A, 1993.

[1721] K. Taugbol, H. H. Zhou, and T. Austad. Low tension polymer flood: the influence of surfactant-polymer interaction. In *Proceedings Volume*, pages 281–294. Rec Adv Oilfield Chem, 5th Royal Soc Chem Int Symp (Ambleside, England, 4/13–4/15), 1994.

[1722] A. S. Taylor. Fluorosilicone anti-foam additive. Patent: GB 2244279, 1991.

[1723] G. N. Taylor. Demulsifier for water-in-oil emulsions and method of use. Patent: EP 696631, 1996.

[1724] G. N. Taylor. Demulsification of water-in-oil emulsions using high molecular weight polyurethanes. Patent: GB 2346378, 2000.

[1725] G. N. Taylor and R. Mgla. Method of demulsifying water-in-oil emulsions. Patent: EP 641853, 1995.

[1726] K. C. Taylor and H. A. Nasr-El-Din. A systematic study of iron control chemicals: Pt 2. In *Proceedings Volume*, pages 649–656. SPE Oilfield Chem Int Symp (Houston, TX, 2/16–2/19), 1999.

[1727] K. C. Taylor, H. A. Nasr-El-Din, and M. J. Al-Alawi. A systematic study of iron control chemicals used during well stimulation. In *Proceedings Volume*, pages 19–25. SPE Formation Damage Contr Int Symp (Lafayette, LA, 2/18–2/19), 1998.

[1728] S. M. Teeters. Corrosion inhibitor. Patent: US 5084210, 1992.

[1729] D. T. Terry, D. D. Onan, P. L. Totten, and B. J. King. Converting drilling fluids to cementitious compositions. Patent: US 5295543, 1994.

[1730] A. B. Theis and J. Leder. Method for the control of biofouling. Patent: US 5128051, 1992.

[1731] M. Thomas and E. Behar. Structure H hydrate equilibria of methane and intermediate hydrocarbon molecules. In *Proceedings Volume*, pages 100–107. 73rd Annu GPA Conv (New Orleans, LA, 3/7–3/9), 1994.

[1732] S. Thomas and Ali. S. M. Farouq. Status and assessment of chemical oil recovery methods. *Energy Sources*, 21(1–2):177–189, January-March 1999.

[1733] T. R. Thomas and K. W. Smith. Method of maintaining subterranean formation permeability and inhibiting clay swelling. Patent: US 5211239, 1993.

[1734] N. E. S. Thompson and R. G. Asperger. Use of tridithiocarbamic acid compositions as demulsifiers. Patent: US 4689177, 1987.

[1735] N. E. S. Thompson and R. G. Asperger. Dithiocarbamates for treating hydrocarbon recovery operations and industrial waters. Patent: US 4864075, 1989.

[1736] N. E. S. Thompson and R. G. Asperger. Methods for treating hydrocarbon recovery operations and industrial waters. Patent: US 4826625, 1989.

[1737] N. E. S. Thompson and R. G. Asperger. Methods for treating hydrocarbon recovery operations and industrial waters. Patent: EP 349681, 1990.

[1738] N. E. S. Thompson and R. G. Asperger. Methods for treating hydrocarbon recovery operations and industrial waters. Patent: US 4956099, 1990.

[1739] N. E. S. Thompson and R. G. Asperger. Methods for treating hydrocarbon recovery operations and industrial waters. Patent: US 5026483, 1991.

[1740] N. E. S. Thompson and R. G. Asperger. Methods for treating hydrocarbon recovery operations and industrial waters. Patent: US 5019274, 1991.

[1741] N. E. S. Thompson and R. G. Asperger. Methods for treating hydrocarbon recovery operations and industrial waters. Patent: US 5013451, 1991.

[1742] N. E. S. Thompson and R. G. Asperger. Method for treating hydrocarbon recovery operations and industrial waters. Patent: US 5089227, 1992.

[1743] N. E. S. Thompson and R. G. Asperger. Methods for treating hydrocarbon recovery operations and industrial waters. Patent: US 5089619, 1992.

[1744] A. Thorhaug, B. Carby, R. Reese, G. Sidrak, M. Anderson, K. Aiken, W. Walker, M. Rodriquez, and H. J. Teas. Dispersant use for tropical nearshore waters: Jamaica. In *Proceedings Volume,* pages 415–418. 12th Bien API et al Oil Spill (Prev, Behav, Contr, Cleanup) Int Conf (San Diego, CA, 3/4–3/7), 1991.

[1745] A. Thorhaug and J. H. Marcus. Preliminary mortality effects of seven dispersants on subtropical/tropical seagrasses. In *Proceedings Volume,* pages 223–224. 10th Bien API et al Oil Spill (Prev, Behav, Contr, Cleanup) Conf (Baltimore, MD, 4/6–4/9), 1987.

[1746] A. Thorhaug, J. McFarlane, B. Carby, F. McDonald, M. Anderson, B. Miller, P. Gayle, and V. Gordon. Dispersed oil effects on tropical habitats: preliminary laboratory results of dispersed oil testing on Jamaica corals and seagrass. In *Proceedings Volume,* pages 455–458. API et al Oil Spill (Prev, Behav, Contr, Cleanup) 20th Anniv Conf (San Antonio, TX, 2/13–2/16), 1989.

[1747] R. M. Tjon-Joe-Pin, H. D. Brannon, and A. R. Rickards. Method of dissolving organic filter cake obtained in drilling and completion of oil and gas wells. Patent: WO 9401654, 1994.

[1748] N. M. To, H. M. Brown, and R. H. Goodman. Data analysis and modeling of dispersant effectiveness in cold water. In *Proceedings Volume,* pages 303–306. 10th Bien API et al Oil Spill (Prev, Behav, Contr, Cleanup) Conf (Baltimore, MD, 4/6–4/9), 1987.

[1749] B. L. Todd, B. R. Reddy, J. V. Fisk, Jr., and J. D. Kercheville. Well drilling and servicing fluids and methods of removing filter cake deposited thereby. Patent: US 6422314, 2002.

[1750] A. A. Toenjes, M. R. Williams, and E. A. Goad. Demulsifier compositions and demulsifying use thereof. Patent: US 5102580, 1992.

[1751] A. A. Toenjes, M. R. Williams, and E. A. Goad. Demulsifier compositions and demulsifying use thereof. Patent: CA 2059387, 1992.

[1752] B. Tohidi, A. Danesh, R. W. Burgass, and A. C. Todd. Effect of heavy hydrate formers on the hydrate free zone of real reservoir fluids. In *Proceedings Volume,* pages 257–261. SPE/Norwegian Petrol Soc Europe Prod Oper Conf (Stavanger, Norway, 4/16–4/17), 1996.

[1753] J. M. Toth. Natural gas odorization and its techniques. In *Proceedings Volume,* pages 170–174. 49th Annu Appalachian Gas Meas Short Course (Coraopolis, PA, 8/15–8/18), 1989.

[1754] P. L. Totten, B. G. Brake, and E. F. Vinson. Retarded acid soluble well cement compositions and methods. Patent: US 5220960, 1993.

[1755] P. L. Totten, B. G. Brake, and E. F. Vinson. Retarded acid soluble well cement compositions. Patent: EP 582367, 1994.

[1756] P. L. Totten, B. G. Brake, and E. F. Vinson. Retarded acid soluble well cement compositions and methods. Patent: CA 2094897, 1994.

[1757] P. L. Totten, B. G. Brake, and E. F. Vinson. Retarded acid soluble well cement compositions and methods. Patent: US 5281270, 1994.

[1758] G. Trabanelli, F. Zucchi, and G. Brunoro. Inhibition of corrosion resistant alloys in hot hydrochloric acid solutions. *Werkstoffe Korrosion,* 39(12):589–594, December 1988.

[1759] A. M. S. Trabelsi and L. S. Al-Samarraie. Fiber content affects porosity, permeability, and strength of cement. *Oil Gas J,* 97(18):108, 110, 112, 114, 1999.

[1760] D. S. Treybig and T. W. Glass. Corrosion inhibitors. Patent: US 4784796, 1988.

[1761] D. S. Treybig and R. G. Martinez. Process for preventing corrosion of a metal in contact with a well fluid. Patent: US 4740320, 1988.

[1762] D. S. Treybig and R. G. Martinez. Compositions prepared from hydrocarbyl substituted nitrogen-containing aromatic heterocyclic compounds, an aldehyde and/or ketone and an amine. Patent: US 4871848, 1989.

[1763] V. F. Trotskij, S. G. Banchuzhnyj, I. G. Zezekalo, and V. I. Tishchenko. Blocking of absorbing strata—by pumping-in specified reagent followed by pumping-in of solution of hydrochloric acid and inhibitor, mixing and leaving for specified time. Patent: SU 1802084-A, 1993.

[1764] A. M. Trushevskaya, L. B. Sklyarskaya, and I. M. Zhurakivskij. Suppressing growth of sulphate(s)—reducing bacteria in stratal injection

water by addition of zinc slurry solutions obtained as waste from filtration stage in production of naphthol-disulphonic acid. Patent: SU 1730502-A, 1992.
[1765] J. P. Tshibangu, J. P. Sarda, and A. Audibert-Hayet. A study of the mechanical and physicochemical interactions between the clay materials and the drilling fluids: application to the boom clay (Belgium) (etude des interactions mecaniques et physicochimiques entre les argiles et les fluides de forage: application a l'argile de boom [Belgique]). *Rev Inst Franc Petrol,* 51(4):497–526, July-August 1996.
[1766] P. F. Tsytsymushkin, S. R. Khajrullin, A. P. Tarnavskij, G. S. Glyantseva, G. G. Iskandarova, and B. V. Mikhajlov. Plugging solution for oil and gas wells—contains plugging cement, modified methylcellulose, mono-substituted sodium phosphate and water. Patent: SU 1740627-A, 1992.
[1767] P. F. Tsytsymushkin, S. R. Khajrullin, A. P. Tarnavskij, P. V. Kovalenko, Z. N. Kudryashova, R. M. Iskhakov, V. N. Levshin, and V. M. Sudakov. Plugging solution for oil and gas wells—contains plugging Portland cement, nitrilo-trimethyl phosphonic acid, hydrazine hydrochloride and water and has corrosion-protective properties. Patent: RU 2002037-C, 1993.
[1768] P. F. Tsytsymushkin, S. R. Khajrullin, A. P. Tarnavskij, Z. N. Kudryashova, P. V. Kovalenko, V. N. Levshin, and B. V. Mikhajlov. Plugging solution for cementing oil and gas wells in salt-bearing strata—contains plugging cement, sodium chloride, calcium chloride, additional preparate bakteritsid, and water. Patent: SU 1803531-A, 1993.
[1769] P. F. Tsytsymushkin, S. R. Khajrullin, A. P. Tarnavskij, Z. N. Kudryashova, V. N. Levshin, and B. V. Mikhajlov. Plugging solution for wells in salt-bearing deposits—contains plugging cement, sodium chloride, water and alkyl-benzyl-methyl-ammonium chloride as additive. Patent: SU 1700201-A, 1991.
[1770] P. F. Tsytsymushkin, S. R. Khajrullin, A. P. Tarnavskij, Z. N. Kudryashova, V. N. Levshin, B. V. Mikhajlov, Y. D. Morozov, and T. B. Kryukova. Plugging solution for oil and gas wells—contains plugging cement, nitrilo-trimethyl-phosphonic acid, product of reacting hexamethylene tetramine with chloro-derivatives unsaturated hydrocarbon(s) and water. Patent: SU 1740629-A, 1992.
[1771] P. F. Tsytsymushkin, S. R. Khajrullin, A. P. Tarnavskij, Z. N. Kudryashova, and B. V. Mikhajlov. Plugging solution contains Portland cement, sodium sulphate, potassium ferricyanide and nitrile-trimethyl-phosphonic acid. Patent: SU 1700204-A, 1991.
[1772] B. I. Tulbovich, L. V. Kazakova, and V. I. Kozhevskikh. Foam-forming composition—contains surfactant, in form of alkali effluent from

caprolactam production, and solvent, in form of e.g. stratal water. Patent: RU 2061859-C, 1996.
[1773] R. G. Udarbe, K. Hancock-Grossi, and C. R. George. Method of and additive for controlling fluid loss from a drilling fluid. Patent: US 6107256, 2000.
[1774] K. H. Ujma, M. Sahr, J. Plank, and J. Schoenlinner. Cost reduction and improvement of drilling mud properties by using polydrill (Kostenreduzierung und Verbesserung der Spülungseigenschaften mit Polydrill). *Erdöl Erdgas Kohle,* 103(5):219–222, May 1987.
[1775] K. H. W. Ujma and J. P. Plank. A new calcium-tolerant polymer helps to improve drilling mud performance and reduce costs. In *Proceedings Volume,* pages 327–334. 62nd Annu SPE Tech Conf (Dallas, TX, 9/27–9/30), 1987.
[1776] V. N. Umutbaev, M. G. Kamaletdinov, B. A. Andreson, R. G. Abdrakhmanov, A. U. Sharipov, and I. V. Utyaganov. Lubricant additive for water-based drilling muds—contains a mixture of phenolic mannich bases, additional phosphoric acid and water. Patent: SU 1799895-A, 1993.
[1777] J. C. Urquhart. Potassium silicate drilling fluid. Patent: WO 9705212, 1997.
[1778] M. A. Urynowicz, R. L. Siegrist, M. L. Crimi, O. R. West, K. S. Lowe, and A. M. Struse. Fundamentals and application of in situ chemical oxidation technologies. In *Abstracts Volume,* page A205. Annu AAPG-SEPM Conv (Denver, CO, 6/3–6/6), 2001.
[1779] M. Vacca-Torelli, A. L. Geraci, and A. Risitano. Dispersant application by hydrofoil: high speed control and cleanup of large oil spills. In *Proceedings Volume,* pages 75–79. 10th Bien API et al Oil Spill (Prev, Behav, Contr, Cleanup) Conf (Baltimore, MD, 4/6–4/9), 1987.
[1780] A. Vadie. Microbial enhanced oil recovery. Internet page: http://www.msstate.edu/dept/wrri/meor/ Accessed 2002.
[1781] R. N. Vaidya and H. S. Fogler. Formation damage due to colloidally induced fines migration. *Colloids Surfaces,* 50:215–229, 1990.
[1782] F. W. Valone. Corrosion inhibiting system containing alkoxylated amines. Patent: EP 207713, 1987.
[1783] F. W. Valone. Corrosion inhibiting system containing alkoxylated amines. Patent: US 4636256, 1987.
[1784] F. W. Valone. Inhibition of microbiological growth. Patent: US 4647589, 1987.
[1785] F. W. Valone. Corrosion inhibiting system containing alkoxylated alkylphenol amines. Patent: US 4867888, 1989.
[1786] F. W. Valone. Corrosion inhibiting system containing alkoxylated amines. Patent: CA 1259185, 1989.
[1787] F. W. Valone. Corrosion inhibiting system containing alkoxylated dialkylphenol amines. Patent: US 4846980, 1989.

[1788] E. Van Oort. Physico-chemical stabilization of shales. In *Proceedings Volume,* pages 523–538. SPE Oilfield Chem Int Symp (Houston, TX, 2/18–2/21), 1997.

[1789] J. P. M. Van Vliet, R. P. A. R. Van Kleef, T. R. Smith, A. P. Plompen, C. A. T. Kuijvenhoven, V. Quaresma, A. Raiturkar, J. M. Schoenmakers, and B. Arentz. Development and field use of fibre-containing cement. In *Proceedings Volume,* volume 4, pages 183–197. 27th Annu SPE et al Offshore Technol Conf (Houston, TX, 5/1–5/4), 1995.

[1790] R. Varadaraj. Polymer-surfactant fluids for decontamination of earth formations. Patent: US 5614474, 1997.

[1791] R. Varadaraj, D. W. Savage, and C. H. Brons. Chemical demulsifier for desalting heavy crude. Patent: US 6168702, 2001.

[1792] Y. V. Vasilchenko and I. G. Luginina. Binder for cementing low-temperature wells—contains modified lignosulphonate(s), potash and cement of specified composition, to increase bending strength of cement rock. Patent: SU 1749199-A, 1992.

[1793] A. Vaussard, A. Ladret, and A. Donche. Scleroglucan drilling mud (boue de forage au scleroglucane). Patent: FR 2661186, 1991.

[1794] A. Vaussard, A. Ladret, and A. Donche. Scleroglucan drilling mud (boue de forage au scleroglucane). Patent: EP 453366, 1991.

[1795] A. Vaussard, A. Ladret, and A. Donche. Scleroglucan based drilling mud. Patent: US 5612294, 1997.

[1796] M. A. Veale, E. Bryan, and R. E. Talbot. A new biocide with respect to industrial water treatment and oilfield applications. In *Proceedings Volume.* Norwegian Soc Chartered Eng Oil Field Chem Conf (Geilo, Norway, 3/19–3/21), 1990.

[1797] R. R. Veldman and D. O. Trahan. Gas treating solution corrosion inhibitor. Patent: WO 9919539, 1999.

[1798] Y. L. Verderevskij, T. G. Valeeva, R. A. Khabirov, V. I. Gusev, Y. N. Arefev, S. N. Golovko, Y. S. Vajsman, D. V. Bulygin, I. V. Mikheeva, V. L. Popova, A. T. Panarin, and G. I. Vasyasin. Development of oil-bearing strata—comprises pumping in spent sulphuric acid, preceded by additional aqueous lignosulphonate(s) to increase efficiency. Patent: SU 1448783-A, 1996.

[1799] S. K. Verma and G. R. Sandor. Corrosion inhibitors for use in oil and gas wells and similar applications. Patent: WO 0146552A, 2001.

[1800] H. A. Videla, P. S. Guiamet, O. R. Pardini, E. Echarte, D. Trujillo, and M. M. S. Freitas. Monitoring biofilms and MIC (microbially induced corrosion) in an oilfield water injection system. In *Proceedings Volume.* Annu NACE Corrosion Conf (Corrosion 91) (Cincinnati, OH, 3/11–3/15), 1991.

[1801] J. P. Vijn. Dispersant and fluid loss control additive for well cement. Patent: EP 1081112-A, 2001.

[1802] J. P. Vijn. Dispersant and fluid loss control additives for well cements, well cement compositions and methods. Patent: US 6182758, 2001.
[1803] J. Villar, J. F. Baret, M. Michaux, and B. Dargaud. Cementing compositions and applications of such compositions to cementing oil (or similar) wells. Patent: US 6060535, 2000.
[1804] J. Villar, J. F. Baret, M. Michaux, and B. Dargoud. Cementing compositions and application of such compositions to cementing oil (or similar) wells. Patent: EP 814067, 1997.
[1805] E. Vindimian, B. Vollat, and J. Garric. Effect of the dispersion of oil in freshwater based on time-dependent daphnia magna toxicity tests. *Bull Environ Contamination Toxicol,* 48(2):209–215, February 1992.
[1806] S. Vitthal and J. M. McGowen. Fracturing fluid leakoff under dynamic conditions: Pt 2: Effect of shear rate, permeability, and pressure. In *Proceedings Volume,* pages 821–835. Annu SPE Tech Conf (Denver, CO, 10/6–10/9), 1996.
[1807] K. von Terzaghi. Die Berechnung der Durchlässigkeitsziffer des Tones aus dem Verlauf der hydrodynamischen Spannungserscheinungen. Sitzungsberichte der Akademie der Wissenschaften in Wien, Mathematisch-Naturwissenschaftliche Klasse, Abteilung 2a, 1923.
[1808] M. A. Vorderbruggen and D. A. Williams. Acid corrosion inhibitor. Patent: US 6117364, 2000.
[1809] K. Waehner. Experience with high temperature resistant water based drilling fluids (erfahrungen beim einsatz hochtemperatur-stabiler wasserbasischer bohrspülung). *Erdöl Erdgas Kohle,* 106(5):200–201, May 1990.
[1810] A. H. Walker and D. R. Henne. The Region III Regional Response Team Technical Symposium on Dispersants: an interactive, educational approach to enlightened decision making. In *Proceedings Volume,* pages 405–410. 12th Bien API et al Oil Spill (Prev, Behav, Contr, Cleanup) Int Conf (San Diego, CA, 3/4–3/7), 1991.
[1811] C. O. Walker. Encapsulated lime as a lost circulation additive for aqueous drilling fluids. Patent: US 4614599, 1986.
[1812] C. O. Walker. Method for controlling lost circulation of drilling fluids with water absorbent polymers. Patent: US 4635726, 1987.
[1813] C. O. Walker. Encapsulated lime as a lost circulation additive for aqueous drilling fluids. Patent: CA 1261604, 1989.
[1814] C. O. Walker. Method for controlling lost circulation of drilling fluids with water absorbent polymers. Patent: CA 1259788, 1989.
[1815] M. L. Walker, W. G. F. Ford, W. R. Dill, and R. D. Gdanski. Composition and method of stimulating subterranean formations. Patent: US 4683954, 1987.

[1816] M. L. Walker, R. L. Martin, and D. J. Poelker. High performance phosphorus-containing corrosion inhibitors for inhibiting corrosion by drilling system fluids. Patent: EP 1076113A, 2001.
[1817] M. L. Walker and C. E. Shuchart. Method for breaking stabilized viscosified fluids. Patent: US 5413178, 1995.
[1818] K. Wall, D. W. Martin, P. W. Zard, and D. J. Barclay-Miller. Temperature stable synthetic oil. Patent: WO 9532265, 1995.
[1819] K. Wall, P. W. Zard, D. J. Barclay-Miller, and D. W. Martin. Amide and imide compounds and their use as lubricant oils. Patent: WO 9530643, 1995.
[1820] K. Wall, P. W. Zard, D. J. Barclay-Miller, and D. W. Martin. Surfactant composition. Patent: WO 9530722, 1995.
[1821] K. Wall, P. W. Zard, D. J. Barclay-Miller, and D. W. Martin. Oil recovery process. Patent: WO 9611044, 1996.
[1822] J. R. Wallace, J. R. Stetter, S. Nacson, and M. W. Findlay, Jr. Odorant analyzer system. Patent: EP 445927, 1991.
[1823] D. Wang. A foam drive method. Patent: WO 9951854, 1999.
[1824] W. J. Wang. The development and applications of cx-18 antichannelling agent. *J Xi'an Petrol Inst,* 11(5):6–7,50–53, 1996.
[1825] Z. Wang. Synthesis of AMPS/AM (acrylamide)/VAC (vinyl acetate) copolymer—a filtration reducer of drilling fluid. *Drilling Prod Technol,* 22(4):4A,55–56, July 1999.
[1826] D. T. Wason. Enhanced oil recovery through in-situ generated surfactants augmented by chemical injection. US DOE Rep DOE/BC/10847-20, Inst Gas Technol, Chicago, IL, August 1990.
[1827] R. M. S. Wat, C. Sitz, M. Rudel, H. Montgomerie, and I. Collins. Development of an oil soluble scale inhibitor. In *Proceedings Volume.* 4th IBC UK Conf Ltd Advances in Solving Oilfield Scaling Int Conf (Aberdeen, Scotland, 1/28–1/29), 1998.
[1828] D. R. Watkins, J. J. Clemens, J. C. Smith, S. N. Sharma, and H. G. Edwards. Use of scale inhibitors in hydraulic fracture fluids to prevent scale build-up. Patent: US 5224543, 1993.
[1829] J. Weaver, K. M. Ravi, L. S. Eoff, R. Gdanski, and J. M. Wilson. Drilling fluid and filter cake removal methods and compositions. Patent: US 5501276, 1996.
[1830] E. Weber, editor. *Molecular inclusion and molecular recognition—Clathrates 1,* volume 140 of *Topics in Current Chemistry.* Springer Verlag, Berlin, 1987.
[1831] C. Wegner and G. Reichert. Hydrogen sulfide scavenger in drilling fluids (schwefelwasserstoff-scavenger in bohrspülungen). In *Proceedings Volume.* BASF et al Chem Prod in Petrol Prod Mtg H_2S—A Hazardous Gas in Crude Oil Recovery Discuss (Clausthal-Zellerfeld, Germany, 9/12–9/13), 1990.

[1832] P. J. Weimer, J. M. Odom, F. B. Cooling, III, and A. G. Anderson. Anthraquinones as inhibitors of sulfide production from sulfate-reducing bacteria. Patent: US 5385842, 1995.
[1833] G. C. West, R. Valenziano, and K. A. Lutgring. Method and composition for sweep of cuttings beds in a deviated borehole. Patent: US 6290001, 2001.
[1834] A. Westerkamp, C. Wegner, and H. P. Müller. Borehole treatment fluids with clay swelling-inhibiting properties (II) (Bohrloch-Behandlungsflüssigkeiten mit tonquellungsinhibierenden Eigenschaften [II]). Patent: EP 451586, 1991.
[1835] D. W. S. Westlake. Microbial ecology of corrosion and reservoir souring. In E. C. Donaldson, editor, *Microbial enhancement of oil recovery: recent advances: Proceedings of the 1990 International Conference on Microbial Enhancement of Oil Recovery*, volume 31 of *Developments in Petroleum Science*, pages 257–263. Elsevier Science Ltd, 1991.
[1836] J. A. Westland, D. A. Lenk, and G. S. Penny. Rheological characteristics of reticulated bacterial cellulose as a performance additive to fracturing and drilling fluids. In *Proceedings Volume,* pages 501–514. SPE Oilfield Chem Int Symp (New Orleans, LA, 3/2–3/5), 1993.
[1837] J. A. Westland, G. S. Penny, and D. A. Lenk. Drilling mud compositions. Patent: WO 9222621, 1992.
[1838] A. J. Wetteman and J. R. Wilson. Operation of large volume odorizers. In *Odorization,* volume 3, pages 251–255. Institute of Gas Technology, Chicago, IL, 1993.
[1839] D. J. White, B. A. Holms, and R. S. Hoover. Using a unique acid-fracturing fluid to control fluid loss improves stimulation results in carbonate formations. In *Proceedings Volume,* pages 601–610. SPE Permian Basin Oil & Gas Recovery Conf (Midland, TX, 3/18–3/20), 1992.
[1840] D. L. Whitfill, E. Kukena, Jr., T. S. Cantu, and M. C. Sooter. Method of controlling lost circulation in well drilling. Patent: US 4957174, 1990.
[1841] J. Wiechert, M. L. Rideout, D. I. Little, D. M. McCormick, E. H. Owens, and B. K. Trudel. Development of dispersant pre-approval for Washington and Oregon coastal waters. In *Proceedings Volume,* pages 435–438. 12th Bien API et al Oil Spill (Prev, Behav, Contr, Cleanup) Int Conf (San Diego, CA, 3/4–3/7), 1991.
[1842] R. D. Wilcox. Surface area approach key to borehole stability. *Oil Gas J,* 88(9):66–80, 1990.
[1843] J. M. Wilkerson, III, D. W. Verstrat, and M. C. Barron. Associative monomers. Patent: US 5412142, 1995.
[1844] A. O. Wilkinson, S. J. Grigson, and R. W. Turnbull. Drilling mud. Patent: WO 9526386, 1995.

[1845] D. A. Williams, J. R. Looney, D. S. Sullivan, B. I. Bourland, J. A. Haslegrave, P. J. Clewlow, N. Carruthers, and T. M. O'Brien. Amine derivatives as corrosion inhibitors. Patent: US 5322630, 1994.
[1846] M. M. Williams, M. A. Phelps, and G. M. Zody. Reduction of viscosity of aqueous fluids. Patent: EP 222615, 1987.
[1847] C. D. Williamson. Chemically modified lignin materials and their use in controlling fluid loss. Patent: GB 2210888, 1989.
[1848] C. D. Williamson and S. J. Allenson. A new nondamaging particulate fluid-loss additive. In *Proceedings Volume,* pages 147–158. SPE Oilfield Chem Int Symp (Houston, TX, 2/8–2/10), 1989.
[1849] C. D. Williamson, S. J. Allenson, and R. K. Gabel. Additive and method for temporarily reducing permeability of subterranean formations. Patent: US 4997581, 1991.
[1850] C. D. Williamson, S. J. Allenson, R. K. Gabel, and D. A. Huddleston. Enzymatically degradable fluid loss additive. Patent: US 5032297, 1991.
[1851] R. Williamson. Well drilling fluids, fluid loss additives therefor and preparation of such additives. Patent: GB 2178785, 1987.
[1852] R. C. Williamson, C. J. Lawrie, and S. J. Miller. Skeletally isomerized linear olefins. Patent: WO 9521225, 1995.
[1853] S. J. Wilson and M. E. Miller. Treatment for shut-in gas well. Patent: US 6302206, 2001.
[1854] H. Wirtz, H. Hoffmann, W. Ritschel, M. Hofinger, M. Mitzlaff, and D. Wolter. Optionally quaternized fatty esters of alkoxylated alkyl-alkylene diamines (Gegebenenfalls quaternierte Fettsäureester von oxyalkylierten Alkyl-Alkylendiaminen). Patent: EP 320769, 1989.
[1855] H. Wirtz, S. P. von Halasz, M. Feustel, and J. Balzer. Copolymers, their mixtures with polymers of esters of methacrylic acid and their use in improving the flowability of crude oil at a low temperature (Neue Copolymere, deren Mischungen mit Poly(Meth)acrylsäureestern und deren Verwendung zur Verbesserung der Fließfähigkeit von Rohölen In der Kälte). Patent: EP 376138, 1994.
[1856] R. Wiser-Halladay. Polyurethane quasi prepolymer for proppant consolidation. Patent: US 4920192, 1990.
[1857] R. R. Wood. Improved drilling fluids. Patent: WO 0153429-A, 2001.
[1858] E. Wrench. Anti-microbial agent. Patent: WO 9006054, 1990.
[1859] E. Wrench. Anti-microbial agent. Patent: GB 2244216, 1991.
[1860] Y. Wu. Chemical problems and costs involved in downstream gas systems. In *Proceedings Volume.* NACE Int Corrosion Forum (Corrosion 90) (Las Vegas, NV, 4/23–4/27), 1990.
[1861] Y. Wu and R. A. Gray. Compositions and methods for inhibiting corrosion. Patent: US 5118536, 1992.

[1862] M. W. Wyganowski. Method of reducing the reactivity of steam and condensate mixtures in enhanced oil recovery. Patent: US 5036915, 1991.
[1863] H. Xiong, B. Davidson, B. Saunders, and S. A. Holditch. A comprehensive approach to select fracturing fluids and additives for fracture treatments. In *Proceedings Volume,* pages 293–301. Annu SPE Tech Conf (Denver, CO, 10/6–10/9), 1996.
[1864] S. S. Yakovlev and E. A. Konovalov. Plugging mixtures on a base of hydrolyzed polyacrylonitrile. *Neft Khoz,* (4):25–27, April 1987.
[1865] T. Yamazaki, K. Aso, H. Okabe, and Y. Akita. Automatic continuous-measuring system of interfacial tension by spinning drop technique. In *Proceedings Volume.* SPE Asia Pacific Conf (Yokohama, Japan, 4/25–4/26), 2000.
[1866] T. Y. Yan. Process for inhibiting scale formation in subterranean formations. Patent: WO 9305270, 1993.
[1867] L. Yang and B. Song. Phosphino maleic anhydride polymer as scale inhibitor for oil/gas field produced waters. *Oilfield Chem,* 15(2):137–140, June 1998.
[1868] V. L. Yashchenko, V. I. Vakulin, A. I. Berdnikov, V. R. Grunvald, V. V. Nikolaev, V. Ya. Klimov, V. I. Nasteka, and N. V. Pikulik. Odor-imparting composition for natural gas—consists of mixture of mercaptan(s) and additionally contains pyridine and α-picoline. Patent: RU 2076137-C, 1997.
[1869] A. A. M. Yassin and A. Kamis. Palm oil derivative as a based fluid in formulating oil based drilling mud. In *Proceedings Volume,* volume 2. 4th SPE et al Latin Amer Petrol Eng Conf (Rio De Janeiro, Brazil, 10/14–10/19), 1990.
[1870] R. R. Yeager and D. E. Bailey. Diesel-based gel concentrate improves Rocky Mountain region fracture treatments. In *Proceedings Volume,* pages 493–497. SPE Rocky Mountain Reg Mtg (Casper, WY, 5/11–5/13), 1988.
[1871] V. Yeager and C. Shuchart. In situ gels improve formation acidizing. *Oil Gas J,* 95(3):70–72, 1997.
[1872] M. H. Yeh. Anionic sulfonated thickening compositions. Patent: EP 632057, 1995.
[1873] M. H. Yeh. Compositions based on cationic polymers and anionic xanthan gum (compositions a base de polymeres cationiques et de gomme xanthane anionique). Patent: EP 654482, 1995.
[1874] T. F. Yen, J. K. Park, K. I. Lee, and Y. Li. Fate of surfactant vesicles surviving from thermophilic, halotolerant, spore forming, clostridium thermohydrosulfuricum. In E. C. Donaldson, editor, *Microbial enhancement of oil recovery: recent advances: Proceedings of the 1990 International Conference on Microbial Enhancement of Oil Recovery,*

volume 31 of *Developments in Petroleum Science*, pages 297–309. Elsevier Science Ltd, 1991.
[1875] Deleted in proofs.
[1876] L. A. Young. Low melting polyalkylenepolyamine corrosion inhibitors. Patent: WO 9319226, 1993.
[1877] S. Young and A. Young. Recent field experience using an acetal based invert emulsion fluid. In *Proceedings Volume*. IBC Tech Serv Ltd Prev of Oil Discharge from Drilling Oper—The Options Conf (Aberdeen, Scotland, 6/15–6/16), 1994.
[1878] A. Zaitoun and N. Berton. Stabilization of montmorillonite clay in porous media by high-molecular-weight polymers. In *Proceedings Volume*, pages 155–164. 9th SPE Formation Damage Contr Symp (Lafayette, LA, 2/22–2/23), 1990.
[1879] P. L. Zaleski, D. J. Derwin, D. J. Weintritt, and G. W. Russell. Drilling fluid loss prevention and lubrication additive. Patent: US 5826669, 1998.
[1880] A. Zaltoun and N. Berton. Stabilization of montmorillonite clay in porous media by high-molecular-weight polymers. *SPE Prod Eng*, 7(2):160–166, May 1992.
[1881] S. M. Zefferi and R. C. May. Corrosion inhibition of calcium chloride brine. Patent: US 5292455, 1994.
[1882] S. M. Zefferi and R. C. May. Corrosion inhibition of calcium chloride brine. Patent: CA 2092207, 1994.
[1883] S. C. Zeilinger, M. J. Mayerhofer, and M. J. Economides. A comparison of the fluid-loss properties of borate-zirconate-crosslinked and noncrosslinked fracturing fluids. In *Proceedings Volume*, pages 201–209. SPE East Reg Conf (Lexington, KY, 10/23–10/25), 1991.
[1884] A. Y. Zekri. Microbial enhanced oil recovery—a short review. *Oil Gas Europe Mag*, 27(1):22–25, March 2001.
[1885] P. Ya. Zeltser. Plugging solution for oil and gas industry—contains plugging Portland cement, lignosulphonate(s), waste from formic acid production obtained at stage of decomposition of sodium formate, and water. Patent: SU 1730431-A, 1992.
[1886] P. Ya. Zeltser, V. I. Chalykh, L. V. Chernyakhovskij, V. N. Smetanin, V. I. Kravchenko, and K. S. Elkin. Light plugging solution production—from mixture of Portland cement, silica-containing dust waste from production of aluminium alloys, sodium hydroxide and water. Patent: SU 1728471-A, 1992.
[1887] Y. B. Zeng and S. B. Fu. The inhibiting property of phosphoric acid esters of rice bran extract for barium sulfate scaling. *Oilfield Chem*, 15(4):333–335,365, December 1998.
[1888] M. J. Zetlmeisl and E. C. French. Corrosion inhibition in highly acidic environments. Patent: US 5169598, 1992.

[1889] M. J. Zetlmeisl and E. C. French. Corrosion inhibition in highly acidic environments. Patent: CA 2067313, 1992.
[1890] C. G. Zhang, M. B. Sun, W. G. Hou, Y. Y. Liu, and D. J. Sun. Study on function mechanism of filtration reducer: comparison. *Drilling Fluid Completion Fluid*, 13(3):11–17, 1996.
[1891] C. G. Zhang, M. B. Sun, W. G. Hou, and D. Sun. Study on function mechanism of filtration reducer: the influence of fluid loss additive on electrical charge density of filter cake fines. *Drilling Fluid Completion Fluid*, 12(4):1–5, 1995.
[1892] G. P. Zhang and H. C. Ye. AM/MA/AMPS terpolymer as non-viscosifying filtrate loss reducer for drilling fluids. *Oilfield Chem*, 15(3):269–271, September 1998.
[1893] H. Zhang, E. J. Mackay, K. S. Sorbie, and P. Chen. Non-equilibrium adsorption and precipitation of scale inhibitors: corefloods and mathematical modelling. In *Proceedings Volume*, page 18 PP. SPE Oil & Gas Int Conf In China (Beijing, China, 11/7–11/10), 2000.
[1894] L. S. Zhang, Y. Q. Wang, and D. Farrar. Polymerisation processes and products. Patent: EP 356242, 1990.
[1895] Y. Zhang and L. Chen. High temperature stabilizer for oil well cement. *Drilling Fluid Completion Fluid*, 10(3):58–61,75–76, May 1993.
[1896] Y. Zhang and Z. Wang. Experiments and applications of low density glass bead slurry. *Petrol Drill Tech*, 17(4):48–51,87, December 1989.
[1897] L. Zhao, Q. Xie, Y. Luo, Z. Sun, S. Xu, H. Su, and Y. Wang. Utilization of slag mix mud conversion cement in the Karamay oilfield, Xinjiang. *J Jianghan Petrol Inst*, 18(3):63–66, September 1996.
[1898] Y. Zhou. Bactericide for drilling fluid. *Drilling Fluid Completion Fluid*, 7(3):2A,10–12, 1990.
[1899] Z. Zhou. Process for reducing permeability in a subterranean formation. Patent: US 6143699, 2000.
[1900] Z. J. Zhou, W. D. Gunter, and R. G. Jonasson. Controlling formation damage using clay stabilizers: a review. In *Proceedings Volume*, volume 2. 46th Annu CIM Petrol Soc Tech Mtg (Banff, Canada, 5/14–5/17), 1995.
[1901] T. Zhu, A. Strycker, C. J. Raible, and K. Vineyard. Foams for mobility control and improved sweep efficiency in gas flooding. In *Proceedings Volume*, volume 2, pages 277–286. 11th SPE/DOE Impr Oil Recovery Symp (Tulsa, OK, 4/19–4/22), 1998.
[1902] V. B. Zhukhovitskij, M. I. Kolomoets, and A. M. Zagrudnyj. Polymeric plugging solution contains urea-formaldehyde resin, expandable resol-phenol-formaldehyde resin containing surfactant and aluminium powder, and maleic anhydride production waste. Patent: SU 1728473-A, 1992.

[1903] C. E. Zobell. Bacterial action in seawater. *Proc Soc Exp Biol Med,* (34):113–116, 1937.
[1904] C. E. Zobell and F. H. Johnson. The influence of hydrostatic pressure on the growth and viability of terestrial and marine bacteria. *J Bacteriol Bioeng,* (57):1979, 1949.
[1905] V. Yu. Zobs, G. V. Belikov, B. F. Sheldybaev, Yu. I. Shaposhnikov, and V. N. Tikhonov. Regulating the density of plugging agents. *Neft Khoz,* (5):29–31, May 1989.
[1906] A. H. Zuzich and G. C. Blytas. Polyethercyclicpolyols from epihalohydrins, polyhydric alcohols and metal hydroxides or epoxy alcohol and optionally polyhydric alcohols with addition of epoxy resins. Patent: US 5286882, 1994.
[1907] A. H. Zuzich, G. C. Blytas, and H. Frank. Polyethercyclicpolyols from epihalohydrins, polyhydric alcohols, and metal hydroxides or epoxy alcohols and optionally polyhydric alcohols with thermal condensation. Patent: US 5428178, 1995.
[1908] C. Zychal. Defoamer and antifoamer composition and method for defoaming aqueous fluid systems. Patent: US 4631145, 1986.

Index

Index of Chemicals

Abietinic acid, **89**
Acetamidoiminodiacetic acid, 106
Acetic acid, 91, 199, 200, 222, 249, **250**, 312
Acetoin, *see* 3-Hydroxy-2-butanon, 222
Acetone, **27**, 94, 330
3-Acetoxy-4-methylthiazol-2(3H)-thione, 74
Acetoxystyrene, **339**
Acrylamide, 5, 12, 15–17, 21, 46, 47, 49–51, **54**, 55–57, 92, 119, 148, 157, 182, 198, 205, 206, 228, 229, 232, 311, 313, 331, 337–339
2-Acrylamido-2-methylpropane sulfonic acid, 16, 46, 47, 49, 50, **52**, **64**, 147, 241, **242**
3-(2-Acrylamido-2-methylpropyldimethyl ammonio)-1-propane sulfonate, 205
Acrylamidopropyltrimethyl ammonium chloride, 337
Acrylic acid, 7, 9, 12, 16, 40, 46, 47, 49, 50, 52, **54**, 55–57, 64, 117, 119, 157, 161, 171, 244, 311, 313, 331, 335, 343
1-Acryloyl-4-methyl piperazine, 337
Adamantane, **176**
Adipic acid, 91
Aldol, 100
Alkylpropoxyethoxy sulfate, 198
Allyl alcohol, **339**
Allylamine, 338
1-Allyloxy-2-hydroxypropyl sulfonic acid, **64**, 241, **242**
Allyloxybenzene phosphonate, 144
Allyloxybenzene sulfonate, 144
Aluminum citrate, 109, 115, 116

Aluminum isopropoxide, 23, 110, 265, **267**, 290
Aluminum oxide, 269
Aluminum sulfate, 110, 142, 291
Aluminum trichloride, 110, 208
Aluminum triisopropyloxide, *see* Aluminum isopropoxide, 23
Amidoiminodiacetic acid, 106
Amino galactomannan, 260
2-Amino-2-hydroxymethyl-1,3-propanediol, 333, **334**
p-Aminobenzoic acid, **310**
Aminoethanol, 100, **188**
N-β-(Aminoethyl)-δ-aminopropyltrimethoxysilane, 270
Aminoethylethanolamine, 48, 55
Aminopyrazine, 99
Ammonia, 169
Ammonium persulfate, 263
Ammonium thioglycolate, 200
Amphoacetates, **212**
Amyl alcohol, 323, **324**
Amylose, **243**
p-tert-Amylphenol, **260**, 268
Aniline, 227
Antimony pentachloride, 330
Antimony tribromide, 100
Arabinose, **99**
Arabitol, **258**
Ascorbic acid, **99**
Aspartic acid, **89**
Azodicarbonamide, 274
Barite, 25, 26, 47, 139
Barium sulfate, 30
Bentonite, 2, 12, 20, 37, 38, 47, 116, 136–138, 205, 284–287, 311

Index **483**

Benzene, 227
Benzene tetracarboxylic acid, 227, **250**
Benzoic acid, 249, **250**, 278
1,2-Benzoisothiazolin-3-one, **18**
Benzothiazyl disulfide, 194
Benzotriazole, **98, 101**, 158, 188, 194
Benzyl trimethyl ammonium hydroxide, 92
Benzylamine, 331, **332**
Benzylsulfinylacetic acid, 102
Benzylsulfonylacetic acid, **101**, 102
Betaine, **91, 267**
2,5-Bis(N-Pyridyl)-1,3,4-oxadiazole, **101**
Bishexamethylenetriamine, **332**
Bisphenol-A, 117, **118**, 333
Boehmite, 269
Boric acid, **254**, 255
2-Bromo-2-bromomethylglutaronitrile, **78**
2-Bromo-4-hydroxyacetophenone, 19, 72
Bromo-2-nitropropane, **77**
Bromo-chorodimethyl hydantoin, 73
1,4-Butanediamine, **90**, 330
Butanol, 334
n-Butanol, 146
trans-2-Butene-1-thiol, 192
Butylene oxide, 45, 57, 63, 341
Butylmercaptan, **164, 195**
tert-Butyl mercaptan, 193
tert-Butylstyrene, 161, **339**
Calcium fluoride, 65
Calcium oxide, 121
Caprolactam, **212**, 284
Carbonic dihydrazide, 107
Carboxymethylcellulose, 2, 5, 10, 12, 16, 17, 37, 39, **43**, 64, 66, 109, 112, 113, 116, 240, 258, 260, 287
Carboxymethylinulin, 107
Cardanol, 342, **343**
Carrageenan, 141
Casein, 144
Castor oil, 228
Cellulose, **243**
Cellulose polyanionic, 17, 37, 39, 56, 63, 64
Cerussite, 25
Cetyltrimethyl ammonium chloride, 163
Chalk, 112, 126, 287
Chlorine dioxide, 73, 74, 76, 97, 102, 199
5-Chloro-2-Methyl-4-isothiazolin-3-one, 74
N-(2-Chloropropyl) maleimide, **312**
Chromium propionate, 115

Cinnamaldehyde, **8**, 87, 92, 99
Citraconic anhydride, 278
Citric acid, 24, 105, 120, 121, 140, 199, **259**, 264, 273, 311
Clean up additives, 264
Cobalt hexacyanide, 226
Cocoalkyl quaternary ammonium chloride, 27
Cocoamphodiacetate, 21
Cocobetaine, 267
2,4,6-Collidine, 94, **95**
Cumarone, 144, **145**, 148
Cupric oxide, 121
Curdlan, 110
Cyamopsis tetragonolobus, 241
Cyclohexane, 334
N-Cyclohexyl maleimide, **95, 312**
Cyclohexylammonium benzoate, **101**, 102
Cysteamine, 200
Cysteine, 200
Daphnia magna, 298
Decanol, 110, 291
1-Decene, 6, **161**
N-Decylacrylamide, 205
5-Decyne-4,7-diol, 96
Dehydroabietinic acid, 89
Di-decyl ether, 9
Diallylamine, 17, 335, **336**
N,N-Diallylcyclohexylamine, 335, **336**
Diallyldimethylammonium chloride, 16, 52, 55, 56, 331, 338
2,4-Diamino-6-mercapto pyrimidine, **101**
2,4-Diamino-6-mercapto pyrimidine sulfate, 102
2,5-Diaminonorbornylene, 95
2,3-Dibromo-1-chloro-4-thiocyanato-2-butene, 77
1,2-Dibromo-2,4-dicyanobutane, **77**
2,2-Dibromo-2-nitroethanol, **77**
2,2-Dibromo-3-nitrilopropionamide, **78**
Dibutyl adipate, 323, **324**
Dicalcium silicate, 127, 128
4,5-Dichloro-2-N-octyl-isothiazolin-3-one, **18**, 74
Dicyclopentadiene, 307
Dicyclopentadiene bis(methylamine) methylenephosphonate, 140
Dicyclopentadiene dicarboxylic acid, **101**, 102, 188
N,N-Didodecylamine, 9

Diethanolamine, 142, 306
Diethyl disulfide, 164
Diethyl ether, 330
Diethylaminoethyl acrylate, 335, **336**
Diethylaminoethyl methacrylate, 335, **336**
Diethyldisulphide, 323, **324**
Diethylene glycol, 18, 66, **186**, 187, 281
Diethylene glycol monobutyl ether, 96, 322, 340
Diethyleneamine, 330
Diethylenetriamine, 93, 95, 96
Di-2-ethylhexyl acetal, 8
Diethylsulfide, **164**, **195**
Dihexyl formal, 8
4,5-Dihydroxy-m-benzenedisulfonic acid, **142**
Dihydroxyisopropylimino-N,N-diacetic acid, 106
2,5-Dimercapto–1,3,4-thiadiazole, **253**
Dimethyl benzene, **145**, 146
Dimethyl diallyl ammonium chloride, 64
Dimethyl disulfide, 164
1,2-Dimethyl-5-nitro-1H-imidazole, **78**
Dimethyl sulfate, 337
Dimethyl sulfide, 193, 194
1,3-Dimethyl-2-thiourea, 158
N,N-Dimethylacrylamide, 39, 46, 50, 265, **267**
Dimethylamino ethyl methacrylate, 49, 55
Dimethylaminoethyl acrylate, 335, **336**
Dimethylaminoethyl methacrylate, 46, 50, **54**, 57, 331, 335, **336**, 338
Dimethylaminopropyl acrylamide, 75, **78**
Dimethylaminopropyl methacrylamide, 17, 75, **78**, 265, **267**, 335, **336**
5,5-Dimethylhydantoin, **78**
N,N-Dimethyl-N′-phenyl-N′-fluorodichloromethylthio)sulfamide, 74
N,N-Dimethylmethacrylamide, 46, **54**
2,5-Dimethylpyridine, 94
Dimethylurea, 14, 19, 48, 55, 74, **78**
4,4-Dimethyl-2-oxazolidinone, **78**
Dimethyl-diallyl ammonium chloride, 112
Dinitrosopentamethylenetetramine, 274
Dioctyl sodium sulfosuccinate, 307
1,4-Dioxan, 330, **332**
Dipropylene glycol, **186**
Dipropylenetriamine, 330
Disodium cocoamphodiacetate, 21

Dithiocarbamic acid, 19
Dithioglycol, **188**
1-Dodecene, **161**
Dodecyl-β-maltoside, 163
Dodecyl diphenyl oxide disulfonate, 163
(N-dodecyl)trimethylammonium bromide, 198
n-Dodecylmercaptan, 93
N-(p-dodecylphenyl)-2,4,6-trimethylpyridinium sulfoacetate, 98
N-dodecylpyridinium bromide, 87
Dodecyl-2-(2-caprolactamyl) ethanamide, 182
Dulcitol, **258**
Epichlorohydrin, 17, 46, 315
Erucyl amine, 253
Erythorbic acid, 199
Erythritol, **258**
Ethanol, 87, 112, 184, **186**, 232, 285
Ethanolamine, 15, **27**, 31, 140, 142, 188, 265, 279
Ethene, **161**
Ethoxylated nonylphenol, **246**
Ethyl acrylate, 11, 313
N-Ethyl maleimide, **312**
Ethyl vinyl ether, **46**, 57
5-Ethyl-5-methylhydantoin, 74
Ethylcellosolve, 281
Ethylene glycol, 18, 112, 181, 184, **185**, **186**, 187–191, 308
Ethylene glycol butyl ether, 133
Ethylene glycol monoacetate, 74, **75**
Ethylene oxide, 14, 45, 57, 63, 329, 330, 332–334, 340, 341
Ethylenediamine, 90, 96, 330, **332**
Ethylenediaminetetraacetic acid, 105, 273, **274**
Ethyleneimine, 330
2-Ethylhexanoic acid, 265
2-Ethylhexanol, 14
2-Ethylhexylacrylate, 244, **245**
Ethylmercaptan, **164**, 192–194, **195**
Ferric acetylacetonate, 109, 113, **114**
Ferrochrome lignosulfonate, 3, 12
Formaldehyde, 5, 14, 46, 48, 51, 55, 72–74, 75, 94, 114, 229, 278, 279, 290, 334
Formamide, 142
Formic acid, 92, 142, 222, **248**, 249, 289

Index **485**

Fructose, 163, 219
Fulvic acid, 315
Fumaric acid, 91, **248**, 249
2-Furaldehyde, **8**, 279
Furfuramide, 282
Furfuryl alcohol, 143, 229, **260**, 270
2-Furfurylidene, **279**
Galactomannan, 240–242, 255, 256
Gibbsite, 269
Gilsonite, 28, 29, 57, 136–138, 147, 150
Gluconic acid, 99, 105, **259**, 264
Glucono-δ-lactone, **260**, 273, **274**
Glucose, 28, 163, 219, 223, 243, **244**
Glucuronic acid, 12, 243, **244**
Glutaraldehyde, 14, 48, 55, 69, 72–74, **75**, 158, 221
Glyceric acid, **259**
Glycerol, 6, 87, 112, 156, 159, 184, 254, 256, **258**
Glycerol diacrylate, 119
Glycerol dimethacrylate, 119
N-Glycerylimino-N,N-diacetic acid, 106
Glycolic acid, 97, 190, **259**
Glyoxal, 14, 48, 55, **256**
Gypsum, 18, 66, 127, 141, 200
Hectorite, 284
Heptane, 334
2-Heptylimidazoline, **212**
1-Hexadecene epoxide, 14
n-Hexadecyltrimethylammonium bromide, 169
1-N-Hexadecyl-1,2,4-triazole bromide, **77**
Hexamethylene tetramine, **176**, 280
Hexamethylenetetramine, 96
Hexane, 334
Hexanol, 110, 155, 265, **266**, 291
1-Hexene, **161**
Hexylene glycol, 97
1-Hexyn-3-ol, 96
Hydrazine chloride, 149
Hydrazine hydrochloride, 280, 281
Hydrochloric acid, 199
Hydrofluoric acid, 107, 199
Hydrogen peroxide, 116, 123, 195, 203, 204, 243
Hydroxamic acid, 19, **101**, 102
1-Hydroxy-4-imino-3-phenyl-2-thiono-1,3-diazaspiro(4.5)decane, 74

1-Hydroxy-5-methyl-4-phenylimidazoline-2-thione, 74
3-Hydroxy-4-methylthiazol-2(3H)-thione, **77**
3-Hydroxy-4-phenylthiazol-2(3H)-thione, 77
3-Hydroxy-2-butanon, 222
Hydroxyacetic acid, 44, 57, 261
2-Hydroxybutyl ether, 21
Hydroxyethyl acrylate, 46, **54**, 57
Hydroxyethylcellulose, 12, 17, 40, 42, **43**, 56, 140, 240, 242, 246, 248, 261, 309
Hydroxyethylethylene diaminetriacetic acid, 273
1-Hydroxyethylidene-1,1-diphosphonic acid, 107
Hydroxyethyliminodiacetic acid, 273
Hydroxyethyl-tris-(hydroxypropyl) ethylene diamine, **259**
Hydroxylamine, 273
o-Hydroxyphenylmethanol, 229
2-Hydroxymethyl-1,3-propanediol, 333
2-Hydroxyphosphono-acetic acid, 102
Hydroxypropyl acrylate, 46, 57, 144, **145**, 147
Hydroxypropylguar, 17, 42, 239–241, **241**, 245, 246, 248, 253, 254, 257, 258, 260
N(2-Hydroxypropyl)imino-N,N-diacetic acid, 106
N-(3-Hydroxypropyl)imino-N,N-diacetic acid, **106**
Ilmenite, 26
2-Imidazolidinethion, **253**
Indene, 144, **145**, 148
Inositol, **258**
Iodoacetone, 74, **75**
Iron carbonate, 25, 88
Iron thiocyanate, 239
Isoascorbic acid, **99**
Isobutanol, 286
Isobutene, 65, 140
Isobutyraldehyde, **8**, 93
Isononylphenol, 199
Isooxazol, **18**
Isophorone diamine, **95**
Isopropanol, 334
Isopropyl benzoate, 228
Isopropyl mercaptan, 193
Isopropylbenzene, 118, 142, 277, 278
Isothiazol, **18**
Itaconic acid, 49, 50, 52, 55, 56, 311
Kaolinite, 269

4-Ketothiazolidine-2-thiol, **253**
Lactic acid, 74, **75**, 199, 222, 223, **259**
Lanthanum oxide, 121
Latex, 117
Lauric acid, 88
Lead carbonate, 25
Levan, 219
Lignosulfonate, 4, 5, 38, 44, 45, 57, 112, 113, 115, 118, 140, 144, 199, 248, 282, 289, 310, 322
Lime, 3, 22, 49, 265
Linoleic acid, 88, **163**
Linolenic acid, 88
Linolic acid, 89
Linseed oil, 308
Lithium aluminum hydride, 92
Magnesium chloride, 134, 191, 326
Magnesium fluoride, 256
Magnesium hydroxide, 256
Magnesium oxide, 12, 121, 134, 145, 256, 269
Magnesium peroxide, 121
Magnesium sulfate, 6
Maleic acid, 14, 48, 55, 140, **250**
Maleic anhydride, 15, 21, 24, 64, 91, 140, 144, **145**, 147, 151, 251, 278, 312
Maleimide, 182, 312
Malic acid, **259**
Malonic acid, 229, **250**
Mandelic acid, **259**
Manganese nitrate, 109, 289
Manganese oxide, 121
Manganese tetroxide, 26
Mannitol, **258**
Mannose, **244**
Melamine, **310**
Mercaptobenzimidazole, 252, **253**
Mercaptobenzothiazole, 188, 194, 252, **253**
2-Mercaptobenzoxazole, **253**
Mercaptoethanol, 200
2-Mercaptothiazoline, **253**
Methacrylamide, 46, **54**, 57, 339
Methacrylamido propyltrimethylammonium chloride, 50, 55
Methacrylic acid, 11, 46, **54**, 57, 64, 117, 161, 171, 313, 331, 335, 339
Methallyl alcohol, **339**
Methane, 175, 176, 178, 214, 217, 220, 223, 234

Methanococcoides euhalobius, 222
Methanol, 27, 87, 112, 155, 162, 163, 181, 184, **186**, 223, 245, 308, 334
2-Methoxyethylimino-N,N-diacetic acid, 106
3-Methoxypropylimino-N,N-diacetic acid, 106
Methyl acrylate, 11
3-Methyl-1-butanethiol, 192
Methyl chloride, 337
n-Methyl diethanol amine, 9
Methyl iodide, 92
2-Methyl-4-isothiazolin-3-one, 74
Methyl styrene sulfonate, 50, 171
Methyl vinyl ether, **46**, 57
Methylbenzoic acid, 227
2-Methylbenzotriazole, 98, **158**
Methylcellulose, 246, 283
α-Methyldecanol, 8
Methylene-bis-acrylamide, 16, **17**, 49
Methylimino-N,N-diacetic acid, 106
Methylisobutylketone, 306
N-Methylmethacrylamide, 46, **54**
N-Methylol acrylamide, 119, **277**, 278
N-Methylol methacrylamide, **119**, **277**, 278
Methylorange, **32**, 33
2-Methylpropene, 65
2-Methylthio-4-tert-butylamino-6-cyclopropylamino-S-triazine, 74
Methyltrimethoxysilane, 20
Monoallylamine, 44, **45**, 331
Montmorillonite, 38, 61, 64, 132
Morpholine, 16, **17**
Myristic acid, 88, **89**
N-Vinyl-N-methyl acetamide, 182
Naphthalene sulfonic acid, 48, 51, 55, 56
Naphthalenediol, 227
Naphthalenedisulfonic acid, 227
Naphthalenesulfonic acid, 227
Naphthalenetrisulfonic acid, 227
Naphthenic acid, 157
1-Naphthol-3,6-disulphonic acid, 72
2-Naphtol, **310**
Napthaleno-sulfonic acid, 310
Neopentylglycol, 15
Nitrilotriacetic acid, 104, 273
Nitrilo-triacetic acid, **274**
Nitrilo-trimethyl phosphonic acid, 280
N-Nitrosophenylhydroxylamine, 119, 277, 278

Index 487

Nonylphenol ethoxylated, 41, 56, 87, 246, 248
Norbornylsiloxane, 323
Novolak, 339
5,9,12-Octadecatrienic acid, 89
1-N-Octadecyl-1,2,4-triazole bromide, 74
Octaethylene glycol decyl ether, 198
Octaethylene glycol dodecyl ether, 198
1-Octene, 6, **161**
n-Octylacrylamide, 232
Oleic acid, 88, 89, **163**
Oxalic acid, 48, 55, 190, **250**
p,p′-Oxybis(benzenesulfonyl hydrazide, 274
Pentaerythrite, 15, 66, 284, **329**
n-Pentane, 227
Pentanedial, *see* Glutaraldehyde
2,4-Pentanedione, **114**
Perfluoromethylcyclohexane, 226
Perfluoromethylcyclopentane, 226
Phenolphthalein, **32**, 33
m-Phenylenediamine, 69
N-Phenyl maleimide, **312**
3-Phenyl-2-propyn-1-ol, **101**
2-Phenylbenzimidazole, 98, **158**
Phosphorous pentoxide, 110, 265, **266**, 290, 291
o-Phthalaldehyde, **72**
Phthalide, **278**
Phthalimide, **53**
Phyllospora comosa, 299
Picoline, **193**
Polyacetoxystyrene, 332, 339
Polyacrolein, 14, 48, 55, 107
Polyacrylic acid, 9, 313
Polyacrylonitrile, 16, 22, 37, 109, 113, 136, 289
Polyaspartate, 89
Polyaspartates, 107
Polyaspartic acid, 90
Polydimethylsiloxane, 189, 318, 321, 324
Polyethyleneimine, 46, 48, 64, 330, 340, 341
Polyethyleneimine phosphonate, 144, 309
Polyisobutylene, 168, 323
Polymethylsiloxane, 282, 322
Polyoxypropylene, 112, **260**, 268, 321
Polyphosphonohydroxybenzene sulfonic acid, 102
Polyvinyl alcohol, 14, 17, 21, 46, 47, 64

Potassium bichromate, 109, 113
Potassium silicate, 64
Propanediamine, 330
2-Propanol, **186**
Propargyl alcohol, 92, **95**, 96
Propargyl ether, 95
Proppants
  light weight, 269
Propylene glycol, 7, 13, 184, 185, **186**, 187, 190
Propylene oxide, 14, 45, 57, 63, 91, 144, 329, 333, **334**, 340, 341
1,2-Propylenediamine, **90**
Pseudozan, 206
Pyrazol, **18**
Pyridine, **195**, 227, 280
N-Pyridineoxide-2-thiol, **253**
Pyruvic aldehyde, 64
Quebracho, 3
Scleroglucan, 17, 28, 43, 57, 140, 143
Sebacic acid, 91, 151, 289
Siderite, 25
Silicium carbide, 269
Smithsonite, 25
Sodium-3-acrylamido-3-methylbutanoate, 205
Sodium-2-acrylamido-2-methylpropanesulfonate, 205
Sodium acrylate, 205, 228
Sodium aluminate, 12
Sodium bichromate, 109, 115
Sodium chlorite, 74
Sodium cyanate, 289
Sodium dodecyl sulfate, 178, 198
Sodium dodecylbenzene sulfonate, 48, 55
Sodium formate, 246
Sodium hypochlorite, 75, 76
Sodium metasilicate, 46, 208
Sodium methallylsulfonate, 312
Sodium orthophosphate, 158
Sodium phosphate, 46, 208
Sodium salicylate, 169
Sodium sulfite, 46
Sodium vinyl sulfonate, 49, 50, 55
Sodium-hexa-metaphosphate, 158
Sorbitan monooleate, 41, 56, **246**, 248, 249, 322
Sorbitol, **258**
Soya bis(2-hydroxyethyl)amine, 163

Starch, 10, 12, 17, 19, 37–39, **40**, 41, 43, 44, 56, 57, 120, 123, 240, 247, 248, 255, 314
Steam flooding, 334
Stearic acid, 27, 88, 91
Styrene, 50, 52, 56, 64, 311, **339**
Styrene sulfonate, 16, 24, 46, 56, 57, 144, **145**, 312
Styrene-butadiene latex, 133
Styrenesulfonic acid, 51, 56
Succinic acid, 27, **250**
Succinimide, **75**
Succinoglycan, 42, 43, 56, 111
Sucrose, 134, 184, 218, 219
Sulfamic acid, **248**, 249
Sulfanilic acid, 310
Sulfoethyl acrylamide, 313
Sulfomethyl acrylamide, 313
Sulfur, 269, 272
Sulfur hexafluoride, 226
Sulfuric acid, 199, 204, 223, 290
Tallow tetramine, 88
Tallow triamine, 88
Tartaric acid, 27, 140, **259**
2,4,8,10-Tetra-oxaspiro-5,5-undecane, 66
Tetracalcium aluminoferrite, 127
Tetradecyl trimethyl ammonium chloride, 198
n-Tetradecyltrimethylammonium bromide, 169
Tetraethylenepentamine, 90, 91
Tetrahydrofuran, 92, 329
Tetrahydrothiophene, 192, **193**, 195
Tetrakis-hydroxymethyl-phosphonium sulfate, 19, 72
Tetramethylammonium chloride, 64, 251
1,3,4-Thiadiazole-2,5-dithiol, 252
Thiazolidines, 102
Thiobacillus denitrificans, 223
Thiocyanomethylthio-benzothiazole, 19, 72
Thioglycerol, 200
2-Thioimidazolidone, 252
2-Thioimidazoline, 252
Thiolactic acid, 200
Thiophene, **193**
Thiourea, 158
Toluene, 334
p-Toluene sulfonyl hydrazide, 274
o-Tolulic acid, **250**, 278
Triazol, **188**

Tributyl phosphate, 282, **283**, 321–323
Tributyltetradecyl phosphonium chloride, 74
Tricalcium aluminate, 127
Tricalcium silicate, 127, 128
Trichloroethane, 214
Tridithiocarbamic acid, 332
Triethanol amine, 9, 142, 289
Triethyl phosphate, 110, **266**, 291
Triethylene glycol, 281
Triethylenetetramine, 95, 330
2,4,6-Trihydroxybenzoic acid, **142**
Trimethylenediamine, **90**
Trimethylolpropane, 15, **329**
2,2,4-Trimethyl-1,6-diaminohexane, 95, **96**
Trimethyl-1-heptanol, 99
Trioxane, 146
TRIS
  see 2-Amino-2-hydroxymethyl-1,3-propanediol, 333
Tris-hydroxy-ethyl-perhydro-1,3,5-triazine, 102
Tris(hydroxymethyl)methylimino-N,N-diacetic acid, 106
Trisodium phosphate, 132
Tritium water, 225
Tylose, 172
Tyrosine, 162, 163
Uintaite, 28, 29, 137
Urea, 75, 114, 175, **176**, 184, 204, 215
Urotropin, 114, 115, **188**, 287
N-Vinyl acetamide, 49, 50
Vinyl imidazole, 50, 56
Vinyl phosphonic acid, **245**
1-Vinyl-2-pyrrolidinone, **93**
Vinyl sulfonic acid, **245**
Vinyl trimethoxysilane, 331, 338
Vinylacetate, 16, 46, 57, 64, **93**, 159, 160
N-Vinylamide, 51
Vinylbenzene sulfonate, 49, 50, 55
N-Vinylcaprolactam, 16
N-N-Vinylformamide, **18**
N-Vinyl-N-methylacetamide, 16, **18**, 50, 52
Vinylphosphonic acid, 55
N-Vinylpyridine, **172**
N-Vinylpyrrolidone, 49–51, 119, **172**
N-Vinyl-2-pyrrolidone, **245**
Vinylsulfonic acid, **172**, 313
Wellan, 45, 140, 205, 243, 248
Wollastonite, 146

Xanthan, 12, 17, 28, 44, 45, 120, 121, 123, 198, 201, 202, 205, 206, 229, 240, 241, 244, 246, 248, 252, 262
Xylene, 334
Xylitol, 258
Zinc carbonate, 25
Zinc dioctyl-phenyl-dithio-phosphate, 66
Zinc oxide, 5, 26, 121
Zirconium oxide, 26
Hydroxyethylcellulose, 121

General Index

Absorption
 drilling solution, 117
 surface active agent, 319
 water, 251, 280, 290
Absorption strata, 118
Acetylenic alcohols, 96
Acid corrosion, 102, 187
Acid fracturing, 234, 271
Acid solubility
 zirconium oxide, 26
Acidizing, 85, 199, 251, 271–273
Adhesive properties
 epoxide resins, 117, 277
Adhesive-coated material
 proppant, 271
Adsorption
 bacteria, 68, 71
 corrosion inhibitor, 87
 polymer, 63
 surfactant, 87
Aerosil, 280
Aging
 drilling mud, 39, 41
 filter cake, 37
 interfacial viscosity, 224
Alkylation
 benzene, 142
 imidazoline, 212
Amphoteric tensides, 212
Anaerobic bacteria, 218, 220, 222
Anti-syneresis, 229
Antifoaming agent, 155, 282, 322
Antioxidant, 158
Aphrons, 23
Asphalt
 blown, 29

 dispersants, 315
 in muds, 2
 precipitation, 228
 processing, 317
 sulfonated, 29, 49, 55, 314, 315
 Uintaite, 29, 137
Asphaltenes, 227, 305, 332
 colloidal dispersion, 227
 dispersants for, 228, 315
 entrained, 155
 precipitation, 227, 228
 precipitation of, 215
 stabilization, 326
 surface tension, 213
 water in oil emulsion, 200
Atomic absorption spectroscopy, 306
Bacteria, 74, 102
 acid-producing, 80
 alkaliphilic, 111
 anaerobic, 218, 220
 biocide
  glutaraldehyde, 73
 biocides, 67, 71
 chemical treatent, 70
 controlled degradation by, 42
 corrosion by, 79
 Corynebacterium petrophilum, 343
 degradation of polymers, 19
 desulfovibrio, 223
 detection, 67, 79
 environmental adaptation, 71
 heterotrophic, 223
 hydrocarbon-oxidizing, 71
 hydrogen sulfide producing, 222
 inoculated, 217
 lactobacillus, 223
 metabolic chemical products, 221
 methanogenic, 222
 microbial enhanced oil recovery, 220
 Nocardia amaraebacteria, 343
 oil-degrading, 218
 pediococcus, 223
 petroleum-oxidizing, 220
 polymer forming, 39
 pseudomonas elodea, 243
 reticulated bacterial cellulose, 243
 Rhodococcus aurantiacus, 343

490   Oil Field Chemicals

Bacteria (*continued*)
  sessile, 70
  slime-forming, 218
  sulfate-reducing, 67, 68, 70, 71, 73, 80, 81, 97, 148, 221
  sulfidogenic, 68
  transport of, 219
  xanthomonas campestris, 244
Bacteria control, 76
Bacteria
  as biodemulsifiers, 343
Bacteria
  biosurfactant, 220
Bacteria control, 19, 69, 72
Bactericides
  cementing, 136
  doses, 68
  for drilling fluids, 19
  for sulfate-reducing bacteria, 68
  fracturing fluids, 69
  in muds, 4
  selection of, 71
Biodegradation, 190, 297
  acceleration by dispersants, 294
  polyaromatic hydrocarbons, 232
  prevention, 19
  tests, 298
Bit bearing, 65
Bitumen, 159, 337, 340
Blast furnace slag, 31, 149, 150, 287
Borehole stabilization, 28, 134
Bridging, 36, 121, 122
  fluid, 121
Bridging agent, 38, 122, 123, 322
Brines, 185
  anti-freeze agent, 185
  bacteria, 223
  corrosive properties, 181
  delayed gelation agents, 113
  determination of corrosion inhibitors in, 86
  hydrate formation temperature, 181
  leak-off rate, 121
  offshore uses, 186
  permeability reduction, 231
  stable gels in, 115
  use of produced, 113
Buffer, 75, 156, 187, 214, 249
Carbon black, 6, 66, 133, 148

Carbon dioxide corrosion, 91, 149
Carbon dioxide flooding, 211, 213
Carbon dioxide injection, 203, 213
Carbonate reservoir, 214
Carbonate rock, 214
Cashew nut shell liquid, 343
Casing
  adhesion, 143, 146, 309
  adhesion of cement, 280
  cementing, 47, 125
  corrosion, 84, 149
  repair, 132, 277, 284
Catalyst
  acid, 335
  alkaline, 322
  aminic adducts, 93
  carbonate rock, 214
  Michael addition, 92
  organometallic compound, 342
Caustic waterflooding, 197, 199, 224, 228
Cement
  accelerators, 141, 142
  adhesion, 280
  anti-freeze, 191
  corrosion, 280, 283
  dispersant, 309–311
  expansion additives, 144
  fiber, 134, 288
  fluid loss, 35, 45, 50, 146
  foam, 139
  furnace slag, 137
  high temperature, 132
  high-alumina, 133
  lightweight, 135, 138
  lost circulation, 23
  low temperature, 133
  magnesian, 133
  oil-based, 23, 132
  permeability, 144, 147, 148, 280
  polymer, 149
  Portland, 126
  pot life, 191
  resin, 130
  retarders, 140
  strength, 285
  thickening time, 46
  thixotropic, 284
  weighting agents, 139
Cement mixer, 113

Cement slurry
  adhesion, 146
  channeling, 148
  specific weight, 129, 137
  thermal thinning, 143
  viscosity, 137, 143, 284, 309, 310
Ceramic particles, 269
Chalk, 226, 232
Chalk reservoirs, 231
Channel inclusion compounds, 175
Chelate, 258, 264
Chelating agent, 24, 105, 109, 111
Chemical resistance, 131, 270
Chromatography
  gas, 80, 226, 301
  liquid, 86
  size exclusion, 205, 239
  thin layer, 86
Clathrates, 174, **175**
Clay stabilization, 20, 58
Clay swelling, 21, 230
Cloud point, 322
Coagulants, 308
Coating, 271
  clay stabilization, 63
  drag reduction, 162, 172
  encapsulation, 263, 264, 272
  proppants, 270
  protective, 83, 158, 159
  water shutoff, 286
Coefficient of friction, 172
Coiled tubing, 13
Colloidal dispersion, 116, 227
Combination flooding, 206
Compressive strength, 126, 130–132, 134, 138, 139, 146, 290
Condensation product
  formaldehyde, 48, 117, 332, 342
  hydroxyacetic acid, 44, 57, 261
Condensation products
  polyalkenylsuccinic acid, 182
Controlled solubility compounds, 264
gen Coral toxicity, 298
Corrosion control, 94, 136, 156
Corrosion inhibitor, 94
  acid-anhydride esters, 91
  acidizing, 85
  alkylaniline-formaldehyde, 94
  analyis, 86

applications, 82
  benzotriazole, 188
  carrier solvent, 93
  carrier solvents, 91
  cinnamaldehyde, 87
  compositions, 94
  Diels–Alder adducts, 91
  efficiency, 85
  imidazoline, 97
  nitrite, 92
  oil-dispersible, 91, 98
  pigs, 84
  polyaspartic acid, 90
  polypeptides, 89
  propargyl alcohol, 95
  surfactants, 91
  table, 102
  water-dispersible, 91
  water-soluble, 97
Corrosion rate, 88, 90, 97
Corrosivity, 87
CROSERF, 298
Crosslinking agents, 40, 116, 235, 255, 288
  bichromate, 112, 113
  boric acid, 255
  chromates, 229
  polyvalent metal cations, 109, 115, 238, 274
Cuttings removal, 12
Deep well drilling, 39
Defoamer, 322–324
Defoaming, 316
Demulsifier
  cactus extract, 343
  diol-polymer, 333
  effectiveness, 327
  phenol formaldehyde, 342
  polyacrylics, 335
  polyalkylene glycol, 333
  polyester, 334
  polysiloxane, 334
Demulsifiers, 331, 332
  interfacial shear viscosity, 342
Desulfurization, 316
Detergent, 307, 313
Dew point, 177
Diatomaceous earth, 136, 138, 141, 179, 309
Differential aeration cell, 80
Differential pressure sticking, 13

Diffusion, 167, 295
  gas, 319, 320
  pore pressure, 60
  simulation, 301
  stability of shales, 60
Dispersant, 51, 55, 56, 144, 228, 311, 338
  analysis, 297
  bentonite suspension, 24, 311
  biodegradation, 298
  cement, 51, 310
  drilling fluid, 313, 314
  effectiveness, 294, 301
  for sulfur, 315
  oil spill, 295
    compositions, 307
    formulation, 308
    spraying, 296
  policies, 292
  toxicity, 299
Displacement
  emulsion stabilizers, 327
  foam, 209
  gas, 164
  oil, 201, 204, 221
  waterflooding, 201
Displacement experiments, 231
Displacement mechanism, 224
Dolomite, 271
Drag reduction, 158, 168, 170
  chain degradation, 167
  elastic macromolecules, 167
  gas, 162
  interpolymer complexes, 170
  measurement, 169
  mechanism, 167
  polymers, 167, 170
  slug flow, 169
  two-phase flow, 168
  viscoelasticy, 167
Drill pipe, 1, 13, 16
Drilling fluid
  anti-freeze, 191
  biodegradable formulations, 10
  filtrate loss, 48
  filtrate reducer, 38
  thermally stable, 47
Drilling fluid disposal, 31
Dynamic tension gradient, 327
Ecotoxicity, 97

Electrical conductivity, 6
Electrochemical cell, 80
Electrochemical impedance, 80
Electrochemical noise, 80
Emulsifier, 275
  natural, 325
Emulsion muds
  inverted, 7, 8
Enhanced recovery, 117, 225
Environmental impact, 6
  odorant, 194
Epoxide resin, 117, 270, 277
Etching, 218, 272
Ethoxylation, 330
Expanding cement, 144
Fermentation, 221, 222, 244
  aerobic, 28, 243
  in-situ bacterial, 217, 218
Fiber cement, 134
Filter cake
  pore size, 37
Filtration rate
  cement slurries, 146
  power law relationship, 34
  test, 32
Fireflooding, 341
Flocculants, 2, 94
Flooding
  acid, 199
  alkali-surfactant, 197
  alkaline, 206
  carbon dioxide, 211
  caustic, 206
  combined, 207
  core, 231
  emulsion, 200
  fire, 230
  foam, 206, 208
  low-permeability, 216
  micellar, 200, 203
  miscible water, 204
  polymer, 206
  polymer water, 205
  steam, 214, 215, 230, 311
  surfactant, 206, 217, 232
Flotation, 26
Flow capacity, 120
Flow model, 231
Flow rate, 35, 296

Index **493**

Fluid loss, 271
  guars, 248
  mechanism, 34, 35
  testing, 36
Fluid loss additive, 34, 247
  acrylics, 50, 51
  asphalt, 147
  brine envrironment, 49
  cellulose based, 39
  cement, 35
  cements, 150
  controlled degradable, 41
  degradation, 247
  electric charge, 36
  electrophoretic mobility, 36
  enzymatically degradable, 247
  gellan, 243
  hydraulic fracturing fluids, 248
  lignite, 48
  lignosulfonate, 45
  pitch, 53, 54
  polyacrylates, 52
  polyhydroxyacetic acid, 44, 261
  polyitaconic acid, 52
  Starch, 40–42
  starch, 247
  sulfomethylated lignins, 45
  sulfonic acids, 46
  table, 55
  tannic-phenolic resin, 45
  testing, 36
  vinyl sulfonate based, 147
Fly ash, 133, 138, 280, 309
Foam, 212, 282, 316
  compressed, 282
  dilatational elasticity, 319
  film thinning, 320
  flow, 210
  fracturing jobs, 267
  Marangoni-effect, 319
  polyurethane, 22
  rag, 326
  stability, 139, 319
  testing, 210
  waterflooding, 198
Foam cement, 139
Foam drilling, 10
Foam film, 282, 322
Foam inhibitors, 189
Foam performance, 213

Foam stability, 209, 320, 322, 323
Foam stabilizer, 282
Formation damage, 20, 37, 231
  clay migration, 231
  filter-cakes, 37
  fracturing jobs, 235
  gas wells, 240
  heat-bank-type flooding, 204
  ion-exchange, 231
  particle deposition, 231
Fracture conductivity, 247, 255, 268, 271
Fracturing, 234
  coal-beds, 268
  efficiency, 274
  fluid leakoff, 237
  fluid loss, 271
  high-permeability, 247
  iron control, 273
  low-temperature, 262
  pressure decline analysis, 234
  stresses, 233
  temporary plugging, 110, 290
Fracturing fluids, 237
  additives, 235, 236
  classification, 234
Friction loss, 15, 169, 247
Friedel Crafts reaction, 278
Fullerene, 160
Garnet type, 13
Gas hydrate, 174
  clathrates, 175
  equilibrium pressure, 178
  inhibition, 180
  inhibitors, 153
  nucleation, 179
  pipelines, 174
  simulation, 180
Gas proportional counter, 225
Gel breaker
  acid, 270
  complexing agents, 272
  oxidative, 239
  sodium acetate, 266
Gel strength, 3, 315
  latex emulsion, 143
Gelation, 3, 112, 114–116, 229
  chelating agents, 109
  delayed, 111, 113, 289
  fibers, 287

Gelling agent, 23, 113, 115, 238, 239, 257, 272
Glass beads, 27, 137
Granules, 263
Gravel packing, 44
Grease, 65
Groundwater
 contaminated, 194
Grouting, 23, 35
Horizontal well, 13, 120
Hydratable clay, 230
Hydrate control, 162, 163, 183
Hydrate formation, 162, 174
 corrosion, 177
 dew point, 177
 induction period, 179
 inhibition, 180
 inhibitor, 181
 kinetics, 178, 179
 nucleation, 179
 pipelines, 174
 water content, 152
Hydrate formation temperature, 181
Hydraulic fracturing, 233, 234
Hydrophilic-lipophilic balance, 25, 51
Hydrosil, see Arosil, 280
Indicator
 acid base titration, 33
Interfacial tension, 133, 199, 200, 202, 207–209, 216, 222, 224, 292, 295, 297, 303, 321, 326, 327, 342
Ion exchange, 38, 199
Kinetics
 of crosslinking, 253, 254
 of degradation, 257, 262
 of hydrate formation, 179, 182
 of silica dissolution, 199
 of swelling, 61
Layer inclusion compounds, 175
Leak Detection, 194
Lightweight cement, 135, 136, 138, 139
Liquefied petroleum gas, 194
Lost circulation, 23, 123, 135, 276
 gelling agents, 108
 glass microspheres, 27
 in-situ polymerization, 119
Lost circulation additives, 22, 23
Lost circulation zone, 23
Lost circulation zones, 284
Lubrication, 14, 29, 65, 66

Mannich reaction, 94
Marsh funnel, 32
Mathematical model, 68, 69, 194, 208
Metabolism, 19, 81, 218, 221, 222
Methane Hydrate
 stability diagram, 177
Micelle, 87, 327
Michael addition, 92
Microcapsule, 173
Microemulsion, 16
Microflora, 70, 71, 217, 307
Microorganisms, 220, 221
 aerobic, 28
 combating, 71, 75, 79
 corrosion, 76, 79, 80
 degradation, 324
 detection, 69, 219
 growth, 219, 325
 hydrogen sulfide, 68
 nitrate reducing, 223
 surfactant producing, 218
Migration
 fines, 214
 gas, 131, 133, 135, 147, 148
Mineral oil, 321
Mineral oil-based muds, 6, 31
Mineral oil substitutes, 7, 8
Mineralization, 71
Mixed metal hydroxides, 12
 horizontal wells, 13
 preparation, 13
 thermally activated, 13
Mobility control, 198, 202, 205–207
Mousse, 326
Muds
 bacterial contamination, 19
 biodegradable, 10
 characterization, 31
 compositions, 4
 conversion to cement, 287
 dispersants, 311
 electric conductive, 6
 filter cake resistance, 36
 freshwater, 3, 4, 37
 inverted emulsion, 7
 lignosulfonate, 3
 lime, 3
 mixed metal hydroxides, 12
 oil-based, 5
 seawater, 4, 49

shale inhibition, 61
synthetic, 6
uncontrolled thickening, 313
water-based, 5
Natural dispersion, 305
Nutrients, 221
  extreme environments, 218
  mineral, 218
  phosphorus, 224
  simulation, 68
  ultramicrobacteria, 223
  water-soluble, 70, 111
Odorant, 192–194
  advanced oxidation technique, 195
  contamination, 194
  deactivating agent, 194
  fade, 194
  partition coefficient, 194
Odorization
  additives in pipelines, 164
Oil coating, 308
Oil spill demulsifiers, 326
Oil spill-treating agents, 292, 293, 295
Oleoresins, 227
Opuntia, 343
Osmosis, 60, 313
Oxalkylation, 330
Oxethylation, 330
Oxygen content
  gas cap, 215
Oxygen scavenger, 23
Paraffin inhibitor, 159
Pipe sticking, 18
Pipeline
  internal corrosion, 84
  oil-in-water emulsions, 191
  pigging, 85, 164, 165
  safety regulations, 192
  scale deposition, 160
  slurry transport, 163
Pipeline flow improvers, 160, 166
Pipeline transportation, 155, 191
Plasticizer
  cement, 282, 284
Plugging agent, 219, 289
Poisson distribution, 330
Pollution, 31, 73, 226, 295, 298, 299
Polymer waterflooding, 205

Portland cement, 31, 126, 149, 284–287, 289
Pour point depressant, 342
Prickly Pear Cactus, 343
Profile control, 196, 229
Redox potential, 80, 221
Rice bran extract, 107
Rotational viscometer, 32
Scale inhibitor, 107, 264
  encapsulated, 105
Scanning electron microscopy, 79, 80
Schulz-Flory distribution, 330
Scintillation counter, 225
Shale control additives, 4
Shear viscosity, 327, 342
Sloughing shales, 28, 29
Sludge, 7, 79, 97, 150, 151, 275, 290, 342
Soil, 83, 232, 297
Soil remediation, 306
Solids control, 6
Spacer fluids, 311
Spotting fluid, 16
Steam flooding, 207, 214, 215, 311
Steam injection, 208, 210, 215
Supercritical carbon dioxide drilling, 10
Surface active agents, 22, 124, 143,
    252, 295
Surface tension, 131, 178, 208, 209, 317,
    319–321
Suspension
  nonaglomerating, 172
Sweep material, 30
Thickener, 11, 28, 41, 65, 66, 205, 236, 240,
    243, 248, 260, 262
Thief zones, 135, 201–203
Thread leaks, 282
Total hardness, 3, 33
Town gas, 152
Tracers, 225
Tunnel sorption, 174
Waste water, 204, 299, 343
Wax granules, 264
Well cements, 31, 50
  classes, 128
  retarded acid soluble, 134
  wastes as additive, 149
Whisker-like hydrate, 174
X-ray photoelectron spectroscopy, 231
Ziegler–Natta catalysts, 171